2nd International Conference on Electronics, Circuits and Information Engineering (ECIE-2022)

AA007069

Journal of Physics: Conference Series Volume 2221

Online
7 - 9 January 2022

ISBN: 978-1-7138-5483-8
ISSN: 1742-6588

Printed from e-media with permission by:

Curran Associates, Inc.
57 Morehouse Lane
Red Hook, NY 12571

Some format issues inherent in the e-media version may also appear in this print version.

This work is licensed under a Creative Commons Attribution 3.0 International Licence.
Licence details: http://creativecommons.org/licenses/by/3.0/.

No changes have been made to the content of these proceedings. There may be changes to pagination and minor adjustments for aesthetics.

Printed with permission by Curran Associates, Inc. (2023)

For permission requests, please contact the Institute of Physics at the address below.

Institute of Physics
Dirac House, Temple Back
Bristol BS1 6BE UK

Phone: 44 1 17 929 7481
Fax: 44 1 17 920 0979

techtracking@iop.org

Additional copies of this publication are available from:

Curran Associates, Inc.
57 Morehouse Lane
Red Hook, NY 12571 USA
Phone: 845-758-0400
Fax: 845-758-2633
Email: curran@proceedings.com
Web: www.proceedings.com

TABLE OF CONTENTS

Preface

In order to meet the health and safety needs of our attendees and staff, 2022 2nd International Conference on Electronics, Circuits and Information Engineering (ECIE 2022) which was scheduled to be held in Xiamen, China was held virtually online during 7th to 9th, January, 2022. Being different from the traditional gatherings we all know, this virtual conference allows us to connect in new ways while keeping expenditure saving and maintaining social distancing. Presentations of presenter from different countries are accessible to hundreds of researchers effectively. There were 110 individuals who attended this online conference, represented many countries including Malaysia, Pakistan, India and China.

ECIE 2022 is an international forum for scientists, engineers, and practitioners to present their latest research and development results in all areas of Electronics, Circuits, and Information Engineering. Conference topics mainly include but are not limited to Software engineering, Power and energy circuits, Optoelectronics, Single-chip microcomputer technology, etc. ECIE 2022 welcomes all high-quality research papers and presentations from related research fields.

During the conference, the conference model was divided into three sessions, including oral presentations, keynote speeches, and online Q&A discussion. In the first part, some scholars, whose submissions were selected as the excellent papers, were given about 5-10 minutes to perform their oral presentations one by one. Then in the second part, keynote speakers were each allocated 30-45 minutes to hold their speeches. In the second part, we invited five professors as our keynote speakers. Prof. Jie Sun, Fuzhou University, China, performed a speech on *Research on micro-LED based displays in China's Mindu Innovation Laboratory.* And then we had Prof. Kai Yang, Huazhong University of Science and Technology (HUST), China. He talked about *New type of direct-drive permanent magnet motor and its robust control technology.* Prof. Weigang Hou, Chongqing University of Posts and Telecommunications, China. He delivered a wonderful speech: *Photonic CNN Accelerator.* Prof. Ghani Albaali, Princess Sumaya University for Tech in Amman, Jordan, performed a speech: *Using Computational Fluid Dynamics in the Processing Engineering.* Lastly, we were glad to invite Prof. Debashis De, Maulana Abul Kalam Azad University of Technology, West Bengal, India as our finale keynote speakers. He shared a speech on *Internet of Drone Things.* Their insightful speeches had triggered heated discussion in the third session of the conference. Every participant praised this conference for disseminating useful and insightful knowledge.

The proceedings are a compilation of the accepted papers and represent an interesting outcome of the conference. Topics include but are not limited to the following areas: VLSI Design and Fabrication, Communication theories and systems, Electrotechnics, Circuits and Systems, Optoelectronics and more related topics. All the papers have been through rigorous review and process to meet the requirements of international publication standard.

We would like to acknowledge all of those who supported ECIE 2022. The help and contribution of each individual and institution was instrumental in the success of the conference. In particular, we would like to thank the organizing committee for its valuable inputs in shaping the conference program and reviewing the submitted papers.

We sincerely hope that the ECIE 2022 turned out to be a forum for excellent discussions that enabled new ideas to come about, promoting collaborative research. We are sure that the proceedings will serve as an important research source of references and knowledge, which will lead to not only scientific and engineering findings but also new products and technologies.

The Committee of ECIE 2022

Content from this work may be used under the terms of the Creative Commons Attribution 3.0 licence. Any further distribution of this work must maintain attribution to the author(s) and the title of the work, journal citation and DOI.
Published under licence by IOP Publishing Ltd

Committee member

Committee Chairs
Prof. Abderrahim Maizate, HASSAN II UNVERSITY, Morocco
Prof. Gheorghe-Daniel Andreescu, Politehnica University Timisoara, Romania

Organizing Committee
A/Prof. Samila binti Mat Zali, Universiti Malaysia Perlis, Malaysia
A/Prof. Akhtar Nawaz Khan, UET Peshawar, Pakistan
A/Prof. Akshaya Kumar Pati, KiiT University, India
Dr. Latifah binti Mohamed, Universiti Malaysia Perlis, Malaysia
Dr. Mohamad Nur Khairul Hafizi Bin Rohani, Universiti Malaysia Perlis, Malaysia
Dr. Mohd Rafi Bin Adzman, Universiti Malaysia Perlis, Malaysia
Dr. Syahrul Ashikin Azmi, Universiti Malaysia Perlis, Malaysia
Dr. Mohd Azrik bin Roslan, Universiti Malaysia Perlis, Malaysia
Dr. Amparo FUSTER-SABATER, Institute of Physical and Information Technologies (CSIC), Spain

Technique Programme Committee
Prof. Ashraf A.M. Khalaf, Minia University, Egypt
Prof. Ghaleb HOBLOS, ÉCOLE D'INGÉNIEURS-ES GÉNÉRALISTES, France
Prof. Kazi Md. Rokibul Alam, Khulna University of Engineering & Technology, Bangladesh
A/Prof. Linquan Bai, University of North Carolina at Charlotte, USA
A/Prof. T. R. Lenka, National Institute of Technology Silchar, India
A/Prof. Dhafer Almakhles, Prince Sultan University, Saudi Arabia
A/Prof. Sushmita Sarkar, R V College of Engineering, India
Dr. Mahajan Sagar Bhaskar, Prince Sultan University, Saudi Arabia

Peer Review Statement

All papers published in this volume have been reviewed through processes administered by the Editors. Reviews were conducted by expert referees to the professional and scientific standards expected of a proceedings journal published by IOP Publishing.

- **Type of peer review:** Single Anonymous
- **Conference submission management system:** Morressier
- **Number of submissions received:** 102
- **Number of submissions sent for review:** 89
- **Number of submissions accepted:** 67
- **Acceptance Rate (Submissions Accepted / Submissions Received × 100):** 65.7
- **Average number of reviews per paper:** 2
- **Total number of reviewers involved:** 17
- **Contact person for queries:**
  Name: Xuexia Ye
  Email: xx.ye@keoaeic.org
  Affiliation: AEIC Academic Exchange Information Centre

Content from this work may be used under the terms of the Creative Commons Attribution 3.0 licence. Any further distribution of this work must maintain attribution to the author(s) and the title of the work, journal citation and DOI.
Published under licence by IOP Publishing Ltd

Design of Memristor-based XOR Logic Gate and Half Adder

Kang Su[1,2,3,a*], Tuo Li[1,2,3], Yanan Du[1,2,3], Xinxin Yuan[1,2,3], Kai Liu[1,2,3], Hongtao Man[1,2,3], Xiaofeng Zou[1,2,3]

[1]State Key Laboratory of High-end Server & Storage Technology, Inspur Group Company Limited, Beijing, China

[2]Shandong Inspur Artificial Intelligence Research Institute Co., Ltd, Jinan, Shandong, China

[3]Inspur Electronic Information Industry Co., Ltd, Jinan, Shandong, China

* Kang Su: [a]email: sukang@inspur.com

Abstract: The non-volatile logic unit based on memristor can build a computing architecture which integrates processor and memory. A novel XOR logic gate as well as half adder with four memristors, one NMOS and one power excitation is proposed in this study. The constraints on the NMOS and power excitation are also presented. The effectiveness of the XOR logic gate proposed is confirmed by the simulation results of LTSPICE. Compared with several prior schemes, the design has better performance in terms of power consumption, operation steps and architecture complexity.

1. Introduction

The traditional computing architecture leads the separation of processor and memory, which results in the energy consumption of data transmission between them and limits the further enhancement of computing performance [1, 2]. In order to support rapid processing of ultra-large amounts of data in the field of big data and artificial intelligence, it is urgent to propose a new computing architecture. The non-volatile logic unit based on memristor has functions of stateful logic operations and storage, which can build a computing architecture integrating processor and memory [3]. It is expected to break through the performance bottleneck of von Neumann architecture.

So far, there are various design schemes for a memristor-based logic gate. Memristor Ratioed Logic (MRL) gate including the memristor and CMOS takes voltage as the logical variable [4]. The circuit design of MRL gates is simple, but a CMOS non-gate is required to elevate the decaying output signal. Besides, MRL gate can only implement logical operations but cannot store the output logical state. A memristor-based material implication (IMPLY) logic gate consists of memristors and one resistor, which takes resistance instead of voltage as the logical variable [3, 5]. Both logical operation and storage can be realized by the IMPLY logic gate where there are yet some disadvantage such as the operation result storing in one of the input memristors and the complicated operations. The memristor aided logic (MAGIC) gate consisting of only memristors also integrates logic operation and storage [6]. In the MAGIC gate, there is a special output memristor to store the operation result. However, the set of MAGIC gates only contains the basic logic gates such as AND, NAND, OR and NOR, and lacks some more complex logic gates such as XOR and XNOR.

This paper proposes a newly designed XOR which is consist of two input memristors, one aided memristor, one output memristor and one NMOS, and requires just one power excitation. Furtherly, the

Content from this work may be used under the terms of the Creative Commons Attribution 3.0 licence. Any further distribution of this work must maintain attribution to the author(s) and the title of the work, journal citation and DOI.
Published under licence by IOP Publishing Ltd

constraints on the NMOS and power excitation are calculated. Additionally, the function of a half adder is realized if the aided memristor is regarded as another output memristor in the XOR logic circuit. Furthermore, the proposed memristor-based logic circuit is simulated and the feasibility to proceed the XOR operation is demonstrated.

2. Memristor-based logic gate design

2.1 The memristor model

At present, various memristor models have been proposed. The memristor model used in this paper [7] is the behavioral memristor model with simple, accurate, generic and easily modifiable characters. In the behavioral memristor model, R_{off} and R_{on} are the maximum and minimum resistances of the memristor, respectively. R_{off} represents logic "0" while R_{on} is logic "1", and R_{off} is much greater than R_{on}. When a positive voltage above V_H is applied to the positive electrode of a memristor, its resistance will increase to R_{off}. On the contrary, its resistance will decrease to R_{on} if a positive voltage below V_L is applied to the negative electrode. The memristor symbol is shown in Fig. 1.

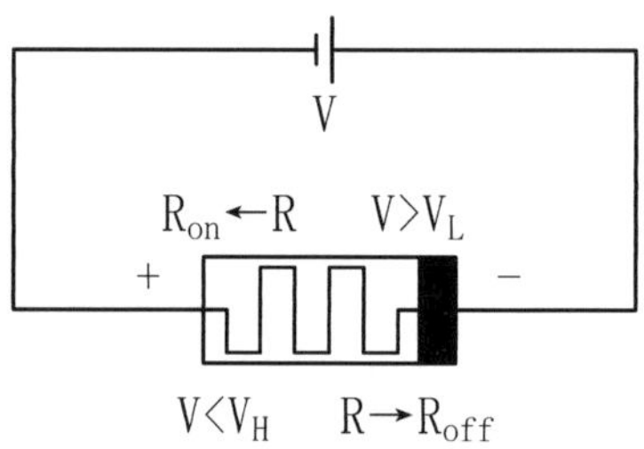

Fig. 1 Memristor symbol

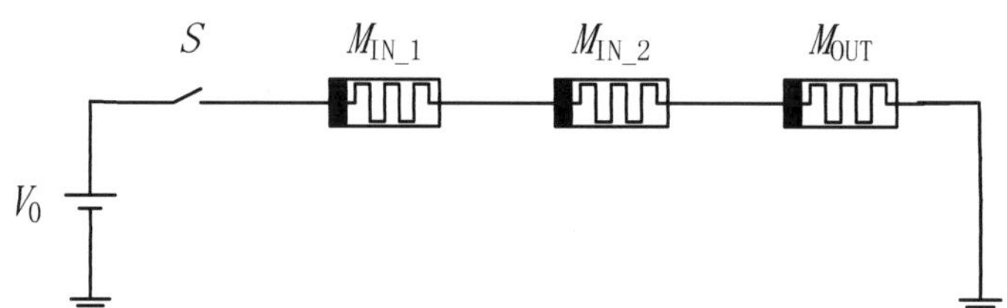

Fig. 2 AND logic circuit in MAGIC model

2.2 XOR gate and half adder circuit

The two-input XOR logic circuit is based on AND logic circuit in MAGIC model. Therefore, the AND logic circuit in [6] is shown in Fig. 2. The AND logic circuit contains two input memristors (M_{in1} and M_{in2}) and one output memristor (M_{out}) connected in series. The AND gate operation requires two sequential steps as follows: (1) the output memristor is initialised to the resistance R_{off} (i.e. logic "0") and the input memristors are loaded; (2) the evaluation is achieved by applying a voltage V_0 ($V_L < V_0 < 2V_L$) at the circuit.

Table 1 The states of memristors and the voltage at the negative endpoint of output memristor in AND gate under different input combinations

Case	M_{in1}	M_{in2}	M_{out}	$V(M_{out}^-)$
1	R_{off} (0)	R_{off} (0)	R_{off} (0)→ R_{off} (0)	$V_0/3$
2	R_{off} (0)	R_{on} (1)	R_{off} (0) → R_{off} (0)	$V_0/2$
3	R_{on} (1)	R_{off} (0)	R_{off} (0) → R_{off} (0)	$V_0/2$
4	R_{on} (1)	R_{on} (1)	R_{off} (0) →R_{on} (1)	$V_0/3$

Table 2 Truth table of XOR gate

Input 1	Input 2	XOR
0	0	0
0	1	1
1	0	1
1	1	0

Table 1 clearly shows the resistance, logic state of memristors and the voltage at the negative endpoint of M_{out} ($V(M_{out}^{-})$)in the AND gate under different input combinations. $V(M_{out}^{-})$ is $V_0/3$ when the input combination is "00" or "11" and $V(M_{out}^{-})$ is $V_0/2$ when the input combination is "01" or "10". According to the truth table of XOR gate, as shown in Table 2, the output state is "0" if the input combination is "00" or "11" and the output state is "1" if the input combination is "01" or "10". Therefore, the output state of the XOR gate can be controlled by the value of $V(M_{out}^{-})$. That is, the output state of the XOR gate is "0" (or "1") when the value of $V(M_{out}^{-})$ is $V_0/3$ (or $V_0/2$). So an NMOS whose threshold voltage (V_{th}) is higher than $V_0/3$ and lower than $V_0/2$ is introduced into the XOR circuit, as shown in Fig. 3.

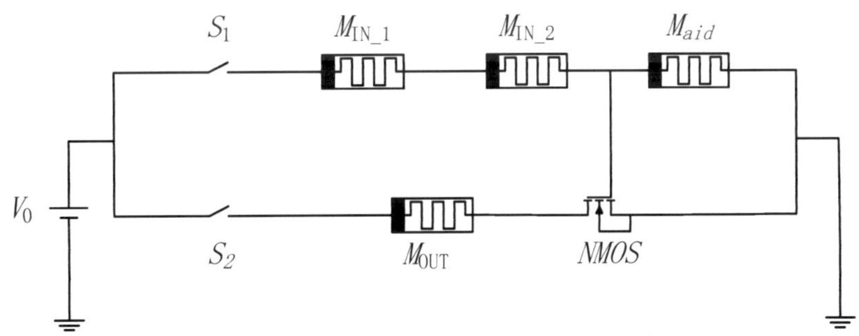

Fig. 3 Schematic of proposed XOR gate and half adder

The top branch of the XOR circuit is the AND logic circuit in MAGIC model mentioned above. The voltage at the negative endpoint of aided memristor (M_{aid}) determines the turn-on or turn-off of the NMOS. The XOR gate operation requires three sequential steps as follows: (1) the aided and output memristors is initialised to the resistance R_{off} (i.e. logic "0") and the input memristors are loaded; (2) whether the NMOS channel is turn-on is determined by applying a voltage V_0 at the top branch of the circuit; (3) the evaluation is achieved by applying a voltage V_0 at the bottom branch of the circuit.

When the logic value combination of the input memristor is logic "00" or "11", the voltage at the negative endpoint of M_{aid} is $V_0/3$ which is lower than the threshold voltage of the NMOS. Hence, the NMOS channel is turn-off and the bottom branch of the circuit is open in step (3). The resistance of M_{out} is keep the original value R_{off} (i.e. logic "0").

When the logic value combination of the input memristor is logic "01" or "10", the voltage at the negative endpoint of M_{aid} is $V_0/2$ which is higher than the threshold voltage of the NMOS. Hence, the NMOS channel is turn-on and the bottom branch of the circuit is close in step (3). The resistance of M_{out} is converted to the R_{on} (i.e. logic "1") by applying a voltage V_0 which is higher than V_L.

During the operation of XOR gate, M_{aid} always stores the result of AND of two input states. According to the logical expressions of the half adder,

$$\begin{aligned} s&=a \text{ XOR } b \\ c&=a \text{ AND } b \end{aligned} , \tag{1}$$

where a, b, c and s are the addend, summand, carry and summary respectively, the values of s and c are obtained in the M_{out} and M_{aid} respectively when the values of a and b are loaded to M_{in1} and M_{in2}. Therefore, the half adder can be performed by the XOR logic circuit.

2.3 Constraints on NMOS and V0

In the XOR logic circuits, when the input combination is logic "01" or "10", the resistance of M_{out} is conversed, which requires the V_0 to be greater than corresponding threshold voltage. But the source-drain saturation current of the NMOS in series with M_{out} determines the maximum value of the current across M_{out}. Therefore, the source-drain saturation current of NMOS must satisfy the following inequality:

$$I_{ds} \cdot R_{off} \geq V_L \tag{2}$$

where, I_{ds} is the source-drain saturation current of NMOS. Furtherly, according to the formula of I_{ds}:

$$I_{ds} = \frac{k_n}{2}\left(V_{gs} - V_{th}\right)^2 \tag{3}$$

where k_n, V_{gs} and V_{th} is the gain factor, gate-source voltage and threshold voltage of NMOS, inequality (2) can be transformed into:

$$V_{gs} - V_{th} \geq \sqrt{\frac{2V_L}{K_n R_{off}}} \tag{4}$$

where,

$$V_{gs} = \frac{R_{off}}{2R_{off}+R_{on}} V_0 \approx \frac{V_0}{2}, \tag{5}$$

$$\frac{V_0}{3} < V_{th} < \frac{V_0}{2}. \tag{6}$$

Combining equations (4), (5) and (6), the limited range of V_{th} and V_0 can be obtained:

$$\frac{V_0}{3} < V_{th} < \frac{V_0}{2} - \sqrt{\frac{2V_L}{K_n R_{off}}}\ , \tag{7}$$

$$V_0 > 6\sqrt{\frac{2V_L}{K_n R_{off}}}\ . \tag{8}$$

Combined with the range of V_0 mentioned above, the constraint on V_0 is

$$\max\left\{V_L, 6\sqrt{\frac{2V_L}{K_n R_{off}}}\right\} < V_0 < 2V_L. \tag{9}$$

Equations (8) and (9) constitute constraints on V_{th} of NMOS and V_0.

3. Simulations

The parameters in [7] are adopted in this paper, of which the threshold voltages V_H and V_L are 1.2 v and -1.2 v respectively, and R_{on} and R_{off} are 100 Ω and 1000 Ω respectively. The LTSPICE tool is used to verify the function of the proposed gate circuits. The V_{th} of NMOS and V_0 are set to 0.7 v and 2 v, respectively. The operation time of each step is set to 0.25 ms. The simulation results of XOR gate are shown in Fig. 4. As Fig. 4 shows, the initial states of M_{aid} and M_{out} both are R_{off} (i.e. logic "0"). The operation result of XOR is obtained at 0.5 ms and stored in M_{out}. The carry and summary outputs stored in M_{aid} and M_{out} of half adder are obtained at 0.25 ms and 0.5 ms. It can be observed from the simulation results that all input-output combinations match the XOR and half adder logic operation. Besides, all input states maintain the initial value during the operation.

4. Comparison

Apart from the proposed XOR logic gate, other structures of memristor-based XOR logic gate are presented, such as IMPLY, MRL, MAD and so on. The comparison results in input–output mode, number of devices, number of operation steps and number of power excitation are collected and listed in Table 3. As Table 3 shows, the input and output variables of XOR gate in MRL are represented by the volatile voltage while those of other logic modes are denoted by the non-volatile resistance of memristor. Obviously, compared with volatile voltage, the non-volatile resistance state is more suitable for recording information. Compared with IMPLY, MRL and MAD, the proposed XOR logic gate does not require resistors, which is beneficial in power consumption. In addition, the proposed XOR logic gate only require one power excitation compared with IMPLY, MAD and XOR logic gate in [8], which reduces the complexity of peripheral circuits. Furthermore, the proposed XOR logic gate also has the function of half adder.

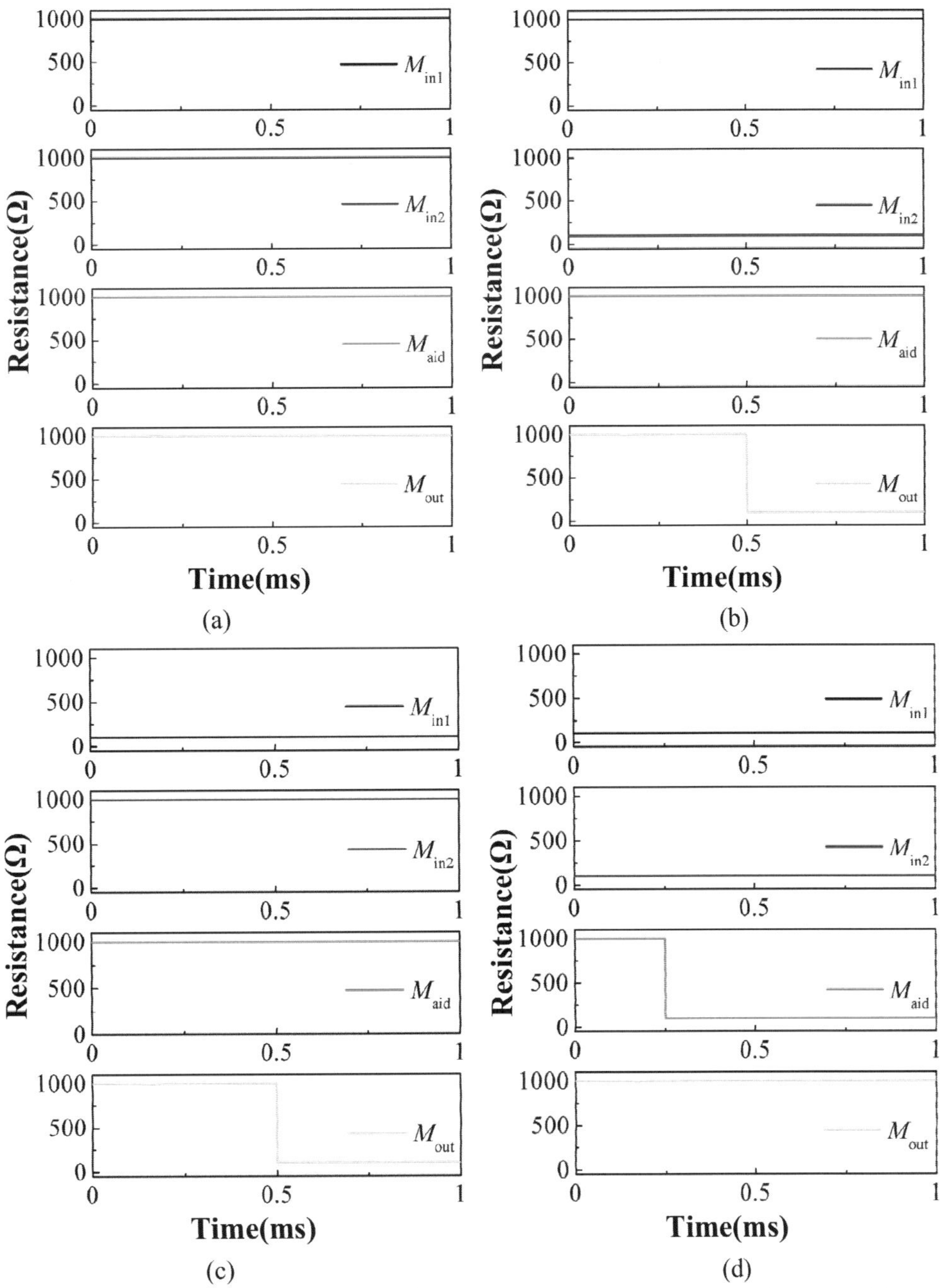

Fig.4 Simulation results of XOR gate under different input combinations: (a) 00, (b) 01, (c) 10 and (d) 11

Table 3 Comparison of different memristor-based XOR logic gate

	IMPLY logic in [9]	MRL in [10]	MAD logic in [11]	Logic in [8]	Proposed logic
Input–output mode	R-R	V-V	R-R	R-R	R-R
No. of devices	4 memristors 1 resistor	4 memristors 6 MOSFETs 1 resistor	3 memristors 3 resistor 2 threshold switchs	4 memristors 2 CNTFETs	4 memristors 1 MOSFET
No. of operation steps	6	1	2	3	3
No. of power excitation	4	1	2	3	1

5. Conclusion

In this paper, a novel XOR logic gate as well as half adder is proposed. The circuit consists of four memristors and one NMOS. Whether the NMOS channel is turn-on is determined by the voltage at the negative endpoint of aided memristor, which further determines the logical value of output memristor. The constraints on threshold voltage of the NMOS and voltage of the power excitation are given. The validity of the proposed logic circuit is confirmed by the simulation results. Besides, the proposed XOR logic gate has better performance in power consumption, operation steps and complexity of peripheral circuit by comparing with prior memristor-based XOR logic gates.

Acknowledgments

This work was supported by Major scientific and technological innovation projects in Shandong Province (Grant No. 2019TSLH0201).

References

[1] Blumenreich, G.A. (2012) The Computer and the Brain. Yale University Press, City of New Haven.

[2] Neumann, V.J. (2002) First Draft of a Report on the EDVAC. IEEE Annals of the History of Computing, 15: 27-75.

[3] Julien, B., Snider, G.S., Kuekes, P.J., J Joshua, Y., Stewart, D.R., and R Stanley, W. (2010) 'Memristive' switches enable 'stateful' logic operations via material implication. Nature, 464: 873-876.

[4] Kvatinsky, S., Wald, N., Satat, G., Kolodny, A., and Friedman, E.G. (2012) MRL — Memristor Ratioed Logic. In: International Workshop on Cellular Nanoscale Networks & Their Applications. Turin. pp. 1-6.

[5] Kvatinsky, S., Satat, G., Wald, N., Friedman, E.G., and Weiser, U.C. (2014) Memristor-Based Material Implication (IMPLY) Logic: Design Principles and Methodologies. IEEE Transactions on Very Large Scale Integration (VLSI) Systems, 22: 2054-2066.

[6] Kvatinsky, S., Belousov, D., Liman, S., Satat, G., and Weiser, U.C. (2014) MAGIC—Memristor-Aided Logic. Circuits & Systems II Express Briefs IEEE Transactions on, 61: 895-899.

[7] Wang, X., Jin, C., Eshraghian, J.K., Iu, H.C., and Ha, C. (2021) A Behavioral SPICE Model of a Binarized Memristor for Digital Logic Implementation. Circuits, Systems, and Signal Processing, 40: 2682–2693.

[8] Luo, L., Dong, Z., Duan, S., and Lai, C.S. (2020) Memristor-based stateful logic gates for multi-functional logic circuit. IET Circuits Devices & Systems, 14: 1-8.

[9] Gan, C., Jiang, X., and Chang, S. (2018) A Two Step Implementation of XOR Logic Unit Based on Memristor. Microelectronics, 48: 596-600.

[10] Kyoungrok, Cho, Sang-Jin, Lee, Kamran, and Eshraghian. (2015) Memristor-CMOS logic and digital computational components. Microelectronics Journal, 46: 214-220.

[11] Guckert, L., and Swartzlander, E.E. (2017) MAD Gates—Memristor Logic Design Using Driver Circuitry. Circuits and Systems II: Express Briefs, IEEE Transactions on, 64: 1-5.

A New Type of Transverse Flux Permanent Magnet Rim Motor and Its Modeling Analysis

Bo Lin[1, a], Jin Gao[1, b*], Yonghui Liu[1], Pengfei Zhan[2]

[1] School of Mechatronic Engineering and Automation, Shanghai University, Shanghai, 201900, China

[2] Shanghai Academy of Spaceflight Technology, Shanghai, 201100, China

[a]email: 944546275@qq.com

[b*]email: gaojin@shu.edu.cn

Abstract. In this paper, a new type of transverse flux permanent magnet rim motor is obtained by combining transverse flux permanent magnet motor with shaftless rim motor. When a sinusoidal current is passed through the stator of the motor, a rotating magnetic field will be generated. The rotating magnetic field interacts with the constant magnetic field generated by the rotor magnet, and the interaction will drive the motor to rotate. The new motor has the advantages of high power, low noise, small vibration and small volume, and is suitable for marine propulsion and other fields. This paper introduces the special magnetic field distribution of the new motor, and uses the 3D finite element analysis method to analyze the electromagnetic field of the motor's no-load and load conditions. The paper will verify the correctness and superiority of the model.

1. Introduction

The traditional permanent magnet motor has a simple structure, the tooth through which the magnetic flux passes and the slot where the armature winding is located are in the same section. The width of the slot and the width of the tooth are mutually restricted. In order to overcome such defects and improve the torque density of the motor, Herbert Wen from Germany proposed a TFPMM (Transverse Flux Permanent Magnet Motor) in 1986[1]. TFPMM has high power density and efficiency[2], and it is suitable for large motor with low-speed, high-torque and direct driven, but it also has the defect of large magnetic leakage[3]. At the same time, in the field of marine motors, the traditional pod propulsion motor has the advantages of low noise and low vibration, but it has many shortcomings such as complex structure and large space occupied by shafting[5]. In order to avoid the problems caused by the motor shafting, a SRDT (Shaftless Driven Rim Thruster) was proposed in 2005[6]. The propeller integrates the motor rotor and the blades, eliminating the shafting structure, and it can effectively save space and reduce overall weight. The propeller has the advantages of low noise and low vibration, because it drives the rim by electric propulsion device[7-8].

At present, TFPMM has a variety of structures such as magnetic convergence, plate plate, switched reluctance and so on, and its development is becoming more and more rapid. In reference [2], a stator bilateral magnetic converging type transverse magnetic field motor structure was designed, and the air-gap flux density diagram and electromagnetic torque results were given by the prototype. However, only the two-phase winding structure was studied, and the motor was a linear motor. In reference [3], TFPMM spatial reluctance was calculated, and inter-pole and inter-tooth flux leakage were analyzed,

Content from this work may be used under the terms of the Creative Commons Attribution 3.0 licence. Any further distribution of this work must maintain attribution to the author(s) and the title of the work, journal citation and DOI.
Published under licence by IOP Publishing Ltd

but the effective measures to solve the flux leakage were not given. In reference [4], bilateral permanent magnet shielding technology was adopted. It placed permanent magnets between stator and rotor poles to reduce magnetic leakage and improve torque. However, this method is only applicable to a constant magnetic field, but not to a rotating magnetic field.

In this paper, TFPMRM (Transverse Flux Permanent Magnet Rim Motor) is proposed by combining TFPMM with SRDT. It changes the existing TFPMM structure and uses three-phase armature winding. When the three-phase sinusoidal current is passed, the stator will produce transverse rotating magnetic field, and the TFPMRM can be applied to high-power Marine motor.

2. TFPMRM

2.1. *Structural characteristics*

Figure 1 is the structure of the TFPMRM. In this picture, the rotor is made of circular iron core, and the stator is composed of concave iron core, the rotor iron core is placed inside the groove of the stator core. The surface permanent magnet is placed on both sides of the rotor axis. The polarity of adjacent permanent magnets is opposite, and the polarity of permanent magnets at the corresponding positions on the axial sides of the rotor is also opposite. The coil windings are wound around the top of the concave stator core, and the number of turns of the coil is determined according to the axial length and the required output torque. The motor is suitable for low speed and high torque occasions. The mutual restriction of slot tooth width is avoided. TFPMRM does not need shafting structure. It occupies less space and has the characteristics of low noise and low vibration.

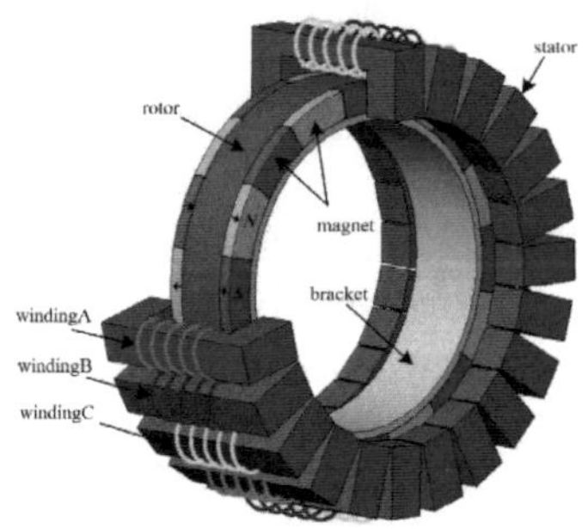

Figure 1. Schematic diagram of three-phase TFPMRM structure

2.2. *Working principle*

When the stator coil is not energized, the magnetic field starts from the N pole at the front end of the rotor axial and is dispersed in three directions inside the rotor core. One part passes through the rotor core and enters the A-phase stator tooth end, then it passes through the top of the A-phase stator tooth and reaches the other end of the stator tooth, a magnetic loop is formed laterally. Another part is dispersed in both directions of the rotor's rotation in the rotor core. It enters the adjacent B phase and C phase stator teeth by the adjacent S pole magnetic steel. The transverse flux path is formed on the stator teeth of phase B and C, which is opposite to the direction of phase A. The paths of flux in rotor core and stator core are shown in Fig. 2 and Fig. 3 respectively.

When the stator coil passes a three-phase sinusoidal current, a rotating magnetic field will be generated on the stator teeth. The stator rotating magnetic field interacts with a constant magnetic field generated by the rotor magnetic, this interaction will drag the rotor rotating synchronously with the stator rotating magnetic field.

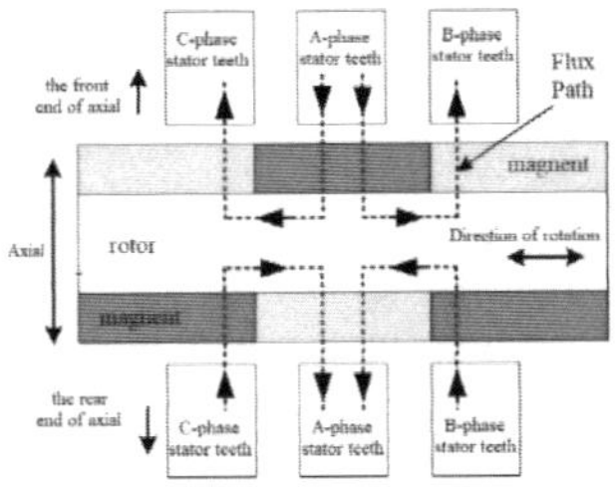

Figure 2. Rotor flux path

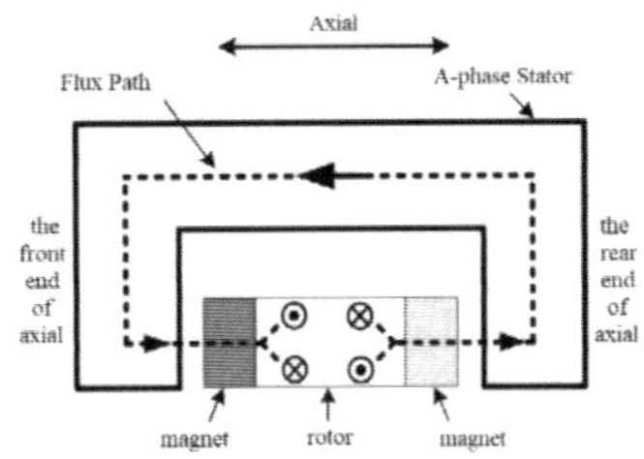

Figure 3. Stator flux path

2.3. *Parameter*

Due to the large positioning torque of general three-phase permanent magnet motors, the parameters of three-phase TFPMRM adopted in this paper are shown as follows. The ratio of poles and slots is 5 : 6.The rotor magnetic steel is 10 pairs of poles, and the stator core is 24, which can effectively reduce the positioning torque. The inner diameter of the motor is 160 mm, and the outer diameter is 215 mm. The axial width is 70 mm, and the thickness of the magnetic steel is 3 mm. The number of turns of coil is 65. The length of air gap between magnetic steel and stator tooth is 1.6mm. The stator and rotor is made of DW-465_50, and the magnetic steel is made of N35SH.

3. Three dimensional finite element analysis

3.1. *Electromagnetic field analysis*

In this paper, the electromagnetic field of TFPMRM is analyzed theoretically and simulated by three-dimensional finite element software which is based on Maxwell's equations. The magnetic dense cloud diagram of TFPMRM stator and rotor is shown in Fig. 4. As can be seen from the figure, the magnetic field distribution of the motor is reasonable. The maximum magnetic flux density at the tooth end of the stator core can reach 1.5T. The magnetic flux density at the top of the stator is low. The iron core does not reach magnetic saturation, and there is magnetic leakage between the stator teeth. Fig. 5 and Fig. 6 show the distribution of magnetic field lines of rotor and stator teeth of the motor respectively. The magnetic flux trend is consistent with the analyzed magnetic flux path, which verifies the rationality of magnetic field distribution and accuracy of magnetic flux loop.

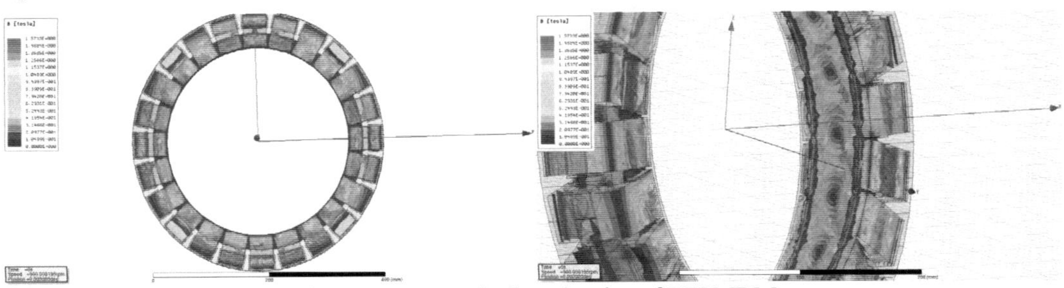
Figure 4. Magnetic flux density of TFPMRM

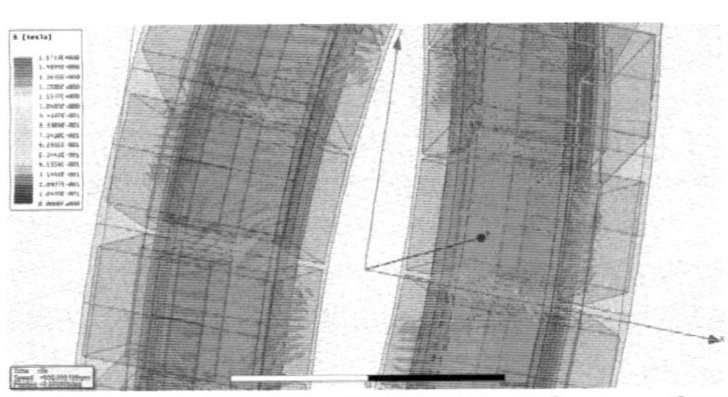
Figure 5. Magnetic field line distribution of rotor of TFPMRM

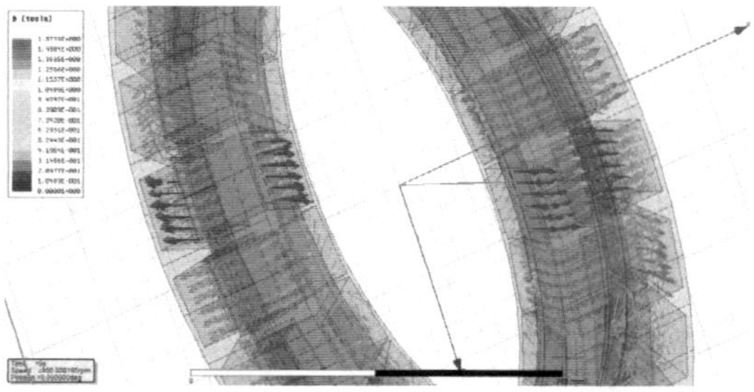

Figure 6. Magnetic field line distribution of stator of TFPMRM

3.2. No-load characteristic

Maxwell was used to simulate the motor model and analyze the no-load characteristics of the motor. In order to reduce the positioning torque of the motor, the arc coefficient of magnetic steel is 0.9， and the axial side of the magnetic steel is inclined 2°. The no-load back potential curve of the motor in two cycles at the speed of 900 r/min is shown in Fig.7. The curve basically presents three-phase sinusoidal waveform which has flat top phenomenon. It indicates that low order harmonic exist in the back potential of TFPMRM. The positioning torque curve of the motor is shown in Fig. 8. The positioning torque is kept within 1 N·m, and its effective value is 371.1195 mN·m, which ensures that the motor can start normally. The larger ripple of torque in the figure indicates that there is a large low harmonic.

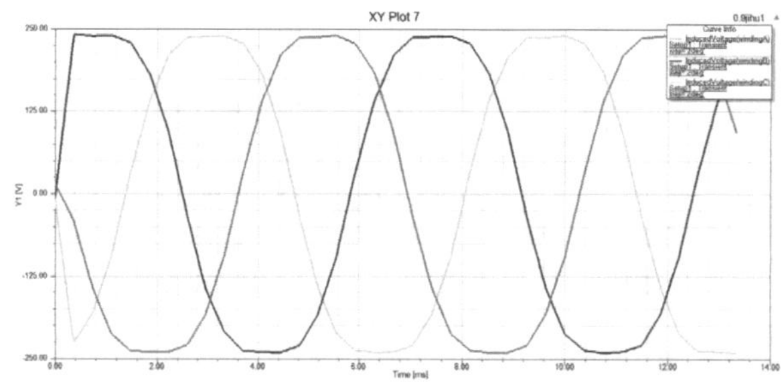

Figure 7. No-load back potential curve

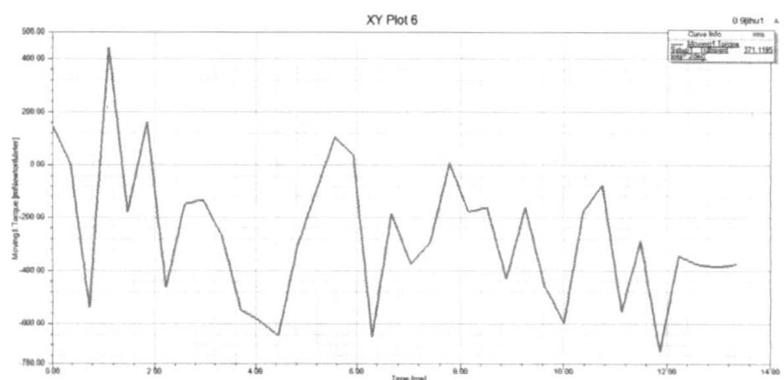

Figure 8. Positioning torque

3.3. Electromagnetic characteristic

The three-phase sinusoidal current which is 5A is passed through the three-phase stator windings to analyze the electromagnetic characteristics of the motor. The initial phase angle of the motor is set as 0°, when the N-pole magnetic steel is facing the A-phase stator of the motor. At the moment, id=0, and the motor torque reaches the maximum value. The voltage under current excitation is shown in Fig. 9, and it can be seen that the voltage is roughly sinusoidal distribution. The electromagnetic torque of the motor under current excitation is shown in Fig. 10. The average torque is about 30.5 N·m.

The output power is shown as follows.

$$P_{out} = \frac{T_e n\pi}{30\times1000} \quad (1)$$

According to Formula (1), the output power of the motor can be calculated as 2.85kW, and the output power is large.

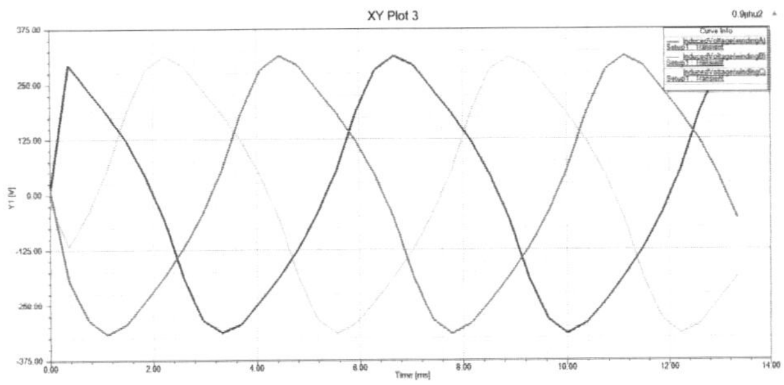

Figure 9. Three-phase voltage under current excitation

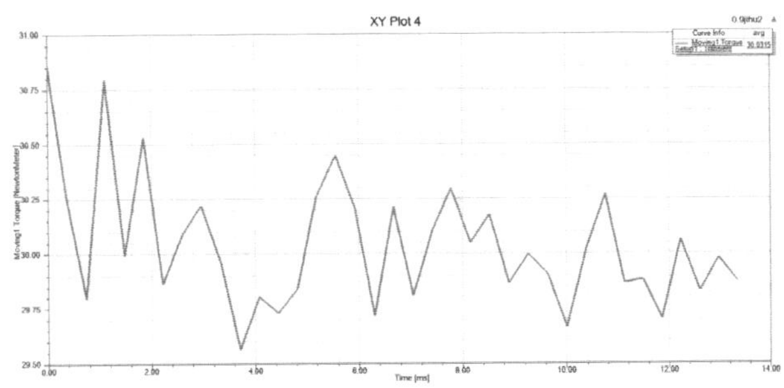

Figure 10. Electromagnetic torque

The magnetic flux trend of TFPMRM stator teeth at 0 s is shown in Fig. 11. It can be seen from the figure that the magnetic flux emanates from the n-pole magnetic steel at the front end of the rotor, enters the front end of the stator tooth, then it passes through the top of the concave stator tooth to the other end of the stator tooth, and forms a transverse magnetic field from the front end to the back end in the stator tooth. The magnetic flux trend of TFPMRM stator tooth at 1/300s is shown in Fig. 12. The stator rotating magnetic field rotates one pole distance, and the rotor also rotates one pole distance. The rotor and stator rotating magnetic field keep in synchronous operation state. The simulation result shows that the stator rotating magnetic field is consistent with the theory, which verifies the feasibility of the motor.

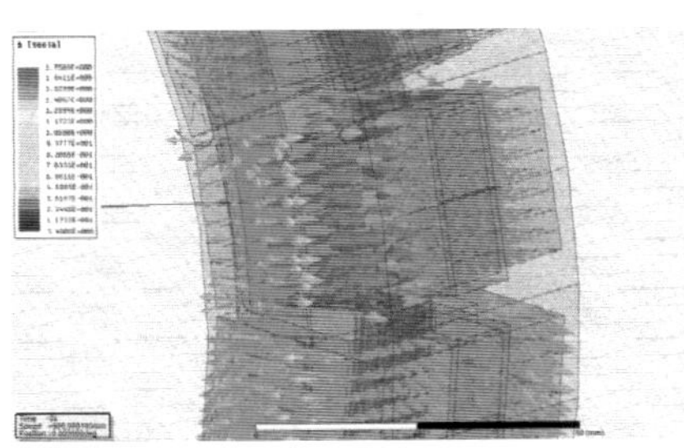

Figure 11. Stator teeth flux trend at 0s

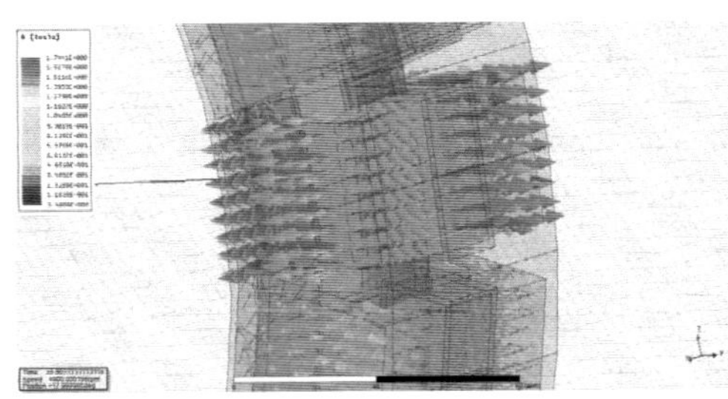

Figure 12. Stator teeth flux trend at 1/300s

4. Motor comparison

The TFPMRM is compared with PMSM (Permanent Magnet Synchronous Motor) of the same size. The optimized electromagnetic torque of a general PMSM is shown in Fig. 13. The electromagnetic torque of the PMSM reaches 71 N·m, and the volume of magnetic steel used is 168007.97692mm^3. At the same size, the volume of magnetic steel used in TFPMRM is 115359.07584mm^3, and the amount of magnetic steel of TFPMRM is smaller than PMSM.

The magnetic flux trend of TFPMRM was analyzed. The stator flux trend diagram of TFPMRM is shown in Fig. 14. It can be seen from the figure that there is a large magnetic leakage phenomenon at the tooth ends of two adjacent stator teeth. The rotor flux trend of TFPMRM is shown in Fig. 15. It can be seen that there is a large flux leakage between the poles in the rotor magnetic steel. The large magnetic flux leakage of TFPMRM results in low magnetic density of transverse magnetic field of stator teeth, which affects the torque and leads to power reduction.

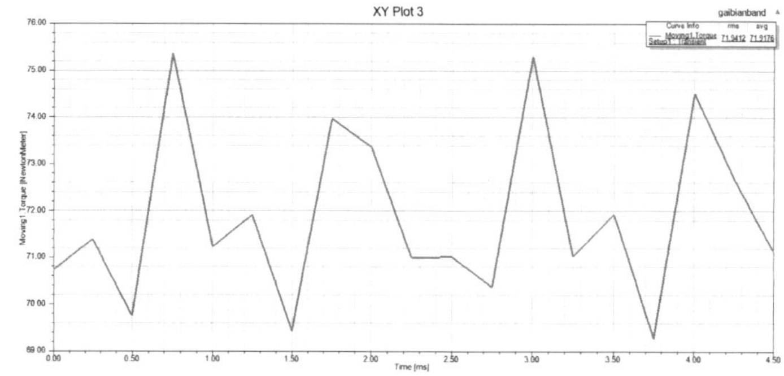

Figure 13. Electromagnetic torque of general PMSM

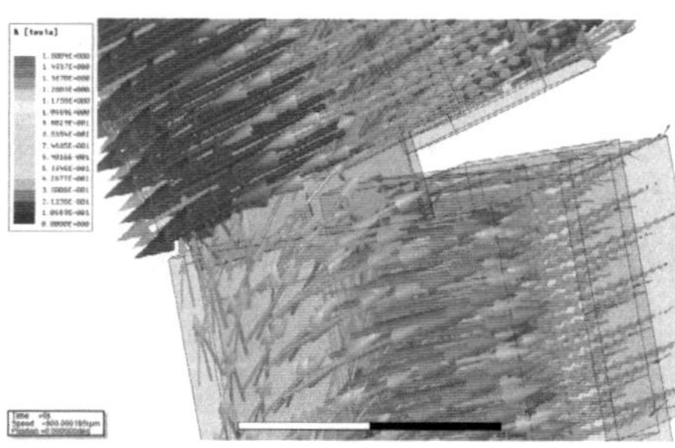
Figure 14. The flux trend of Stator

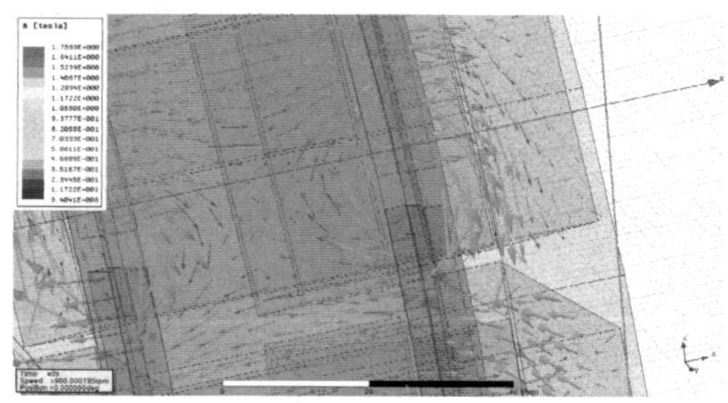
Figure 15. The flux tend of Rotor

5. Conclusion

In this paper, a new type of TFPMRM is proposed. The special structure of the new type of motor is emphasized. Different from the existing transverse magnetic field motor, it can generate stator rotating magnetic field and drive the rotor to run synchronously. The no-load characteristics and electromagnetic characteristics of the motor are analyzed. The correctness and validity of motor operation principle are verified. It is suitable for the large marine motor with low speed and high torque, and it has practical value. Compared with the general PMSM, the new motor has serious magnetic leakage, so it is necessary to analyze the magnetic leakage of the motor in the future to solve the magnetic leakage problem and improve the power density and torque density of the motor.

References

[1] Weh,H. (1986) Achievable force densities for permanent magnet excited machines in new configurations. International Conference on Electrical Machines, German.

[2] Li,Y.B., Yuan,Q. (2003) Research on a novel transverse flux permanent-magnetic machine with a concentrated flux rotor. Transactions of China Electrotechnical Society, 05:46-49+53.

[3] Wu,D., Gu,C.L. (2010) 3D magnetic circuit analysis of a novel transverse flux permanent magnet machine. Transactions of China Electrotechnical Society, 33:90-95.

[4] Li,Y.B., Gao,J., Jiang,J.Z.(2003) Research on a new transverse-flux switched reluctance machine with PM shields. Transactions of China Electrotechnical Society, 11:168-172.

[5] Hu,H.B., Gao,H.B., Lin,Z.G. (2019) Simulation of mixed eccentricity fault of shaftless rim propulsion motor.Journal of Dalian Maritime University, 04:117-122.

[6] Tan,W.Z. (2015) Research status and prospect of shaftless rim propulsion system. Journal of Wuhan University of Technology (Transportation Science & Engineering), 03:601-605.
[7] Song,B.W., Wang,Y.J. (2015) Open water performance comparison between rim-type and rim-less rim driven thrusters based on CFD method. Ocean Engineering, 103: 55-63.
[8] Yan,X.P., Liang,X.X., Wu,O. (2017) A review of progress and applications of ship shaft-less rim-driven thrusters. Ocean Engineering, 144: 142-156

Digital Control System for Stepping Motor Driver Chip

Song Wang [1,a], Xiaoning Xin [2], Jian Ren [3], Guofa Guo [4]

[1]School of Information Science and Engineering, Shenyang University of Technology, Shenyang China

[2]School of Information Science and Engineering, Shenyang University of Technology, Shenyang China

[3]School of Information Science and Engineering, Shenyang University of Technology, Shenyang China

[4]School of Information Science and Engineering, Shenyang University of Technology, Shenyang China

[a]Email: imwang_song@163.com

Abstract—Stepper motor is a common execution unit in electromechanical control system. In order to realize the high precision and effective control of the stepper motor, a digital control scheme integrated in the stepping motor driver chip is proposed. On the ADMS platform, the digital control system is designed by Verilog HDL language, and the digital control system is connected with the H-bridge circuit. The control of the two-phase micro-stepping motor is realized by Eldo and Modelsim simulation. It can support SPI bus to set working mode, attenuation time, micro-stepping work mode and other parameters. The intelligent harmonic scheme can effectively reduce the ripple current, and the ripple current can be controlled. The control scheme adopts modular design, the coding is simple and easy, and has high application value.

1. Introduction

The device for converting the pulse signal of the stepping motor into the corresponding angular displacement is one of the actuators of the electromechanical control system[1][2]. It has the characteristics of low price, easy control, stable performance and no error accumulation[3~4]. The case of non overload motor speed and stop position just depends on the pulse signal frequency and pulse numbers, not affected by load change[5].At the same time, in the control process, the stepping motor does not need position sensor as feedback, which greatly reduces the difficulty of control system design, so it is widely used in a variety of automatic control situations[6]. However, because the stepping angle is large, the stepping motor is difficult to be used in situations with high precision. In order to further improve the accuracy, it is necessary to subdivide the motion process of the stepping motor[7~8]. In the control system of traditional stepping motor, the controller with MCU and H-bridge drive is mostly adopted, which increases the operation time greatly. Therefore, the digital control part will be integrated in the driver chip.

2. The Control principle of H Bridge

A single H-bridge consists of four high voltage NMOSFET, which controls and stabilizes the current flowing through the motor by controlling the state of the four NMOSFET[9~10].

Content from this work may be used under the terms of the Creative Commons Attribution 3.0 licence. Any further distribution of this work must maintain attribution to the author(s) and the title of the work, journal citation and DOI.
Published under licence by IOP Publishing Ltd

2.1. Four states of H Bridge

The driving state of the H-bridge is the charging of the motor winding, which increases the current flowing through the motor winding. Therefore, the state is the conduction of the high and low bridge arms on different sides in accordance with the direction of the current. Taking the current flowing through the winding from left to right as an example, the driving state of the H-bridge can be shown in Fig 1.

The attenuation state of H-bridge is the discharge of motor winding, which can be divided into two kinds: slow attenuation state and fast attenuation state. The low bridge arm is turned on at the same time so that the current in the motor winding is slowly lost, so this attenuation mode is also known as slow attenuation, as shown in Fig 1.

The fast attenuation mode of the H bridge is that the arms on different sides of the bridge, which are opposite to the direction of the current flowing through the winding, are turned on at the same time, as shown in Fig 1.

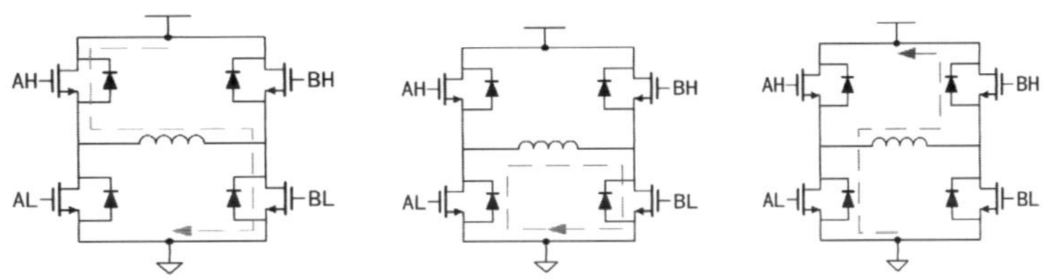

Figure 1. Three states of H bridge

2.2. Intelligent Harmonic Mode

This mode requires a lower limit under the target current value. When the target current value does not change, the current is in a slow attenuation, and decreases slowly. When the target current value is lower than the current value, the current is in a rapid attenuation, in order to rapidly reduce the current to the vicinity of the target current value. Compared with other traditional continuous current modes, the harmonic mode has more than a pair of comparators, but the steady-state current variation amplitude is greatly reduced and the dynamic change speed is greatly improved, and the current accuracy is also relatively higher. The specific work processor is as follows: the current increases to I_{tar} in the rising phase, the slow attenuation mode is used in the continuous phase, the current decreases to I_{edg}, the ripple current is small, and the fast attenuation mode is used in the descending stage, and the current drops rapidly to I_{edg}. The schematic diagram of intelligent harmonic continuous flow is shown in Fig 2:

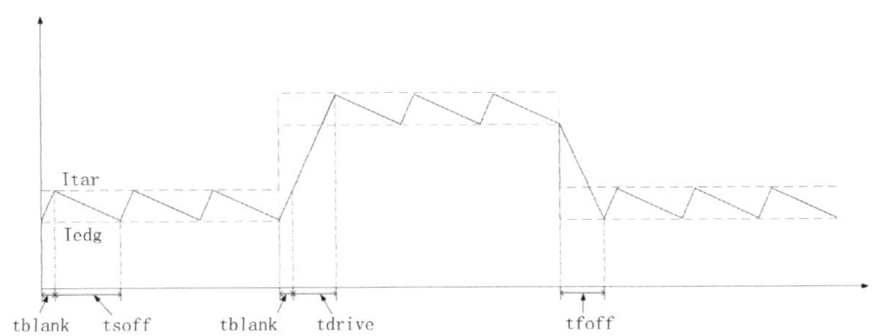

Figure 2. Intelligent harmonic example

3. Micro-stepping of Two-phase Stepping Motor

The micro-stepping motor divides a sinusoidal period equally, and each bisection point becomes a step [11~12]. The phase difference of the two-phase current is 90°.The current in the H bridge as follow:

$$I_a = I_{max}\sin(step * 360°/n) \quad (1)$$

$$I_b = I_{max} \cos(step * 360°/n) \quad (2)$$

Where I_a is A phase current, I_b is B phase current, I_{max} is the full range current, step is the current number of steps, and n is the micro-step mode

4. Digital Controller

The digital controller consists of a system controller, two H-bridge controllers, an input buffer and a data interface SPI. The overall architecture is shown in Fig 3.

4.1. System controller

The system controller module is responsible for the transition of the working state of the system, and schedules the work of other modules according to the working state. The main flow chart is shown in Fig 4.

The system controller module has four states: 'IDLE', 'ADJS', 'WORK' and 'STOP'. When the system is reset, the system controller enters the 'IDLE'. The signal to be reset jumps up, and the system controller changes to the 'ADJS'. When the system is in 'ADJS', the system controller outputs 'switch_adjs'; the system controller outputs short pulses at a fixed frequency, which makes the internal current of the B-phase H-bridge controller rise to the maximum. When the internal current of the B-phase H-bridge controller reaches the current maximum, the ready signal will be given to make the system controller switch to the 'WORK'. When the system is in 'WORK', the system controller outputs 'switch_en'; and the system controller will output the corresponding switch_step according to the input step short pulse. When the system is in the working stage, that is, when the system is in 'ADJS' and 'WORK', and the error signal fault is logical '1', the system controller will enter the 'STOP'. When the system is in 'STOP', 'switch_step', 'swtich_en', and 'switch_adjs' are set to logical '0'. At this point, all the digital control parts are locked and no other actions can be performed. When the user eliminates the error after troubleshooting and sets the fault signal to logic '0', the system controller switches to 'IDLE'. Alternatively, the user makes a global reset of the overall system, and the system controller then switches to a 'IDLE'.

4.2. H-bridge controller

The H-bridge controller module is mainly responsible for controlling the H-bridge and controlling the breaking of the H-bridge according to the working state. H-bridge controller mainly consists of three parts: counter, state controller and switch control module. The counter converts the pulse into steps under the clock pulse, that is, the counter counts once when the pulse rises. In order to place the effect of the burr, the rising edge of the pulse needs to maintain at least four clock cycles. The steps are added by one when turning forward and minus one when reversed, and the steps are transmitted to the outside world through the decoder as the digital input of current rudder DAC.

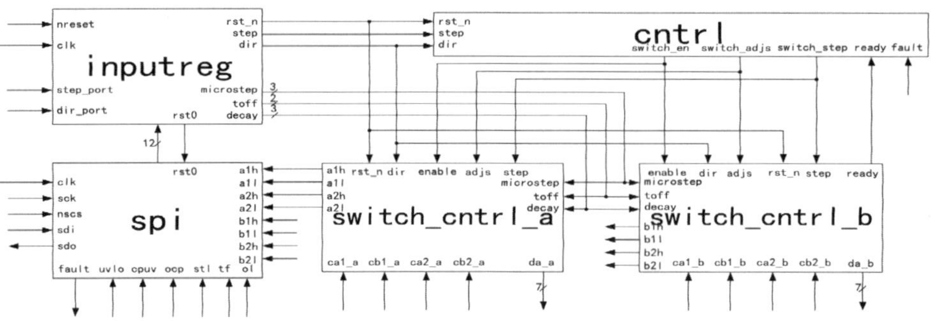

Figure 3. Digital controller

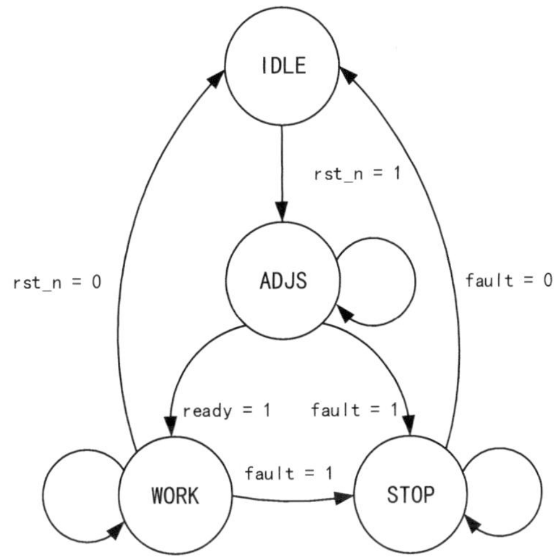

Figure 4. system controller flow chart

The state controller mainly controls the breaking of H-bridge, and its flow chart is shown in Fig 5.

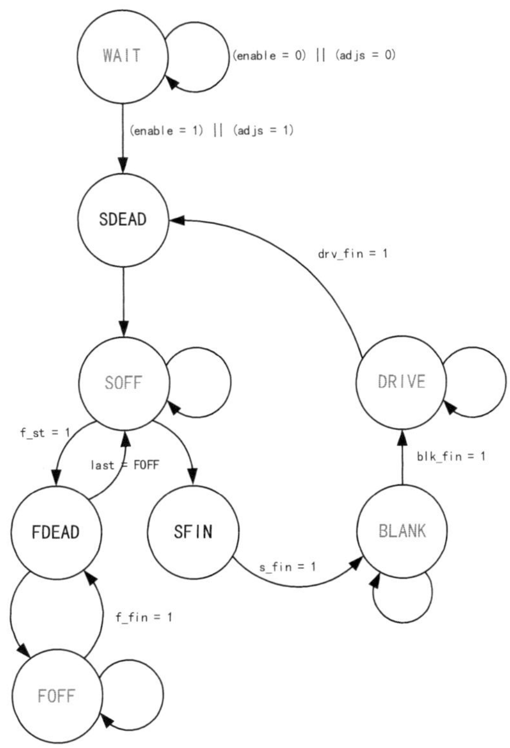

Figure 5. State controller in H-bridge controller

Among the various step modes and attenuation modes, there are three basic working states of H-bridge: driving stage, slow attenuation and fast attenuation. Therefore, in this design, the basic state is subdivided into eight types: 'WAIT'(Wait for the system to start), 'BLANK'(Blank detection H-bridge time), 'DRIVE'(Drive state of H-bridge), 'SDEAD'(Dead time of state transition), 'SOFF'(Slow attenuation state of H bridge), 'SFIN'(Dead time of state transition), 'FDEAD'(Dead time of state transition) and

'FOFF'(Fast attenuation state of H bridge). Because the high back EMF of the motor winding will lead to the misjudgment of the system, it is necessary to avoid 'BLANK' state for this period of time.

In this design, due to the large inductance of the motor winding, the reaction electromotive force is too large, it is possible to break down the chip, the 'dead zone' state is inserted between the working states, and the short circuit caused by opening the same side bridge arm at the same time is avoided. At the same time, because the reaction EMF generated by the motor winding will interfere with the comparator in the process of state transition, in order to prevent the impact on the normal work flow, the 'BLANK' is inserted before the 'DRIVE'. Take the harmonic attenuation mode as an example, the work flow is as follows: when the system is reset, the state converter is in 'WAIT'. When the system starts to work, the state converter changes from 'WAIT' to 'SDEAD', and then to 'SOFF'. The slow attenuation mode can be maintained at the current voltage with smaller ripples. When the voltage attenuates to the lower limit, the state transition logic sends out a s_fin signal. The state converter changes from 'SOFF' to 'SFIN', and then to 'BLANK' and 'DRIVE'. In this design, in order to eliminate the interference of the comparator, the 'BLANK' will maintain the 7ns. When the state converter is in 'DRIVE', the state transition logic will send out a 'drv_fin' signal when the H-bridge current reaches the sinusoidal current value. The turntable converter changes from 'DRIVE' to 'SDEAD', and then to 'SOFF'. Repeat the previous steps. When the step signal is provided from the outside and is in the rising phase, the state converter will be in the 'DRIVE' for a long time until the H-bridge current reaches the corresponding sinusoidal current value, and the driving state time is recorded. When the state converter is in the descending phase, the state converter passes from 'SOFF', 'FDEAD' to 'FOFF', and the attenuation counter starts to work. When the attenuation counter coincides with the recorded time, the state transition logic sends out a f_fin signal, and the state converter transitions to a 'SOFF' via a 'FDEAD'. Repeat the state transition of the previous phase.

The function of the switch control module is to control the breaking of the high voltage nmos tube on the four arms of the H bridge according to the 'dir' signal of the rotation direction of the motor and the state of the state converter.

5. Overall simulation results

In order to verify the performance, the Hspice Netlist of the peripheral drive circuit is translated into a Verilog file which can be read by the digital simulation period in the ADMS environment, and the digital part and the peripheral model are called in the testbench file for simulation.

The simulation results in 32 step mode are shown in Fig 6.

When using the device, the rotation direction of the motor can be changed.

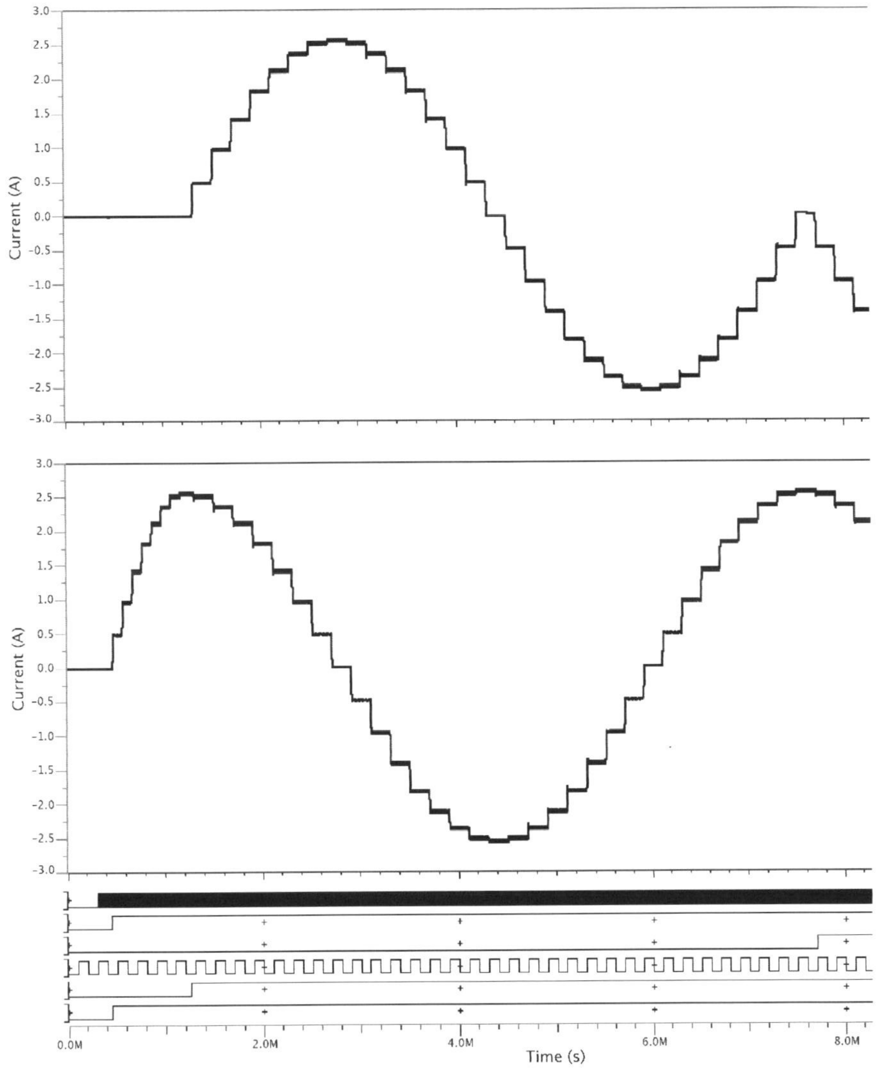

Figure 6. current waveform of 32 step mode

As can be seen from the above figure, the current ripple is relatively small during the rotation of the stepper motor, and there is no loss of steps during the rotation, and the rotation direction can be changed arbitrarily during use.

6. Conclusions

The principle of stepper motor subdivision drive is analyzed, and a highly reliable two-phase stepper motor subdivision drive digital control system is designed, which can be integrated in the driver chip. By using the intelligent harmonic mode, the current ripple can be effectively reduced, up to 32 subdivision control, and the user can configure the corresponding subdivision mode, rotation direction, attenuation time, attenuation mode and other parameters through the spi interface.

References

[1] Carrica, D. O. , SA González, and M. Benedetti . "A high speed velocity control algorithm of multiple stepper motors." Mechatronics 14.6(2004):675-684.

[2] Matsui, N. , M. Nakamura , and T. Kosaka . "Instantaneous torque analysis of hybrid stepping motor." IEEE Transactions on Industry Applications 32.5(1996):1176-1182.
[3] Kruse, R. , G. Pfaff , and C. Pfeiffer . "Transverse flux reluctance motor for direct servodrive applications." CONFERENCE RECORD- IEEE INDUSTRY APPLICATIONS ANNUAL MEETING IEEE, 1998.
[4] Wangdan, Zhang Shuochengt , and Q. W. Jinglan . "Step Motor Control System." Imp & Hirfl Annual Report 00(2003):106-106.
[5] Jiang, Z. , et al. "A novel hysteresis current control for active power filter with constant frequency." Electric Power Systems Research 68.1(2004):75-82.
[6] Mizutami, K. , S. Hayashi , and N. Matsui . "Modeling and control of hybrid stepping motors." Industry Applications Society Meeting IEEE, 1993.
[7] H.-S., et al. "A Study on Micro-step of 2-phase Hybrid Type Linear Stepping Motor." TRANSACTIONS- KOREAN INSTITUTE OF ELECTRICAL ENGINEERS B (2000).
[8] Jung-Ho, et al. "High-speed micro-step drive of Stepper motor using Current-Controller." INFORMATION AND CONTROL SYMPOSIUM (2010).
[9] Betin, F. , and D. Pinchon . "Fuzzy logic applied to speed control of a stepping motor drive." Industrial Electronics IEEE Transactions on 47.3(2000):610-622.
[10] Yang, S. M. , F. C. Lin , and M. T. Chen . "Micro-stepping control of a two-phase linear stepping motor with three-phase VSI inverter for high-speed applications." 38th IAS Annual Meeting on Conference Record of the Industry Applications Conference, 2003 IEEE, 2003.
[11] Lu, Y. , D. Meng , and Y. Xu . Subdivision Controller Design of Stepping Motor Based on FPGA. 2010.
[12] Wang, H. B. , et al. "Stepper motor SPWM subdivision control circuit design based on FPGA." Conf. Computer and Information Science (ICIS), Wuhan, China, 2017, pp.889-893.

On Magnetic-coupled Matrix Rectifier Applied to Accelerator Power Supply

Bin WANG [1,a], Yang SI [2,b], Zhiwen ZHOU [3,c]

[1] School of Physics and Electronic-Electrical Engineering, Ningxia University, Yinchuan, Ningxia, China

[2] New Energy (Photovoltaic) Industry Research Center, Qinghai University, Xining, Qinghai, China

[3] School of Electrical Engineering, Northwest Minzu University, Lanzhou, Gansu, China

[a]wangbin053@nxu.edu.cn, [b]siyang@qhu.edu.cn, [c]zzw407@qq.com

Abstract: The power supply designing of superconducting magnet is vital for magnetic field system in the development of accelerator. Magnet power supply can still be improved in grid-side power factor, and better cost-effective solutions can be sought. For release or consumption of the huge energy stored in the future superconducting magnet, so that it is urgent to carry out related research to solve the energy feedback problem. Utilizing the characteristics of adjustable power factor and bi-directional energy transfer of matrix converter, a novel magnetic-coupled matrix rectifier topology is presented, which is suitable for accelerator power supply. An improved control strategy is proposed to realize sinusoidal waveform of input current, unity input power factor as well as output current tracking the instruction current waveform. Prototype test shows that the matrix converter can realize DC output and the control strategy can achieve expected requirements. Load disturbance experiment of quasi-trapezoidal wave output is preliminarily verified. The matrix converter studied in the paper, as a kind of feasible attractive choice of accelerator power converter topology, has a favorable research prospect.

1. Introduction

Along with the development of the modern accelerator technology, requirements of beam quality and stability are increasingly enhanced. Requirement of magnet power supply is consequently improved[1]. In the national twelfth five-year major science and technology infrastructure project - High Intensity Heavy-ion Accelerator Facility (HIAF), superconducting coil is considered to be applied to part of dipole magnets and the magnet power supply is required to work in pulse operation mode, in order to meet the demands of HIAF. When current pulse falls, energy stored in the superconducting magnets need to be transferred from the magnet in an appropriate way[2]. Therefore, it is necessary to adopt new technology to realize energy feedback to power grid.

In addition, most of the magnet power supplies work in traditional phased-controlled rectifier or non-controlled rectifier model, which brings about a great impact and harmonic pollution to grid. And multiple series-parallel switch converter and auxiliary switch phase shifting drive are usually applied to increase the equivalent switching frequency, which causes the system complexity and high costs.

Content from this work may be used under the terms of the Creative Commons Attribution 3.0 licence. Any further distribution of this work must maintain attribution to the author(s) and the title of the work, journal citation and DOI.
Published under licence by IOP Publishing Ltd

Therefore current power supply can still be improved in grid-side power factor, and better cost-effective solutions can continue to be sought.

As a significant technique of the power supply, matrix converter and its control strategies have been studied extensively [3-7]. Aiming at energy feedback and power factor problem, the feasibility of applying the matrix converter to magnet power supply of heavy ion accelerator is studied in this paper. With matrix converter worked as rectifier, the device can realize sinusoidal input current so as to achieve high power factor. Matrix converter can realize bi-directional flow of energy, which can transmit excessive energy of magnet back to the grid. In HIAF superconducting magnet power supply project, matrix transformation technology is expected to play an important role.

2. Matrix converter topology

2.1 Matrix converter

Matrix converter, whose switch network is composed of four-quadrant bi-directional switches, is a kind of flexible, general, green power converter. Compared with the traditional converter, matrix converter has the following distinctive features: single-stage frequency transformation, without large capacity of energy storage component, bi-directional electricity flow, sinusoidal input and output, controllable input power factor, etc. [8]. Three-phase input three-phase output (33MC) is as shown in figure 1.

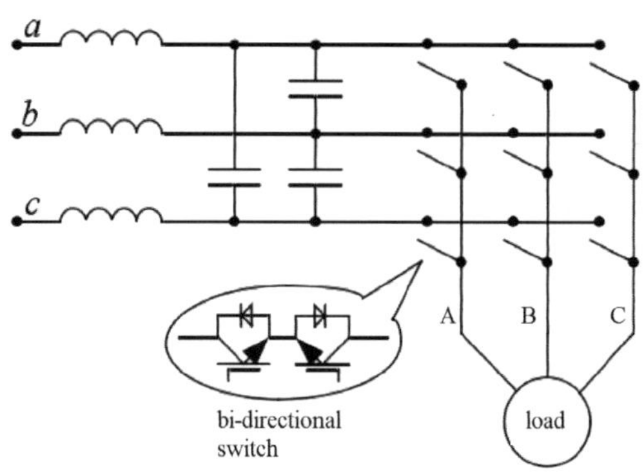

Fig. 1 Three-phase input three-phase output of matrix converter (33MC)

If three bi-directional switches in one output phase is removed from 33MC and expected output frequency is set to zero so as to get a DC output, the matrix converter can be called Matrix Rectifier (MR).

2.2 Magnetic-coupled matrix rectifier

Two types of AC/DC converter circuit can be evolved from 33MC, one is magnetic-coupled MR, the other is direct-coupled MR. Magnetic-coupled MR studied in this paper, can not only realize electrical isolation between source and load, but also effectively reduce the volume and weight of the transformer with larger output voltage range.

Typical magnetic-coupled matrix converter is shown in figure 2. MR is expected to realize energy feedback to grid so as to be energy saving if it is controlled to be worked as an inverter. The experimental results show that steady current output, energy flow from the grid side to load side. To realize the bi-directional flow of energy control strategy is more complex, which will be studied in the following work.

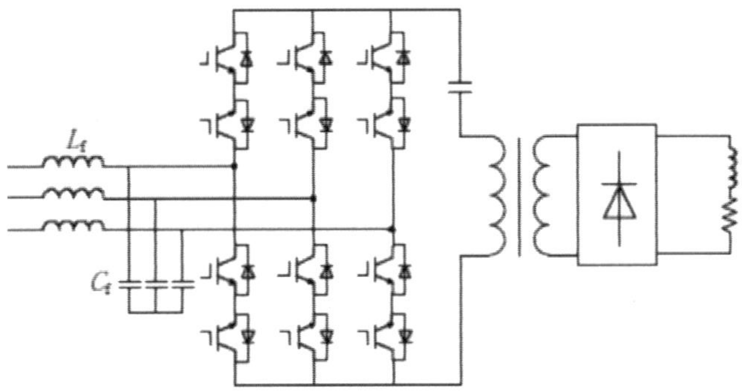

Fig. 2 Magnetic-coupled matrix converter topology

3. Control strategy

3.1 Preceding stage space vector modulation principle

Appropriate control strategy needs to be applied at both preceding stage and post stage. In figure 2, the matrix converter on the left of the transformer is preceding stage, in which input current space vector modulation is applied in order to obtain sinusoidal input current waveform and high input power factor [9-10].

Preceding stage can be virtual as rectifier and inverter. A three-phase input voltage is set as the formula (1), where ω_i refers to input voltage frequency.

$$\begin{pmatrix} u_a \\ u_b \\ u_c \end{pmatrix} = u_{im} \begin{pmatrix} cos(\omega_i t) \\ cos(\omega_i t - 2\pi/3) \\ cos(\omega_i t - 4\pi/3) \end{pmatrix} \tag{1}$$

Six bi-directional switches of rectifier part can synthesize six effective input current space vectors, as shown in figure 3. In figure 3, the letters in brackets respectively represent the connection status of switches between three-phase bridge arm (a, b, c) and upper and lower switch and DC terminals (P, N). Such as "ac" refers to the switch between phase a and terminals P is turned on, and the switch between phase c and terminals N is also turned on, the rest of the switches are all turned off. Input current space vector is zero vector, when the upper and lower switches in one arm are both turned on [11].

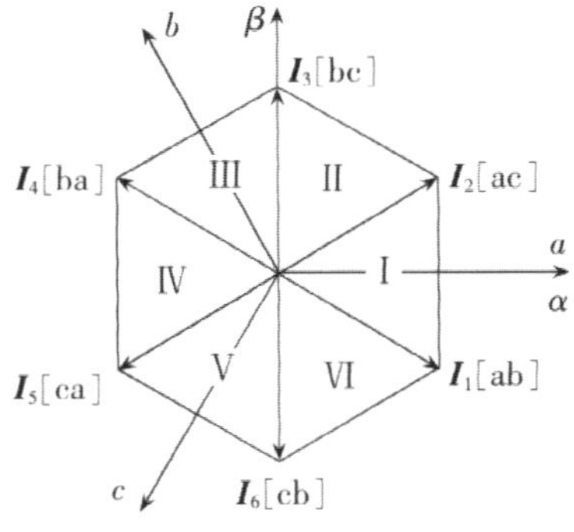

Fig. 3 Space vector diagram of switch current states

By Park transformation the three-phase instantaneous input current can be synthesized in time domain into a spin current space vector, namely reference current vector. Reference current vector consists of static vectors, I_{ref} can be synthesized by the adjacent two effective space vector I_k, $I_{k+1}$and zero vector I_0, in the phase sector where the input current space vector I_{ref} is located.

Given that d_k, d_{k+1} and d_0 respectively refer to duty ratio of switch I_k, I_{k+1} and I_0, then

$$I_{ref}=d_k I_k + d_{k+1} I_{k+1} + d_0 I_0 \tag{2}$$

The calculation formula of switch duty ratio is

$$d_k = \frac{2\rho\sin(60°-\theta_r)}{\sqrt{3}\,|I_k|} = m_c \sin(60°-\theta_r)$$
$$d_{k+1} = \frac{2\rho\sin\theta_r}{\sqrt{3}\,|I_{k+1}|} = m_c \sin\theta_r$$
$$d_0 = 1 - d_k - d_{k+1} \tag{3}$$

$$|I_k| = |I_{k+1}| = \sqrt{2} I_{dc} \tag{4}$$

where m_c （$0 \le m_c \le 1$） refers to current modulation coefficient, $\theta_r = \omega_i t - \phi_i + 30^\circ$, φ_i refers to the phase difference between input current and input voltage.

If the input current space vector is located in the first sector, the average input current in a switching period is,

$$\overline{I_i} = \begin{pmatrix} \bar{i}_a \\ \bar{i}_b \\ \bar{i}_c \end{pmatrix} = \begin{pmatrix} d_k + d_{k+1} \\ -d_k \\ -d_{k+1} \end{pmatrix} \cdot I_{dc} = m_c \cdot \begin{pmatrix} cos(\omega_i t - \phi_i) \\ cos(\omega_i t - \phi_i - 2\pi/3) \\ cos(\omega_i t - \phi_i - 4\pi/3) \end{pmatrix} \cdot I_{dc} \tag{5}$$

By comparing formula (1) and (5) it is concluded that, without considering the effect of high frequency harmonic, the input phase current changes according to the sine curve at the same frequency of corresponding input phase voltage with phase difference φ_i. Input power factor can be controlled by changing φ_i.

Output DC voltage V_{dc} of virtual rectifier part is input of virtual inverter part, virtual inverter part generates single-phase high-frequency bipolar pulse output, whose reference voltage vector synthesis is similar as and rectifier part.

3.2 Overall control strategy of accelerator power supply

No matter accelerator power supply works in DC mode or pulse mode, the control target of the power is load current. a series of indicators of the load current waveform are focused on, such as stability of DC power supply, the tracking error of pulse power supply, etc. It is very important for power supply's normal operation and satisfactory performance obtaining to choose suitable closed-loop control method based on topology model of power supply.

Considered the requirements of DC magnet power supply, meanwhile ensured that load current keep track of instruction current waveform, based on analysis of control strategy of matrix converter, current space vector modulation (SVM) of preceding stage and proportional integral (PI) feedback control of load current is combined as a feasible solution, its schematic diagram as shown in figure 2. In order to eliminate deviation and make the output following the given value, control variable is used to revise rectifier modulation ratio of preceding stage. Input current can be improved by using current space vector modulation (SVM), as well as output current can keep good track of instruction current by using PI regulating. The method is simple and with good robustness.

As pulse power supply is concerned, since the direct-coupled MR is actually a three-phase one-phase matrix converter(31MC) in terms of topology, with frequency of output phase set as the corresponding harmonic frequency and output phase processed, the pulse output of quasi-trapezoidal wave can be obtained through high frequency synthesis. however the complexity of algorithm will be significantly increased.

Magnetic-coupled MR, like direct-coupled MR, can meet the requirements of ion accelerator magnet power supply, such as high current, high stability, low ripple and high dynamic performance.

For pulse power supply, the given value of the closed-loop control of the post stage can be modified to a non-constant value, and the instruction value can be continuously issued according to the discrete value of the pulse waveform, and the desired pulse output current can be obtained on the load.

As for preceding regulated matrix rectifier (MR), pulse output can be realized as long as the constant-voltage output of the preceding rectifier is maintained and the appropriate modulation strategy is adopted for post stage.

4. Prototype experiment

A prototype of MR DC magnet power supply is constructed with adopting magnetic-coupled matrix AC/DC converter topology structure. In the prototype, IGBT K30H603HAJ108 is selected to constitute bi-directional switches, DSP2812 is selected as system control devices, CPLD EPM240T100C5N is selected to generate drive pulses and realize the commutation logic. Since the prototype is expected to conduct original validation experiments, currently only experiment with impedance load is carried out, whose actual power is small, with output current less than 5A, voltage less than 50V.

Output current and voltage waveform when load resistance of MR switches from 10.3Ω to 13 Ω are shown in figure 4. It is illustrated that although output voltage has a corresponding change when simulated load switches, output current roughly remains stable at the given value after a transient closed loop regulation.

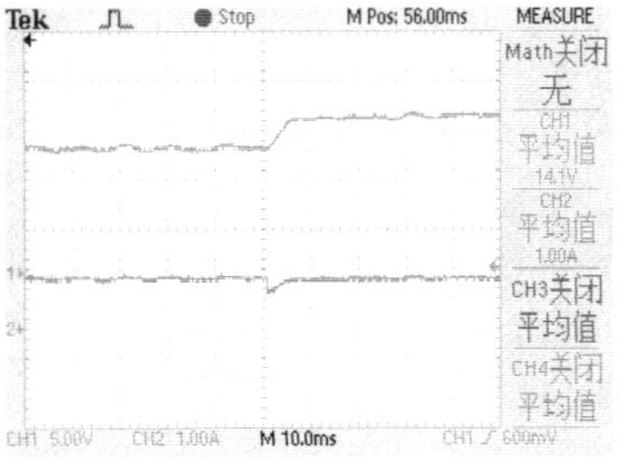

Fig. 4 Waveform of closed-loop output current and voltage when switching to larger load

Without changing control procedures and related parameters MR can output constant current according to the given value in the case of load switching, with rapid dynamic regulation time and high grid side power factor. Consequently feasibility of the topology applied to accelerator power supply and validity of the control strategy are both verified by prototype experiments.

The left figure in Fig.5 shows the input voltage and current waveform under open-loop control. According to the waveform, the input current is basically sinusoidal and slightly ahead of the input voltage. The capacitiveness characteristic is due to the input filter.

The right figure in Fig. 5 shows the input voltage and current waveform of grid side under closed-loop control. It can be observed that under closed-loop control, the input voltage(yellow) and current(blue) of RMC are basically in the same phase, with high power factor, high utilization rate of grid and good performance of grid side.

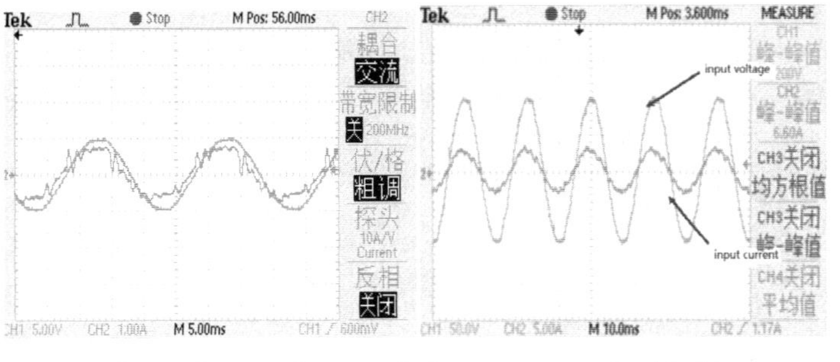

Fig. 5 Phase voltage and phase current of grid side

Finally, load disturbance experiment is carried out and tracking error of pulse output is analyzed. In order to ensure the synchronization of the accelerator cooling storage ring (CSR), the control system will provide the reference waveform signal I_{ref} to the power supply within a specified time, I_{ref} is shown on the left in Fig. 6.

The output current I_{out} of the power supply must follow the reference waveform I_{ref} in real time, as shown on the right in Fig. 6. The deviation between I_{out} and I_{ref} in amplitude and time, is known as tracking error, which is the smaller the better. Thus power supply is required to have a quite fast response speed. A similar closed loop test scheme is used for magnetic-coupled matrix rectifier (MR).

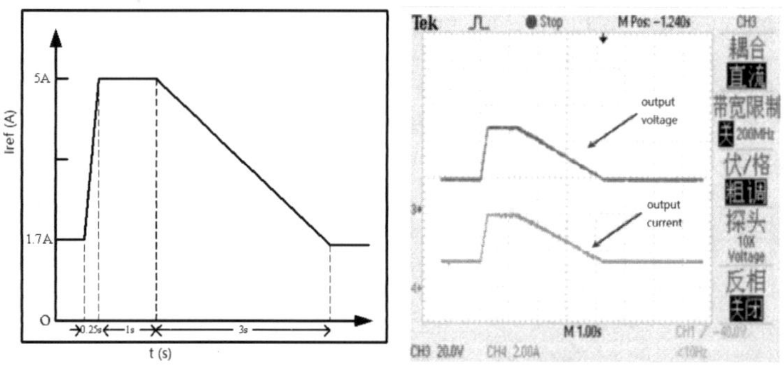

Fig. 6 Reference current I_{ref} waveform(left) and actual current waveform (right)

From experimental results, it is shown that in the case of low current and low current, it can be stable and timely follow the given waveform in principle, and achieve predetermined closed-loop effect. Because of the low power, the correctness of the program is verified more in principle.

5. Conclusion

Through prototype experiment, magnetic-coupled MR topology is validated basically suitable for DC power supply, the output current of which can overcome certain disturbance to realize constant current control with high grid side power factor. Also for magnetic-coupled MR topology, the feasibility of its application to pulse power supply has been discussed. Eventually matrix converter has an excellent research prospect, as a kind of potential attractive choice of accelerator power converter topology.

Acknowledgments

This work is supported by National Natural Science Foundation (NNSF) of China (Grants 11805156, 61867005), Natural Science Foundation of Qinghai Province (Grant 2019-ZJ-948Q) and Natural Science Foundation of Ningxia Province (Grant 2020AAC03068).

References

[1] Huang Y. Z., Chen Y. X., Zhou Z. Z., et al., "Research and design of digital power supply for HIRFL-CSR sexupole magnet", Nuclear Physics Review, 2011, 28 (3): 296-299. (In Chinese)

[2] Li H., Ding H., Yuan Y., "Design of a 60T dual-coil quasi-continuous magnetic field system with a hybrid power supply", IEEE Transactions on Applied Superconductivity, 2014, 24(3): 1-5.

[3] Deng H., Peng T., Dan H., et al., "Error-voltage based open-switch fault diagnosis strategy for matrix converters with model predictive control method", IEEE Transactions on Industry Applications, 2017: 1-7.

[4] Reddy G. P. R., Kumar M. V., "Two level versus matrix converters performance in wind energy conversion systems employing DFIG", Journal of the Institution of Engineers, 2017(2): 1-13.

[5] Trentin A., Zanchetta P., Empringham L., et al., "Experimental comparison of devices thermal cycling in direct matrix converters (DMC) and indirect matrix converters (IMC) using SiC MOSFETs", Energy Conversion Congress and Exposition, 2017:1-7.

[6] Mengoni M., Zarri L., Tani A., et al., "Modulation strategies for three-phase AC-DC matrix converters: A comparison", Energy Conversion Congress and Exposition, 2017: 1-7.

[7] Garcia C. F., Rivera M. E., Rodríguez J. R., et al., "Predictive current control with instantaneous reactive power minimization for a four-Leg indirect matrix converter", IEEE Transactions on Industrial Electronics, 2017, 64(2): 922-929.

[8] Nguyen T. D., Lee H. H., "Modulation strategies to reduce common-mode voltage for indirect matrix converters", IEEE Transactions on Industrial Electronics, 2012, 59(1): 129-140.

[9] Su M., Wang H., Sun Y., et al., "AC/DC matrix converter with an optimized modulation strategy for V2G applications", IEEE Transactions on Power Electronics, 2013, 28(12): 5736-5745.

[10] Dabour S. M., Hassan E. W., Rashad E. M., "Analysis and implementation of space vector modulated five-phase matrix converter", International Journal of Electrical Power & Energy Systems, 2014, 63(8):740–746.

[11] Klumpner C., Wijekoon T., Wheeler P., "New methods for the active compensation of unbalanced supply voltages for two-stage direct power converters", IEEJ. Transactions on Industry Applications, 2006, 126(5): 589-598.

A humidity compensation method for a temperature and humidity sensor

Yuchuan Yang[1], Xiuxiu Liang[2], Ting Hong[3] and Xin Li[4]

[1]College of Information Science and Engineering, Shenyang University of Technology, Shenyang, Liaoning,110020, China.

[2]College of Information Science and Engineering, Shenyang University of Technology, Shenyang, Liaoning,110020, China.

[3] Shandong Huake Semiconductor Research Institute Co. LTD, ,100000, China.

[4]College of Information Science and Engineering, Shenyang University of Technology, Shenyang, Liaoning,110020, China.

[1]Corresponding author's e-mail: fs_yyc@163.com

Abstract. This paper introduces the method of humidity compensation in the temperature and humidity sensor to improve the measurement accuracy of humidity. The nonlinearity of humidity and humidity affected by temperature are compensated in circuit and data processing respectively. On the circuit, the sampling principle and the structure of the operational amplifier in the design are introduced. The data processing is mainly polynomial fitting of the chip test data by MATLAB software, so as to improve the humidity measurement accuracy. After the final fitting, the humidity error can reach ±1.5RH in the range of 10 °C to 50 °C.

1. Introduction

People's lives are closely related to humidity, and with the development of The Times, it is difficult to find a field that has little to do with humidity. Therefore, how to accurately obtain the humidity in the environment is very important [1]. Now the humidity sensor is mainly divided into two kinds: resistive and capacitive. The principle of capacitive humidity sensor is that the dielectric constant of humidity sensitive material changes with the change of humidity, which is further shown as the capacitance value changes. Because the capacitive humidity sensor is easy to integrate, high sensitivity, so its principle is widely used in the integrated humidity sensor. Because humidity is closely related to temperature, it is generally a sensor integrated with temperature and humidity.

This paper mainly introduces how to compensate the humidity of the temperature and humidity sensor to improve its humidity accuracy. The structure of this paper is arranged as follows: The second section mainly introduces the basic principle of circuit design in this paper, the third section is the nonlinear compensation method of humidity under constant temperature, the fourth section introduces the nonlinear compensation method of humidity under temperature change, and the last section is the summary of the whole paper.

Content from this work may be used under the terms of the Creative Commons Attribution 3.0 licence. Any further distribution of this work must maintain attribution to the author(s) and the title of the work, journal citation and DOI.
Published under licence by IOP Publishing Ltd

2. Basic principles of circuit design

2.1 Sampling circuit principle:

The sampling part of the circuit uses the switching capacitor sampling circuit, Vin is the input voltage[2], S is the switch, and its principle is as follows:

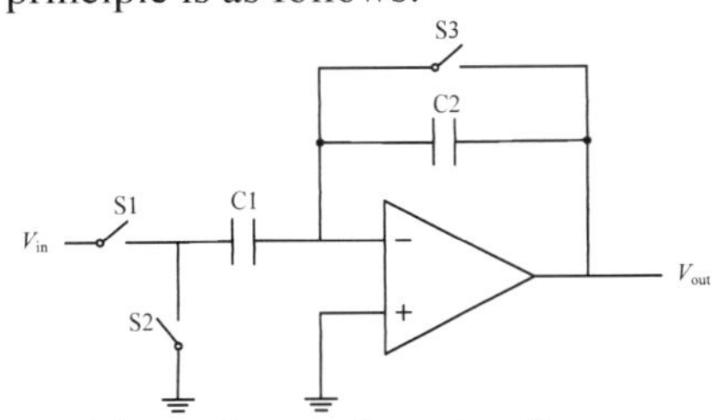

Figure 1 sampling circuit

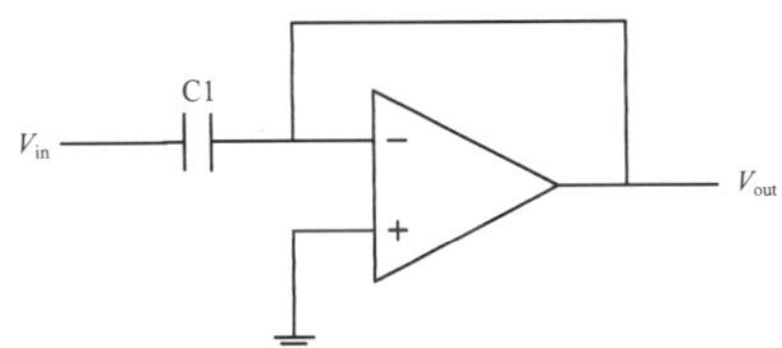

Figure 2 The sampling period

Sampling stage: switches S1 and S3 are closed, S2 is disconnected, and the charge quantity of capacitor C1 Qc1 is:

$$Q_{C1} = V_{in} * C1 \tag{1}$$

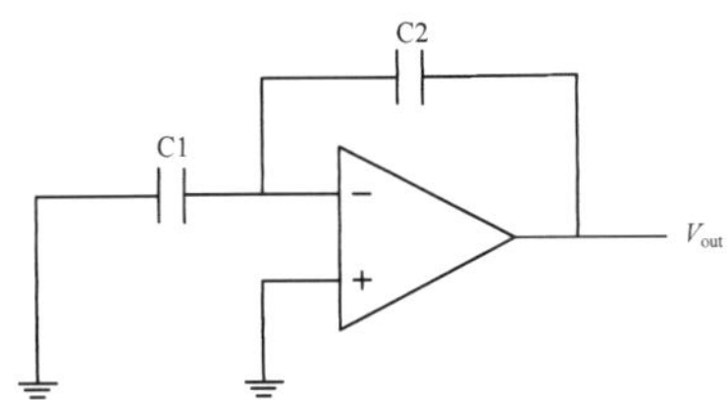

Figure 3 The sampling period

Integration stage: switch S2 is closed, S1 and S3 are disconnected. Since the voltage at both ends of capacitor C1 is 0V[3], the capacitor on capacitor C1 is transferred to capacitor C2, so the output voltage Vout is:

$$V_{out} = \frac{Q_{C2}}{C2} = \frac{V_{in} * C1}{C2} \tag{2}$$

When humidity is measured, the input voltage Vin is the reference voltage output by the bandgap reference circuit, which is used to charge the humidity sensitive capacitor. When humidity changes[4], the capacitance value of the humidity sensitive capacitor changes. According to Formula (3), the transferred charge changes [5], so as to realize the sampling and integration of humidity signal. The principle of temperature measurement is basically the same, except that the input voltage proportional to temperature charges the sampling capacitor[6].

2.2 Gain enhanced operational amplifier

In this design, a first-order Sigma-Delta ADC structure is adopted. The circuit used mainly affects the performance of the integrator, that is, the performance of the operational amplifier. The gain-enhanced operational amplifier [7] adopted in this design improves the output impedance through two auxiliary operational amplifiers, thus improving the gain.

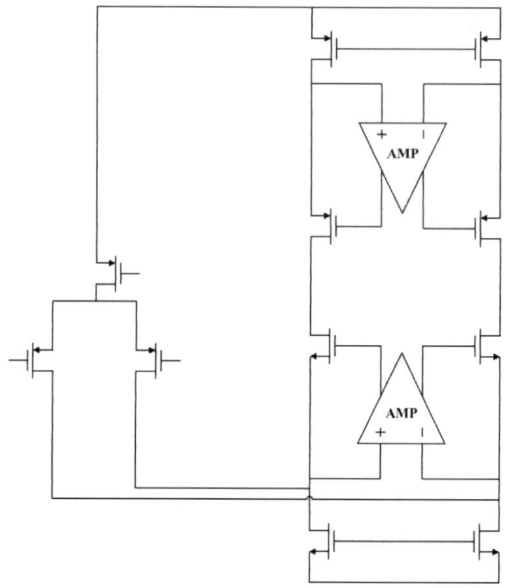

Figure 4. Gain enhanced operational amplifier

3. Constant temperature humidity compensation

After a lot of tests, it is found that there is a nonlinear relationship between humidity and hygrocapacitance at constant temperature, as shown in Figure 5[8]. It is difficult to compensate the nonlinearity from the analog circuit, but the accuracy of humidity can be significantly improved by processing the measured data from the perspective of data processing.

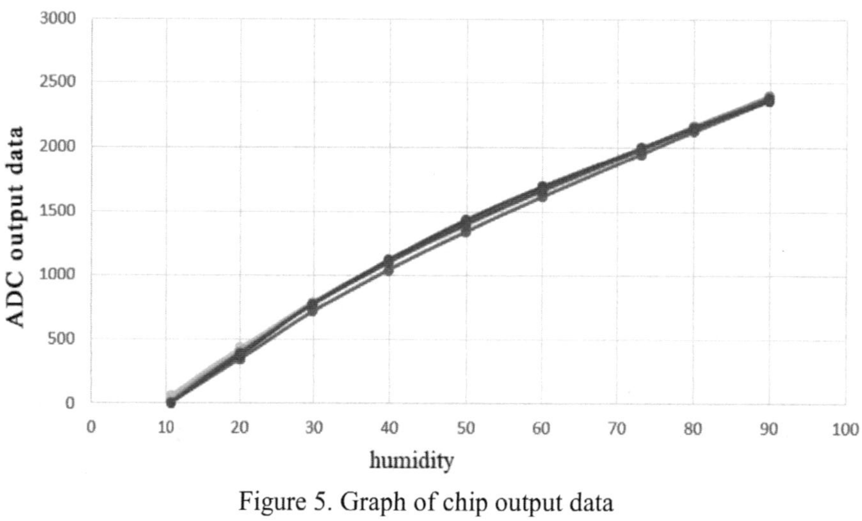

Figure 5. Graph of chip output data

Figure 6 and Figure 7 show the curve and humidity error of the chip after polynomial fitting of data by MATLAB software.

The steps of data fitting using MATLAB are as follows:

1. Use the readtable function to extract data from Excel files
2. Obtain the sub-term coefficients after data polynomial fitting by polyfit function
3. Obtain the fitted value through polyval function

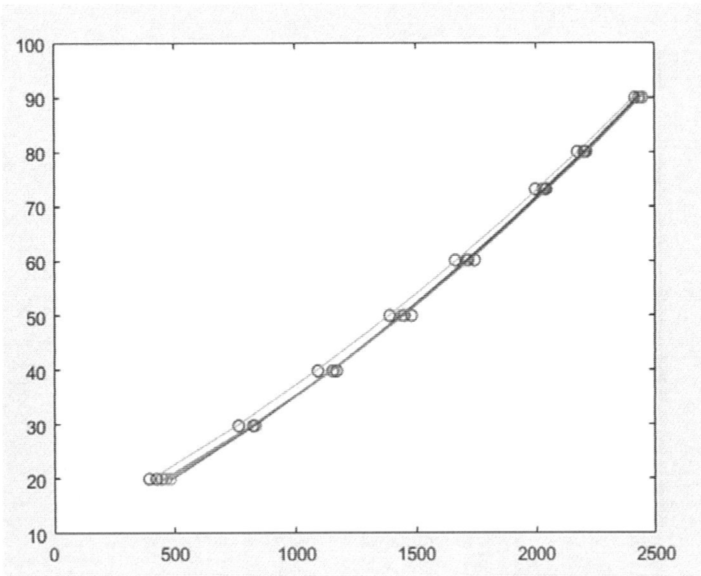

Figure 6 Fitting curve of test data of multiple chips

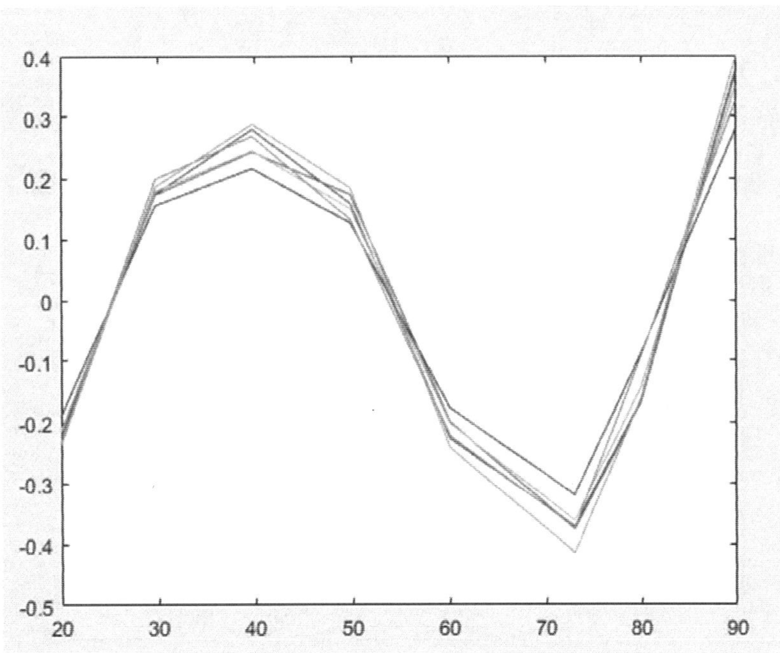

Figure 7 Humidity error curve after fitting

4. Compensation for temperature characteristics of humidity

The input of the humidity sensor will also be affected by different temperatures, especially under high humidity conditions. The specific test data is shown in Figure 8.

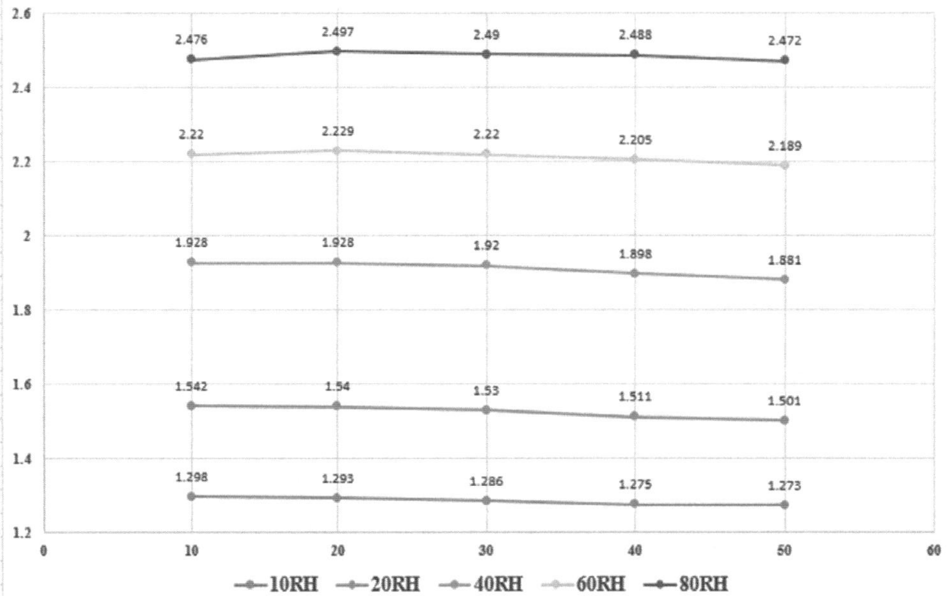

Figure 8 Humidity sensitive capacitance curve with temperature

It can be found from the curve that the humidity sensitive capacitance and temperature change can be approximately negative correlation. The compensation is mainly carried out from two aspects[9]. One is through circuit modification. When humidity sampling is carried out, the input PTAT voltage is compensated by appropriate coefficient, as shown in Figure 9[10]. Second, polynomial fitting was carried out for the output humidity data and temperature data. Finally, the humidity error after fitting was shown in Figure 10, which could be within ±1.5 within the range of 10°C to 50°C[11].

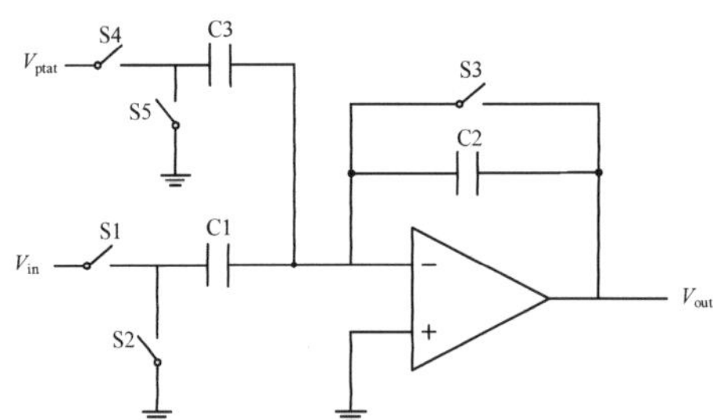

Figure 9. Principle of temperature characteristic compensation circuit for humidity

$$V_{out} = \frac{V_{ptat} * C3 + V_{in} * C1}{C2} \quad (3)$$

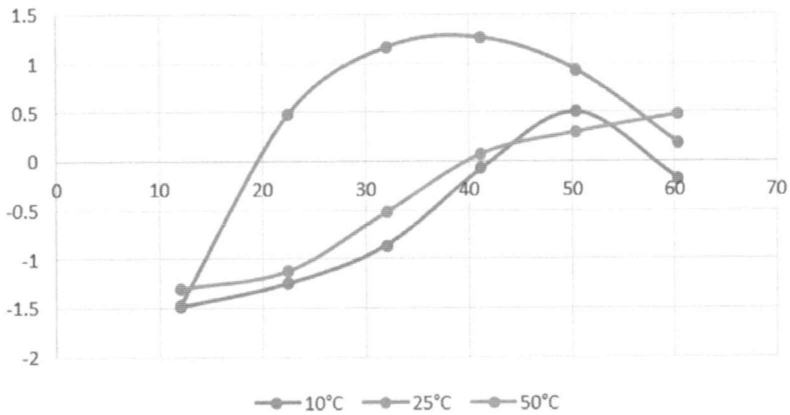

Figure 10. Humidity error curve after temperature compensation and fitting

5. Conclusion

This paper presents a reliable method to improve the accuracy of temperature and humidity sensor, and introduces the basic principle of switching capacitance sampling and the structure and principle of gain enhanced operational amplifier.At the same time, the nonlinear problem of humidity in the temperature and humidity sensor is analyzed, and the nonlinear problem of humidity itself and the influence of temperature on humidity are analyzed and the compensation method is proposed.The results show that this method can improve the accuracy of humidity sensor and has reliability.

References

[1] Yang Bin, Ying Xiumei, Yang Huazhong. A High-Speed HighResolution Sample-and-HoldCircuit[J]. Chinese Journal of Semiconductors.Vol.28, No.10,Oct.2007.

[2] Rohana Musa, Yuzman Yusoff, Tan Kong Yew . Design of SingleStage Folded-Cascode Gain Boost Amplifier for 100mW 10-bit 50MS/s Pipelined Analog-to-Digital Converter.ICSE2006Proc.2006.

[3] T. V. Burmas, K. C. Dyer, P. J. Hurst, and S. H. Lewis, "A second-order double-sampled delta-sigma modulator using additive-error switching," IEEE J. Solid-State Circuits, vol. 31, no. 3, pp.284–293, Mar. 1996.

[4] F. Ueno, T. Inoue, K. Sugitani, M. Kinoshita, and Y. Ogata, "An oversampled sigma–delta A/D converter using time-division multiplexed integrator," in IEEE Proc. 33rd Midwest Symp. Circuits Syst., Aug.1990, vol. 2, pp. 748–751.

[5] P. J. Hurst and W. J. McIntyre, "Double sampling in switched-capacitor delta-sigma A/D converters," in Proc. 1990 IEEE Int. Symp. Circuits Systems, New Orleans, LA, May 1990, pp. 902–905

[6] S. S. Bazarjani, "Mixed analog-digital design considerations in deep sub-micron CMOS technologies," Ph.D. dissertation, Carleton University,Ottawa, Ont., Canada, 1996

[7] H. Yoshizawa and G. C. Temes, "High-linearity switched-capacitor circuits in digital CMOS technology," in Proc. IEEE Int. Symp. Circuits and Systems, 1995, pp. 1029–1032.

[8] H. Yoshizawa, "High-linearity switched-capacitor circuits in digital CMOS technologies," Ph.D. dissertation, Oregon State University, Cor-vallis, OR, 1997.

[9] K. Haug, G. C. Temes, and K. Martin, "Improved offset compensation schemes for switched capacitor circuits," in Proc. IEEE Int. Symp.Circuits and Systems, 1984, pp. 1054–1057.

[10] H. Yoshizawa, Y. Huang, and G. C. Temes, "MOSFET-only switched-capacitor circuits in

digital CMOS technologies," in Proc. of the IEEE Int. Symp. on Circuits and Systems, 1997, pp. 457–460.
[11] B. E. Boser and B. A. Wooley, "The design of sigma-delta modulation analog-to-digital converters," IEEE J. Solid-State Circuits, vol. 23, pp.1298–1308, Dec. 1988.

Performance analysis of congestion control protocols over 5G mmWave links

Yi Luo[1,a], Min Wang[1,b*], Deteng Zhao[1,c], Chun Liu[1,d]

[1]College of Computer Science and Information Technology Yunnan Normal University,Kunming, P.R. China

[a]email: cqjtu307@outlook.com

[c]email: 89089201@qq.com

[d]email: 785604676@qq.com

* Corresponding author: Min Wang, [b]email: honghewangmin@sina.com

Abstract. Millimeter wave technology is an important basic technology in 5G applications. It has higher bandwidth and lower delay. But high dynamics of the channel poses a major challenge to the design of congestion control protocols. Based on ns3, this paper conducts extensive experiments on the TCP congestion control protocols, and analyzes the transmission performance of different protocols over 5G millimeter wave links in detail. It will lay a foundation for further research on congestion control protocols suitable for 5G millimeter wave communications.

1. Introduction

Currently, 5G (the fifth-generation technical standard for cellular networks) is becoming more and more commercialized. Millimeter wave technology is an important basic technology in 5G applications. Because of the higher communication frequency, it has higher bandwidth and lower delay. But it also brings higher signal attenuation and more serious diffraction problems, which results in high dynamics of the channel.

In order to study the transmission performance of the existing TCP congestion control protocol under 5G millimeter wave network, this paper simulates three 5G millimeter wave network scenes through ns3, and selects six congestion control protocols for comparison and analysis. These protocols respectively represent packet loss-based, delay-based, and bandwidth-based congestion control mechanisms. The simulation experiment shows the congestion window, round-trip delay and throughput changes of each protocol under 5G millimeter wave, and analyzes the reasons for the different performance of the protocol.

The rest of this article is organized as follows. In the next section, the main control logic and characteristics of the selected protocol are introduced, the relevant experimental analysis work is exemplified and their conclusions are summarized. Then it lists the simulated experiment environment and experiment parameters carried out in this article. Finally, according to different simulation experimental environments, the experimental results of throughput, congestion window and round-trip delay with time are obtained, and the evaluation results of different types of protocols are obtained through comparative analysis of the experimental results.

Content from this work may be used under the terms of the Creative Commons Attribution 3.0 licence. Any further distribution of this work must maintain attribution to the author(s) and the title of the work, journal citation and DOI.
Published under licence by IOP Publishing Ltd

2. Related Work

After decades of development of the TCP congestion control protocol, various algorithms have emerged. Based on their different ways of judging congestion, they can be roughly divided into loss-based (such as NewReno), delay-based (such as Vegas), and bandwidth-based (such as BBR). We select 6 congestion control protocols from these three categories, which are NewReno, Cubic, HighSpeed, Vegas, Westwood, and BBR. The following is a brief overview of these six protocols in turn.

On the basis of Reno, NewReno solves the problem of multiple halving of the congestion window caused by multiple packet loss in one congestion. Cubic designs a cubic function in the congestion avoidance phase to achieve RTT fairness and fast convergence of the cwnd. HighSpeed uses formulas 1 and 2 to change cwnd. When cwnd is small, the behavior is similar to that of standard TCP. When cwnd exceeds the threshold, a(w) and b(w) also change to keep cwnd in a larger range.

$$cwnd = cwnd + a(w)/cwnd \tag{1}$$

$$cwnd = \left(1 - b(w)\right) * cwnd \tag{2}$$

The improvement of the previous three algorithms to classic TCP focuses on avoiding excessive reduction of the cwnd when encountering congestion and how to quickly restore the cwnd after reduction. Vegas judges the link status and sets cwnd through the change of RTT. And Vegas adopts a two-stage RTT reciprocating growth method in the slow-start stage, the first period increases exponentially, and the latter period observes the changes of RTT.

Westwood will estimate the link bandwidth and use it to set cwnd and ssthresh when congestion is detected. BBR uses formulas 3 and 4 to estimate the appropriate BDP of the link, and uses it to set cwnd and pacing rate.

$$\widehat{RTprop} = RTprop + \min(\eta_t) = \min(RTT_t) \tag{3}$$

$$\widehat{BtlBw} = \max(deliveryRate_t) \tag{4}$$

The current evaluation of various congestion control protocols under 5G networks is not perfect. Reference [1-4] mainly analyzes the throughput, delay, cwnd, fairness and other performance indicators of different TCP congestion control algorithms under wired links. Reference [1-2] evaluated the behavior, performance and fairness of each protocol under wired links under different speed, simulation time, number of traffic sources and other parameters. Reference [3-4] focus on various performance tests of the BBR algorithm and the fairness issues between BBR and loss-based algorithms. These studies have conducted relatively sufficient tests on various protocols and their behaviors, but they are all carried out under reliable wired links, and their behavior and stability under lossy links have not been proven yet.

Reference [5-9] evaluated various performance indicators of TCP congestion control protocol in wireless network environment. Reference [5] and [9] evaluated the performance of TCP under 5G millimeter wave networks under different variables, including server location, network connection mode, congestion control protocol, data packet size, and RLC buffer space, but its selection of congestion control protocols is limited. Reference [6] evaluated the cellular uplink, and combined with NewReno, Cubic, and BBR to analyze the behavior and performance of the protocol in stationary and mobile scenes, but congestion usually occurs in the downlink. Reference [7] Analyzed the behavior of several popular congestion control algorithms in the real-time LTE network static and moving away from the base station. Reference [8] studied the performance fairness of BBR and Cubic streams in Wi-Fi networks. This type of research mainly tested the performance and fairness of loss-based congestion control protocol and BBR under lossy links such as Wi-Fi, and did not fully test various protocols.

3. Experiments and Results

3.1. Topology

The 5G millimeter wave network simulation environment is constructed in ns3 using the modules provided by Reference [10]. Scene one is an environment without any obstacles for testing the highest throughput that each protocol can achieve. The user in scene two is completely obscured by a building to test the performance of the protocol under diffraction. In scene three, the user keeps moving at a constant speed to test the performance of the protocol under frequent signal changes. Basically, the performance of each protocol under various conditions can be detected through three experiments. In Figure 1, the left picture shows the topology of the experimental scene, and the right picture shows the change in the ratio of signal to interference plus noise ratio.

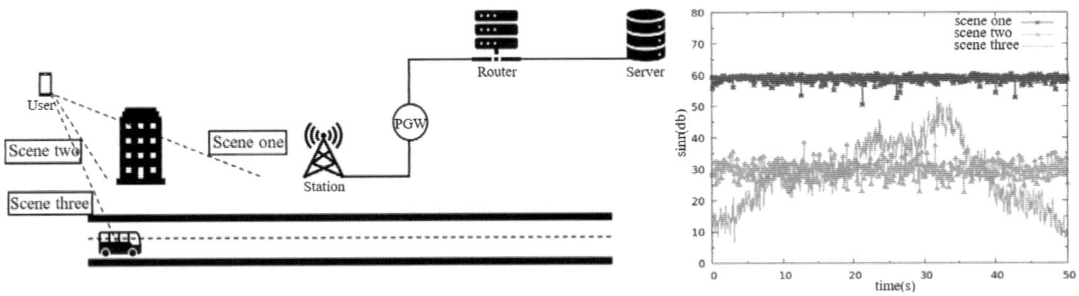

Figure 1 Simulation experiment scene

3.2. Scene one

It can be seen from Figure 2 that in the case of obstacle-free interference, each protocol basically has a stable and high throughput, but there are also some problems. Vegas is limited by latency and is more cautious about the growth of the congestion window, resulting in too low throughput. Westwood's estimate of bandwidth is too conservative, which reduces the congestion window too much and cannot quickly restore throughput, so it is not as high as NewReno. HighSpeed is too aggressive, it will not easily reduce the congestion window excessively after congestion occurs, and the recovery is extremely fast. BBR needs to actively reduce the throughput every 10s to measure the minimum delay.

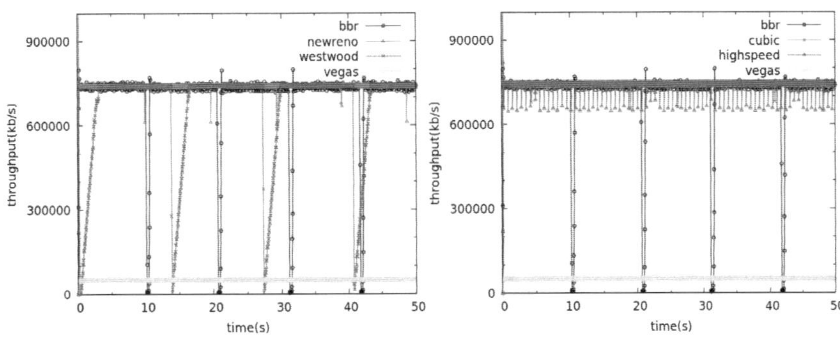

Figure 2 Scene one graph of throughput over time

3.3. Scene two

3.3.1. throughput and cwmd

In scene two, the signal is transmitted by diffraction due to the occlusion of the building, which increases the probability of data packet errors. Combining Figures 3 and 4, the BBR can still accurately detect the available bandwidth, and then set an appropriate congestion window, so it can maintain the throughput as in scene one, but some fluctuations inevitably occur. Vegas is far from reaching the throughput that should be affected, so no analysis is done. The packet loss-based congestion control protocol cannot distinguish between congested packet loss and non-congestion packet loss, so it significantly increases the action to reduce the congestion window, resulting in a significant change in throughput. Because HighSpeed is too aggressive, it can quickly recover even if the throughput is reduced many times. NewReno and Westwood algorithm are similar, the difference is that Westwood uses the estimated bandwidth to set the congestion window after detecting congestion. Therefore, the congestion window and throughput changes of the two are similar, but it can be clearly seen that the estimated bandwidth of Westwood in 5G millimeter wave is smaller, which results in a slightly worse recovery than NewReno. Surprisingly, Cubic's congestion window and throughput are significantly reduced compared to Scene one. Therefore, Cubic seems to be unable to infer the appropriate cubic function in the 5G millimeter wave network under certain interference, and there is also a reduction in the congestion window caused by non-congestion, which makes it extremely difficult to improve throughput.

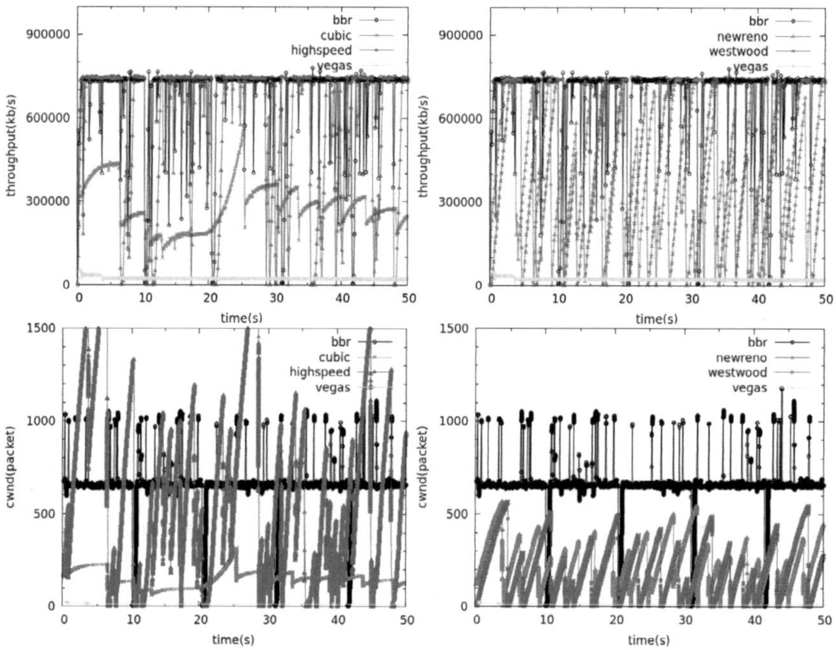

Figure 3 Scene two throughput and cwnd over time

3.3.2.rtt

It can be seen from Figure 4 that the round-trip delay of basically all protocols fluctuates sharply. Among them, HighSpeed is too aggressive, data packet queuing is serious, and delay is high. The remaining packet loss-based protocols failed to reach the peak rate as in Scene one, and the congestion window was reduced due to non-congested packet loss, so the average delay was lower.

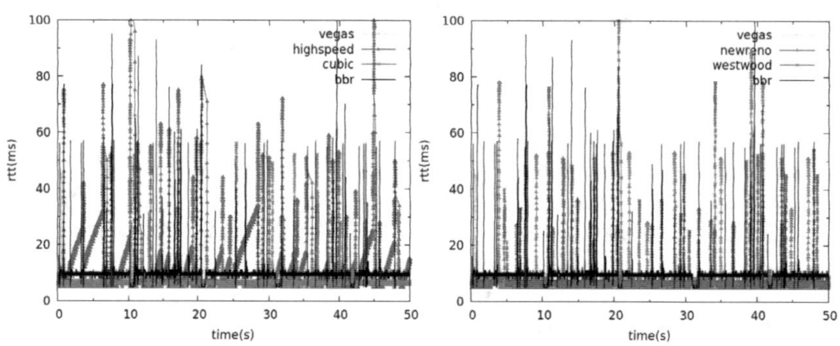

Figure 4 Scene two rtt over time graph

3.4.Scene three

3.4.1.throughput and cwnd

The performance of Vegas in mobile scene is the same as scene one and two, so no analysis is done. In the time periods of 0-25 seconds and 35-50 seconds, the signal is weak and the interference is large. The congestion window of the congestion control protocol based on packet loss is difficult to rise to a larger value, but BBR can still detect several times of high bandwidth in the interference and set a higher congestion window. However, the susceptibility to interference of 5G millimeter waves causes the throughput of BBR to fluctuate drastically. However, when the available bandwidth becomes smaller, the BBR still uses a larger value to set the sending rate, which will inevitably cause a delay to cause the BBR to enter the queue emptying stage. Such repetition will produce violent fluctuations. In the protocol based on packet loss, the more aggressive the protocol, the greater the fluctuation of the throughput.

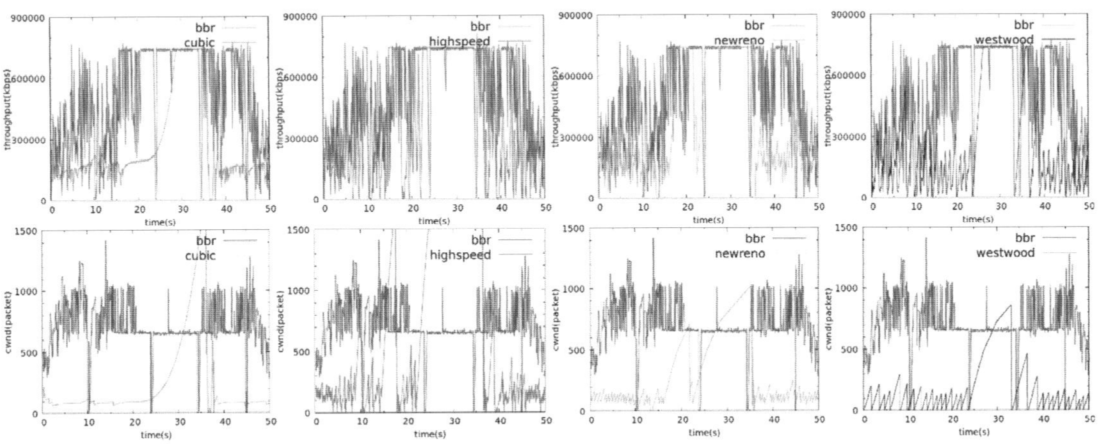

Figure 5 Scene three throughput and cwnd over time

3.4.2.rtt

It can be seen from the foregoing that BBR estimates the BDP based on the measured maximum bandwidth, and sets the congestion window and sending rate based on this. However, it can be seen from Figure 6 that the BBR has a huge time delay and drastic changes. It shows that the BDP measured by BBR at this time is different from the actual network, which does not conform to the current network environment, nor does it conform to the theory of congestion control at the knee point proposed by BBR. According to the different degree of aggressiveness of other protocols, the degree of delay change is also different. But overall, compared with BBR, there is a lower latency. In the case of low noise, BBR can recover a relatively stable and low delay, and the delay of a protocol that judges congestion based on packet loss will gradually increase.

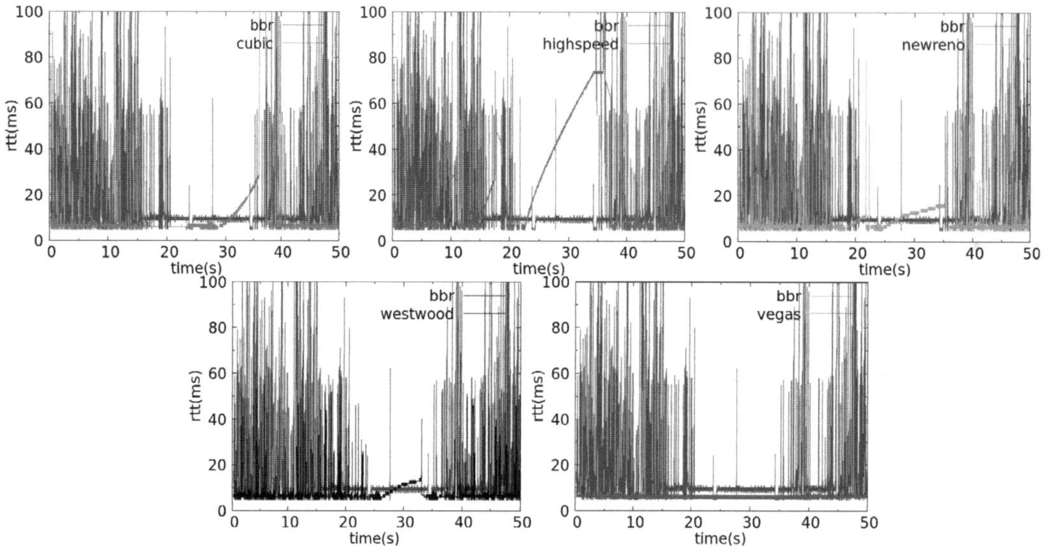

Figure 6 Scene three graph of rtt over time

4. Conclusions

In order to study the transmission performance of the existing TCP congestion control protocol under 5G millimeter wave network, six protocols are simulated in 5G millimeter wave network scenes. The experimental results show that all protocols achieve different performance in different scenes. For the protocols based on packet loss, their throughput and delay are relatively stable when the noise is large. However, they often make misjudgments on congestion to reduce the congestion window, resulting in lower throughput. When the noise is small and the signal is strong, they will increase the congestion window aggressively, resulting in an increase of delay. The delay-based protocols, such as Vegas, are too sensitive to delay. Therefore, they maintain extremely low and stable throughput and delay in static or dynamic environments. BBR, a representative protocol based on link bandwidth, has excellent performance when the signal is strong and the noise is small. However, BBR's estimation of BDP is too aggressive when the 5G signal is interfered by strong noise, resulting in great fluctuations in throughput and delay. This does not conform to the theory of congestion control at the knee point proposed by BBR. In the case of low noise, the throughput of BBR still fluctuates greatly.

Acknowledgments

This work was supported by the National Natural Science Foundation of China (61561055), the National Natural Science Foundation of China (61562093), Key Project of Applied Basic Research Program of Yunnan Province (2016FA024), and the Scientific Research Fund Project of Education Department of Yunnan Province (2021J0429).

References

[1] N. I. A. Altahir and H. A. Ali.(2016) Performance evaluation of TCP congestion control mechanisms using NS-2. In: 2016 Conference of Basic Sciences and Engineering Studies (SGCAC). Khartoum, Sudan. pp. 146-151.

[2] M. Hock, R. Bless and M. Zitterbart. (2017) Experimental evaluation of BBR congestion control. In: 2017 IEEE 25th International Conference on Network Protocols (ICNP). Toronto, ON, Canada. pp. 1-10.

[3] K. Sasaki, M. Hanai, K. Miyazawa, A. Kobayashi, N. Oda and S. Yamaguchi. (2018) TCP Fairness Among Modern TCP Congestion Control Algorithms Including TCP BBR. In: 2018 IEEE 7th International Conference on Cloud Networking (CloudNet). Tokyo, Japan. pp. 1-4.

[4] P. Hurtig et al. (2018) Impact of TCP BBR on CUBIC Traffic: A Mixed Workload Evaluation.

In: 2018 30th International Teletraffic Congress (ITC 30). Vienna, Austria. pp. 218-226.
[5] M. Zhang et al. (2016) Transport layer performance in 5G mmWave cellular, In: 2016 IEEE Conference on Computer Communications Workshops (INFOCOM WKSHPS). San Francisco, CA, USA. pp. 730-735.
[6] A. Parichehreh, S. Alfredsson and A. Brunstrom. (2018) Measurement Analysis of TCP Congestion Control Algorithms in LTE Uplink. In: 2018 Network Traffic Measurement and Analysis Conference (TMA). Vienna, Austria. pp. 1-8.
[7] M. Akselrod and M. Fidler. (2020) TCP Congestion Control Performance on a Highway in a Live LTE Network. In: 2020 IEEE 92nd Vehicular Technology Conference (VTC2020-Fall). Victoria, BC, Canada. pp. 1-7.
[8] K. Miyazawa, S. Yamaguchi and A. Kobayashi. (2020) Performance Evaluation of TCP BBR and CUBIC TCP in Smart Devices Downloading on Wi-Fi. In: 2020 IEEE International Conference on Consumer Electronics - Taiwan (ICCE-Taiwan). Taoyuan, Taiwan. pp. 1-2.
[9] M. Zhang et al. (2019) Will TCP Work in mmWave 5G Cellular Networks?. IEEE Communications Magazine, vol. 57, no. 1, pp. 65-71
[10] M. Mezzavilla et al. (2018) End-to-End Simulation of 5G mmWave Networks. IEEE Communications Surveys & Tutorials, vol. 20, no. 3, pp. 2237-2263

Modeling and Simulation of Dual-Active-Bridge Based on PI Control

Xiaorui Ai[a], Dongsheng Zuo[b*], Ye Zhang[c]

School of Electrical Engineering, Shanghai Dianji University, Shanghai, China

[a]email: aisuperfood@qq.com , [c]email: 253891469@qq.com

[b]email: zuods@sdju.edu.cn

Abstract: Dual-active-bridge (DAB) is a DC/DC converter,which is commonly used in solid-state-transformer (SST) and electric vehicle (EV).In order to obtain the expected output voltage,the converter needs to be modeled and controlled.Firstly,the working modes in different time intervals of the switching cycle under single-phase-shift (SPS) modulation are analyzed,and the mathematical models of output voltage,current stress and phase-shifting duty cycle are constructed.Then,the simulation model is built on Simulink,and the PI controller is used for closed-loop voltage control,The accuracy of the mathematical model is verified.

1. Introduction

Dual-active-bridge (DAB) circuit has gradually become the main circuit topology of energy storage charging pile because of its characteristics of high power density, input and output magnetic isolation, two-way energy flow, wide soft switching range, easy modularization and multi-level realization [1-4].

DAB is usually modulated by SPS at high transmission power level[5].By changing the duty cycle of phase shift between bridges,the magnitude and direction of transmission power can be changed. When voltage transfer ratio changes,the current stress of the converter will also change simultaneously[6-8].Therefore, it is necessary to build the model of transmission power and current stress and use appropriate control to obtain ideal output results.

Firstly,this paper analyzes the working modes of DAB in different time intervals under SPS modulation.According to the mathematical expression of inductance current in steady state,the mathematical model of transferred power and current stress is established by integrating the power in sections.Then,the main circuit model is built on MATLAB/Simulink,the output voltage is sampled,and the PI controller is used to realize the negative feedback of voltage loop,Finally,the effectiveness of the model is verified by the simulation results under the closed-loop.

2. Analysis of DAB

2.1. The Topology of DAB

The main circuit of DAB converter is shown in Figure 1.The full bridge at primary side is composed of switches S_1-S_4,and the full bridge at secondary side is composed of switches Q_1-Q_4.

V_1 and V_2 represent the DC voltage on the primary side and secondary side respectively,C_1 and C_2 represent the capacitance parallel on the DC voltage side,which plays the role of voltage stabilization and filtering.The transformation ratio of high-frequency transformer is n:1,ignoring the current on its excitation inductance branch,L is composed of transformer leakage inductance and external auxiliary

Content from this work may be used under the terms of the Creative Commons Attribution 3.0 licence. Any further distribution of this work must maintain attribution to the author(s) and the title of the work, journal citation and DOI.
Published under licence by IOP Publishing Ltd

inductance.v_p and v_s represent the voltage between the midpoint of the two bridge arms at the primary side and the secondary side respectively.i_L is the inductance current,and the voltage transmission ratio is M,M = nV_2/V_1. The load is represented by resistance R.

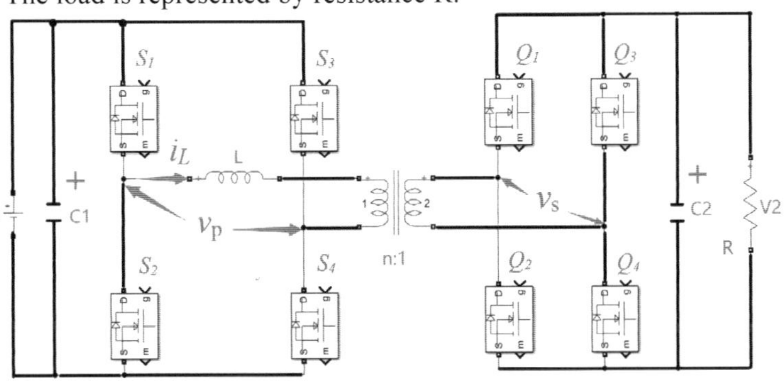

Figure 1. The Main Circuit of DAB

2.2. The Driving Signal of DAB

The driving signal sequence of each switch and the main working waveforms of DAB converter are shown in Figure 2.For simplicity,the symbols S_1-S_4 and Q_1-Q_4 of each switch represent the driving signal,the high level represents the device on,the low level represents the device off,T represents half of the switching cycle value T_s,f represents the switching frequency,T_s =1/f,and D_0 represents the inter bridge phase shift duty cycle.

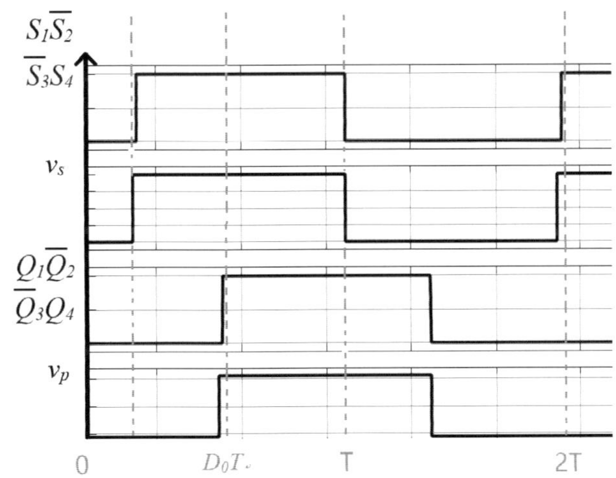

Figure 2. The Driving Signal of DAB

3. Modeling

3.1. Modeling of Current Stress

Here, take the situation of the transferred power from the primary side to the secondary side and M<1 for an example.When the converter works in steady state,the inductance current and the full bridge output waveformS are shown in Figure 3.

The current is the largest at time t_2,that is,the current is the current stress by the converter.By observing the waveform,we can get some infermation from it:

$$\begin{cases} t_1 - t_0 = D_0 T \\ t_2 - t_1 = (1 - D_0)T \end{cases} \tag{1}$$

According to the current direction in Figure 3,the voltage at both ends of the inductor exists:

$$u_L(t) = L\frac{\Delta i_L(t)}{\Delta t} \tag{2}$$

Combined with the switching modes of DAB converter in different time periods, according to Kirchhoff Voltage Laws(KVL):

$$\begin{cases} t_0 \sim t1 : u_L(t) = V_1 + nV_2 \\ t_1 \sim t2 : u_L(t) = V_1 - nV_2 \end{cases} \tag{3}$$

According to equations (1)-(3),it can be obtained that:

$$i_{L\max}(D_0, M) = \frac{TV_1}{2L}(2MD_0 - M + 1) \tag{4}$$

3.2. Modeling of Transferred Power

Although a switching cycle is divided into four different operating modes,due to the symmetry of the waveform,the power in the 0~T interval is equivalent to the power in the switching cycle.The transferred power can be obtained by piecewise integration in two intervals.

$$P(D_0, M) = \frac{1}{T}\left(\int_{t_0}^{t_1} v_s(t)i_L(t)dt + \int_{t_1}^{t_2} v_s(t)i_L(t)dt\right) \tag{5}$$

According to equations (1)-(5),it can be obtained that:

$$P(D_0, M) = \frac{V_1^2 T}{L} MD_0(1 - D_0) \tag{6}$$

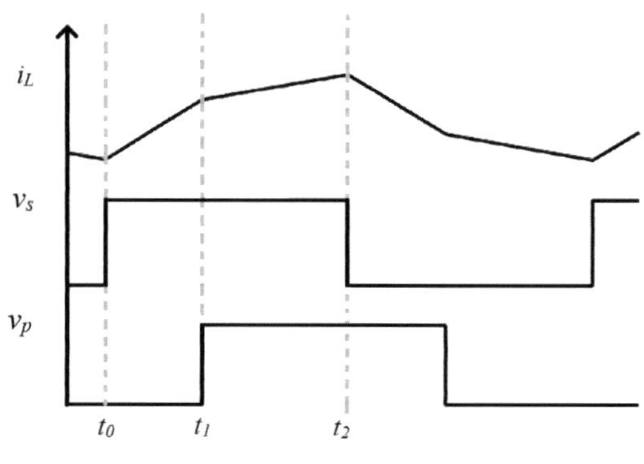

Figure 3. The Main Waveforms of DAB

4. Simulation and Results

4.1. PI Control and Simulation

Based on equation (6) and power formula,it can be found that:

$$V_2 = \frac{V_1 TR}{L} D_0(1 - D_0) \tag{7}$$

The mathematical model of output voltage and phase shift duty cycle is obtained by equation (7).In this case,we can build PI controller to realize negative feedback of voltage loop.

Figure 4 depicts the PI control block diagram.In practical application,it also needs driving circuit,sampling circuit,protection circuit and auxiliary power supply.In the diagram,S-fun convert the following output voltage to the corresponding phase-shifting duty cycle,from the equation (7),the expression of S-fun can be derived:

$$D0 = \frac{1 \pm \sqrt{1 - \frac{4LV_2}{V_1TR}}}{2} \tag{8}$$

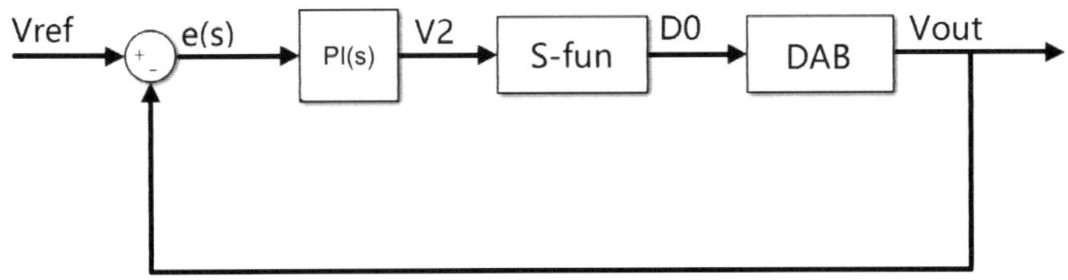

Figure 4.PI Control Block Diagram

In order to verify the correctness of the model and the control strategy,a simulation model on the Matlab/Simulink was constructed.The parameters of the model are shown in Table 1.

Table 1.Parameters of The DAB Model

Input Voltage V_1	200V
Reference Voltage Vref	100V
Transformer Turn Raito n	1:1
Switching Frequency f	20kHz
Load R	16Ω
Voltage Stabilizing Capacitor C1	1000uF
Filter Capacitor C2	1000uF
Inductor L	100uH

4.2. Results

The waveforms of output voltage are illustrated in Figure 5.

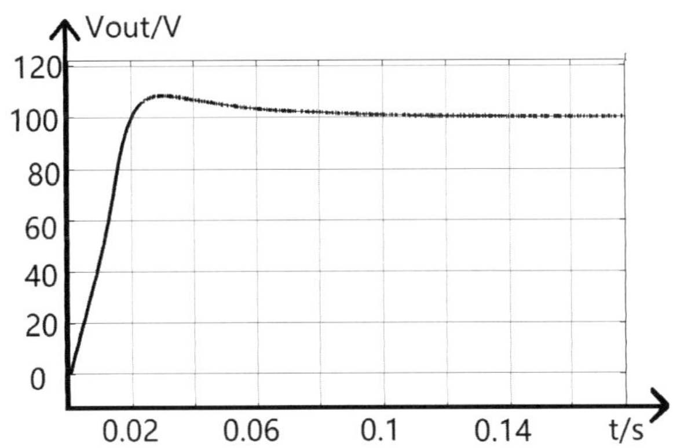

Figure 5.The Output Voltage

It can be seen from Figure 5 that the rise time is about t = 0.018s,the system has good dynamic response and the steady-state error is 0.The dynamic response speed can be further improved by appropriately modifying the parameters of the PI controller.

5. Conclusions

By analyzing the working waveform of DAB under SPS modulation,the mathematical model of output voltage and current stress about phase shift duty cycle and voltage transmission ratio is obtained.Then, by building Simulink simulation model,the correctness of the proposed control model is verified under voltage closed-loop control with PI controller, which has a certain engineering practice value.

Reference

[1] Shen, K., Tong, A., Hang, L., Li, G., et al. (2019) Modulation strategy of dual active bridge converters applied in renewable energy systems. Automation of Electric Power Systems., 43: 154-159.
[2] Zhao, B., Song, Q., Liu, W., et al. (2014) Overview of dual-active-bridge isolated bidirectional DC-DC converter for high-frequency-link power-conversion system. IEEE Transactions on Power Electronics., 29: 4091-4106.
[3] Guan, M., Shen, J., Lou, P. (2020) Design scheme of DC substation based on modular dual-active-bridge converter. Automation of Electric Power Systems., 44: 172-178.
[4] Zhou, J., Wang, H., Zhang, J., et al. (2020) MMC-based solid state transformer's submodule capacitance optimization method based on fluctuation power delivery. Proceedings of the CSEE., 40: 3990-4003.
[5] Huang, J., Wang, Y., Li, Z., et al. (2016) Optimized modulation scheme of dual active bridge DC-DC converter based on triple-phase-shift control. Proceedings of the CSEE., 36: 1658-1666.
[6] Zhao, B., Yu, Q., Sun, W. (2012) Extended-phase-shift control of isolated bidirectional DC-DC converter for power distribution in microgrid. IEEE Transactions on Power Electronics., 27: 4667-4680.
[7] Zhao, B., Yu, Q., Sun, W. (2012) Bi-directional full-bridge DC-DC converters with dual-phase-shifting control and its backflow power characteristic analysis. Proceedings of the CSEE., 32: 43-50.
[8] Zhao, B., Song, Q., Liu, W., et al. (2013) Current-stress-optimized switching strategy of isolated bidirectional DC/DC converter with dual-phase-shift control. IEEE Transactions on Industrial Electronics., 60: 4458-4467.

Research and design of low-power, high-performance processor based on RISC-V instruction set architecture

GuangTang[1], Jiancheng, Li[1*]

School of Physics and Optoelectronic Engineering, Xiangtan University, Xiangtan, Hunan, 411105, China

The first author's e-mail: guangtang1858@163.com

Corresponding author's e-mail: lijc2016@qq.com

Abstract: In embedded IoT applications, in order to meet the needs of small area, low power consumption, and high performance, this paper designs a 32-bit low-power, high-performance microprocessor based on the RISC-V instruction set architecture. It uses a 3-stage pipeline structure with static branch prediction function, supports RV32IM instruction set, and adopts AHB bus as the control bus. The logic function of the processor was verified under the iverilog environment, and the simulation results showed that the processor can run normally. The CoreMark score of 2.38 CoreMark/MHz was measured by running the CoreMark running program, and it supports FreeRTOS transplantation. The analysis and verification results show that the design can be well applied to small-area, low-power embedded application scenarios.

1 Introduction

With the rapid development of electronic communications and integrated circuit industry, the characteristics of embedded microprocessors with high performance and low power consumption have been widely used in the IoT industry[1], and the ARM series products represented by STM32 account for a large market share in the field of embedded products. Although its development is relatively mature and perfect, its instruction set is relatively redundant and complex, and it requires expensive commercial authorization, which is not affordable by small and medium-sized companies and enterprises, bringing many problems to the development of the integrated circuit industry. As a new completely open source and free simplified instruction set architecture, RISC-V has a strong competitive advantage with low cost, low power consumption, support for third-party extensions, and security and reliability[2-4]. Since the birth of RISC-V, many universities and enterprises at home and abroad have conducted research based on the RISC-V instruction set. The RISC-V R&D team at the University of California, Berkeley took the lead in launching a processor called Rocket Chip; In 2018, Xinlai Technology launched the Hummingbird E203 processor; Zhaoyi Innovation also cooperated with Xinlai Technology to launch a customized RISC-V processor Bumblebee; Ali Pingtou released the high-performance processor Xuantie 910 based on the RISC-V architecture, and the first RISC-V China Summit was successfully held at ShanghaiTech University in June 2021.

RISC-V's flexible scalability and high customization make it have greater potential in high-end new markets (such as embedded IoT, artificial intelligence, etc.) that are less dependent on the ecosystem. In view of the demand for processor area and power performance of embedded devices, this paper uses the memory structure of Harvard structure based on the RISC-V processor architecture, uses a 3-stage pipeline structure, and designs a 32-bit low-power, high-performance RISC-V processor

Content from this work may be used under the terms of the Creative Commons Attribution 3.0 licence. Any further distribution of this work must maintain attribution to the author(s) and the title of the work, journal citation and DOI.
Published under licence by IOP Publishing Ltd

with single-transmit, sequential execution and out-of-order write-back, and adopts AHB bus interface to realize the interconnection of various devices.

2. RISC-V instruction set analysis

RISC-V is itself a directive set standard specification, like common C/C++ language specifications, POSIX and other system standards[5]. The modular design of the RISC-V architecture attempts to meet a variety of application scenarios through a unified architecture, which is not available in ARM and x86 architectures, and can achieve a balance between cost, power consumption, performance and so on. In particular, in the field of low-power small-area processors, it has the following obvious advantages:

(1) Completely open source: Open source adopts a relaxed BSD protocol, which is completely free for enterprises to use free of charge, and also allows enterprises to add their own instruction set expansion without having to open and share to achieve differentiated development;

(2) The architecture is simple: the RISC-V basic instruction set is only more than 40, plus other modular extension instructions for a total of dozens of instructions, without backward compatibility;

(3) Modular design: RISC-V its different parts are organized together in a modular manner, and users can flexibly choose different module combinations to achieve the needs of their own customized equipment;

(4) Complete toolchain: The RISC-V community has provided a complete toolchain, and chip design companies no longer worry about toolchain problems, just focus on chip design.

The RISC-V instruction format is a typical three-operand, 7-bit opcode instruction format. The RISC-V instruction set has six basic instruction formats (R/I/S/B/U/J), as shown in Figure 1: Where opcode represents a 7-bit instruction operator code, its role is to distinguish between different instructions; funct3 represents a 3-bit function code, funct7 represents a 7-bit function code, which can help distinguish different kinds of instructions; rs1 and rs2 represent two 5-bit source registers; rd is a 5-bit destination register, and the result of the instruction operation is stored in rd; imm represents an immediate number of different lengths, which can be used directly as an operand.

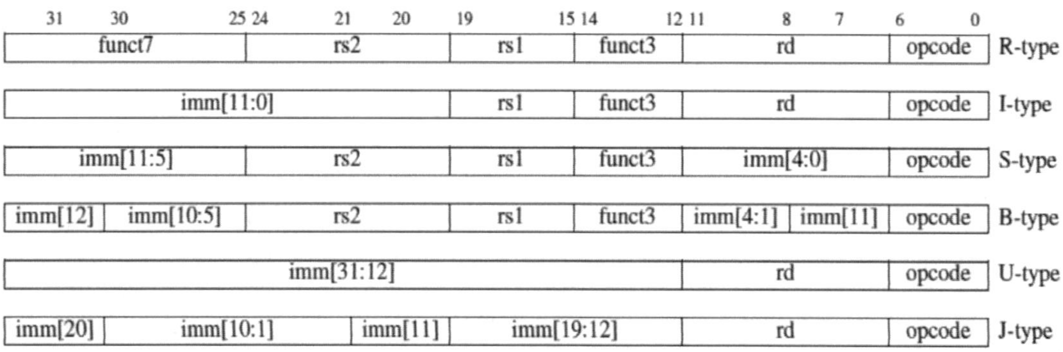

Figure 1. Six basic instruction formats of the RISC-V instruction set

3. RISC-V processor core design

3.1 The overall structure of RISC-V microprocessor

This article mainly designs a RISC-V microprocessor based on the characteristics of embedded software development, which mainly consists of a processor core, a bus, a debug unit, and some peripherals. The processor core adopts a 3-stage pipeline design, single-emitting, sequential execution, single-core 32-bit RISC-V microprocessor. Not only supports the RV32I basic instruction set module, but also supports the RV32M extended instruction set module. The overall structure of RISC-V SoC is shown in Figure 2:

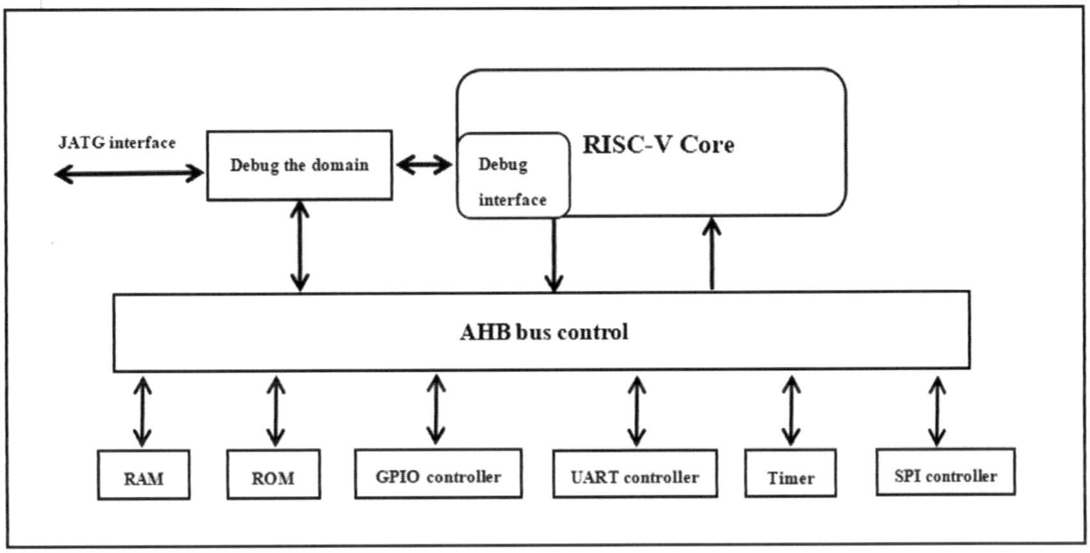

Figure 2. RISC-V SoC overall structure diagram

3.2 Pipeline design

An assembly line can essentially be understood as a means of exchanging area for performance and space for time[6]. However, the deeper the pipeline, the better, and the depth of the assembly line depends on the specific design requirements. Based on the processor performance and RISC-V instruction set characteristics, the design depth is a pipeline with a depth of 3, as shown in Figure 3. There are registers between each level of the pipeline in the classical MIPS five-level pipeline (fingering, decoding, execution, access, and writeback), and each pipeline series has its own combinatorial logic data path inside, and there is no reuse of resources between each other, so its area overhead is relatively large. The 3-level pipeline greatly reduces the number of hardware logic to a certain extent, and while ensuring basic functions, it not only reduces the hardware cost but also greatly reduces the system power consumption.

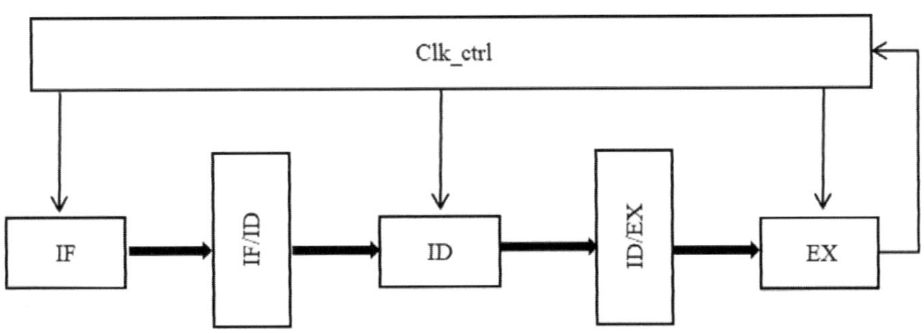

Figure 3. Three-stage pipeline structure of RISC-V processor

The first stage of the pipeline is the instruction fetch module: the PC fetches the instruction and accesses the instruction memory, and stores the fetched instruction in the register. The instructions fetched in the process of fetching can be divided into two types: branch jump instructions and non-branch jump instructions. If the address of a non-branch jump instruction is fetched, the fetch module can continuously fetch instructions from the ROM in sequence, and use the current PC+4 value as the address of the next instruction. The instruction fetch unit generates the PC value of the next instruction to be fetched every cycle to ensure that instructions can be continuously fetched from the instruction ROM in each cycle, thereby improving processor performance. In addition, during the fetch stage, a simple branch prediction is performed on the jump instruction. The type of jump instruction is shown in Table 1. The code for judging whether it is a jump instruction is as follows: assign prdt_taken = (dec_jal|dec_jarl|(dec_bxx & dec_bjp_imm[31]))

For the unconditional direct jump instruction, there is no need to judge the direction of the jump because it does not need to judge that the condition will definitely jump. Its target jump address can be calculated directly from the immediate number in the instruction code.

For conditional direct jump instructions, a simple static branch prediction is used, and the jump address can be obtained directly by adding the PC value and the immediate number according to the definition of the instruction.

For an unconditional indirect jump instruction, since it is bound to jump, there is no need to predict the direction of its jump, and for the target jump address, the base address required for the jump target calculation of the instruction needs to be calculated from the operand of the register index.

Table 1. Branch jump instructions in RV32I

Grouping	Instruction	Number of digits
Unconditional direct jump/branch instructions	Jal	32
Unconditional Indirect Jump/Branch Instructions	Jal	32
Conditional direct jump/branch instructions	beq,ben,blt,bltu,bge,bgeu	32

The second stage of the pipeline is the decoding module: the main function is to decode the instruction fetched from the instruction fetch unit to obtain the instruction type, operation information, etc, access the general register, obtain the register value to be read, and send the relevant information to Execution unit. RISC-V's instruction encoding format is very regular, so it is more convenient to read relevant information such as operands and instruction types from the register during decoding, which greatly improves efficiency.

The final phase is also an important and complex part of the design, the execution phase: since the processor of this design has only 3 levels of pipeline, the fetch and writeback phases will also be completed in this module. According to the different instruction types, the execution module is mainly divided into "ALU", "AGU", "MDV", "CSR" and so on. ALU is mainly responsible for the execution of logical operations, addition and subtraction, shift and other instructions, AGU is mainly responsible for the address generation of Load, Store and "A" extension instructions, MDV is mainly responsible for the execution of multiplication and division instructions, and CSR is mainly responsible for the execution of CSR instructions.

One of the most important functions of the execution phase is to maintain and resolve pipeline conflicts. In order to test the correlation between WAW and RAW data of long instructions (including Load, Store, and multiply and divide instructions), an OITF (Outstanding Instructions Track FIFO) module is added to this design[7]. As shown in Figure 4, OITF is essentially a FIFO with a depth of 2 table entries, first in, first out. At the dispatch point of the pipeline, each time a long instruction is dispatched, an entry is assigned in the OITF, in which the source operand register index and the result register index of the long instruction are stored. At the Write-Back point of the pipeline, after writing back a long instruction in order each time, the table entry of this instruction in the OITF is removed, that is, it withdraws from its FIFO to complete the mission.

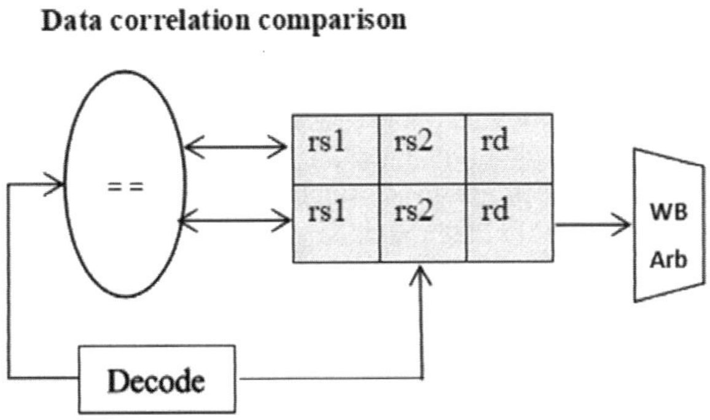

Figure 4. OITF micro-architecture diagram

4. Simulation and verification

This article uses the hardware description language Verilog HDL to implement the functions of each module in code to form RTL code. The processor is functionally verified through the instruction set architecture test cases in riscv-tests, and before the test cases are run in test platform TestBench, they also need to compile their Verilog code, and can be simulated after successful compilation. The simulation tool uses the open source and free iverilog, combined with the gtkwave tool to view the simulation waveforms. Finally, the simulation results of all instructions show that all the instructions of the RV32I in the processor designed this time have passed the simulation test. The verification process is shown in Figure 5:

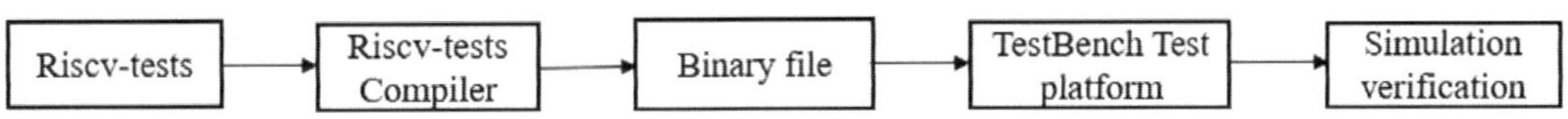

Figure 5 Instruction set verification process

In order to evaluate the energy efficiency information such as area and power consumption of this design, the Cortex M-series processor and the Hummingbird E203 processor were logically synthesized. Using Synopsys' Design Compiler tool, the SIMC 110nm process was used to logically synthesize the processor designed in this article, and the processor was tested by the CoreMark running score program, and the comprehensive running score test results were compared with other processors as shown in Table 2. In terms of area and power consumption, the Cortex M3 processor contains more instructions and functions, and has greater power consumption and area compared to the design; compared with Cortex M3, the performance is slightly lower, but the power consumption is reduced 0.55mW, the area is reduced by about 58.7%; compared with the Hummingbird E203 processor, the power consumption and area are slightly higher, but the performance is improved by about 12%; this design includes a multiplication and division unit, so the area is slightly higher than Cortex M0 processing Device.

Table 2 Comparison of performance parameters of different processors

Processor	ARM Cortex-M0	ARM Cortex-M3	Hummingbird E203	This design
Process/nm	SMIC 110nm	SMIC 110nm	SMIC 110nm	SMIC 110nm
The number of logical gates	14.7k	48k	18.5k	19.8k
Power consumption/mW	0.17	0.75	0.145	0.20
CoreMark (CoreMark/MHz)	2.33	3.32	2.14	2.38

5. Conclusion

This paper designs a 32-bit small area, low power, high-performance RISC-V processor for the small area and low power consumption characteristics of embedded applications. The pipeline depth is 3 levels, with a Harvard structure, single launch, sequential execution, out-of-order write-back, and parallel access to instructions and data via a separate AHB bus. The verification results show that the microprocessor is functionally simulated through the TestBench test platform in the iverilog environment, and the functions are correct; The SIMC 110nm process is used to synthesize logic and run score tests on a 32-bit RISC-V processor in a DC environment, which consumes 0.20mW, has an area overhead of 19.8k logic gates, and has a command execution speed of 2.38CoreMark/MHz. The test results show that the processor's low power consumption and high performance characteristics can be well used in the Internet of Things, artificial intelligence, and embedded devices.

References

[1] Li Jiemin, Zhang Shancong. Research and performance optimization of five-stage pipeline RISC-V processor[J/OL]. Microelectronics and Computer: 1-7[2021-12-26].DOI: 10.19304/J .ISSN1000-7180.2021.0856.

[2] Raveendran A, Patil VB, Selvakumar D, et al. A RISC-V instruction set processor-micro-architecture design and anal ysis[C]//2016 International Conference on VLSI Systems, Architectures, Technology and Applications (VLSI-SATA), 2016: 1-7

[3] Banakar R, Steinke S, Lee B S, et al. Scratchpad memory; design alternative for cache on-chip memory inembedded systems[C]//2015 International Conference on Hardware/Software Codesign and System Synthesis (CODES+ISSS),2002:73-78.

[4] Keller B.A RISC-V processor SoC with integrated power management at submicrosecond timescales in 28 nmFD-SOI[J].IEEE Journal of Solid-State Circuits, 2017, 1:1863-1875.

[5] Patil V, Raveendran A, Sobha PM, et al. Out of order floating point coprocessor for RISC V ISA[C]//201519th International Symposium on VLSI Design and Test, 2015:1-7.

[6] Rao J, Ao T, Xu S, et al. Design exploration of SHA-3 ASIP for IoT on a 32-bit RISC-V processor[J].IEICE Transactions on Information and Systems, 2018 (11): 2698-2705.

[7] Hao Zhenhe,Jiao Jiye,Li Yuqian.Design and implementation of RISC-V microprocessor based on AHB bus[J].Computer Engineering and Applications,2020,56(20):52-58.

Journal of Physics: Conference Series **2221** (2022) 012009 doi:10.1088/1742-6596/2221/1/012009

Power transformer life analysis based on Lambert W function

Jingyang Wang[1], Liyou Fu[2]

[1] Business College, Shanghai Dian Ji University, Shanghai, 201306, China

[2] Business College, Shanghai Dian Ji University, Shanghai, 201306, China

Liyou Fu: e-mail: fuly@sdju.edu.cn

Abstract—As the core equipment in many construction power fields, the reasonable life management of power transformer plays an important role in ensuring the safe production and economic benefits of the public and enterprises. Based on lambert-w function, the exact solution is obtained by calculating the total annual allocated cost. The variable values that may have errors in the function model are deeply analyzed. Combined with the inherent characteristics of power transformer, the factors affecting the economic life in the results are analyzed. The reliability of the method is verified by an example. Compared with the traditional tabular method, this method is convenient to further analyze the influencing factors and change laws of economic life. At the same time, through the analysis of specific problems, this method can also be used to study the economic life of other equipment. It has popularization value.

1. Introduction

Transformers are widely used in almost all electronic products, especially power transformers in modern industrial enterprises. As one of the core equipment of the enterprise, the transformer needs to be fully guaranteed in terms of equipment installation, stable operation, safety and reliability. The research on the economic life of power transformer is a powerful guarantee for these factors. By studying the economic life and formulating an economic and reasonable maintenance and renewal plan, it will more effectively promote the utilization of enterprise core assets and bring greater economic benefits to the enterprise.

There is no specific mandatory standard for the service life of the transformer. At present, most studies at home and abroad focus on economic evaluation, and there is no unified definition of the economic life of transformer. At the same time, there is little research on the operation and management of power equipment in China, and there is a lack of sufficient information to form a scientific and reliable retirement decision-making criterion. In recent years, CiteSpace has little analysis and Research on the economic life and reference frequency of transformers. Literature[3] comprehensively considers the impact of the life cycle cost of active transformer, the economic performance of the selected reference transformer and the random fuzzy uncertainty of parameters on the economic life of active transformer, but there are few quantitative data and more qualitative components; The economic life defined in reference[4] refers to the year in which the net benefit value obtained by comprehensive economic indicators such as depreciation, energy conservation and loans is the largest; Literature[5] is a comparison of maintenance and replacement based on the principle of maximum annual average income; Reference[6] considers the investment, maintenance and fault loss of transformer in the whole life cycle; Reference[7] puts forward the economic life model of power transformer based on the lowest equivalent cost of transformer life year; The above documents do not

Content from this work may be used under the terms of the Creative Commons Attribution 3.0 licence. Any further distribution of this work must maintain attribution to the author(s) and the title of the work, journal citation and DOI.
Published under licence by IOP Publishing Ltd

provide accurate solutions for economic life, nor do they combine sufficient influencing factors. Reference[9] proposes an economic life solution for general equipment, but it is not completely applicable to any equipment. The power transformer needs to be analyzed in detail.

Combined with the advantages of the above literature, the exact solution of the economic life of power transformer is derived by using Lambert W function. By analyzing the factors affecting the economic life of transformer and the variation law of equipment economic life, it is of great significance to improve the economic life and economic benefits of transformer.

2. Mathematical Model of Economic Life of Power Transformer

The method for calculating the economic feasibility of energy transformer shall consider the procurement cost, maintenance cost and residual value of equipment. Considering these factors, there are other reasons why the machine cannot be reused before the end of the service life of the power transformer. This is the loss caused by excessive damage caused by frequent use of equipment in the later stage of the transformer. With the increase of operation period, the annual investment will decrease, but on the contrary, the cost of maintaining and repairing equipment will increase in the opposite direction, resulting in a sharp increase in the annual allocation of the lowest total cost.

Therefore, in order to reduce the cost as much as possible, according to the annual cost method, the economic feasibility of the equipment is the operation period with the lowest average cost per unit time, and the lowest value is determined.

This document describes the idea of establishing the model used in literature[9]. Combined with the characteristics of transformer, the original model is improved, and the model is reconstructed to improve the accuracy of the model.

It is assumed that the annual allocation of procurement cost and residual value is C_1. C_1 is the purchase cost multiplied by the present value of the pension minus the residual value multiplied by the final value of the pension. Experience has shown that the residual value of transformer closure is usually 30% to 40% of the original value. Only 30% here.

$$\begin{aligned} C_1 &= K\left(A / P, i, n\right) - 0.3K\left(A / F, i, n\right) \\ &= Ki\frac{(1+i)^n - 0.3}{(1+i)^n - 1} \end{aligned} \tag{1}$$

In formula (1), K represents the purchase fee, the equivalent annual value of the fund, P represents the capital limit, f represents the final value of the fund, I represents the annual interest rate, and N represents the year.

If the annual allocation of control and maintenance costs is C2, you can get:

$$\begin{aligned} C_2 &= \begin{pmatrix} C(1)(P/F,i,1) + C(2)(P/F,i,2) \\ + \cdots + C(n)(P/F,i,n) \end{pmatrix} \\ (A/P,i,n) &= \frac{i}{(1+i)^n - 1}\sum_{j=1}^{n} C(j)(1+i)^{n-j} \end{aligned} \tag{2}$$

In formula (2), j is one year. It can be assumed that the annual growth of control and maintenance costs is, and C_2 can be changed to

$$C_2 = \frac{i}{(1+i)^n - 1}\sum_{j=1}^{n}\left[C(1) + (j-1)\Delta C\right](1+i)^{n-j} \tag{3}$$

Through simplification, the final term C2 can be as follows:

$$C_2 = \Delta C\frac{i}{(1+i)^n - 1}\cdot\frac{(1+i)^n - ni - 1}{i^2} + C(1) \tag{4}$$

Assuming that the total annual cost is allocated as CT, we can get

$$C_T(n) = C_1 + C_2 = Ki\frac{(1+i)^n - 0.3}{(1+i)^n - 1} \tag{5}$$

$$+\Delta C\frac{i}{(1+i)^n - 1}\cdot\frac{(1+i)^n - ni - 1}{i^2} + C(1)$$

Make

$$\frac{\mathrm{d}C_T(n)}{\mathrm{d}n} = 0 \tag{6}$$

$$\frac{\Delta C - \Delta C(1+i)^n - (1+i)^n(0.7Ki - n\Delta C)\ln(1+i)}{\left(-1+(1+i)^n\right)^2} = 0 \tag{7}$$

As can be seen from formula (7), if the denominator on the left side of the equation is equal to 0, we can get

$$n\ln(1+i) - 1 - \frac{0.7Ki\ln(1+i)}{\Delta C} = -\frac{1}{(1+i)^n} \tag{8}$$

Multiply both sides at the same time

$$e^{n\ln(1+i)-1-\frac{0.7K_0 i\ln(1+i)}{\Delta C}} \tag{9}$$

Obtain

$$\left(n\ln(1+i) - 1 - \frac{0.7Ki\ln(1+i)}{\Delta C}\right) \tag{10}$$

$$\cdot e^{n\ln(1+i)-1-\frac{0.7Ki\ln(1+i)}{\Delta C}}$$

$$= -\frac{1}{(1+i)^n}\cdot e^{n\ln(1+i)-1-\frac{0.7Ki\ln(1+i)}{\Delta C}}$$

$$= -\mathrm{e}^{-1-\frac{0.7Ki\ln(1+i)}{\Delta C}}$$

You can see that formula (10) is in the form of Lambert W function and can be described as follows

$$W(z) = n\ln(1+i) - 1 - \frac{0.7Ki\ln(1+i)}{\Delta C} \tag{11}$$

$$z = -e^{-1-\frac{0.7Ki\ln(1+i)}{\Delta C}} \tag{12}$$

The exact solution of the economic life is as follows:

$$n^* = \frac{0.7Ki}{\Delta C} + \frac{W\left(-e^{-1-\frac{0.7Ki\ln(1+i)}{\Delta C}}\right) + 1}{\ln(1+i)} \tag{13}$$

In formula (11), the cost and annual percentage I of purchasing K are greater than 0, and the Z range in the function Lambert W should be (-e-1, 0). According to the Lambert W function, it can be seen from this property that in (z), there must be a solution of the actual number, and the exact solution of the Lambert W function can be obtained by Mathematica. Mathematics draws a curve, as shown below.

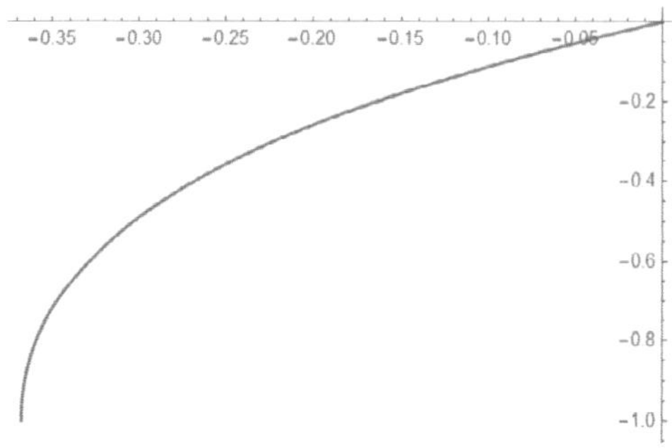

Fig 1. Lambert W function graph

As shown in Figure 1, the horizontal axis represents Z and the vertical axis represents Z. W (z) increases with the increase of E, and Z increases monotonically (- e-1,0).

3. Analysis of Factors Affecting Economic Life of Power Transformer

It is included in the cost allocation model of the original model, but does not consider the accurate economic solution. Therefore, its value does not affect the determination of economic life of power transformer. According to the value of Lambert W function of the actual solution, the function image increases monotonically. With its growth, Z will decrease and in W(z) will decrease accordingly. At the same time, it will also decrease, resulting in the reduction of economic life. Therefore, with the increase of annual control and maintenance costs, the economic life of energy transformer will be shortened. In the case of purchase fee K, the increase of purchase fee K will increase Z, and in W(z) will increase accordingly. At the same time,. Therefore, the economic life of transformer increases with the increase of order cost.

For the annual percentage I, literature[9] mentioned that the higher the annual percentage I, the greater the annual value of procurement cost allocation. As the annual percentage I increases, the annual value of control and maintenance cost allocation also increases. Therefore, the impact of annual percentage I on the economic life of energy transformer can not be proved only by equating the economic life with the capital recovery formula. The influence of I value on the economic life of transformer is affected by K / H ratio. It is necessary to use professional mathematical software to evaluate variable control methods. The following figure shows that the economic life change trend of annual percentage energy transformer I is 0-0.3 when the remaining variables remain unchanged.

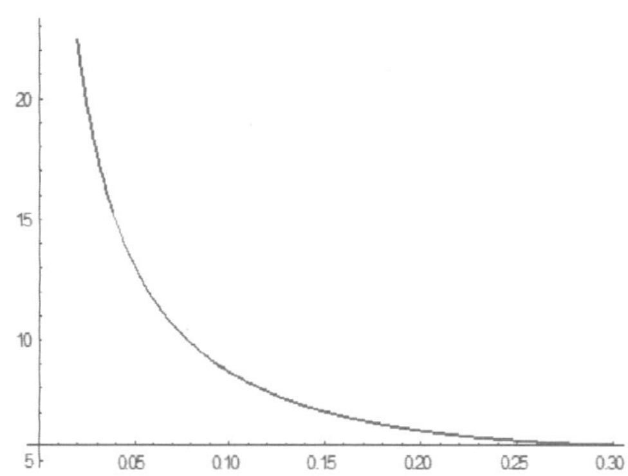

Fig 2. Economic life curve of power transformer with annual interest rate I in the range (0-0.3)

As shown in Figure 2, with the increase of annual percentage I, the economic life of energy transformer is still declining, which shows that the annual common living cost is an important part in solving its own economic problems. This also confirms that with the increase of business period, the annual allocation of control and maintenance costs continues to increase, while the annual allocation of contracts becomes less and less.

In the process of solving the procurement cost K and annual percentage I, it is easy to obtain, but in terms of value, it is usually difficult to obtain the control and maintenance cost that increases almost every year. Therefore, C2 needs further analysis.

The annual allocation of control and maintenance costs shall be:

$$C_2(t) = \frac{r(t)C_{MAD}(t) + (1 - r(t))C_{MAL}(t)}{(1+\delta)^t} \tag{14}$$

In equation(14), $C_2(t)$ represents the control and maintenance cost in year t, $r(t)$ represents the function of power transformer fault coefficient. Power transformers installed in different years have different fault coefficient functions, $C_{MAD}(t)$ represents the loss and maintenance cost caused by power transformer fault this year. $C_{MAL}(t)$ represents the temporary control fee and δ represents the discount rate for the current year.

According to this formula, if the maintenance cost must be amortized and increased every year, the maintenance cost must be deducted to achieve annual growth, and the best solution can be found by analyzing the relationship.

Based on the analysis of the above influencing factors, we should start with these influencing factors in order to improve the economic life of energy transformer. The annual percentage rate cannot be controlled. For energy transformers that have been purchased and put into use, the purchase fee K cannot be changed, i.e. The annual allocation of control and maintenance costs should be reduced. Economical and reliable maintenance methods can be adopted to reduce maintenance costs. For example, the main internal components of a transformer are iron core, winding and transformer oil. If there is a problem, replacing these parts will greatly increase the maintenance cost. When repairing the iron core, first remove the oil stain on the iron core to ensure that dust and impurities will not enter the transformer oil tank during disassembly and assembly, and then check the shape and position of the iron core (mainly including color, purity, connection tightness, grounding, dye foil, etc.). If there is a problem, it shall be adjusted at any time. The initial steps of repairing moisture are the same as those of repairing iron core, flushing, controlling appearance position and aging pressure. If the pressure is very fine, the insulating packaging must be replaced. The number of transformers and oil in power transformers is relatively large. Considering the high oil cost, the most economical method is to absorb the oil and pollutants at the bottom of the transformer through the oil absorption device.

Although the control and maintenance costs of energy transformers are minimized, considering that the design durability of energy transformers defined in the national standard is 20 years, few energy transformers can operate beyond this limit, and the increase of control and maintenance costs is only taking a batch of thin insulated aluminum tube transformers produced around 1976 as an example. You has issued documents on early elimination, Therefore, innovation in infrastructure and technology continues to have a significant impact.

4. Case analysis

The analysis and calculation shall be taken as an example of substation power transformer. The voltage level of the energy transformer is 1000% of the voltage used in the 1980s. The annual percentage rate is 5%, the discount rate is 5%, the average error cost of the first transformer is 1000 yuan / time, and the conventional control cost is 1500 yuan. Try to analyze the economic life of power transformer (the error coefficient function is% / unit 1833; year).

Pictures showing losses and maintenance costs caused by transformer failure.

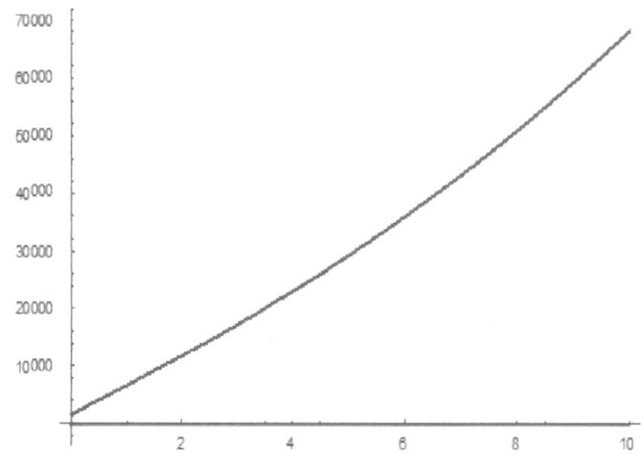

Fig 3. Loss caused by power transformer failure and maintenance cost chart

As shown in Figure 3, the horizontal axis represents the year and the vertical axis represents the loss and maintenance costs. Although the annual loss and maintenance cost increase with the increase of operation period, the curve in the figure shows an approximate linear increase. Well, the trend is about 5000 yuan.

Enter this value into the Lambert W function and analyze it using the product logarithm template.

$$\mathrm{ProductLog}\left[-e^{-1-\frac{0.7*80000*\ln(1.05)}{5000}}\right] = -0.2825$$

Which is

$$W\left[-e^{-1-\frac{0.7*80000*\ln(1.05)}{5000}}\right] = -0.2825$$

Then, the economic life of the transformer is accurately solved as

$$n^* = \frac{0.7*80000*0.05}{5000} + \frac{-0.2825+1}{\ln(1+0.05)} = 15.26 \text{ years}$$

The economic life is much shorter than the design life. The best economic life does not mean that the longer the economic life, the better. In fact, as long as the maximum annual average net income is lower than its control and maintenance costs until it reaches positive income, the power transformer can continue to be used when it reaches its economic life. In the design process, it is difficult to realize the feasibility of power transformer. Different reasons may reduce the economic feasibility of the transformer during use, thus shortening the service life.

In this example, based on the comprehensive analysis of the operation state of the transformer, it is recommended to complete the optimal operation years of the transformer within 15-18 years.

5. Conclusion

In this paper, the Lambert W function is used to establish the analysis model to solve the economic life problem of energy transformer. Although the original model is based on the same conditions as the annual growth of control and maintenance costs, through a thorough analysis of the annual breakdown of control and maintenance costs, the average annual growth of control and maintenance costs is obtained. The design of the complete model is persuasive and accurate. The results obtained by lambert-w function can be applied to practice. This is a new method to calculate the economic life of energy transformer, not just the economic life of transformer. It has high reference value for planning other types of equipment and even substations in the future.

References

[1] Wu Z.H. (2018) Economic life model of power transformer and its application example. Technology Wind, 98-99.
[2] Wang W.B., Bai W.G., Shi L.L. (2019) Comprehensive life cycle economic physical life assessment method for power transformer. Power System Protection and Control, 91-98.
[3] Li R., Han B., Lu Y. (2014) Transformer's economic life assessment based on random and fuzzy theory. Power System Protection and Control, 42: 9-16.
[4] Yu J.L., Wang C.F., Zhang B. (2010) Economic Life Evaluation of Power Transformer in Service. Proceedings of the CSU-EPSA, 86-90.
[5] Liu Y.W., Ma L., Wu L.Y., Zhou Y., Lian C. (2012) Economic Life Model of Power Transformer and Its Application. Power System Technology, 235-240.
[6] Luo X.C., Li L., Wei Z.L., Ge J.B. (2011) Applications of Life Cycle Cost Theory in Decision-Making of Investment for Distribution Transformer Renovation. Power System Technology, 207-211.
[7] Wang H.F., Zhao W.F, Du Z.D. (2015) Economic Life Prediction of Power Transformers Based on the Lifetime Data. Power System Technology, 39: 810-816.
[8] Hu B., Hu W.P., Jia Z.H. (2010) Life cycle cost and economic life analysis of transformer. The 9th Symposium of equipment life cycle cost Committee of China Equipment Management Association, Shanghai. 227-230.
[9] Wu S.H., Liu X.D., He B. (2016) Economic life analysis for weapon equipment based on Lambert W function. Systems Engineering and Electronic, 38: 844-851.

Research on the reliability of Micro LED high-density solder joints under thermal cycling conditions

Mingyang Liu[1,a], Junyang Nie[2], Yifeng Liu[1], Jie Sun[1,b]*, Min Li[3,c]*, Chunli Yan[4], Qun Yan[1]

[1]National and Local United Engineering Laboratory of Flat Panel Display Technology, College of Physics and Information Engineering, Fuzhou University, Fuzhou 350108, China

[2]Faculty of Electronic and Information Engineering, Xi'an Jiaotong University, Xi'an 710049, China

[3]Fujian Science & Technology Innovation Laboratory for Optoelectronic Information of China, Fuzhou 350116, China

[4]Department of Information and Automation, Library of Fuzhou University, Fuzhou 350116, China

[a]email: mingyang_l97@foxmail.com

*Corresponding authors: [b]email: jie.sun@fzu.edu.cn, [c]email: admi123@163.com

Abstract. For the purpose of studying the effect of high-density, small-size solder joints on the thermal fatigue life of display devices, a 3D model of the corresponding device structure is established, using finite element analysis methods, and using Anand constitutive equations and Coffin-Manson life prediction models. The equivalent stress and strain change law of an indium solder joint with a diameter of 10 μm under thermal cycling loading conditions is explored, and the key position of the solder joint damage is inferred based on the simulated strain cloud diagram. In addition, the influence of different bump height, bump contact area and Under-Bump Metallization (UBM) thickness on the thermal fatigue life of solder joints is studied separately. The final simulation results show that the location where the solder joint may be damaged first is at the point on the outer side where the solder joint is connected to the UBM on the edge away from the center point. The height of the bump and the contact area of the bump have a significant effect on its thermal fatigue life. In the actual experiment, we should focus on the impact of this aspect.

1. Introduction

Compared with Organic Light Emitting Display (OLED) and Liquid Crystal Display (LCD), Micro Light Emitting Diode (LED) is expected to become the core of next-generation display technology by virtue of its characteristics of high brightness, high resolution, high color gamut, low power consumption, and long time. It is the technological high ground in the industry. As far as the current development status of Micro LED is concerned, however, it still faces many technical challenges: miniaturization process technology, transfer technology, bonding technology, defect reparation and detection technology, full color technology, etc [1,2]. Due to the limitation of epitaxial growth technology at this stage, the Micro LED chips and Complementary Metal Oxide Semiconctor (CMOS)

Content from this work may be used under the terms of the Creative Commons Attribution 3.0 licence. Any further distribution of this work must maintain attribution to the author(s) and the title of the work, journal citation and DOI.
Published under licence by IOP Publishing Ltd

driver chips have to be grown on different substrates, typically GaN/sapphire and silicon. That is, the two are made separately on two substrates [3]. Thus, for the driving control of the Micro LED displays, how to better realize the electrical and mechanical interconnection is one of the issues that need to be addressed in the bonding technology. One of the mainstream bonding technologies for high-density and small-size solder joints in this research field is the bump interconnection technology, which is well adapted to the development trend of chip miniaturization, integration, small size and multi-function. However, as a bonding technology, it needs to consider how to extend the service life of the device and improve the overall reliability, which is currently studied very little in literature.

At present, the main reason for most of the damage of electronic components is the loss of functionality due to the failure of solder joints. According to the statistics of the U.S. Air Force Electronics Industry Department, 55% of the failure of electronic components is caused by temperature changes, 20% is caused by vibration, 19% is caused by moisture, and the other 6% is caused by dust [4], as shown in Fig. 1.

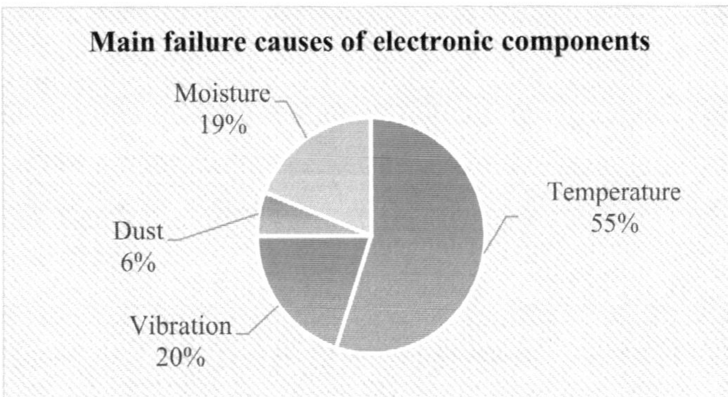

Fig.1 Main failure causes of electronic components.

Temperature influence is the core cause of the device solder joint failure, and the failure mechanism is usually due to the difference in thermal expansion coefficient between the solder joint material and the substrate. In the changing temperature environment, the solder joint itself is stimulated to generate thermal stress, and plasticity, creep, and elastic strain behavior occur. The accumulation of plastic deformation and creep damage for a long time leads to the generation of voids or cracks in the solder joint, which leads to the failure of the solder joint. Meanwhile, at the present stage, the micron-sized bump repairation technology under flip-chip welding is imperfect and the changes in the device working environment tend to be even worse. Therefore, studying the reliability of thermal fatigue of solder joints should be one of the main research directions to extend the life of electronic components [5]. However, there is little research on the reliability of related Micro LED micro bumps, most of which being aimed at large-size bumps on the chip. Wenzhong Zhao et al. studied the thermal fatigue life of solder joints with a diameter of 64 μm under different chip sizes, and summarized the influence of chip size on the temperature and stress/strain distribution. It was found that the stress and strain increase with the increase in size [6]. Xiaoqin Lin et al. studied the influence of the shape of the 100 μm solder joint on the failure surface and fatigue life, and explored its stress-strain law and failure mechanism. The corners of the chip bear the greatest tensile stress [7]. Jiang Han et al. studied the influence of solder joint volume and diameter on the thermal fatigue behavior of Through Silicon Via (TSV) solder bumps under power cycling [8].

The research on thermal fatigue of solder joints is to explore the thermal stress distribution and the accumulation of plastic and creep strain damage of solder joints with temperature changes. Due to the influence of the small size and large quantity of Micro LED micro bumps, it is practically difficult for current testing instruments to fully reflect the thermal stress distribution and strain change trends of the device bumps. However, the finite element simulation can better reflect the thermal stress distribution and strain behavior of the bumps, and can greatly save the cost of sample preparation and sample inspection time. In recent years, the method of studying the reliability of micro devices through finite element simulation has been proved to be feasible and powerful [9-11]. Therefore, the finite element analysis method is used here to establish a three-dimensional model to analyze the thermal fatigue life of the Micro LED micro bump array. The equivalent stress-strain change rule of indium bumps with 10 μm diameter and 4 μm height under thermal cycling is systematically studied. It is found that the stress and strain value spreads from the center to the edge, and the analysis shows that the dangerous point appears at the edge of the solder joint and the UBM contact point. The influence of the basic morphology of the bump diameter, height and the thickness of the UBM on the life change of the micro-bump are respectively explored, and the contact area of the double-bump is used instead of the diameter of the bump as a key factor in the simulation, because the contact area can more directly reflect the changes in the bonding quality. The Anand constitutive model is used to simulate the accumulation of viscoplastic strain in each thermal cycle, and the Coffin-Manson model is used to predict the thermal fatigue life of the micro bumps. Although this article has indium solder joint research as a representative, it also has important reference value for other chip bump bonding.

2. Finite element model

2.1 Parameters used in the model

The construction of the 5×5 microbump array 3D model refers to the structure and parameters of the 19.2 μm Micro LED as shown in Fig. 2. Due to the limitation of simulation calculation capacity and the main research object of thermal fatigue life being the solder joint, the entire structure is not completely simulated, and only a local array is selected to study the law of stress and strain, and derive the change of the entire structure. The material from top to bottom is GaN, Au, In, Au, Si (LED main material, UBM, solder, UBM, substrate), the upper and lower substrate size is 100 μm ×100 μm, the height is 10 μm, the UBM is 12 μm ×12 μm, the height is 0.5 μm. The micro solder joint adopts the round table model, the top and bottom radius are 4 μm, 5 μm, and the height is 4 μm, as shown in Fig. 3.

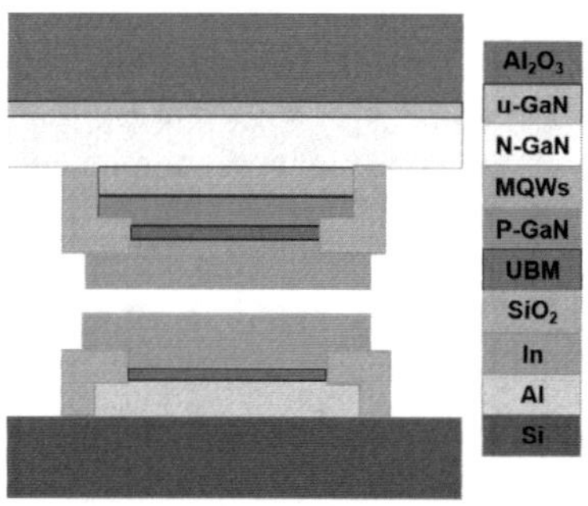

Fig.2 19.2 μm Micro LED structure diagram.

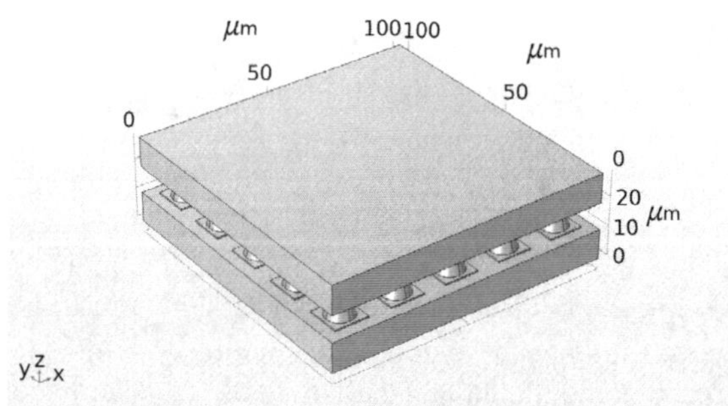

Fig.3 3D model of the local array of micro bumps.

2.2 Material parameters

Several basic material parameters required for each structural material in the finite element model in this paper are shown in Table 1.

Table 1. Material properties of various materials.

Material	Poisson's Ratio [1]	Young's Modulus [GPa]	Thermal Conductivity [W/(m×K)]	Density [kg/m³]	CTE [1/K]
Si	0.28	170	130	2329	$2.6e^{-6}$
GaN	0.2	290	130	6070	$3.17e^{-6}$
In	0.45	11	81.6	7290	$32.1e^{-6}$
Au	0.44	70	317	19300	$14.2e^{-6}$

The main research data of the Coffin-Manson life prediction model is to explore the accumulation of viscoplastic strain damage of the solder joints. The micro bump material used is a low melting point metal. Under thermal cycling loading conditions, elastic strain and plastic strain mainly occur. Therefore, except for microbump materials, other structural materials are set to isotropic linear elastic materials, and microbump materials are set to viscoplastic materials. The Anand viscoplastic constitutive model is used to describe the viscoplastic changes of microbump materials. The Anand constitutive model equation has 9 key parameters, as shown in Table 2.

Table 2. The parameters of the indium bump Anand model.

A [1/s]	$1.49e^7$	m [1]	0.303	h_0 [N/m^2]	2640.75
Q [J/mol]	90046	s_0 [MPa]	80.42	a [1]	1.34
ξ [1]	11	$\hat{s}$ [MPa]	56.33	n [1]	0.0231

2.3 Load setting and structure optimization

The thermal cycle loading condition setting refers to the thermal cycle standard JESD22-A104-B. The cycle temperature is set to -55～125 °C, the cycle period is 40 min, the heating rate and cooling rate are both 18 °C/min, and the maximum temperature and minimum temperature maintenance time are both 10 min (see Fig. 4). The life prediction adopts the plastic strain Coffin-Manson life prediction model based on the Anand constitutive equation. The stress and strain distribution in the fourth cycle of the thermal cycle is selected as the initial value, and the stress and strain in the fourth cycle tends to be stable, increasing the model accuracy of prediction. Some assumptions defined by the simulation model are: Each bump is completely evenly distributed; All materials are ideally connected; The role of the internal light-emitting structure and the insulating layer of the LED is ignored; The residual stress and strain that may be generated during the preparation process are ignored.

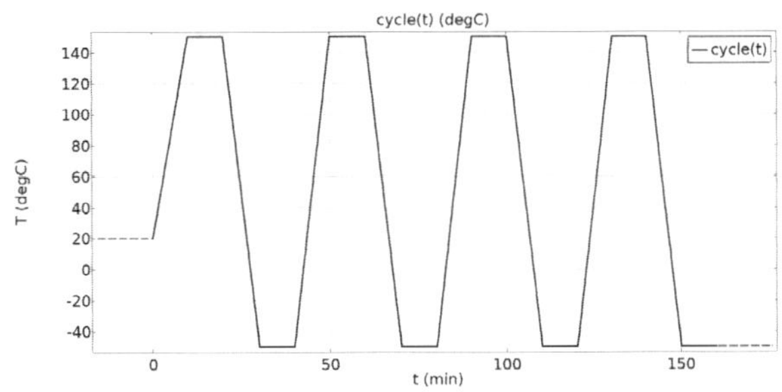

Fig.4 Thermal cycle graph of 4 cycles.

3. Simulation analysis

3.1 Stress and strain analysis of micro bumps

Under normal circumstances, tens of thousands of solder joints will not be damaged at the same time, and often local solder joints or individual key points are greatly damaged, which leads to device failure. Studying the damage of these key solder joints is the focus of prolonging the life of the device. The key position of the key solder joint damage can be found through the simulation stress and strain situation of three angles of point, line and plane.

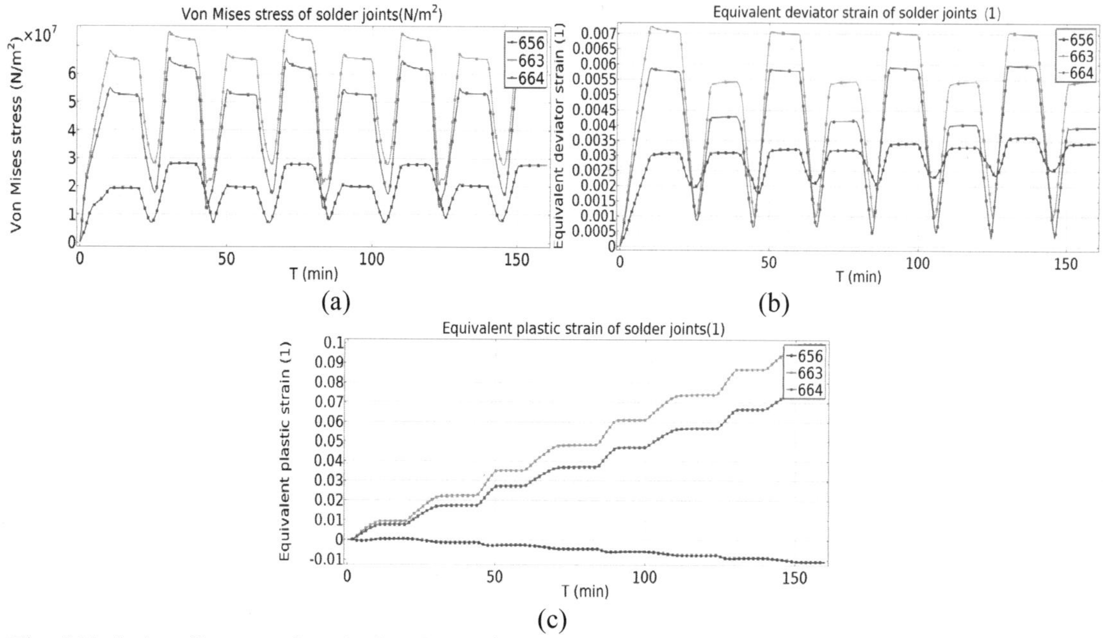

Fig. 5 Variation diagram of equivalent Von Mises stress, strain, and viscoplastic strain accumulation at different points of micro bumps over time (664: bump and upper UBM contact point, 663: bump and lower UBM contact point, 656: bump and bump contact point).

By comparing the equivalent Von Mises stress, equivalent strain, and thermal cycle change diagrams at different positions of the micro bumps, the equivalent stress and strain in the four cycles in Fig. 5a-b all change periodically. The equivalent stress and strain of the four cycles has tended to change steadily, focusing on the analysis of the equivalent stress and strain change law in the fourth cycle, it can be found that the equivalent stress value and the equivalent strain value change with time during the heating stage, and show an upward trend, the stress-strain values reach the maximum value at the end of the heating phase, the change rule of the cooling phase is slightly different from that of the heating phase, the stress and strain first drop and then rise, and the change is more dramatic. Due to the difference in thermal expansion coefficients of various materials, the expansion or contraction behavior cannot occur at the same time under the same temperature change. The equivalent stress and strain changes in the high temperature maintenance phase and the low temperature maintenance phase are not obvious, and the equivalent stress decreases slowly with time due to the effect of stress relaxation. According to the change law of the equivalent stress and strain of the micro bumps with time, the sudden temperature change stage has a greater impact on the equivalent stress and strain of the micro bumps, and it is inferred that the micro bumps are prone to failure at this stage. The equivalent stress and strain and the accumulation of viscoplastic strain at the contact position of the bump and the upper and lower interfaces are significantly greater than the contact position of the double bump. The place where cracks are likely to occur first is the contact position between the bumps and the upper and lower interfaces. It is necessary to focus on the key points of the dangerous solder joints.

The Coffin-Manson model based on plastic deformation is used in this solder joint life prediction, and it is based on the generalized fatigue damage law of metal materials. It is suitable for low-cycle fatigue analysis caused by temperature loads. The basic formula is as follows [12]:

$$\frac{\Delta\varepsilon_p}{2} = \varepsilon_f'\left(2N_f\right)^c \quad (1)$$

In the formula: ε_f' is the fatigue toughness coefficient; $\Delta\varepsilon_p$ is the plastic strain range; c is the fatigue toughness index; N_f is the number of failure cycles.

The key data is the accumulation of equivalent viscoplastic strain. According to Fig. 5c, the accumulation of equivalent viscoplastic strain steadily increases with the increase of the thermal cycle period. Once the accumulation of plastic strain reaches or exceeds the tensile strength of the material, cracks or fractures may occur.

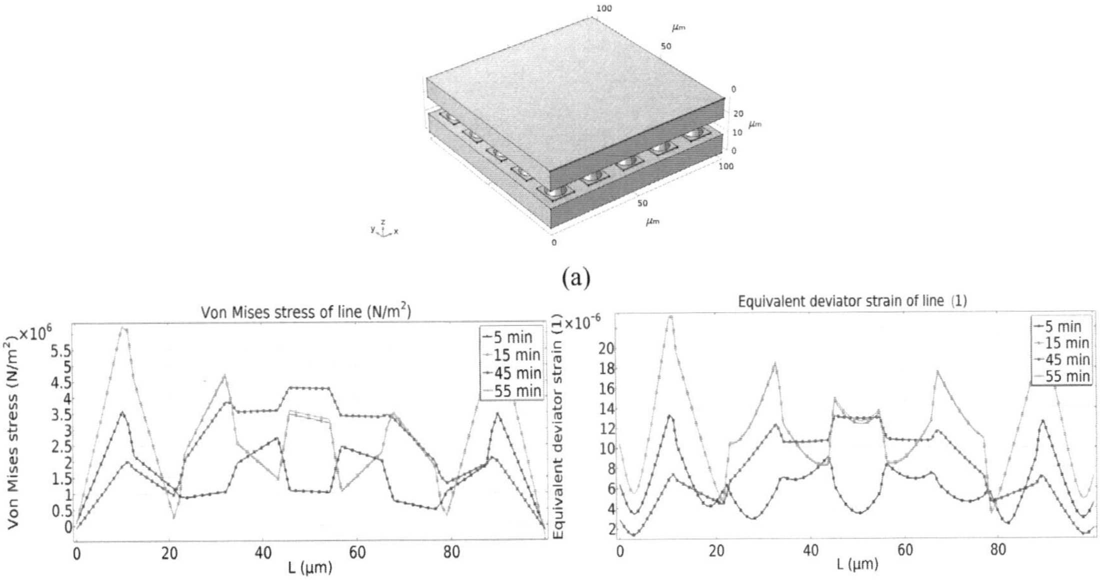

Fig.6 The stress and strain of the edge line segment of the substrate during the heating and high temperature maintenance stage.

Fig. 6a shows the selected line segment position. The stress and strain peak position in Fig. 6b-c corresponds to the solder joint distribution position. During the heating stage and the high temperature maintenance stage, the equivalent stress and strain at the edge of the substrate is obviously greater than the middle part, and the thermal stress generated by the thermal expansion of the bumps spreads to both sides of the edge significantly, the difference between the maximum stress and the minimum stress at the edge is the largest. The huge sudden stress and strain will cause greater damage, and the pull-up deformation at the edge of the solder joint interface is greater than the deformation at the center position. It can be judged that the bump at the edge position is a dangerous solder joint, and the key damage point is inferred to be the contact point between the bump and the upper and lower interfaces, the following re-analyze and confirm the position of the dangerous solder joint according to the oblique section cloud diagram of stress and strain.

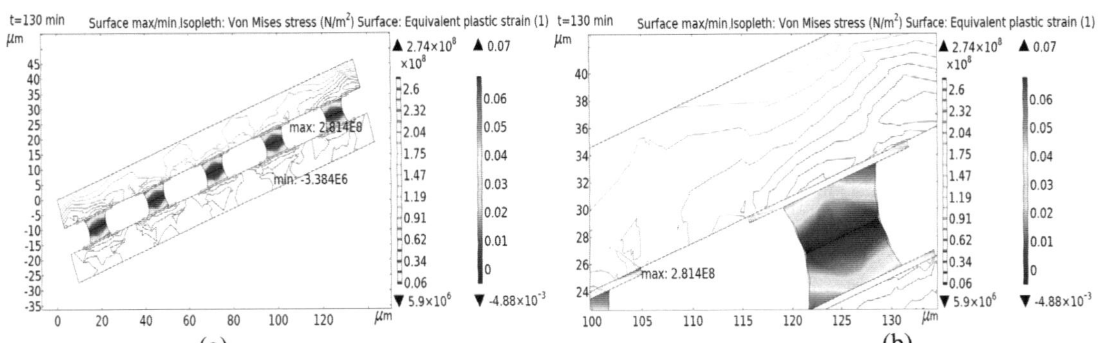

Fig.7 Overall equivalent viscoplastic strain cloud diagram and equivalent stress contour diagram of the oblique section.

The stress cloud diagram in Fig. 7 selects the end of the fourth cycle of heating, from the equivalent stress contour distribution in Fig. 7b. It can be clearly observed that the stress distribution area is mainly at the edge position of the upper substrate and the edge position of the pad UBM. The maximum equivalent stress appears on the contact surface between the upper substrate and UBM at the second solder joint on the edge, the minimum appears on the middle of the lower substrate. The overall distribution law conforms to the analysis of the equivalent stress spreading to the edge, and the equivalent stress and strain of the solder joints at the edge is greater than that of the solder joints near the center [13]. The accumulation of equivalent viscoplastic strain increases outwardly from the center of the solder joint, the high-level distribution mainly appears at the corner contact position between the UBM and the solder joint. The viscoplastic accumulation of the contact point toward the edge is greater than the accumulation toward the center, and the development direction of the crack may change from the outside to the inside. The stress and strain change law of the local bump array is deduced to the entire Micro led bump array, and multi-faceted analysis confirms that dangerous solder joints appear at the edge of the substrate. The location where damage may occur first under long thermal cycles is the contact point between the solder joint and the outer side of the UBM. In the actual preparation process, more consideration can be given to the protection of the edge, similar to the filling treatment or the edge sealing treatment [14].

3.2 Influencing factors of thermal cycle life of micro bumps

The chip bump interconnection structure is basically the same. The thermal cycle life of each structure under the same material is mainly the impact of its own package size on its life, followed by the configuration of the structure directly connected to the bump. Therefore, exploring the influence of different bumps' topography and size and UBM thickness on their thermal cycle life provides a reference analysis for the actual packaging structure and the preparation of interconnect bumps. The following analysis is based on the maximum equivalent stress and strain values that occurred in 4 cycles, the accumulation of equivalent viscoplastic strain at the end of the 4th cycle, and the number of lives predicted by the Coffin-Manson model as comparison data. Explore the degree of influence on thermal fatigue of solder joints.

3.2.1 Influence of the height of micro bumps on the thermal fatigue life of solder joints

In order to explore the influence of the height of the micro bumps on the thermal fatigue life of the solder joints, the heights of the micro bumps are respectively set to 2, 3, 4 μm, the bottom radius of the bump is 5 μm, the top radius is 4 μm, and the UBM thickness is 0.5 μm. The other conditional parameters are kept unchanged, the corresponding finite element analysis models are established respectively. It can be seen from Table 3 that as the height of the micro bumps increases, the maximum equivalent stress value increases gradually, the maximum equivalent strain value decreases approximately at the same amplitude, and the viscoplastic strain accumulation value gradually decreases, and the increase is considerable. According to the changes in the simulation data of different bump heights, it shows that the increase of the micro bump height can effectively increase the number of thermal fatigue life predicted by the micro bump.

Table 3. Data table of equivalent stress, strain and viscoplastic strain accumulation for different bump heights.

Bump height [μm]	Maximum stress [N/m^2]	Maximum strain [1]	Equivalent viscoplastic strain accumulation [1]	Thermal fatigue life times [1]
2	1.73328e^8	0.013889	0.15224	1001
3	1.8022e^8	0.011426	0.12163	1548
4	2.1468e^8	0.0092869	0.10384	2093

3.2.2 The influence of microbump contact area on the thermal fatigue life of solder joints
Set the contact area of the micro bumps to 4π, 9π, 16π μm^2, and other condition parameters remain unchanged, respectively establish the corresponding finite element analysis model, compare the simulation data Table 4 of different bump contact surfaces, increasing the maximum equivalent stress, strain, and viscoplastic strain accumulation of the micro-bump contact area increases, and the corresponding thermal fatigue life shows a downward trend. According to data analysis, reducing the bump contact area can indeed increase the thermal fatigue life. However, if the bump contact area is too small, it will affect the shear performance of the solder joint, resulting in a decrease in the anti-collision performance. The comparison of the contact area of 4π and 9π μm^2, and 9π and 16π μm^2 is carefully analyzed. The increase of the equivalent stress and strain value of the latter is obviously greater than that of the former. Therefore, in the actual preparation, a suitable bump contact area should be selected appropriately to avoid excessively large or small impacts on the reliability of the overall device.

Table 4. Data table of equivalent stress, strain and viscoplastic strain accumulation of different bump contact areas.

Bump contact area [μm^2]	Maximum stress [N/m^2]	Maximum strain [1]	Equivalent viscoplastic strain accumulation [1]	Thermal fatigue life times [1]
4π	1.7566e^8	0.0074324	0.096876	2396
9π	1.7551e^8	0.0086906	0.10666	2160
16π	1.8022e^8	0.01142	0.12163	1548

3.2.3 The influence of UBM thickness on the thermal fatigue life of solder joints
UBM is a structure that directly contacts the bumps, and its morphology and material settings may affect the thermal fatigue life of the solder joints. Therefore, the UBM thickness is set to 0.5, 1.5, 2, 2.5 μm, and the other condition parameters are kept unchanged, and the corresponding finite element analysis models are established to explore the degree of influence. According to the simulation data Table 5, as the thickness of UBM increases, the maximum equivalent stress gradually weakens, and the maximum strain presents an oscillating change pattern, which has a relatively weak effect on the thermal fatigue life. The main purpose of UBM is to act as an infiltration layer and barrier layer between different materials. The appropriate thickness can be selected according to the characteristics of the material. Its UBM thickness will not have a large impact on the thermal fatigue life.

Table 5. Data table of equivalent stress, strain and viscoplastic strain accumulation of different UBM thicknesses.

UBM thickness [μm]	Maximum stress [N/m^2]	Maximum strain [1]	Equivalent viscoplastic strain accumulation [1]	Thermal fatigue life times [1]
0.5	2.1468e^8	0.0092869	0.10384	2093
1.5	1.9302e^8	0.011317	0.10449	1930
2	1.7997e^8	0.010724	0.10370	2079
2.5	1.8166e^8	0.0093029	0.10411	2070

4. Conclusions

This paper aims at the thermal fatigue reliability research of Micro LED high-density solder joints, and establishes the corresponding 3D model through finite element analysis of the equivalent stress and strain change law of the micro bump with a diameter of 10 μm and the main factors affecting its life. According to the simulation results, the following conclusions are drawn:

1. The equivalent stress and strain distribution law of the micro bump array, the stress and strain distribution gradually spreads and increases from the center position to the surroundings, and the equivalent stress and strain of the bump to the edge side is greater than the stress and strain to the center. The solder joints at the edge of the array are dangerous solder joints.

2. According to the Anand viscoplastic constitutive model to analyze the viscoplastic strain accumulation of the bumps, the viscoplastic accumulation of the upper and lower ends of the micro bumps in contact with the UBM is significantly greater than the middle position of the bumps, and the contact points toward the edge are generally distributed at the highest value. The key damage location of the micro bumps is likely to occur at the point where the UBM is in contact with the edge side.

3. The impact of bump height, bump contact area, and UBM thickness on the thermal fatigue life of micro bumps is explored, and the main influence is the ranking bump height>bump contact area>UBM thickness. In particular, the impact of the height of the bump and the contact area of the bump is more significant, and the actual preparation should consider its impact.

Acknowledgments

We thank the support from Mindu projects (Nos. 2021ZZ122, 2020ZZ110), Fujian provincial projects (Nos. 2021HZ0114, 2021J01583, 2021L3004), and National Natural Science Foundation of China (No. 62175032).

References

[1] Guo, W., Tai, J., Liu, J., & Sun, J. (2019) Process optimization of passive matrix GaN-based micro-LED arrays for display applications. Journal of Electronic Materials, 48(8): 5195-5201.

[2] Du, Z., Li, D., Guo, W., Xiong, F., & Sun, J. (2021) Quantum Dot Color Conversion Efficiency Enhancement in Micro-Light-Emitting Diodes by Non-radiative Energy Transfer. IEEE Electron Device Letters. 42(8): 1184-1187.

[3] Fu, Y., Sun, J., Du, Z., Guo, W., & Yan, Q. (2019) Monolithic integrated device of GaN micro-LED with graphene transparent electrode and graphene active-matrix driving transistor. Materials, 12(3): 428.

[4] Guo, Q., Zhao, M., & Wang, H.F. (2005) SMT solder joint's semi-experimental fatigue model. Mechanics Research Communications, 32(3): 351-358.

[5] Zhang, K., Liu, Y., Kwok, H.S., & Liu, Z. (2020) Investigation of electrical properties and reliability of GaN-based micro-LEDs. Nanomaterials, 10(4): 689.

[6] Zhao, W., Yao, Q., Cao, Y., Lin, P., & Lu, F. (2018) Effect of chip size on thermal fatigue of flip-chip bump. In: 2018 International Conference on Electronics Technology. pp. 69-73.

[7] Lin, X., & Luo, L. (2008) The reliability study of sub 100 microns SnAg flip chip solder bump on FR4 substrate under thermal cycling. In: International Conference on Electronic Packaging Technology & High Density Packaging. pp. 1-5.

[8] Jiang, H., Liang, S.B., Yuwen, H.H., & Zhang, X.P. (2016) Finite element simulation of the size effect on thermal fatigue behavior of solder bump joints in TSV structure. In: International Conference on Electronic Packaging Technology. pp. 947-952.

[9] Jiang, L., Zhu, W., & He, H. (2017) Comparison of darveaux model and coffin-manson model for fatigue life prediction of BGA solder joints. In: International Conference on Electronic Packaging Technology. pp. 1474-1477.

[10] Gui, X., & Sheng, L. (2011) Experiments and numerical simulation on the performance of LED subjected to thermal shock. In: International Conference on Electronic Packaging

Technology & High Density Packaging. pp. 1-4.

[11] Wang F., & Yun, C. (2012) Modeling study of thermosonic flip chip bonding process. Microelectronics Reliability, 52(11): 2749-2755.

[12] Long X., Liu, Y., Jia, F., Wu, Y., Fu, Y., & Zhou, C. (2019) Thermal fatigue life of sn-3.0ag-0.5cu solder joint under temperature cycling coupled with electric current. Journal of Materials Science, 30(8): 7654-7664.

[13] Amalu, E.H., & Ekere, N.N. (2012) High-temperature fatigue life of flip chip lead-free solder joints at varying component stand-off height. Microelectronics Reliability, 52(12): 2982-2994.

[14] Shantaram, S., Sakib, A., & Lakhera, N. (2018) Coupled Thermal and Thermo - Mechanical Simulation for Flip-chip Component Level Copper Pillar Bump Fatigue. In: Intersociety Conference on Thermal and Thermomechanical Phenomena in Electronic Systems. pp. 1381-1386.

A low resource consumption Arbiter PUF improved switch component design for FPGA

Jingfu Yang[a], Xiaoping Yu[b], Rikun Wei[c]

School of Mechanical and Electrical Engineering, Chengdu University of Technology, Chengdu, Sichuan, China

[a]email: yangjingfu@stu.cdut.edu.cn, [b]email: yxpsc@126.com, [c]email: 516033225@qq.com

Abstract. Aiming at the problems of poor uniqueness and high resource consumption of slice after the implementation of Arbiter physical unclonable function (Arbiter PUF) on field programmable gate array (FPGA), a new improved scheme of switch component structure is proposed. The switch component structure of the improved scheme adopts the parallel-connected mode to improve the uniqueness of the circuit, which avoids the cross-connected mode unable to achieve symmetrical layout on FPGA. At the same time, the improved scheme uses lookup tables to construct the structure of programmable delay line (PDL) with a multiplexer (MUX), which can reduce the internal resource consumption of slice while receiving the same 64-bit challenges. The improved scheme was tested on FPGA boards, the uniqueness and steadiness of different switch component schemes are compared and analyzed, and the feasibility of the improved scheme is verified. The results show that in the generation of Arbiter PUF, compared with the conventional scheme, the improved scheme reduces the resource consumption and improves the uniqueness by 22.2%; Compared with the MUX + MUX scheme, the improved scheme saves 50% of resource consumption while maintaining good uniqueness.

1. Introduction

Key storage security is an important part of hardware security. Traditional key storage often stores keys in nonvolatile memory. Since the data will not disappear after the nonvolatile memory is powered off, the key stored in nonvolatile memory is vulnerable to physical attack [1] or side channel attack [2], so it is difficult to ensure its security. The manufacturing process deviation of integrated circuits will lead to uncontrollable physical randomness. Physical unclonable function (PUF) uses this randomness to realize the mapping from challenges to responses. The key mechanism based on PUF does not need to store the key locally, which avoids the possibility of theft and attack. Moreover, the physical process deviation of each chip is unique, the chip cannot be copied, and has certain security [3], so PUF can be used for IP protection, and device authentication [4].

Arbiter PUF is a typical delay based PUF, which generates response by using the difference of two completely symmetrical path delays caused by process deviation during chip manufacturing. Arbiter PUF was proposed by Lim D et al. [5,6] in 2005, but the uniqueness of this conventional Arbiter PUF in FPGA environment is quite poor [7,8]. At present, many researchers have proposed different methods to solve this problem. For example, literature [9] uses look-up table (LUT) to realize programmable delay line (PDL) to improve the uniqueness of Arbiter PUF, but when the number of challenges is large, the resource consumption is too large; Literature [10,11] proposed Double Arbiter

Content from this work may be used under the terms of the Creative Commons Attribution 3.0 licence. Any further distribution of this work must maintain attribution to the author(s) and the title of the work, journal citation and DOI.
Published under licence by IOP Publishing Ltd

PUF (DA PUF) to improve the uniqueness of Arbiter PUF, but the uniqueness of DA PUF is not stable, and the resource consumption increases exponentially.

In order to improve the uniqueness of Arbiter PUF in FPGA environment and reduce resource consumption, an improved switch component scheme using the structure of PDL overlay multiplexer (MUX) is proposed. The switch component structure adopts parallel-connection to improve the uniqueness and reduce the consumption of internal resources of slice when receiving 64-bit challenges.

2. Conventional Arbiter PUF

2.1. Implementation Structure

The conventional Arbiter PUF circuit structure is shown in Figure 1(a). The scheme has two perfectly symmetrical signal transmission delay paths in the upper and lower layouts. When the pulse arrives, the path direction within the switch component is selected according to the input challenges [c0, c1, ..., c63].

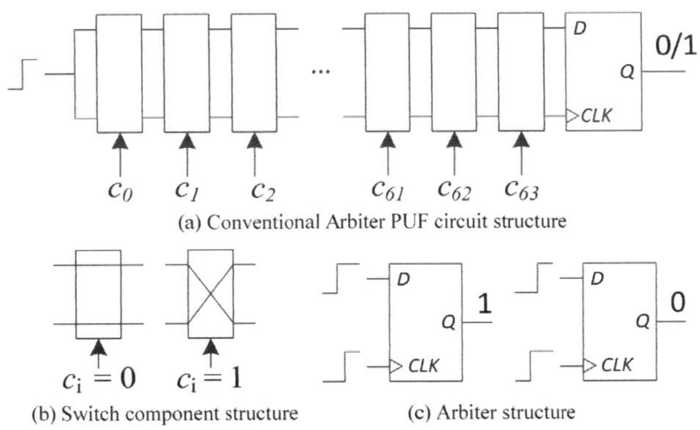

Figure 1. Conventional Arbiter PUF design.

Each switch component has two input terminals, one control terminal and two output terminals, as shown in Figure 1 (b), when the control signal ci = 0, the signal passes horizontally; When ci = 1, the signals cross over. The switch components corresponding to the upper and lower paths share a control signal, that is, the signal path structure of the upper and lower paths is the same. Ideally, the pulse signals on the two paths will reach both input ends of the arbiter at the same time. However, the delay of the two paths will be slightly delayed due to the existence of process bias in the actual implementation. That is, the signals on the two paths do not arrive at the arbiter at the same time. Finally, the arbiter determines that the output is 0 or 1 according to its sequence. As shown in Figure 1 (c), when the pulse signal from port D of the arbiter arrives first, the output response is 1; When the CLK port pulse signal of the arbiter arrives first, the output response is 0. Since this process bias is completely random, the resulting delay in signal transmission on both routes is also uncertain, and the output of the final arbiter is unpredictable.

2.2. Analysis of Design Principle

The basic principle of arbiter PUF design is to symmetrically layout and route the two paths to make them have the same theoretical delay of routing path, and the final response is generated according to the random delay deviation in the switch component caused by the manufacturing process deviation. Since the location of logic resources and routing resources on the FPGA has been fixed and cannot be changed at will [10], it is impossible to achieve accurate and symmetrical layout and routing effect, and due to the asymmetric routing of the two paths, the Arbiter PUF has quite poor uniqueness on the FPGA. Finally, if it is much larger than the delay deviation caused by the manufacturing process, the delay deviation caused by the asymmetric layout and wiring determines the output response value.

When the layout and wiring of different chips are the same, the output response value is basically the same, which cannot reflect the difference between different chips.

3. Improvement scheme of Arbiter PUF Circuit

In order to adapt to the development environment of FPGA platform, improve uniqueness and avoid the situation that cross-connection cannot realize symmetrical wiring, the improvement scheme changes the connection mode of switch module in conventional Arbiter PUF structure from cross-connection to parallel-connection. At the same time, the switch component adopts the structure of PDL overlay MUX, when 64-bit challenges is also received, the consumption of internal resources of slice can be reduced.

One of the influencing factors of path symmetrical layout and routing is the connection mode of switch components on the path, which is usually divided into cross-connection and parallel-connection.

3.1. Cross-connected switch component

The switch component of conventional Arbiter PUF is composed of MUX and adopts cross-connected mode. Its model is shown in Figure 2 (a). The switch component with cross-connected mode is more suitable for ASIC implementation, and its model is shown in Figure 2 (b). The delay model of the cross-connected switch component implemented on FPGA is shown in Figure 2 (c). Due to the fixed resource location on FPGA, the cross-connected line cannot realize symmetrical wiring on FPGA.

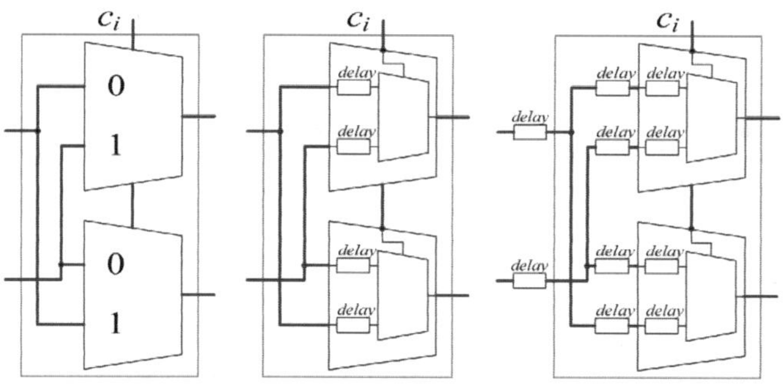

(a)Cross-connected switch component (b)Cross-connected ASIC model (c)Cross-connected FPGA model

Figure 2. Cross-connected switch component model.

3.2. Parallel-connected switch component

The parallel-connected switch component includes MUX type switch component and PDL type switch component. The upper and lower paths of MUX switch component are in parallel structure, and the delay model is shown in Figure 3 (a), which solves the problem that cross-connected model cannot realize symmetrical wiring on FPGA; The upper and lower paths of PDL switch component are also parallel structures, and the difference in delay time is only related to the internal structure of LUT. The delay model is shown in Figure 3 (b). There is no cross wiring, and the symmetrical wiring of Arbiter PUF structure can be realized on FPGA.

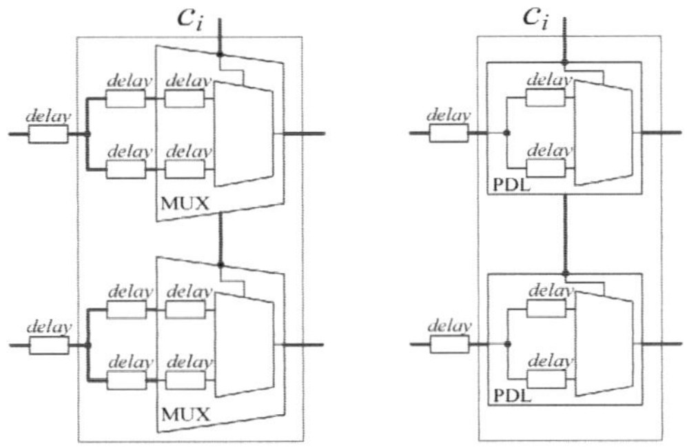

(a)Parallel-connected MUX switch component model (b)Parallel-connected PDL switch component model

Figure 3. Parallel-connected switch component model.

3.3. Improvement principle and method

Compared with MUX switch component, PDL switch component has more relaxed design constraints, higher delay accuracy and less logic resource consumption. An input LUT can be configured as a PDL with n-1 control ports. PDL has only one input signal path, and the delay difference is achieved by different propagation path delays of input signals with different control signals. Simultaneous interpreting delay is only related to the internal structure of LUT. The truth table is listed first, as shown in Table 1. Regardless of the value of the strobe signal, the output O is the value of the input signal I0. Only when the values of I2 and I1 are different, the path of the input I0 signal inside the look-up table is long and short, that is, the propagation of the input signal is sometimes different.

Table 1 Truth table for configuring the three-input LUT to PDL.

I2	I1	I0	O	SRAM
0	0	0	0	0
0	0	1	1	1
0	1	0	0	0
0	1	1	1	1
1	0	0	0	0
1	0	1	1	1
1	1	0	0	0
1	1	1	1	1

A three input LUT is configured as a PDL. The input I0 as the path option determines the output read from the SRAM. The initial value of the SRAM is assigned corresponding to the value of I0. I1 and I2 as strobe signals will not affect the output result, but their values will affect the propagation path of the input signal in the look-up table.

The improved switch component scheme uses four six-input LUTs and three multi-selectors inside slice simultaneously, that is, a PDL constructed from the LUT with a multiplexer (PDL + MUX). Using the structure of the above principle, four six input LUTs (A6LUT, B6LUT, C6LUT, D6LUT) in slice can be configured into four PDLs which can receive 5-bit challenges. As shown in Fig. 4 (a), I0 is used as signal input, I1, I2, I3, I4 and I5 are selected signals, that is, challenge signals. For example, when I5I4I3I2I1 = 00000, the signal propagation will have the shortest path; When I5I4I3I2I1 = 11111, the signal will have the longest path. The signal propagation delay of LUT when the challenge signal is all zero is less than that when the challenge signal is all one. The function of PDL can be implemented so that a PDL can receive 5-bit challenges signal input.

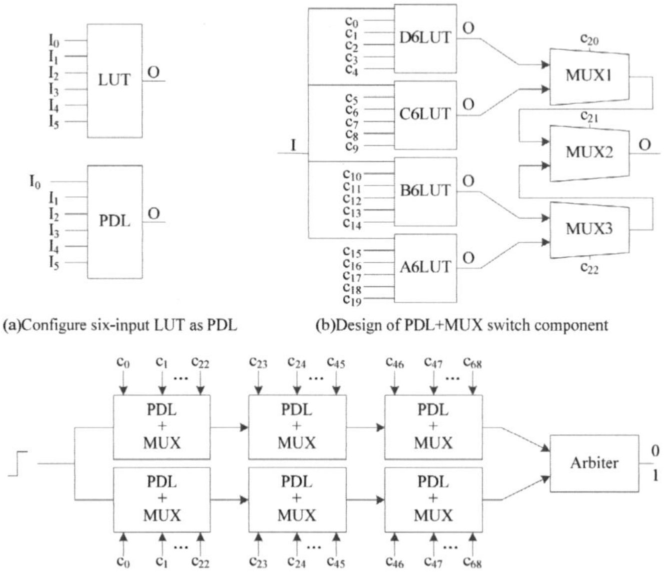

(a)Configure six-input LUT as PDL (b)Design of PDL+MUX switch component
(c)Improved Arbiter PUF circuit

Figure 4. Improved Arbiter PUF switch component design.

The pulse signals are connected to the signal inputs of four PDLs at the same time. In order to ensure path symmetry, connect the outputs of D6LUT and C6LUT at the upper position in the slice with the input of MUX1, connect the outputs of B6LUT and A6LUT at the lower position in the slice with the input of MUX3, and then connect the outputs of MUX1 and MUX3 with the input of MUX2 at the middle position, as shown in Fig. 4 (b), the challenges that can be received by each slice is increased to 23 bits.

Three PDL + MUX switch components are cascaded, and six slices can be used for the upper and lower paths to form an Arbiter PUF circuit that can receive 69-bit challenges. As shown in Fig. 4 (c), theoretically, the connection mode is parallel, which improves the uniqueness compared with the conventional Arbiter PUF cross-connected switch component, and makes full use of the multiplexer other than LUT in the slice, when receiving the same number of incentives, the consumption of slice resources can be reduced.

4. Experiment and analysis

In order to verify the design of the improved scheme, the improved Arbiter PUF switch scheme is implemented on four XC7Z020-2CLG400I SOC chips of the same Xilinx ZYNQ 7000 series. In the experiment, the challenge signals are 69 bits. 5 bits are selected randomly and fixed to 0 and 1. Only 64 bits of them are used. The circuit includes Upper computer, UART module, Challenge generation circuit, Pulse generation circuit and Arbiter PUF delay path.

4.1. Scheme validation and performance comparison

The performance index of the PUF circuit determines whether the design is feasible or not. The uniqueness and steadiness of the PUF performance can usually be considered [12]. The uniqueness of PUF is usually measured by the inter Hamming distance, that is, the ratio of different response bits to the total response bits in the response signal of two different PUF circuits to the same challenge. The closer the inter Hamming distance is to 50%, the better the uniqueness of PUF. The formula for calculating inter Hamming distance between slices is as follows

$$\mu_{inter} = \frac{2}{m(m-1)} \sum_{p=1}^{m-1} \sum_{q=p+1}^{m} \frac{HD(X_p, X_q)}{k} \times 100\% \tag{1}$$

In the formula, *m* denotes the number of PUFs with the same structure, *k* denotes the number of bits per response, *Xp* and *Xq* denote the response of the *p* and *q* chips to the same challenge, and *HD* denotes the Hamming distance.

The steadiness of PUF is usually measured by the intra Hamming distance, that is, the same challenge signal is repeatedly input into the same PUF circuit, and the ratio of different response bits in the two response signals to the total response bits. The closer the intra Hamming distance is to 0%, the better the steadiness of PUF. The formulas for calculating the intra Hamming distance in a slice are as follows

$$\mu_{intra} = \frac{1}{s}\sum_{t=1}^{s}\frac{HD(X_i, X_{i,t})}{k} \times 100\% \quad (2)$$

In the formula, *k* represents the number of bits of each response, k-bit PUFs receives *s* challenges, each challenge measures *t* group, *Xi* represents the reference response value of the *ith* chip, *Xi,t* represents the response generated by the *t* group of the *ith* chip.

The test environment temperature is 21 ℃, and 10,000 bits of response is generated each time. Each 20 groups of four FPGA boards are tested. According to the uniqueness and steadiness calculation formula, the inter and intra Hamming distances can be obtained by collecting the output response of the software through the serial port of the upper computer, as shown in Figure 5. Among them, the maximum intra Hamming distance is 0.431%, the minimum is 0.286%, and the mean is 0.357%. The maximum inter Hamming distance between slices is 47.2%, the minimum is 43.1%, and the mean is 45.2%.

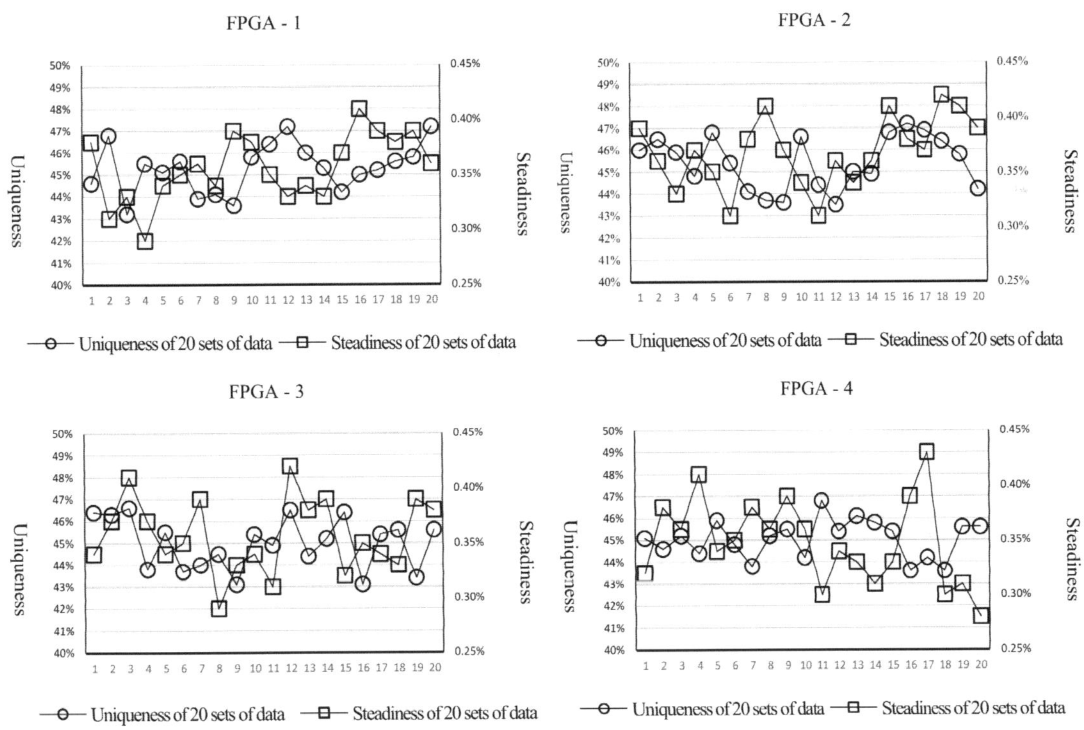

Figure 5. Test results of Uniqueness and Steadiness.

The uniqueness and steadiness of conventional Arbiter PUF, MUX + MUX and improved PDL + MUX switch component can be obtained by comparing the test results. As shown in Table 2, the uniqueness and steadiness values are averaged.

Table 2 Parameters comparison of different Arbiter PUF switch component implementations.

Comparative item	Uniqueness	Steadiness
Ideal value	50.0 %	0.000 %
Conventional Arbiter PUF	23.0 %	0.700 %
MUX + MUX	45.7 %	0.276 %
PDL + MUX	45.2 %	0.357 %

The results show that the improved Arbiter PUF switch component adopts a parallel-connected structure, and the uniqueness of the output response is improved by 22.2% compared with the conventional Arbiter PUF. Compared with the MUX + MUX scheme, the improved Arbiter PUF design reduces uniqueness by 0.5%, steadiness by 0.081%, and the numerical value does not change much. The feasibility of the improved scheme is verified.

4.2. Resource comparison and analysis

In terms of resource consumption, there are four LUTS in a slice, now it is necessary to realize 64-bit challenges signal input. If the MUX switch component implementation scheme is used, each LUT can be configured as a 4-1 MUX that can receive 2-bit challenges. The four LUTs can be cascaded to receive 8-bit challenges, and the upper and lower paths need to occupy a total of 16 slices; If the PDL scheme is used, each LUT can be configured as a PDL capable of receiving 5-bit challenges. Four LUTS can be cascaded to receive 20-bit challenges, requiring a total of 8 slices; If the MUX + MUX switch component improvement scheme is used, each slice can be configured into a structure capable of receiving 11-bit challenges, requiring a total of 12 slices; If the improved PDL + MUX switch component implementation scheme is used, each slice can be configured into a structure that can receive 23-bit challenges, requiring a total of 6 slices. Through comparison, it can be found that the number of slices consumed by the improved PDL + MUX scheme to receive 64-bit challenges is reduced by 62.5% compared with MUX scheme, 25% compared with PDL scheme and 50% compared with MUX + MUX scheme, further reducing the occupation of logical resources. The comparison of resource consumption is shown in Table 3.

Table 3 Comparison of resource consumption of different switch component implementations.

Schemes	The number of challenges each slice can receive	The number of slices consumed to receive 64-bit challenges
MUX	8 bits	16
PDL	20 bits	8
MUX + MUX	11 bits	12
PDL + MUX	23 bits	6

5. Conclusion

Due to its non clonable characteristics, PUF has application prospects in the field of information security, and resource consumption, uniqueness and steadiness are important indicators to evaluate the performance of PUF. Here, an improved switch component design of low resource consumption Arbiter PUF for FPGA is proposed. When receiving the same 64-bit challenges, the uniqueness and steadiness are compared with different switch component schemes, the feasibility of the improved scheme is verified. The improved scheme not only improves the uniqueness of Arbiter PUF in FPGA environment, but also reduces the consumption of logic resources, which is in line with the expected design. In the future, the research will focus on the adjustment circuit and adjustment algorithm to improve the randomness of PUF.

References

[1] Dofe, J., Pahlevanzadeh, H. & Yu, Q. A Comprehensive FPGA-Based Assessment on Fault-Resistant AES against Correlation Power Analysis Attack. J Electron Test 32, 611–624 (2016). https://doi.org/10.1007/s10836-016-5598-9

[2] Madhavapandian, S., MaruthuPandi, P. System Level Protection Against Side-Channel Attack Using High Performance Virtual Secure Circuit for Cryptographic Processor. Wireless Pers Commun 117, 2667–2677 (2021). https://doi.org/10.1007/s11277-019-06930-w
[3] Rahman M T, Rahman F, Forte D, et al. An Aging-Resistant RO-PUF for Reliable Key Generation[J]. IEEE Transactions on Emerging Topics in Computing, 2016, 4(3):335-348.
[4] Huang, Z., Wang, Q. A PUF-based unified identity verification framework for secure IoT hardware via device authentication. World Wide Web 23, 1057–1088 (2020). https://doi.org/10.1007/s11280-019-00677-x
[5] Lee J W, Lim D, Gassend B, et al. A technique to build a secret key in integrated circuits for identification and authentication applications[C]// VLSI Circuits, 2004. Digest of Technical Papers. 2004 Symposium on. IEEE, 2004.
[6] Daihyun, Lim, Lee, et al. Extracting secret keys from integrated circuits[J]. Very Large Scale Integration Systems IEEE Transactions on, 2005.
[7] Maiti A, Gunreddy V, Schaumont P. A Systematic Method to Evaluate and Compare the Performance of Physical Unclonable Functions[M]. Springer New York, 2011.
[8] Hori Y, Katashita T, Kobara K. Performance Evaluation of Physical Unclonable Functions on Kintex-7 FPGA[J]. Ieice Technical Report, 2013, 113.
[9] Devadas S, Kharaya A, Koushanfar F, et al. Automated Design, Implementation, and Evaluation of Arbiter-based PUF on FPGA using Programmable Delay Lines. 2014.
[10] Machida T, Dai Y, Iwamoto M, et al. A new mode of operation for arbiter PUF to improve uniqueness on FPGA[C]// 2014 Federated Conference on Computer Science and Information Systems. IEEE, 2014.
[11] Machida T, Dai Y, Iwamoto M, et al. Implementation of double arbiter PUF and its performance evaluation on FPGA[C]// Design Automation Conference. IEEE, 2015.
[12] Takanori Machida, Dai Yamamoto, Mitsugu Iwamoto, et al. A New Arbiter PUF for Enhancing Unpredictability on FPGA[J]. The Scientific World Journal, 2015, 2015.

Apple sugar content non-destructive detection device based on near-infrared multi-characteristic wavelength

Weihui Liu[1,a], Yanan Han[1,b], Na Wang[1,c], Zhen Zhang[1,d], Qingguo Wang[1,e], Yongping Miao[1,f*]

[1]Shandong University of Science and Technology, College of Electronic and Information Engineering, Qingdao, Shandong, China

[a]email: liuweier@126.com, [b]email: 1134016594@qq.com, [c]email: 859815526@qq.com, [d]email: 1661979865@qq.com, [e]email: 646910026@qq.com

[f*]email: miaoyongping78@126.com

Abstract. Knowing the sugar content of apples in time and grading the internal quality of apples can meet the requirements of different consumers for apples. In this paper, Yantai Red Fuji apples are used as the research object, a low-cost and portable non-destructive testing device for apple sugar content is designed based on near-infrared detection technology. To realize the quantitative detection of apple sugar content, The relationship between the voltage value of diffuse reflection light and the sugar content of apples at multiple characteristic wavelengths is studied, and the multiple linear regression method is used to establish the apple sugar content prediction model. The correlation coefficient predicted by the model is 0.7708, and the minimum relative error is 0.89%. The experimental results show that within a certain error range, the device can realize fast and non-destructive detection of the sugar content of apples. This article provides a reference for the realization of a low-cost, portable non-destructive testing device for Apple.

1. Introduction

Sugar content is one of the important indicators to measure the internal quality of apples, and timely understanding of the sugar content information of apples is of great significance for apple quality sorting [1]. The traditional apple sugar content detection equipment mainly uses the refractive characteristics to perform destructive detection of apples. The operation process is complicated and can only be used for random inspection [2]. At present, many researchers at home and abroad use near-infrared spectroscopy technology to collect near-infrared spectroscopy information of apples through spectrometers for non-destructive testing [3-5]. However, the cost of spectrometer equipment is high, it is difficult to popularize, and the high-precision spectrometer is large in size and not easy to carry [6]. Therefore, the study of a portable fruit testing equipment that can detect the internal quality of fruits with high precision and non-destructively, but also with low cost and low power consumption, is of great significance to promote the development of the fruit industry [7]. This article is based on the principle of near-infrared, taking Yantai Red Fuji apple as the research object, using near-infrared LED and photodiode to design a detection device to realize low-cost, non-destructive inspection of apples.

Content from this work may be used under the terms of the Creative Commons Attribution 3.0 licence. Any further distribution of this work must maintain attribution to the author(s) and the title of the work, journal citation and DOI.
Published under licence by IOP Publishing Ltd

2. Design principles and methods

2.1. Mechanism of interaction between apples and near-infrared light

Scan apple samples in the near-infrared spectrum, when the vibration frequency of the hydrogen-containing groups (C-H, O-H, N-H, etc.) in the apple sugar is the same as the frequency of a certain wavelength of the incident light, resonant absorption will occur and a characteristic absorption peak is formed. The intensity is determined by the number of light-absorbing molecules in the sample, that is, the concentration [8]. In diffuse reflectance spectroscopy analysis, the degree of apple sugar content is different, the selective absorption of the spectrum is different, so the diffuse reflection light signal can be detected by photoelectric conversion to convert it into an electrical signal, and the relationship model between voltage and sugar content can be established.

2.2. Multiple linear regression

The sugar in apples contains sucrose, fructose, and glucose. A single absorption peak does not fully represent the sugar content of apples, so multiple linear regression methods are used. The multiple linear regression analysis studies the influence of multiple independent variables on the dependent variable. This article mainly studies the relationship between the photovoltage value and the sugar content of apple sugar at three characteristic wavelengths, namely:

$$y = a_0 + a_1 u_1 + a_2 u_2 + a_3 u_3 + \varepsilon \quad (1)$$

In the formula, y is the sugar content value; u_1, u_2, u_3 is the diffuse reflection light voltage value with wavelengths of 820nm, 880nm, 940nm, respectively, a_1, a_2, a_3 is the regression coefficient, a_0 is the regression constant.

The model is evaluated by the coefficient of determination R^2 and the root mean square error RMES. The calculation formulas for them are as follows:

$$R^2 = 1 - \frac{\sum_{i=1}^{n} (y_i - \hat{y})^2}{\sum_{i=1}^{n} (y_i - \overline{y})^2} \quad (2)$$

$$RMSE = \sqrt{\frac{1}{n} \sum_{i=1}^{n} (y_i - \hat{y})^2} \quad (3)$$

3. Detection device design

The detection device is mainly composed of a case, a detection array, and a microprocessor circuit board. The square shell is made of opaque material. A black sponge is embedded at the bottom of the shell to facilitate fixing of fruits. Four detection arrays are evenly distributed on the inside of the shell. Each detection array is composed of a photodiode and six LEDs with three wavelengths; the microprocessor circuit board is designed with buttons and display modules on the top of the housing. The structure of the detection device is shown in Figure 1:

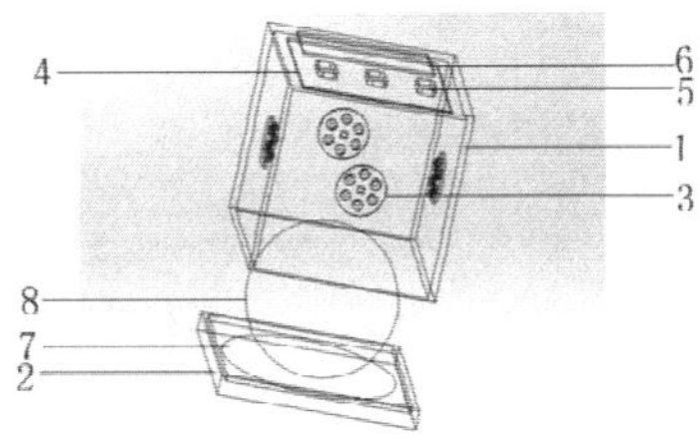

Figure 1 Structure diagram of detection device (1 Case, 2 Bottom of the shell, 3 Detection array, 4 Microprocessor circuit board, 5 Button, 6 Display screen, 7 Black sponge, 8 Apple)

3.1. Hardware Design

According to the system function, the hardware part of the detection device mainly includes LED drive circuit, signal processing circuit, power supply circuit, microcontroller and its peripheral circuits.

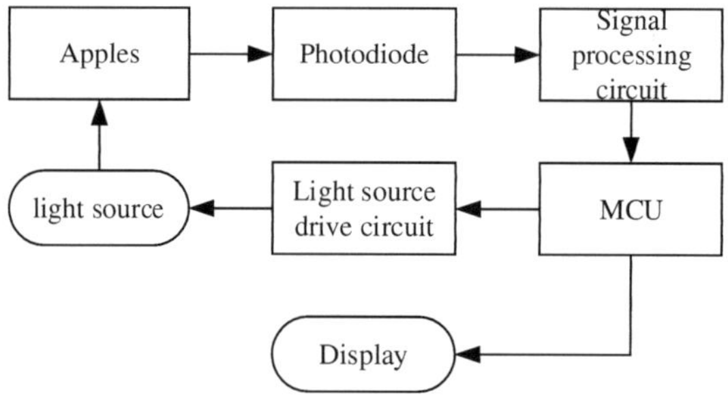

Figure 2 System design process

3.1.1. LED Drive Circuit. Considering the low-cost and low-power design of the device, a near-infrared LED light-emitting diode is selected as the light-emitting source of the experimental system. Compare and analyze the relevant performance parameters of LED light sources on the market, and compare prices and wavelengths, and finally select LEDs with center wavelengths of 820nm, 880nm, and 940nm. The working state of the LED greatly affects the performance of the detection device. The MT7285 constant current drive chip is selected to provide constant current drive for the LED, so that the LED can obtain a more stable working performance. The LED current is set by the external detection resistor Rcs, the formula:

$$ILED = \frac{205}{Rcs(ohm)}(mA) \quad (4)$$

3.1.2. Signal Processing Circuit. The diffuse reflection light signal carrying the information of the apple sugar content is received by the photodiode and converted into an electrical signal. The integrated operational amplifier AD825 is used to amplify the weak electrical signal. AD825 can maintain stable gain under various load conditions. The unique input stage has ultra-low input bias current and input current noise.

The design of the amplifying circuit directly affects the detection accuracy of the device. For the amplifying circuit, the selection of a stable resistor can reduce the temperature drift error, but the resistance of the resistance above 1MΩ has poor stability. In this paper, the T-type network structure is designed, and a resistor with a smaller resistance value is selected to achieve a higher gain, which reduces the measurement error caused by the excessive resistance of the resistor. The specific circuit is shown in the figure. Connect the photodiode to U5 in the figure. When light is detected, the photodiode converts the light signal into a current I, which is amplified by AD825 to convert the output voltage U. The output voltage is:

$$U = -I(1+\frac{R_8}{R_7})R_6 \quad (5)$$

3.1.3. Microprocessor and its peripheral circuit. In this design, GD32F103ZET6 is used as the main control chip, and external 8M and 32.768k passive crystal oscillator are used. The chip, reset circuit, and program download circuit constitute the smallest system of the single-chip microcomputer. The device is connected with an external 5V power supply, the voltage conversion from 5V to -5V is realized through the ICL7660 chip, and the conversion from +5V to +3.3V is realized by BL1117. To realize the display of sugar content information, 2.8 inch TFTLCD is used.

3.2. Software Design

The software part of this work is mainly divided into main control module, AD module, data processing algorithm module, display module and so on based on hardware function module. The software is written using Keil uVision5 tool. The overall structure of the software is mainly composed of the following block diagram, which is mainly to initialize each module and call each module to complete the corresponding task.

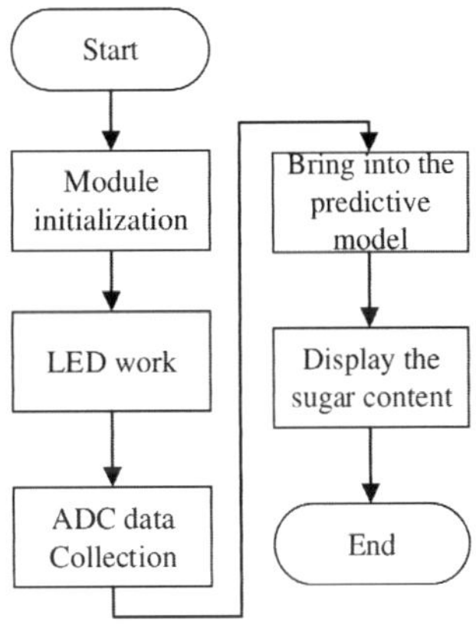

Figure 3 Software design flow chart

4. Results and Analysis

4.1. Methods and procedures

In the supermarket of Shandong University of Science and Technology, the same batch of Yantai Red Fuji apples with uniform shape and fruit diameter of 80~85mm were selected as experimental samples. The sample selection should cover the range of apple sugar content as much as possible. After washing with clean water, number the samples and let them stand at room temperature at 20°C for 24 hours. Put the apple sample to be tested inside the detection device, power on the system, record the voltage value at different wavelengths, and use the Lichen Technology LC-DR-53B digital display brix meter to measure the actual value of the sugar content of the apples.

4.2. Analysis of results

Data processing uses MATLAB software, based on experimental data, finds the relationship between voltage value and fruit sugar content through multiple linear regression methods and establishes a sugar content prediction model. The results are as follows:

$$y = 43.2182 - 11.6145u_1 + 1.4202u_2 - 1.5761u_3 \tag{6}$$

In the above formula, y represents the predicted value of apple sugar content, and u_i represents the value of photovoltage.

In order to verify the accuracy of the device, seven apples sample under the same standard are selected to measure the sugar content of the apple under the same conditions with the device and compared with a digital sugar meter.

Table 1 Model prediction parameter table

Predictive value	Actual value	Absolute error	Relative error
11.95701	10.6	1.357014	12.80%
11.96768	10.7	1.267679	11.85%
9.697634	9.8	-0.10237	-1.04%
10.19262	10.1	0.09262	0.92%
10.2909	10.2	0.090898	0.89%
11.29592	10.8	0.495916	4.59%
11.38014	10.3	1.080143	10.49%

It can be seen from the above table that the minimum relative error of the apple sugar content measured by this device is 0.89%, and the fitting relationship between the predicted value and the true value is shown in the figure below:

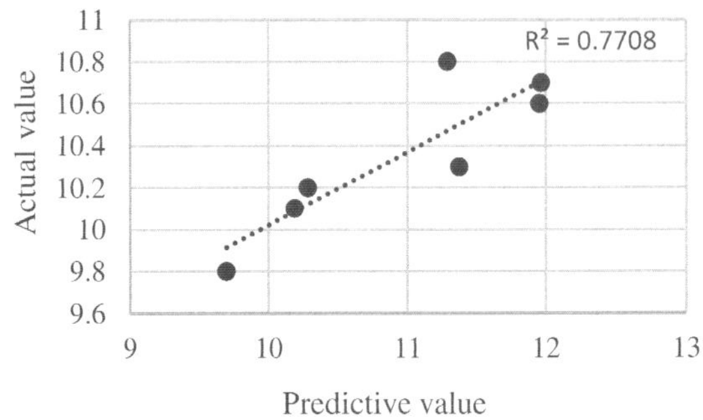

Figure 4 Fitting curve between predicted and actual values

As can be seen from the above figure, the prediction model determination coefficient is 0.7708, and the prediction root mean square error is 0.8356.

5. Conclusions

A quantitative analysis model of apple sugar content was established using the principle of near-infrared, and hardware circuits and software were independently designed and debugged, and a non-destructive testing device for apple sugar content was successfully built. Experimental results show that the use of photodiodes and LEDs with characteristic wavelengths of apple sugar content components instead of traditional spectrometers greatly reduces the cost of the devices. Within a certain error range, the device can realize fast and non-destructive testing of apple sugar content. Due to the small number of samples, the model prediction accuracy is not high. The number of samples can be increased later to optimize the model. This article provides a reference for the realization of a low-cost, portable fruit quality inspection instrument.

References

[1] Liu, Y.D., Zhu, D.N., Sun, X.D., Wu, M.M., Han, R.B., Ma, K.R., Ye, L.Y., Zhang, B.C. (2017) Experimental study on portable determination of soluble solids in Apple. J. Spectroscopy and Spectral Analysis, 37(10): 3260-3265.

[2] Xu, S., Lu, H.Z., Qiu, G.J., Wang, C., Liang, X. (2020) Progress and application of nondestructive testing for fruit quality. J. Guangdong Agricultural Sciences, 47(12): 229-236.

[3] Tan, B.H., Xiao, T.F., Liu, Q.L., Li, G., Li, G., Huang, C.X. (2021) Test and Analysis of Near-infrared Diffuse Reflectance Detection of Fruit Sugar Content. J. Journal of Central China Normal University (Natural Science Edition), 55(02): 211-220.

[4] Wang, D., Wu, J.Z., Han, P., Wang, K. (2021) The application progress of the screening of key spectral variables in the non-destructive testing of agricultural products and food quality. J. Spectroscopy and Spectral Analysis, 41(05): 1593-1601.

[5] Li, X.N., Huang, J.C., Xiong, Y.J., Zhou, J., Tan, X.Y., Zhang, B.H. (2018) Determination of soluble solid content in multi-origin 'Fuji' apples by using FT-NIR spectroscopy and an origin discriminant strategy. J. Computers and Electronics in Agriculture, 155: 23-31.

[6] Huang, Y.W., Li, H., Wang, R.L. (2017) Progress in the application of micro near infrared spectrometer in agricultural products detection. J. Grains and fats, 30(07): 1-3.

[7] Liu, Y.D., Xu, H., Sun, X.D., Jiang, X.G., Rao, Y., Zhang, Y. (2020) Research on a universal model for online detection of sugar content of different varieties of apples by near-infrared spectroscopy. J. Spectroscopy and Spectral Analysis, 40(03): 922-928.

[8] Yuan, L.M., Cai, J.R., Sun, L., Han, E. (2016) Nondestructive measurement of soluble solids content in apples by a portable fruit analyzer. J. Food Analytical Methods, 9(3): 785-794.

CFAR Detector Based on the Identification of Sea Clutter Distribution Characteristics

Bowen Li[a], Xiaoming Liu[b*], Shasha Du[c], Wanhan Li[d]

Information Science and Technology College, Dalian Maritime University, Dalian, Liao Ning, China

[a]email: 1017255076@qq.com, [c]email: dushasha872@163.com,

[d]email: liwh1127@163.com

[b]email: lxmdmu@dlmu.edu.cn

Abstract. To improve the capability of radar detection in a time-varying maritime environment, an algorithm was proposed and analyzed in this paper that can be applied to determine the shape and scale parameters of the K-distributed model using the radar parameters and environmental information under the context of K-distributed sea clutter, and improved the original algorithm to improve the accuracy of the parameter estimation algorithm. Furthermore, a CFAR detector capable of identifying the characteristics of sea clutter distribution was developed. It can estimate the parameters of K-distributed model for sea clutter in real time using both radar parameters and environmental perception information, thus improving the accuracy of clutter background estimation. When the false alarm rate reaches a certain level, the adaptive threshold factor can be obtained for the CFAR detector proposed in this paper, thus improving the effectiveness of control on the CFAR detector. As suggested by the simulation results, the proposed method outperforms the CFAR detector with fixed parameters in the capability of detection and robustness under the context of K-distributed clutter.

1. Introduction

When a radar is used to detect maritime targets, it is common for the existence of sea clutter to cause severe interference with radar detection. Not only is sea clutter affected by the inherent properties of radar, such as radar parameters, it also shows a complex relationship with such environmental factors as wind speed, wind direction, and maritime state. According to the recent research, the improvement of radar resolution and the reduction in grazing angle contribute to a longer tail of the amplitude distribution of sea clutter. Besides, not only can the K-distributed model meet the characteristics of amplitude distribution for the sea clutter under observation, it is also indicative of the relationships between pulses, which makes it a classic model for the description of sea clutter [1].

Given different environments of clutter background, there have been multiple improved CFAR detectors proposed by researchers. Despite some literature where a comparative analysis was conducted of several common CFAR detectors[2], the analysis of most CFAR detectors is premised on the conditions of different noise interference, with the corresponding threshold factor T obtainable through reference. Consequently, it lacks adaptability to non-stationary maritime clutter environment. In order to address this problem, it is necessary to identify the characteristics of sea clutter distribution in real time. To obtain the parameters of clutter distribution model in real time, there are a series of clutter model estimation methods proposed by scholars, such as maximum likelihood estimation method[3] and

Content from this work may be used under the terms of the Creative Commons Attribution 3.0 licence. Any further distribution of this work must maintain attribution to the author(s) and the title of the work, journal citation and DOI.
Published under licence by IOP Publishing Ltd

moment estimation method[4]. However, they all require the parameters of the clutter distribution model to be estimated using the measurement data. In case of limited measurement data on sea clutter, it is necessary to obtain K-distributed model parameters using other methods.

Based on radar parameters and environmental information, the estimation algorithm of reflected power σ^0 per unit sea surface was improved in this paper to determine the shape parameter v and scale parameter b for the K-distributed model. Then, the K-distributed model estimation algorithm was validated. With the OS-CFAR detector as an example, a CFAR detector based on the identification of clutter distribution characteristics was put forward. When the false alarm rate reached a certain level, the optimal threshold factor T of the CFAR detector was obtained through self-adaptation. In this way, the CFAR detector could achieve higher detection probability and better robustness while maintaining the false alarm rate as a constant.

2. K-distributed statistical distribution model parameter estimation

The amplitude distribution characteristics of K-distributed sea clutter can be expressed as probability density function (PDF) below[5].

$$f(x;v,b)=\frac{2b}{\Gamma(v)}(\frac{bx}{2})^{v}K_{v-1}(bx) \tag{1}$$

Where x represents the sea clutter amplitude; v indicates the shape parameter, which determines the sharpness of clutter; b denotes the scale parameter. The larger the scale parameter, the smaller the clutter intensity; $\Gamma(v)$ refers to the Gamma function; K_v stands for the v-order second modified Bessel function.

According to the properties of Bessel function, the distribution function (CDF) of clutter amplitude x is written as

$$F(x)=\int_0^x f(x)dx=1-\frac{2b}{\Gamma(v)}(\frac{bx}{2})^{v}K_{v}(bx) \tag{2}$$

The shape parameter v is a determining factor in the degree to which the K-distributed deviates from Rayleigh distribution. When v approaches 0, there is a relatively long tail; when it approaches ∞, it nearly conforms to Rayleigh distribution. Meanwhile, the smaller the value of v, the more significant the deviation and the longer the tail. Therefore, the false alarm probability will increase when the same detection threshold is adopted for detection [6].

To monitor the clutter environment in real time and impose control on the CFAR detector, it is necessary to determine the functional relationship of radar and environmental parameters with shape and scale parameters, thus improving the outcome of detection.

2.1 shape parameter v estimation model

Based on an in-depth study on the sea clutter data obtained, researchers have made attempt to summarize the empirical model of K-distribution parameters, which can be directly estimated using environmental and radar parameters. Currently, the shape parameter v is commonly estimated using Ward[7], Watts[8], Ryan[9] models, all of which are based on the assumption that the influencing factors in shape parameter v include radar incidence angle, wave direction, and the mode of radar polarization.

Ward constructed an empirical model to estimate shape parameter v by analyzing 300 sets of airborne radar experimental data, and this model performs well within the range of 5-35 GHz[7]. The Ryan model adopted in this paper is an improved version of the Ward model, with the width of radar pulse introduced. Within the X-band range of common marine radar, the shape parameter v can be expressed as[9]:

$$v=10\exp[(2/3)\lg\varphi_{gr}+(5/8)\lg A_c-(1/3)\cos(2\theta_{sw})-k+\lg(\tau/30)\lg(50/\varphi_{gr})\lg(5.5\varphi_{gr})^{0.8}] \tag{3}$$

Where θ_{sw} represents the wind direction angle, indicating the included angle formed between the radar line of sight and the upwind direction; k denotes the parameter related to the polarization mode (HH was polarized to 2.09, VV was polarized to 1.9); τ refers to the radar pulse width; A_c stands for the area of the radar resolution unit, and its value is

$$A_c = c\tau r\Delta\theta_B/2\cos\varphi_{gr} \tag{4}$$

Where c represents the speed of light; r indicates the distance from the radar to the clutter center, $\Delta\theta_B$ denotes the antenna 3 dB beam width; φ_{gr} is referred to as the incident residual angle. The general value of φ_{gr} falls within the range of $0.1°\sim10°$, and it can be expressed as

$$\varphi_{gr} = \arcsin\left(h/r - r/2\alpha\right) \tag{5}$$

Where h represents the height of the radar antenna (unit: meter); α refers to the effective radius of the earth.

2.2 Scale parameter b estimation model

The scale parameter b is correlated with the power characteristics of sea clutter. The smaller the scale parameter, the stronger the echo power of the sea surface. Conversely, the larger the scale parameter, the more gentle the trend of distribution for clutter amplitude. According to Equation (1), the secondary moment of the clutter amplitude can be obtained, while the clutter power can be expressed as follows:

$$P_c = 4v/b^2 \tag{6}$$

According to the radar equation, the power of clutter can be expressed as

$$P_c = \left[P_t G^2\lambda^2/(4\pi)^3 r^4 F^4\right]\sigma^0 A_c \tag{7}$$

Where P_t represents the pulse power transmitted by radar; G indicates the antenna gain; λ denotes the incident wavelength of radar; F refers to the loss factor; and σ^0 stands for the reflected power per unit sea surface. By combining Equations (4), (6) and (7), the scale parameter b can be expressed as

$$b = 16\sqrt{2v\pi^3 r^3\cos\left(\varphi_{gr}\right)/P_t G^2\lambda^2 F^4 c\Delta\theta_B\sigma^0} \tag{8}$$

2.3 Clutter reflection coefficient model

According to Equation (8), the reflected power per unit sea surface σ^0 must be obtained to solve the scale parameter b. For this reason, there are several estimated per unit sea surface reflected power models deduced by researchers, including GIT model[10], TSC model[11], HYB model[12] and NRL model[13].

When the environment of communication is unknown, it is considered better to adopt the TSC model than others, because it is closer to the average of all communication environments[14]. Meanwhile, the model is applicable when radar frequency falls within the range of 0.5-30 GHz and the ground scraping angle varies from $0°\sim90°$, which is consistent with the working environment required by the radar studied in this paper. In addition, all wind directions from downwind to upwind are covered. Therefore, this paper focuses mainly on the study of the TSC model, in which the calculation formula of reflected power per unit sea surface is expressed as follows:

$$\begin{cases} \sigma_{HH}^0 = 10\lg\left[1.7\times10^{-5}\varphi_{gr}^{0.5}G_u G_W G_A/(3.2808\lambda+0.05)^{1.8}\right] \\ \sigma_{VV}^0 = \begin{cases} \sigma_{HH}^0 - 1.73\ln(8.255\sigma_z+0.05)+3.76\ln\lambda \\ +2.46\ln\left(\sin\varphi_{gr}+0.0001\right)+24.2627; f<2\text{GHz} \\ \sigma_{HH}^0 - 1.05\ln(8.255\sigma_z+0.05)+1.09\ln\lambda \\ +1.27\ln\left(\sin\varphi_{gr}+0.0001\right)+10.945; f\geqslant 2\text{GHz} \end{cases} \end{cases} \tag{9}$$

2.3.1 Small gazing angle factor G_A

$$\begin{cases}\sigma_Z=0.03505S^{1.95}\\ \sigma_\alpha=4.5416\varphi_{gr}\left(3.2808\sigma_Z+0.25\right)/\lambda\\ G_A=\sigma_\alpha^{1.5}/\left(1+\sigma_\alpha^{1.5}\right)\end{cases} \tag{10}$$

Where σ_Z represents the surface height standard deviation (meter); λ denotes the radar wavelength (meter); φ_{gr} refers to the gazing angle (rad); S stands for the Douglas sea state grade.

2.3.2 Wind speed factor G_W

$$\begin{cases}U=3.189S^{0.8},Q=\varphi_{gr}^{0.6}\\ A_1=\left[1+\left(\lambda/0.00914\right)^3\right]^{0.1},A_2=\left[1+\left(\lambda/0.00305\right)^3\right]^{0.1}\\ A_3=\left[1+\left(\lambda/0.00914\right)^3\right]^{Q/3},A_4=1+0.35Q\\ A=2.63A_1/\left(A_2A_3A_4\right),G_W=\left[\left(1.9438U+4\right)/15\right]^A\end{cases} \tag{11}$$

Where U represents the wind speed (m/s), with other parameters carrying the same meanings as Equation (10).

2.3.3 Wind direction factor G_u

$$G_u=\begin{cases}1;\ \theta_{sw}=\pi/2\\ \exp\left[0.3\cos\theta_{sw}\exp\dfrac{\left(-\varphi_{gr}/0.17\right)}{\left(\lambda^2+0.005\right)^{0.2}}\right];\theta_{sw}<\pi/2\end{cases} \tag{12}$$

Where θ_{sw} represents the wind direction angle, indicating the included angle between the radar line of sight and the upwind direction (rad), while other parameters are of the same meanings as Equation (10).

Since the model is suitable for a low incidence angle, there is a certain error between the wind direction factor G_u and the measured data as can be known from the previous experiment. The minimum of wind direction factor G_u of the TSC model appears in the downwind direction, while the minimum of measurement data is supposed to appear in the crosswind direction. On this basis, the wind direction factor is corrected in this paper as follows:

$$G_u=\begin{cases}1-A\sin^2\theta_{sw};\ \theta_{sw}=\pi/2\\ \left(1-A\sin^2\theta_{sw}\right)\exp\left[0.3\cos\theta_{sw}\exp\dfrac{\left(-\varphi_{gr}/0.17\right)}{\left(\lambda^2+0.005\right)^{0.2}}\right];\ \text{others}\end{cases} \tag{13}$$

Maintaining the original form of the TSC model, the corrected wind direction factor G_u was modulated by sine function to obtain a relative minimum in the crosswind direction. Besides, the change rate reached its minimum when it was 0° and 180°, respectively. With $1-A\sin^2\theta_{sw}$ multiplied by the wind factor G_u in the TSC model, a minimum of wind factor G_u will result at the wind angle of 90° without affecting the upwind and downwind values.

Fig. 1 shows the changes in the modulated wind direction factor as parameter A changes from 0.2 to 0.8. It can be seen that there is a 3 dB difference in the wind factor between the downwind and crosswind in Nathanson data, which is consistent with the value of A=0.63. Therefore, in this paper, A=0.63.

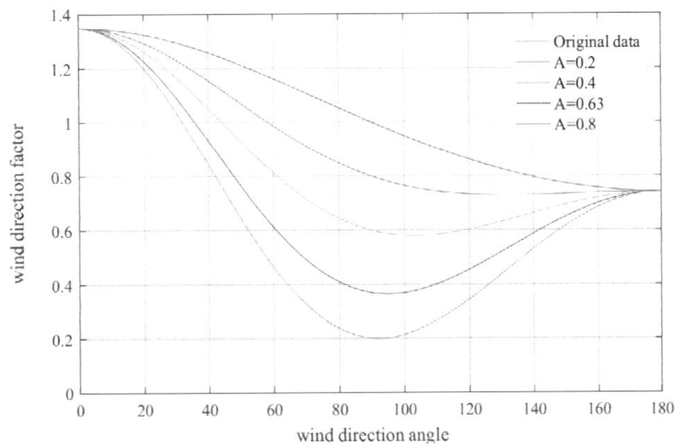

Fig. 1 Variation in wind direction factor with parameter A

2.4 K-distribution parameter estimation of sea clutter

Based on radar and clutter environmental parameters, the process of estimating K-distributed shape parameter v and scale parameter b is detailed as follows:

According to the distance and angle between the radar and the clutter center, the wind direction as determined by the wind speed sensor, as well as the preexisting information like radar parameters, the shape parameter v can be obtained through Equations (3)~(5).

The wind direction as determined by the wind speed sensor and the maritime meteorological and hydrological information buoy is referenced to calculate the average wave height. Then, the Douglas sea state level can be obtained by looking up the table. Combined with the preexisting information such as radar parameters, the reflected power per unit sea surface can be obtained through Equations (9)~(13).

Based on the shape parameter v and reflected power per unit sea surface σ^0 as obtained above, as well as such preexisting information as radar parameters, the scale parameters can be obtained through Equations (6)~(8).

3. CFAR detector based on the identification of sea clutter distribution characteristics

In this paper, the OS-CFAR detector was exemplified to conduct experiment, which led to the result that this method still performs well when applied on other CFAR detectors. Figure 2 shows the schematic diagram of the CFAR detector based on the identification of clutter distribution characteristics. In this figure, D indicates the detection unit, while M denotes the protection unit. There are R reference units on both sides, with the length of the reference sliding window at the front and back denoted as N. First of all, the reference unit sample is sorted in descending order. Secondly, the kth sorted sample is applied as the detector to estimate the clutter power level Z. T represents the threshold factor obtained through self-adaptation according to the distribution characteristics of sea clutter and the set false alarm probability. Compared with conventional OS-CFAR, the CFAR detector proposed in this paper on the basis of identifying clutter distribution is fitted with a parameter estimator and a threshold factor calculator, so as to obtain the optimal adaptive threshold factor T that can better match the environment.

The judgment criterion is $D \underset{H_0}{\overset{H_1}{\gtrless}} T \bullet Z$, where H_1 indicates the existence of target and H_0 indicates otherwise.

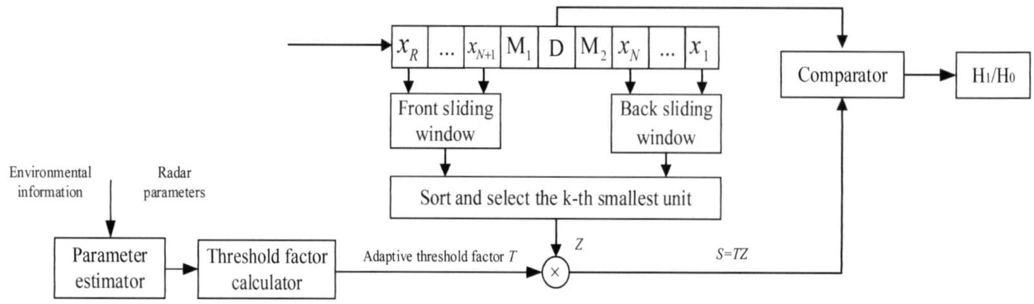

Fig. 2 Schematic diagram of OS-CFAR detector based on the identification of sea clutter distribution characteristics

The random variable conforms to K-distribution, the density function of which is determined by Equation (1). Based on the assumption that R reference units are independent and identically distributed random variables, the probability density function of the kth sample can obtained as follows after the reference units are ranked from small to large:

$$f_k(x) = k\binom{R}{k} f(x) F(x)^{k-1} (1-F(x))^{R-k} \tag{14}$$

Based on Equations (1), (2) and (14), it can be inferred that the false alarm rate of the OS-CFAR detector under the context of K-distributed clutter is

$$\begin{aligned} P_f &= P(X \geq TZ) \\ &= \int_0^\infty 2bk\binom{R}{k}\left[\frac{2(bx)^v}{\Gamma(v)}\right]^{R-k+1} \bullet K_{v-1}(2bx) \\ &\bullet K_v^{R-k}(2bx) \bullet \left[1-\left(\frac{2(bx)^v}{\Gamma(v)}\right)K_v(2bx)\right]^{k-1} \\ &\bullet \left[1-\frac{2(Tbx)^v}{\Gamma(v)}K_v(2Tbx)\right]dx \end{aligned} \tag{15}$$

Let $z = bx$, then Equation (15) can be transformed into:

$$\begin{aligned} P_f &= \int_0^\infty 2k\binom{R}{k}\left[\frac{2(z)^v}{\Gamma(v)}\right]^{R-k+1} \bullet K_{v-1}(2z) \\ &\bullet K_v^{R-k}(2z) \bullet \left[1-\left(\frac{2(z)^v}{\Gamma(v)}\right)K_v(2z)\right]^{k-1} \\ &\bullet \left[1-\frac{2(Tz)^v}{\Gamma(v)}K_v(2Tz)\right]dz \end{aligned} \tag{16}$$

It can be seen from Equation (16) that when the shape parameter v is known, the false alarm probability P_f is irrelevant to the scale parameter b, with the former related solely to the threshold factor T and shape parameter v, indicating that the CFAR detector remains capable to maintain the false alarm characteristics when the clutter intensity varies.

The CFAR detector proposed in this paper on the basis of identifying sea clutter distribution characteristics can be applied to estimate the shape parameter v of the K-distributed model in real time using radar parameters and the information about maritime environment, with the corresponding adaptive threshold factor made obtainable by the threshold factor estimator T under the context of varying false alarm probability P_f.

4. Experimental analysis

4.1 Performance analysis as to the K-distributed parameter estimation of measured sea clutter

First of all, the measurement data of sea clutter was used to evaluate the performance of the proposed method in estimating shape parameter v and scale parameter b for the sea clutter distribution model. The measurement data used in this section is the sea clutter data obtained by using IPIX radar sourced from McMaster University in Canada in 1993. The operating parameters of IPIX radar are shown partially in Table 1. In this paper, the influencing factor for two-way communication is 0.02.

Table 1 IPIX radar parameters

Parameters	Parameter value
Radar frequency	9.39GHz
Height of radar antenna	30m
Peak power of transmitter	8kW
Antenna gain	45.7dB
Horizontal beam width	0.9°
Pulse width	0.2us
PRF	1kHz
Range resolution	30m

IPIX radar data revealed 14 sets of resident mode data, four of which suffered severe truncation effect. For this reason, only 10 sets of data were available for use. Due to the limited amount of data available for IPIX, the experimental results might be rendered accidental to some degree. To avoid this issue, the IPIX data will be segmented next.

Table 2 IPIX data target distribution

Serial number	Unit of main target	Unit of secondary target
#17	9	8,10,11
#26	7	6,8
#30	7	6,8
#31	7	6,8,9
#40	7	5,6,8
#54	8	7,9,10
#280	8	7,9,10
#310	7	6,8,9
#311	7	6,8,9
#320	7	6,8,9

In this experiment, there are 10 sets of data available to use for the IPIX radar, with each set including 14 range cells, each of which comprised of 131,072 pulses. As the cooperative target drifts left and right while fluctuating with the wave, the energy expands in distance and pulse dimension. Table 2 details the distribution of a specific target. Take # 17 data as an example. First of all, with the four units removed where the main and second targets are located in the distance dimension, the 10 pure sea clutter distance units remained. With pulse accumulation set as $N = 26000$, the post-segmentation data can conform to the time correlation, which means the data in each distance unit was divided into 26000 × 5 data segments. Then, 10 pure sea clutter distance units were divided into 50 data segments (# 26, #30 two groups of data could be divided into 55 data segments). Under HH and VV polarization modes, #17 data was divided into 100 data segments (#26, #30 two groups of data could be divided into 110 data segments). Likewise, the IPIX radar data was divided into 1020 groups of data segments. The above algorithm was applied to estimate the parameters of the sea clutter K-distributed model for the processed data segments. The spatial correlation conforms to an exponential correlation, while the power spectral density corresponding to time correlation is Gaussian spectrum. The TSC model was adopted for the clutter reflection coefficient model, the Ryan model was used for shape parameter estimation, and the SIRP method based on the clutter scattering mechanism was applied to simulate the K-distributed clutter.

In this section, the results of K-S test are treated as the measurement standards to verify the above-mentioned sea clutter K-distributed parameter estimation algorithm for its robustness under different maritime conditions and different radar polarization modes when α=0.1. The experimental results are listed in Table 3. Under the context of HH polarization, there are 428 pieces of data conforming to the theoretical value, with a coincidence rate of 83.92%. Given VV polarization, there are 442 pieces of data conforming to the theoretical value, with a coincidence rate of 86.67%. From this, it can be found out that in the K-S test carried out using the sea clutter parameter estimation method as proposed in this paper and the pure sea clutter data of IPIX radar, there are 870 data segments (1020 data segments in total) which conform to the theoretical values, accounting for 85.29% of the total. Therefore, it is believed that there is an excellent robustness possessed by the improved estimation algorithm for the K-distributed shape parameter v and scale parameter b of sea clutter.The data in the table is the ratio of the theoretical data volume to the total data volume.

Table 3 K-S test data of IPIX pure sea clutter data

Serial number	HH polarization	VV polarization
#17	36/50	40/50
#26	52/55	50/55
#30	51/55	53/55
#31	42/50	43/50
#40	42/50	44/50
#54	43/50	43/50
#280	41/50	42/50
#310	39/50	42/50
#311	40/50	41/50
#320	42/50	44/50
Total	428/510 83.92%	442/510 86.67%
	870/1020 85.29%	

4.2 Relationship between false alarm probability and threshold factor

According to Equation (16), the relationship between false alarm probability P_f and threshold factor T can be determined using the numerical integration method when the shape parameter v is known, as shown in Fig. 3. When the false alarm probability is constant, the curve of relationship between the threshold factor T and shape parameter v is also obtainable, as shown in Fig. 4. As for the shape parameter v of clutter, it can be estimated using radar parameters and environmental information, so as to determine the corresponding adaptive threshold factor T .

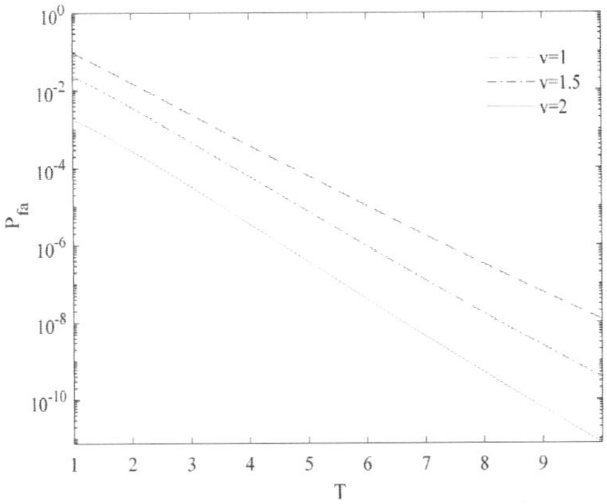

Fig. 3 Relation between P_f and T

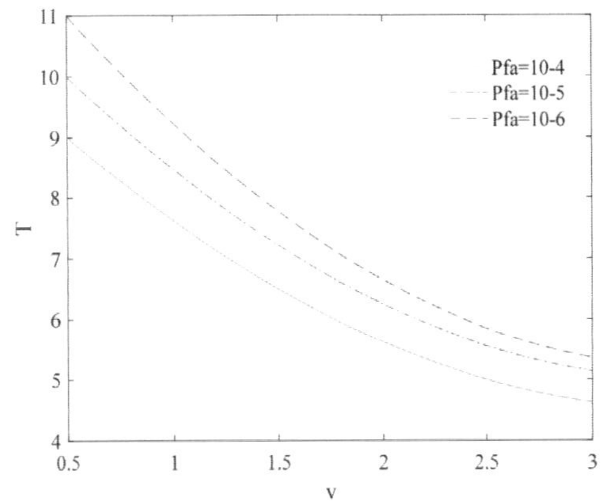

Fig. 4 Relation between T and v

4.3 Performance analysis of CFAR detector under K-distributed clutter background

In order to verify the better performance of the algorithm proposed in this study than the conventional OS-CFAR detection algorithm, a Swerling type 1 target was introduced into the pure sea clutter background of #54 and #310 of the IPIX radar, respectively (#54 represents low sea condition and #310 represents high sea condition). There were 10 reference units and 2 protection units of the OS-CFAR detector set based on the identification of clutter distribution characteristics with the Monte Carlo experiments performed 10^7 times and the probability of false alarm is 10^{-5}.

As shown in Fig. 5, in the presence of K-distributed sea clutter, the CFAR detector as proposed in this paper based on the identification of clutter distribution characteristics evidently outperforms the conventional OS-CFAR detector with fixed parameters. When the detection probability is 0.8, the requirement on the input signal-to-clutter ratio of the CFAR detector based on the identification of clutter distribution characteristics is improved by 1dB compared with the conventional OS-CFAR detector with fixed parameters. The CFAR detector based on the identification of sea clutter distribution characteristics showed similarity in detection performance under two different maritime conditions, which is attributed to the CFAR detector based on the identification of clutter distribution characteristics integrating radar parameters with the real-time information about environmental parameters. As a result, the level of robustness improves when compared with the conventional CFAR detector.Figure 5(a) is the #54 data detection performance comparison under low sea conditions, Figure 5(b) is the #310 data detection performance comparison under high sea conditions.

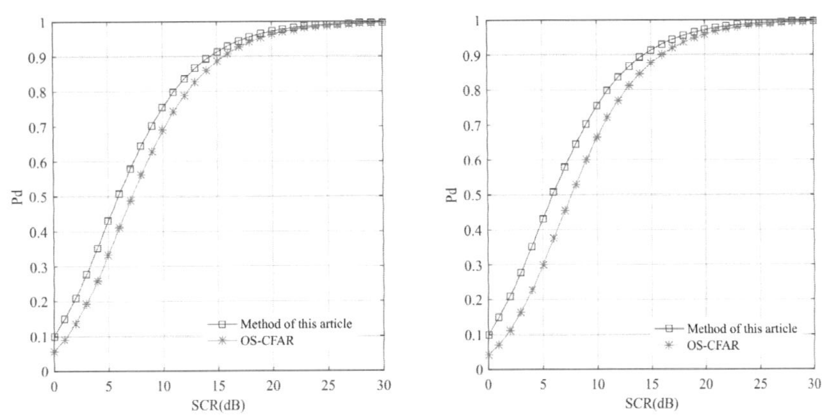

Fig. 5 Relationship between detection probability and signal-to-clutter ratio of the CFAR detector under different maritime conditions

5. Conclusion

The main content of this paper is detailed as follows. On the one hand, an analysis was conducted to determine the relationship of environmental information and radar parameters with the shape parameter v and scale parameter b of the K-distributed model under the context of K-distributed clutter. On the other hand, the maritime clutter shape parameter v as estimated using the above algorithm was used to solve the adaptive threshold factor T of the OS-CFAR detector for better control on the CFAR detector.

As revealed by the above analysis, with changes occurring to radar parameters or the maritime environment, the proposed CFAR detector based on sea clutter identification can be applied to obtain the threshold factor T adapted to the environment in real time through self-adaptation. Moreover, this method proves applicable to both the OS-CFAR detector and other CFAR detectors. Furthermore, the simulation data was used to verify the performance of the algorithm. According to the results, the proposed CFAR detector based on sea clutter cognition outperforms the traditional CFAR detector algorithm without identifying clutter characteristics in detection and robustness.

References

[1] Liu N.B, Jiang X.Y, Ding H, Guan J. (2021) Research Overview of Sea Clutter Characteristics of Big Radar Gazing Angle. Journal of Electronics & Information Technology, 201,43:2771-2780.

[2] Xiong X, Yu G.W, Guo S.F, Wu Z.L. (2018) An Adaptive Constant False Alarm Detection Method Against Multiple False Target Interference. Journal of Air Force Early Warning Academy,32:391-396.

[3] Lu Y, Ying S.J, Sun Z, Zou X.P. (2012) Maritime Radar Target Detection Algorithm Based on Maximum Likelihood Estimation. Journal of Shanghai Maritime University, 33:1-4.

[4] Schoenecker, Steven, Willett, et al. (2016) The Effect of K-Distributed Clutter on Trackability. IEEE Transactions on Signal Processing: A publication of the IEEE Signal Processing Society, 2016. 64:475-484.

[5] Zhang K, Shui P.L, Wang G.H. (2020) Incoherent accumulation Constant False Alarm Detection Method under the Background K-distributed Sea Clutter of Correlative Radar. Journal of Electronics & Information Technology, 42:1627-1635.

[6] Meng X.W. (2020) Performance Analysis of Quantized Rank Nonparametric CFAR Detector in Clutter Edge. Journal of Electronics, 48:384-389.

[7] Watts S , Wicks D C . (1990) Empirical models for detection prediction in K-distributed radar sea clutter. In: IEEE International Radar Conference. Arlington, 189-194.

[8] Ryan J , Johnson M . (1992) Radar performance prediction for target detection at sea. In: Radar 92 International Conference. Brighton, 13-17.

[9] Horst M.M, Dyer F.B, Tuley M.T. (1978) Radar Sea Clutter model. Antennas and Propagation. London, 6-10.

[10] Leonard T.P , Antipov I , Ward K.D . (2002) A comparison of radar sea clutter models. RADAR 2002. Edinburgh, 429-433.

[11] Reilly J P , Dockery G D . (2002) Influence of evaporation ducts on radar sea return. IEE Proceedings F - Radar and Signal Processing, 137:80-88.

[12] Gregers-Hansen V , Mital R . (2012) An Improved Empirical Model for Radar Sea Clutter Reflectivity. Aerospace & Electronic Systems IEEE Transactions on, 48:3512-3524.

[13] Rohling H . (1983) Radar CFAR Thresholding in Clutter and Multiple Target Situations. IEEE Transactions on Aerospace and Electronic Systems, 19:608-621.

[14] Deng S.Q, Jin L, Liang H. (2017) Simulation of Spatiotemporal Coherence Correlation K-distributed Sea Clutter. Electronic Measurement Technology, 40:61-65.

Design of high voltage electrostatic protection device for CAN bus

Xuebing Su [1,3,a], Yang Wang [1,3,b*], Xiangliang Jin [2,3,c], Hongjiao Yang [1,3,d]

[1] School of Physics and Optoelectronics, Xiangtan University, Xiangtan, Hunan, China

[2] School of Physics and Electronics, Hunan Normal University, Changsha, Hunan, China

[3] Hunan Engineering Laboratory for Microelectronics, Optoelectronics and System on a Chip, Xiangtan, Hunan, China

[a]email: sxbxtu@163.com, [b*]email: wangyang@xtu.edu.cn, [c]email: jinxl@hunnu.edu.cn, [d]email: yanghongjiao2004@xtu.edu.cn

Abstract: The Dual-Direction Silicon Controlled Rectifier (DDSCR) device has dual-direction electrostatic protection function and strong current discharging ability, which is widely used in ESD on-chip protection. In this paper, a high performance symmetric high voltage Dual-Direction Silicon Controlled Rectifier with floating P+ (HVDDSCR_FP+) is designed for CAN bus under 0.18 μm BCD process. In order to predict and verify the ESD performance of the device, TCAD two-dimensional device simulation and Transmission Line Pulse (TLP) test system were used. The experimental results show that the HVDDSCR_FP + structure has not only symmetrical positive and reverse I-V curves, but also the characteristics of high holding voltage and high failure current. Compared with the traditional HVDDSCR, HVDDSCR_FP+, at 25V forward and reverse trigger voltages (V_t), has increased its holding voltage (V_h) from 12V to 20V, failure current (I_{t2}) from 5A to 35A. In the actual tape-out process, it is found that the width of the N-well has a great influence on the performance of the device. After analyzing the reason and improving it, the leakage current of the device is reduced from μA level to nA level.

1 Introduction

With the rapid development of electronic technology, automobile electronization is increasingly growing faster and higher. Harsh application environments and complex electronic systems bring great challenges to the design of on-chip electrostatic discharge (ESD) protection [1-3]. ESD protection of automotive communication ports usually requires high holding voltage (V_h) to avoid circuit latch-up and high failure current (I_{t2}) to have good ESD robustness [4-6]. In practical electrostatic protection schemes, transient voltage suppressor (TVS) devices integrated on the chip are generally used. TVS devices have the advantages of high temperature resistance, high voltage resistance, small parasitic capacitance and fast response speed in electrostatic protection. Among them, the silicon controlled rectifier (SCR) is a very attractive choice, and the dual-directional silicon controlled rectifier (DDSCR) provides bidirectional ESD protection for circuits in compact devices. However, its high trigger voltage makes SCR prone to core damage and low holding voltage leads to latch-up issue , which is an urgent problem to be solved in the industry [7]. The holding voltage of high voltage SCR is usually dependent on the doping concentration and depth of substrate, P-epi and NBL, but this adjustment has

Content from this work may be used under the terms of the Creative Commons Attribution 3.0 licence. Any further distribution of this work must maintain attribution to the author(s) and the title of the work, journal citation and DOI.
Published under licence by IOP Publishing Ltd

great limitations. At present, the proposed holding voltage improvement technology includes the insertion of floating N+ in the N-well [8] and segmented emitter layout [9][10]. These technologies improve the holding voltage (V_h) to a large extent, but significantly reduce the failure current (I_{t2}), which also increases the difficulty of design. Literature [11] analyzed and verified that the holding voltage (V_h) and failure current (I_{t2}) of DDSCR could be significantly increased by adding N-type buried layer. Literature [12] introduces an asymmetric DDSCR structure, in which a floating P+ is inserted into the P-well of the device, and its holding voltage and failure current are greatly improved. This provides a great design scheme for designers to optimize the holding voltage and failure current of devices.

In this paper, symmetric HVDDSCR devices were studied under 0.18 μm BCD process. In order to achieve the design purpose of high holding voltage and high failure current, the structure and working mechanism of HVDDSCR_FP+ are proposed and discussed on the basis of traditional symmetric HVDDSCR devices, and the effect of the width of N-well in the structure on the performance of HVDDSCR_FP+. The theoretical analysis is carried out by using TCAD two-dimensional simulation software. The performance of the device was verified by tape-out and transmission line pulse test (TLP).

2 Device structure and working mechanism

The cross-section view of a conventional symmetric HVDDSCR device is shown in Fig.1 (a). Either the cathode or the anode has the lateral parasitic transistor NPN ($NPNL_F$ or $NPNL_R$) with N-well as collector and the vertical parasitic transistor NPN ($NPNV_F$ or $NPNV_R$) with N-type buried layer (NBL) as collector. Forward SCR path is composed of P-well resistors R1 and R2, parasitic transistor PNP, $NPNV_F$ and $NPNL_F$. The equivalent circuit of forward SCR is shown in Fig.1 (c). The reverse SCR path consists of P-well resistors R2 and R1, parasitic transistor PNP, $NPNV_R$ and $NPNL_R$, and the equivalent circuit of reverse SCR is shown in Fig.1 (d). The distance between PW and NW was denoted as Sp, the distance between the N+ and the breakdown surface was denoted as Wb, and the width of the N-well in the middle of the device was denoted as X.

The symmetrical HVDDSCR is completely symmetrical in structure and can discharge current in both positive and negative directions. As shown in Fig.1 (a), when the anode of HVDDSCR receives ESD forward pulse and reaches the breakdown voltage at the breakdown surface, avalanche breakdown will occur in N-well and P-epi, which will generate a large number of electron-hole pairs and the holes will reach the cathode P+ of the device through P-epi resistance R2. As the number of holes increases, the junction voltage V_{be} derived from the base and emitter of $NPNV_F$ and $NPNL_F$ will increase gradually. When V_{be} increases to 0.7V, $NPNV_F$ and $NPNL_F$ will turn on. The parasitic transistor $NPNV_F$ and $NPNL_F$ provide current to the base of the parasitic transistor PNP after opening, making the parasitic transistor PNP turning on. A positive feedback mechanism will be formed between the two kinds of parasitic transistor, making the device in the latch-up state, and the device begins to release the current. Current discharge path P1 is composed of the $NPNL_F$ and the parasitic transistor PNP, and path P2 is composed of the $NPNV_F$ and the parasitic transistor PNP, forming forward SCR paths P1 and P2, as shown in Fig.1 (c). The working mechanism of the reverse path is the same as that of the forward path.

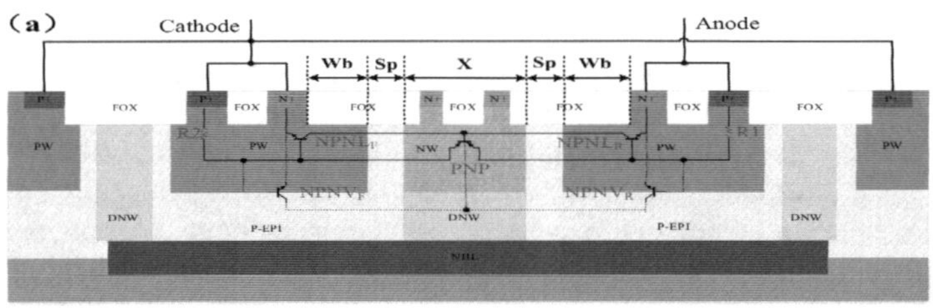

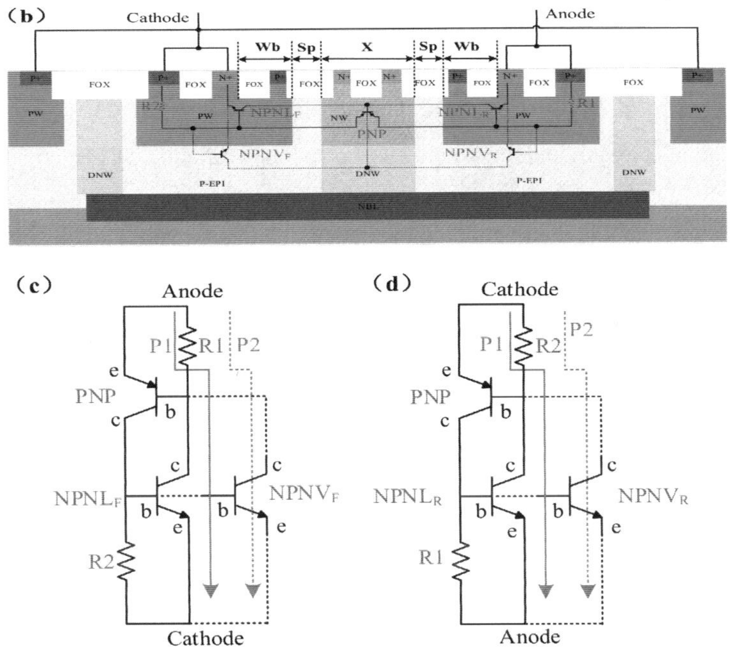

Fig.1 (a) Cross-section view of the HVDDSCR ; (b) Cross-section view of the HVDDSCR_FP+; (c) Forward equivalent circuit; (d) Reverse equivalent circuit.

In the traditional symmetric HVDDSCR structure, because of the deep N-type buried layer (NBL), the width of the base of the vertical parasitic transistor NPNV is much wider than that of the lateral parasitic transistor NPNL, so the current amplification factor β of the parasitic transistor on the lateral path is larger. The current discharge path P1 is the main discharge path, and P2 is the second path. Widening the distance of the Wb can increase the lateral parasitic transistor NPNL base of effective width, inhibite the lateral current amplification factor β, nominate the discharge current path (P2) as main discharge path. It makes the discharge path longer and hence holding voltage higher. The wider the Wb gets, however, the larger the chip area will be, which contributes to the device improperly working.Therefore, the symmetric HVDDSCR_FP+ structure studied in this paper is to insert floating P+ at the Wb position of the traditional symmetric HVDDSCR structure. The purpose is not to increase the lateral size of the device, make P2 the main current release path, and improve the performance of the device. The cross-section view of symmetric HVDDSCR_FP+ is shown in Fig. 1(b), and its equivalent circuit diagram is the same as that of traditional symmetric HVDDSCR, as shown in Fig. 1(c) and (d).

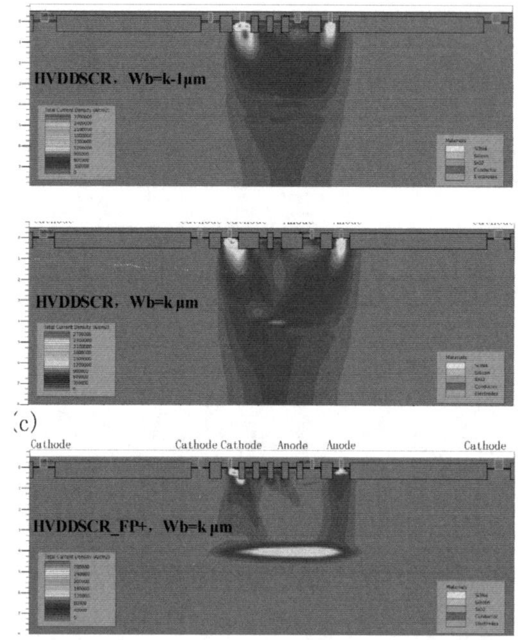

Fig.2 TCAD simulated forward current density in the HVDDSCR, Wb=k-1μm (a) and the HVDDSCR, Wb=k μm (b) and the HVDDSCR_FP+, Wb=k μm (c), under 0.1A forward current pulse.

3 TCAD simulation analysis

3.1 Performance simulation of HVDDSCR and HVDDSCR_FP+ devices

Through transient simulation of TCAD, HVDDSCR with Wb as K-1 μm and Wb as K μm and HVDDSCR_FP+ with Wb as K μm were simulated to study the working mechanism of the device. A square wave current pulse similar to TLP was applied to the anode with a pulse amplitude of 0.1A and a pulse duration of 100ns, and the cathode was grounded. Fig. 2 (a) When the Wb of HVDDSCR is K-1 μm, the current is mainly emitted by the SCR path P1 formed by the $NPNL_F$ and the parasitic transistor PNP. Fig. 2 (b) When the Wb of HVDDSCR is K μm, the current begins to go deep into the buried layer of the device, and the SCR path P2 is released by the $NPNV_F$ and the parasitic transistor PNP. Fig. 2 (c) shows the HVDDSCR_FP+ with Wb as K μm. It can be observed more significantly that a large amount of current is concentrated in the buried layer.

The Widened Wb increases the effective base width of the NPNL and inhibits the current amplification factor β of the NPNL. The current discharge move toward the vertical parasitic path inside the device rather than toward the surface of the device. However, when floating P+ is inserted into the device, the current path walk deep inside the device, causing the main current discharge path to change from P1 to P2. P2 has a longer current path and thus the device obtains a higher holding voltage. Its mechanism can be explained from the following two aspects:

1) Under the action of transient forward electrostatic pulse, because floating P+ and Ground have the same potential, floating P+ will block part of the electron flow from the emitter (N+) to the collector (NW) in the $NPNL_F$, and the current will flow deeper into the device, resulting in a longer current path and a higher holding voltage.

2) When floating P+ is inserted into the P-well, the base width and the base doping concentration of the NPNL increase, resulting in a decrease of the current amplification factor β of the NPNL, and the current discharge path P2 becomes the main path, resulting in a higher holding voltage of the device [13].

3.2 Effect of N-well width X in HVDDSCR_FP+ on device performance

It can be seen from the cross-section view of HVDDSCR_FP+ device that the N-well and the deep N-well in the middle of the device (the width of the N-well and the deep N-well is the same) play a role of isolation in the structure. In ideal scenario, no current should flow from anode to cathode until the device has reached the trigger voltage, and in practical applications, the leakage current of the device is generally required to be nA level. TCAD two-dimensional device simulation software was used to simulate HVDDSCR_FP+ structure X is t μm and 2t μm respectively. Fig. 3 shows the current density distribution of the device when the N-well is not broken down. When the width of the N-well X is t μm, due to the depletion of PN junction, a small amount of current will pass through the N-well before breakdown, resulting in large leakage current of the device, which leads to the device cannot be used properly. The position of device penetration due to depletion is marked with red circles in Fig. 3 (a). When the width of the N-well X is 2t μm, the device is not penetrated when the N-well is not broken down, which is also marked with red circle in Fig. 3(b). For HVDDSCR_FP+, the smaller the X distance, the narrower the base width of the parasitic transistor PNP, the larger the current amplification factor β, and the SCR formed with the parasitic transistor NPN is easier to open, but the

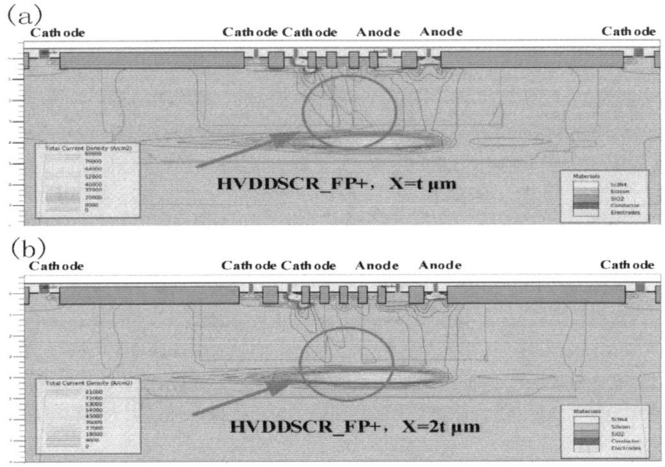

Fig.3 TCAD simulated forward current density in the HVDDSCR_FP+ (a) X=t μm (b) X=2t μm, under 0.1A forward current pulse.

simulation results show that the small X will make the device punch through, and large leakage current. Secondly, in order to improve the device holding voltage, it is feasible to widen X distance appropriately. The mechanism is not only to increase the resistance along the discharge path, but also to reduce the current amplification factor β of the parasitic transistor PNP, and weaken the positive feedback of SCR. However, the N-well width X should not be too wide, because too wide X will make the SCR to hardly open.

4 TLP test verification

4.1 TLP test of HVDDSCR and HVDDSCR_FP+

The traditional symmetric HVDDSCR and the symmetric HVDDSCR_FP+ with floating P+ were fabricated and their performance were verified under 0.18μm process. The layout of each structure mentioned in this paper was drawn with finger length of 50μm and eight-finger pattern, Fig. 4 is the image of HVDDSCR_FP+ under metallographic microscope. The main application scenario of the 20V device is the vehicle network CAN bus, and the normal operating voltage of each port is ±12V. Thus, in the TLP test, the leakage current was measured at 13.2 V DC voltage (1.1 times the operating voltage).

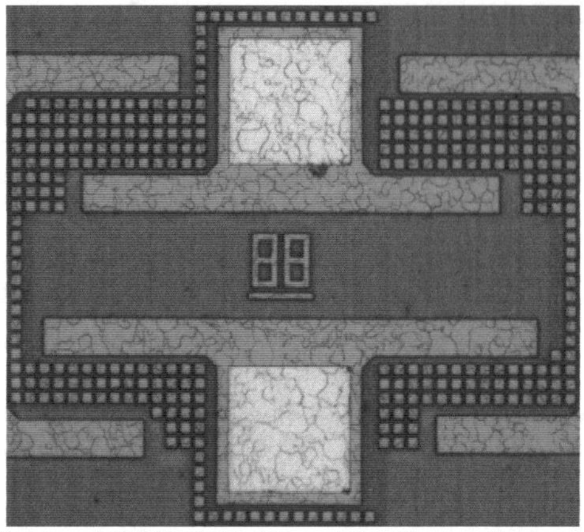

Fig.4 HVDDSCR FP+ eight-finger diagram.

TLP test is a method to measure the I-V characteristics of ESD protection devices under high current stress by using rectangular short pulses. It is also an important test method to evaluate the ESD protection technology of integrated circuits. Fig. 5 shows the TLP test results of HVDDSCR with Wb as K-1 μm and Wb as K+1 μm, and HVDDSCR_FP+ with Wb as K μm. The specific data are shown in Table 1. The diagram shows the structure of HVDDSCR. When Wb is K-1 μm, the forward and reverse holding voltages are 12.5V and 12.3V, the forward and reverse failure currents are 4.8A and 5.9A, and the forward and reverse leakage currents are about 20nA. When Wb is k+1 μm, the forward and reverse holding voltages of the device are 21.5V and 20.6V, and the forward and reverse failure currents are 35.3A and 33.6A respectively. Both the holding voltage and failure current of the device increase greatly, but the forward and reverse leakage currents are 53nA and 134nA. In the HVDDSCR_FP+, Wb is K μm, the forward and reverse holding voltage and failure current are higher than 22V and 35A, respectively, and the forward and reverse leakage currents are less than 20nA. Through comparison, it can be found that the performance of the device can be improved by widening Wb and inserting floating P+, but the method of inserting floating P+ can ensure that the holding voltage and failure current can be improved under the condition that the size of the device is basically unchanged.

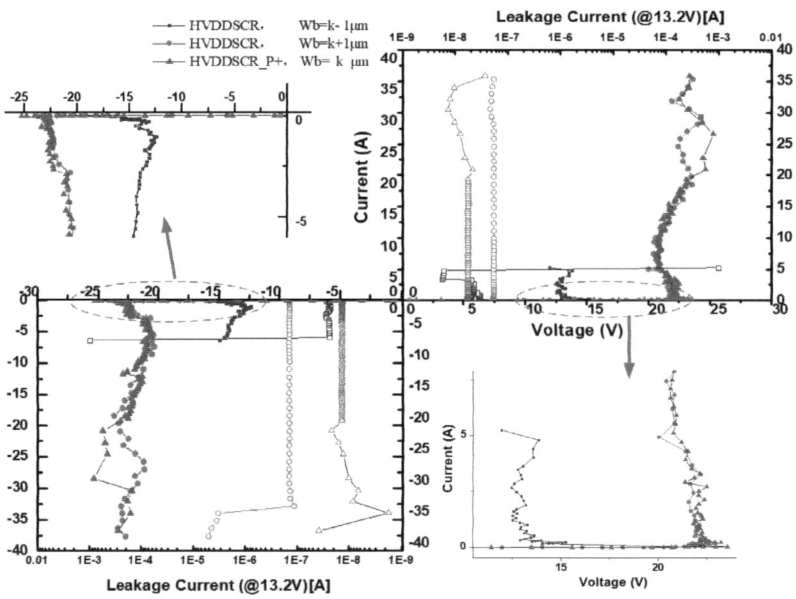

Fig.5 TLP test curves of traditional HVDDSCR with Wb=K-1 μm and Wb=K+1 μm and HVDDSCR_FP+ with X= k μm respectively.

Tab.1 TLP test data for HVDDSCR and HVDDSCR FP+

Device name		V_t (V)	V_h (V)	I_{t2} (A)	$I_{leakage}$(nA)
HVDDSCR Wb=k-1 μm	Forward	21.5	12.5	4.8	21.0
	Reverse	20.6	12.3	5.9	26.2
HVDDSCR Wb=k+1 μm	Forward	22.6	20	35.5	53.0
	Reverse	23.7	20.6	33.6	134.0
HVDDSCR_FP+ Wb=k μm	Forward	23.5	22.3	35.8	17.4
	Reverse	25.2	22.8	36.8	13.0

4.2 TLP test of N-well width X in HVDDSCR_FP+

The performance of HVDDSCR_FP+ device with three sizes of t μm, 2t μm and 4t μm was verified by tape-out. Fig. 6 shows the TLP test results of the three structures respectively, and the specific data are shown in Table 2. It can be seen from the diagram that when X is t μm, the forward and reverse holding voltage and failure current of the device reach 16V and 15A respectively. However, under the operating voltage of 13.2V, the leakage current is more than 1μA before the device is turned on, and the device cannot be used normally. When X is 2t μm, the device performance is great, the forward and reverse holding voltage and failure current of the device are increased to 22V and 35A, respectively, and the forward and reverse leakage currents are less than 20nA. When X is 4t μm, the failure current of the device is very low. The main reason is that the SCR is not open normally and this curve is not shown. The measured TLP data of the three structures also verified the correctness of the TCAD simulation results in Section 3.2.

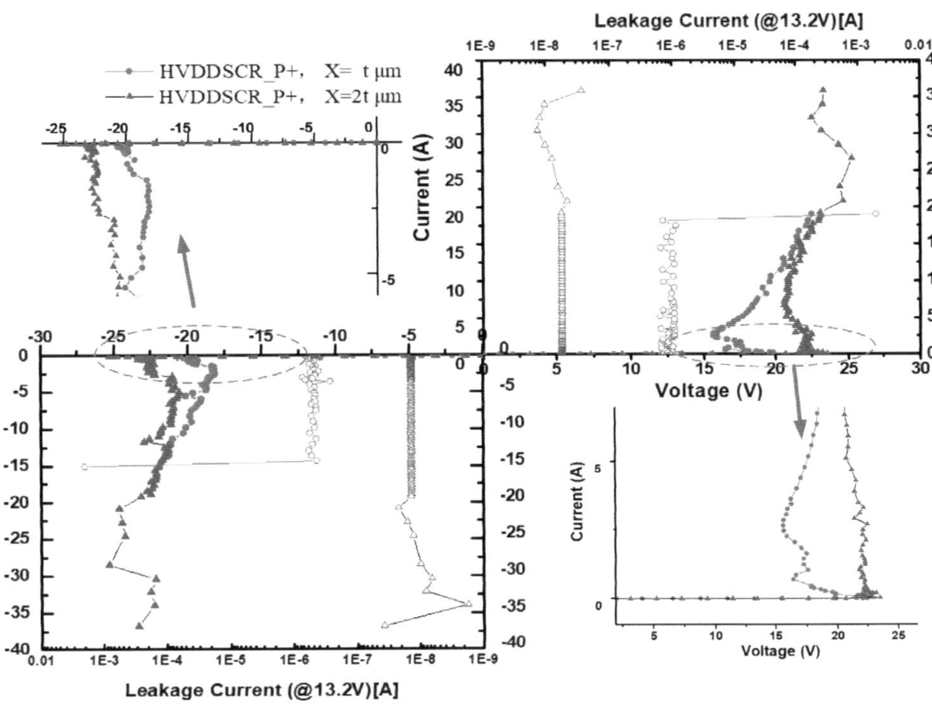

Fig.6 TLP test curves of HVDDSCR_FP+ structure with X = t μm and X= 2t μm.

Tab.2 TLP test data of HVDDSCR_FP+ structure with X = t μm and X= 2t μm.

Device name		V_t (V)	V_h (V)	I_{t2} (A)	$I_{leakage}$(nA)
HVDDSCR X=t μm	Forward	22.7	16.4	18.2	1030.0
	Reverse	21.9	18.1	15.0	1070.0
HVDDSCR_FP+ X=2t μm	Forward	23.5	22.3	35.8	17.4
	Reverse	25.2	22.8	36.8	13.0

5 conclusions

Under 0.18μm BCD process, theoretical analysis, TCAD simulation and tape-out verification show that the floating P+ can greatly improve the holding voltage and failure current of symmetric HVDDSCR devices. Secondly, the effect of N-well width X in symmetric HVDDSCR devices on device performance is discussed. The work in this paper has a certain guiding significance for the design of HVDDSCR electrostatic protection devices.

Acknowledgments

This work is supported by National Natural Science Foundation of China (Grant No.61774129, 61827812), and by Excellent youth funding of Hunan Provincial Education Department (Grant No. 19B557), and by Degree & Postgraduate Education Reform Project of Hunan Province(QL20210141).

References

[1] Chuang C H , MD Ker. (2017) System-Level ESD Protection for Automotive Electronics by Co-Design of TVS and CAN Transceiver Chips. IEEE Transactions on Device and Materials Reliability., PP:1-1.

[2] Frédéric L, Priscila F. (2014) ESD performance analysis of automotive application based on improved Integrated Circuit ESD model. 2014 International Symposium on Electromagnetic Compatibility., pp: 494-499.

[3] Stefan D, Danielle G. (2019) ESD Design Considerations for Ultra-Low Power Crystal Oscillators in Automotive Products. 2019 41st Annual EOS/ESD Symposium (EOS/ESD)., pp:1-6.

[4] Zhou Z, Jin X, Wang Y. (2021) New DDSCR structure with high holding voltage for robust ESD applications. Chinese Physics B., 30(03):610-620.

[5] Wang Y, Jin X, Peng Y. (2020) Analysis of High-Failure Mechanism Based on Gate-Controlled Device for Electro-Static Discharge Protection. IEEE Access., 8:1-1.

[6] Wang Z, Qi Z, Liang L. (2020) Novel HighHolding Voltage SCR with Embedded Carrier Recombination Structure for Latch-up Immune and Robust ESD Protection. Nanoscale research letters.,14:175.

[7] Li J, Wang Y, Jia D ,Wei W, Dong P. (2020) New embedded DDSCR structure with high holding voltage and high robustness for 12-V applications. Chinese Physics B, 29(10):635-640.

[8] Ko J H , Kim H G , Jeon J S . (2013) Gate bounded diode triggered high holding voltage SCR clamp for on-chip ESD protection in HV ICs. 2013 35th Electrical Overstress/Electrostatic Discharge Symposium., pp. 1-8.

[9] Huang X , Liou J J , Liu Z. (2016) A New High Holding Voltage Dual-Direction SCR With Optimized Segmented Topology. IEEE Electron Device Letters., 37(10):1311-1313.

[10] Liu Z , Jin H , Liou J J. (2012) Segmented SCR for high voltage ESD protection. 2012 IEEE 11th International Conference on Solid-State and Integrated Circuit Technology., pp: 1-4.

[11] Wang Y , Chen X , Jia D. (2020) The Influence of N-Type Buried Layer on SCR ESD Protection Devices. IEEE Transactions on Device and Materials Reliability., 20(4): 658-666.

[12] Chen X , Wang Y , Jin X. (2019) An ESD robust high holding voltage dual-direction SCR with

symmetrical I-V curve by inserting a floating P+ in PWell. Solid-State Electronics., 160:107627.
[13] Jin X , Zheng Y , Wang Y. (2017) ESD robustness improving for the low-voltage triggering silicon-controlled rectifier by adding NWell at cathode. Solid-State Electronics., 2017, 139(jan.):69-74.

A Design of MIPI DSI Interface for LCD Display Driver

Yeming Liu[1,a], Chengyue He[2,b]

[1]XIDIAN-WUHU RESEARCH INSTITUTE, Wuhu, Anhui, China

[2]XIDIAN-WUHU RESEARCH INSTITUTE, Wuhu, Anhui, China

[a]email:ymliu@stu.xidian.edu.cn, [b]email:cyhe19@stu.xidian.edu.cn,

Abstract: With the development of HD display field, the display interface needs a standard, stable, scalable, low-power, and high-speed solution. Traditional serial interfaces such as TTL, LVDS and RGB have some shortcomings in terms of power and speed. In addition, parallel interfaces need to involve more than 10 channels of signals, and the display-related applications have up to 20 channels. A large number of signals increase the difficulty of wiring, and limit the frequency and transmission distance of the signal, and also restrict the miniaturization of the product, so it has been gradually eliminated by the new technologies. The development direction of the interface technology is mainly low-voltage, high-speed, serial, and miniaturization to meet the demanding requirements of the mobile market for the cost, volume, weight, power consumption, performance and reliability. This paper proposes a solution of serial interface applied to display technology, which is compatible with DPI, DBI and DCS interface. The MIPI DSI receives the pixel information from the host in serial mode. There is an independent protocol in the communication filed, and the interface can support the design of 1080p/60fps display driver.

1. Introduction

The display driver is the control device of the display panel. It displays images by controlling the brightness and color signals of the screen. With the development of display technology, improving display performance has become a trend of the display driver, and the design of high-speed interface is particularly important. The traditional display interfaces such as LVDS, RGB and TTL have their own disadvantages in high-speed transmission. RGB and TTL interfaces have poor anti-interference ability and low rate. LVDS can only transmit high-speed video data, but cannot transmit control instruction data[1]. So a design of MIPI DSI interface based on MIPI DSI protocol is proposed.

The MIPI DSI interface receives pixel information in a serial manner and reads commands from the host. The received serial data are converted into the form of original packets, and accompanied by packet detection and parsing mechanism, and finally output by DPI and DBI interfaces. MIPI DSI uses the PHY layer to transmit serial data, which greatly improves the rate[2]. Moreover, compared with TTL and RGB digital signals, the analog signals transmitted by the D PHY layer have stronger anti-interference ability. LVDS interface does not have the characteristic to transmit command signals, while MIPI DSI interface can transmit control instructions and other data in low power mode[3]. The maximum rate of a single channel is 1Gbps in the high-speed mode, and its rate is 10Mbps in the low-power mode [4]. The clock channel works in the high-speed mode and is used to transmit synchronized signals. Data channel 0 and 1 have the same working mechanism in the high-speed mode[5]. Therefore, MIPI DSI interface is suitable for display driver. This paper discusses the MIPI DSI interface of the LCD display driver. We need to design an interface to support 1080p/60fps display requirements.

Content from this work may be used under the terms of the Creative Commons Attribution 3.0 licence. Any further distribution of this work must maintain attribution to the author(s) and the title of the work, journal citation and DOI.
Published under licence by IOP Publishing Ltd

Journal of Physics: Conference Series **2221** (2022) 012015 doi:10.1088/1742-6596/2221/1/012015

2. Design architecture of MIPI DSI interface

The Display Serial Interface Specification defines protocols between a host processor and peripheral devices that adhere to MIPI Alliance specifications for mobile interfaces. The DSI specification builds on existing specifications by adopting pixel formats and command set defined in MIPI03. As shown in the Figure 1, the MIPI DSI consists of one differential clock channel and four differential data channels. The working mode of the whole interface can be divided into high-speed mode and low-power mode. The high-speed mode transmits pixel information, and the low-power mode transmits control information. Data channel 0 can work in the mode of bidirectional transmission.

According to the function, there are four functional modules: PHY Layer, Lane Management Layer, Low Level Protocol Layer and Application Layer. The functions of each layer are described in the figure1.

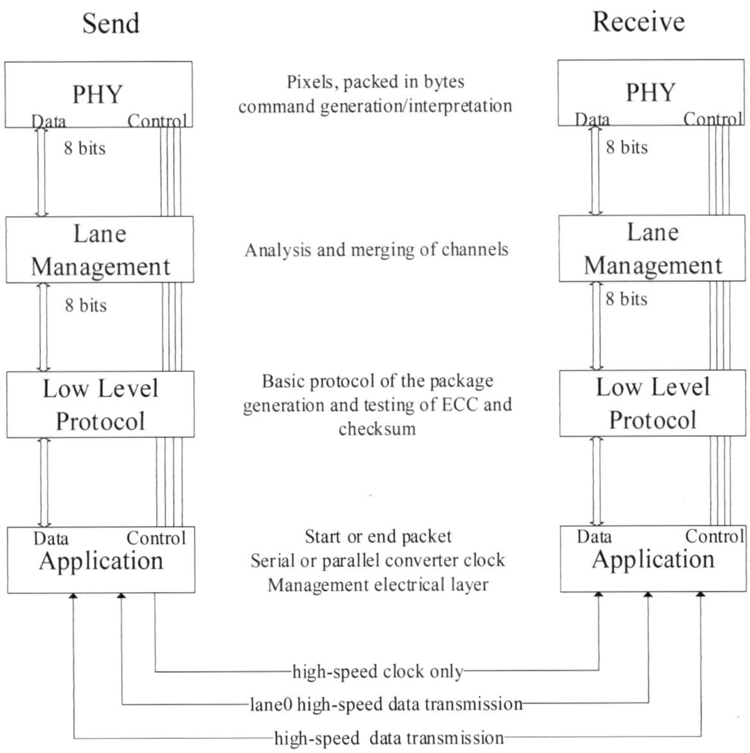

Figure 1 The circuit architecture of MIPI DSI interface

3. The concrete realization of MIPI DSI interface

According to the MIPI DSI protocol specification shown in the figure 1, the module of MIPI DSI interface is designed, as shown in the figure 2. This picture illustrates the function of each module.

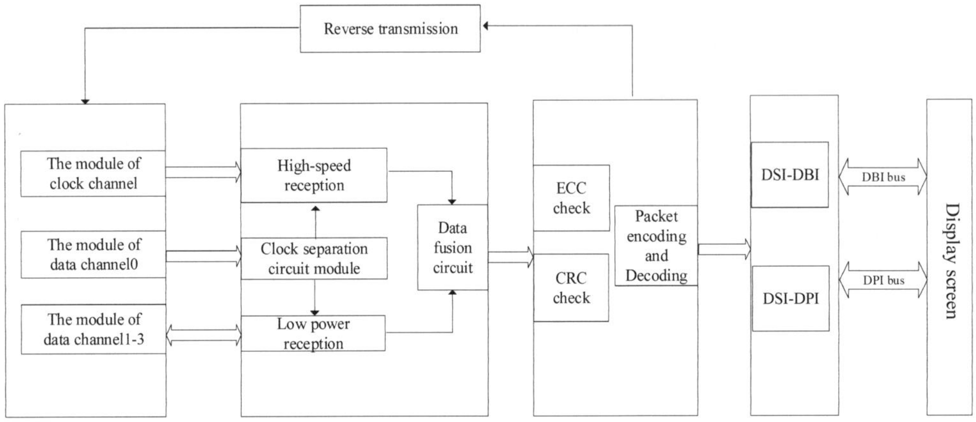

Figure 2 The module of MIPI DSI interface

3.1 PHY Layer

According to the MIPI DSI specification, the D PHY layer specifies electrical characteristics for various transmission modes. High-speed mode and low power mode use the different level to transport, and high-speed mode uses differential clock and data signals. High-speed mode mainly transmits image information and synchronized information. It transmits command and control information in the low power mode. The electrical characteristics of PHY Layer are shown in the figure 3.

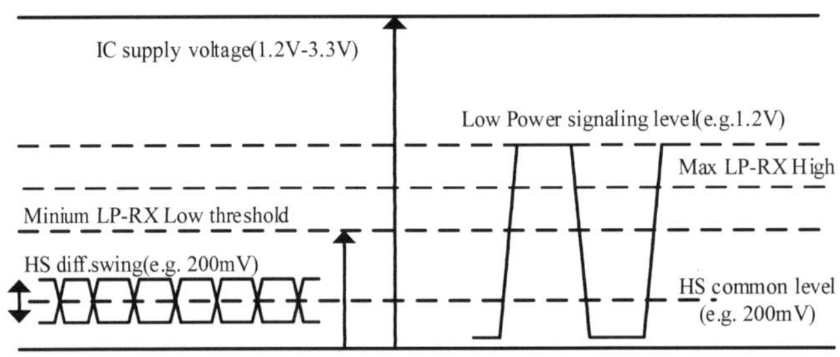

Figure 3 The electrical characteristics of PHY Layer

D PHY layer triggers and generates valid signals in different modes under the state jump mechanism, and converts the serial data into parallel data.

3.2 Lane Management Layer

The data packets sent from the host are transmitted in a certain order in the four channels. The channel management layer starts to receive data from the D PHY layer after identifying the valid signal. According to the MIPI DSI specification, for the transmission requirements of 1080p/60fps, the theoretical bandwidth can be calculated under the number of channels, as shown in the following table. The maximum bandwidth of each channel is 1.5 Gbps under the D PHY1.3 specification. The parameters calculated according to the theoretical bandwidth are shown in Table 1. The number of channels can be configured according to the theoretical value when the different transmission modes are selected.

The channel management layer needs to rearrange the byte order of each channel according to the order of the original packet to restore the correct packet. The theoretical bandwidth corresponding to different channels is shown in the table 1.

Table 1 Theoretical bandwidth required for different channel numbers

Pixel / Lane /Gbps	RGB-24bpp	RGB-18bpp	RGB-16bpp
1 lane	3.244	2.433	2.163
2 lane	1.622	1.216	1.081
3 lane	1.081	0.811	0.721
4 lane	0.811	0.608	0.541

3.3 Low Level Protocol Layer

The Low Level Protocol Layer resolves and verifies the packets recovered by the Lane Management Layer. Data packets in the channel can be divided into long packets and short packets. The transmission state of the short packet is shown in the figure 4, data packets have the starting SoT and ending EoT packets. In short packets, when SOT is detected, data are received. The first four bytes of the packet header are received, and the ECC check value is calculated according to the first three bytes. Compared with the transmitted ECC value, a certain byte in the header can be verified and corrected. In a long packet, the check mode of the packet header is the same as that of a short packet. The packet data of a long packet is the information containing image data. After the number of bytes carried by the packet is resolved, the image data is parsed and received, and CRC check is performed in the image data. When the wrong data information is identified in transmission, the packet is discarded and the host is requested to resend the packet.

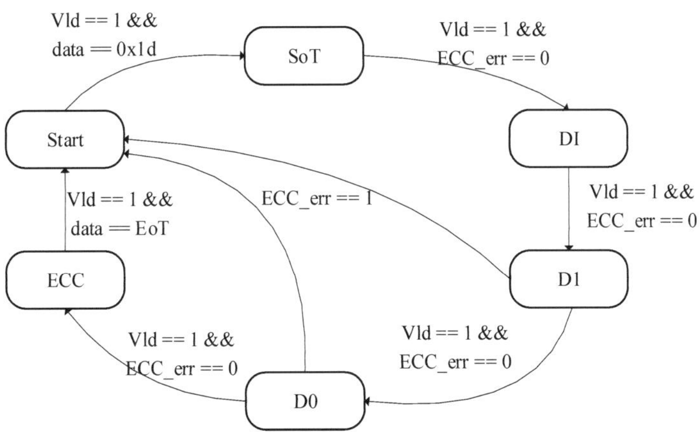

Figure 4 The state diagram of receiving short packets

3.4 Application Layer

The MIPI DSI has two output interfaces: DBI interface and DPI interface. DBI is mainly the output interface of control instructions, including some enable signals, read and write control signals, read and write commands and other state information. The data in the short packet is generally output from the DBI interface. The parsed command information is stored in the packet header, including turn off/on color mode command, turn off/on the mode command, DCS read and write command. The interface circuit is shown in figure 5.

DPI is the interface of the pixel data. DSI-DPI module mainly outputs the pixel information from the protocol layer in the format of the display sequence. In the DSI-DBI module, the output is generally long data packet parsed by the protocol layer, including RGB565, RGB666 and RGB888 output formats. After determining the output format, it detects and output signals of the Hsync, Vsync, DPI data and others.

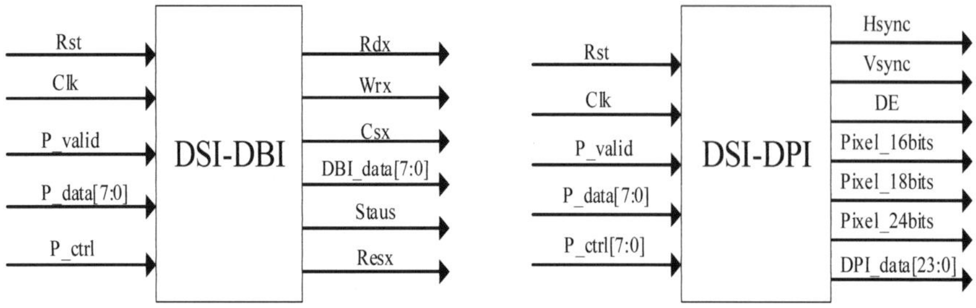

Figure 5 The circuit diagram of DBI and DPI interface

4. The simulation verification of MIPI DS interface

The MIPI DSI interface is simulated and verified in modelsim. A frame of 1080p grayscale is converted into RGB data as the input data of MIPI DSI's host, and the functions of Lane Management Layer, Low Level Protocol Layer and Application Layer are simulated and verified.

As shown in the figure 6, the number of configured channels are 4, including rx_hs_data_0, rx_hs_data _1, rx_hs_data _2 and rx_hs_data _3. The four channels poll and transmit the received data in turn, and output data through the signal of data out.

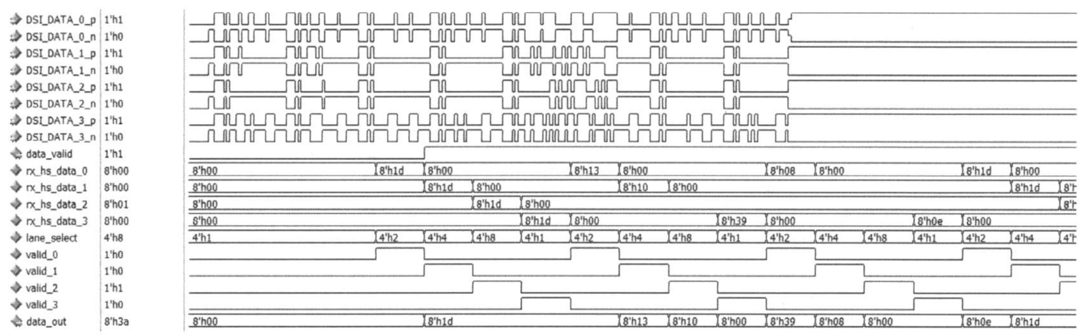

Figure 6 The simulation results of Channel management layer

The data path of the low protocol layer is shown in the figure 7, In the simulation results, packet header is correctly resolved in the figure. The signals of data_identification_rec,head_d0_rec , head_d1_rec, and others are processed through the ECC check. After ECC calibration, the signal of data_identification_ecc, head_D0_ECC, and head_d1_ECC are displayed.

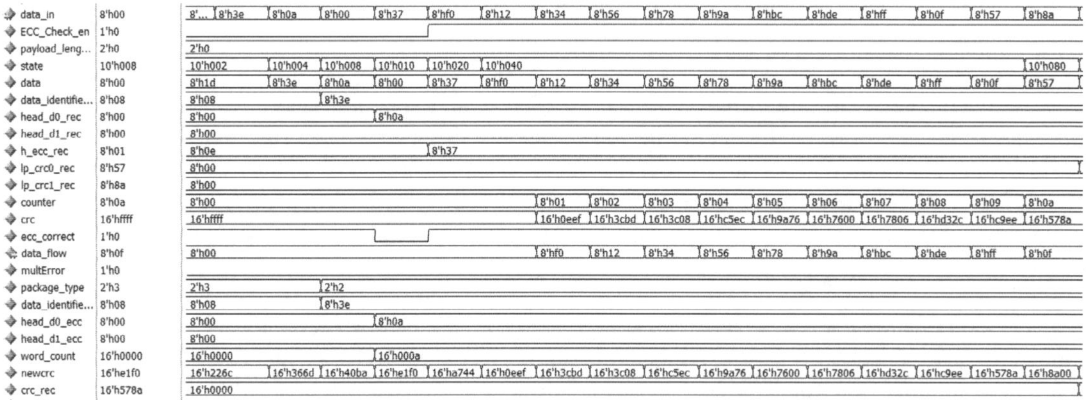

Figure 7 The simulation results of Low Level Protocol

During the transmission of long data packets, CRC check is performed in the payload of long data, and the generated checksum is stored in the new CRC register, as shown in the figure7. All bytes are correctly parsed.

The transmission path of the application layer is shown in the figure 8, and DSI is converted into DBI format for output. DBI data is parsed and output correctly. The data of 0x13_0x10_0x00_0x39 is in the packet header. WRX is correctly parsed and pulled down, and the signal of RDX is enabled. The command is output through the signal of DBI_data.

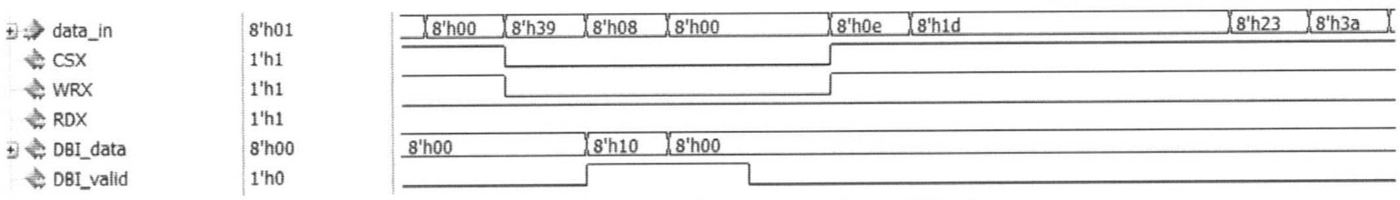

Figure 8 The simulation results of DBI

As shown in the figure 9, DPI data is parsed and output correctly. Pixel stream is in the payload of long data packets. The format of RGB888 is correctly output, the signal of Hsync, Vsync, DE and other enable signals can also be correctly output.

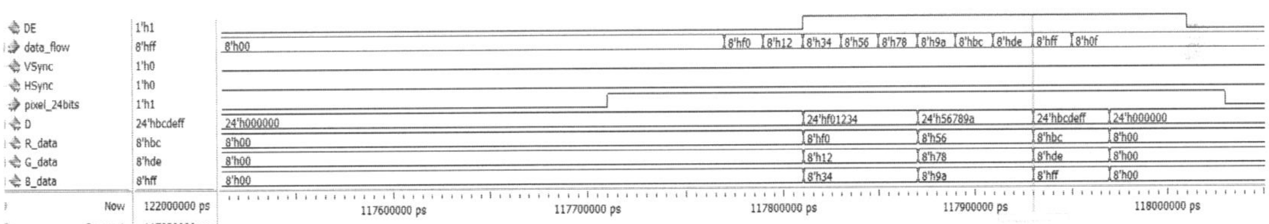

Figure 9 The simulation results of DPI

5. Conclusion

This paper discusses the MIPI DSI interface of the LCD display driver. MIPI DSI interface includes analog and digital parts. The analog logic is PHY Layer, and the digital logic is Management Layer, Low Level Protocol Layer and Application Layer. The transmission of 1080p/60fps picture data, using SMIC 90nm process, each channel can reach 0.8Gbps theoretical bandwidth. Its working frequency can reach 150Mhz in the high-speed mode. It can support display driver well, and the DPI and DBI functions meet requirements of the market. This design has been used in the specific project of display driver, which can well support the transmission of 1080p/60fps data, and has a good application value in the field of consumer electronics.

References

[1] Qiang Zhou. Research on the application of MIPI DSI interface[J]. China Integrated Circuits,2020,29(10):38-43.
[2] Zhen Zhang. A kind of MIPI protocol high-speed interface D-PHY physical layer circuit design[D]. Dalian University of Technology, 2021.
[3] Tae-Jin Kim,Jong-Il Hwang,Sangkyu Lee,Sengsub Chun,Seong-Young Ryu,Soo-Joo Lee,Hyunwoo Cho,Wonho Tak,Yong-Jae Kim,Hoomin Lee,Hansu Pae,Hyun-Wook Lim,Jae-Youl Lee. 39-2: Distinguished Paper: A 14-Gb/s Dual Mode Receiver with MIPI D-PHY and C-PHY Interfaces for Mobile Display Drivers[J]. SID Symposium Digest of Technical Papers,2020,51(1):
[4] Synopsys and Hardent Deliver Interoperable Display IP Solutions[J]. Wireless News,2015,:
[5] PARADE DEBUTS MIPI DSI TO EDP VIDEO CONVERTERS[J]. Electro Manufacturing,2014,27(4):

Design and Implementation of a Direct Memory Access ControllerBased on Microcontroller Unit

Chao Lu*, **Haosen Yang, Qi Wu**

School of Microelectronics, Xidian University, Xi' an, Shaanxi, China

email: chlu@stu.xidian.edu.cn

Abstract: With increasing IoT technology, the application of MCU is becoming more and more extensive, and higher requirements are put forward for the functions and data processing capabilities of MCU. This trend means the amount of data that the MCU needs to process in the system has increased sharply, so this paper proposes a hardware design scheme for a DMA controller based on the AHB-Lite protocol to replace the MCU to transfer data to improve system performance. The controller supports 8 transfer channels and adopts the method of fixed priority level arbitrating to select channels. The channel source and destination data width and burst type are configurable. In the process of data burst transfer, the condition of crossing the 1K address boundary is taken into consideration. At the same time, the design adopts the method that 8 transfer channels share a buffer FIFO to optimize the area of the controller.

1. Introduction

As the diversification of application scenarios, and the functions of MCUs have become more complex, the amount of data that to be processed in the chip is increasing sharply. The efficiency of data transfer among various modules has become a bottleneck in improving the performance of the MCU. In order to improve system performance, it is necessary for a DMA (Direct Memory Access) controller to replace the CPU of completion a large amount of data transfer operations [1].

DMA is a kind of technology that data can transfer among memory, input and output devices [2]. DMA plays an important role in the system, for it can independently perform data transfer operations without CPU participation. At the same time, the CPU can perform other arithmetic operations, which can improve the parallelism of the system, and increase the data throughput rate. All these factors can enhance the system performance [3]. The DMA controller is essentially a special interface circuit [4], the function is to generate read and write addresses. Then it reads the data from the source device to the internal FIFO, the data of which will be transferred to the destination device. It exerts its role as a bridge for data transfer between internal memory and peripherals [5].

This paper introduces a design technology of a DMA controller based on MCU. The controller supports 8 transfer channels, 4 transfer methods between memory and peripheral, data width of the source and destination configured as byte, halfword and word 3 types. It also supports burst types of transfer configured as single, burst4, burst8 and burst16. The controller can achieve channel switching after accomplishing a DMA transfer (once source transfer and destination transfer). Furthermore, this paper proposes an approach for crossing the 1K address boundary during multi-beat burst transfer and a processing method for the remaining data that does not meet the configured burst transfer type.

2. Design architecture of DMA controller

The overall architecture of the DMA controller is demonstrated in Figure 1. It is mainly composed of

Content from this work may be used under the terms of the Creative Commons Attribution 3.0 licence. Any further distribution of this work must maintain attribution to the author(s) and the title of the work, journal citation and DOI.
Published under licence by IOP Publishing Ltd

AHB-Lite Slave module, channel request module, arbitration module, transfer control module, interrupt generation module, handshake module, AHB-Lite master module and other sub-modules.

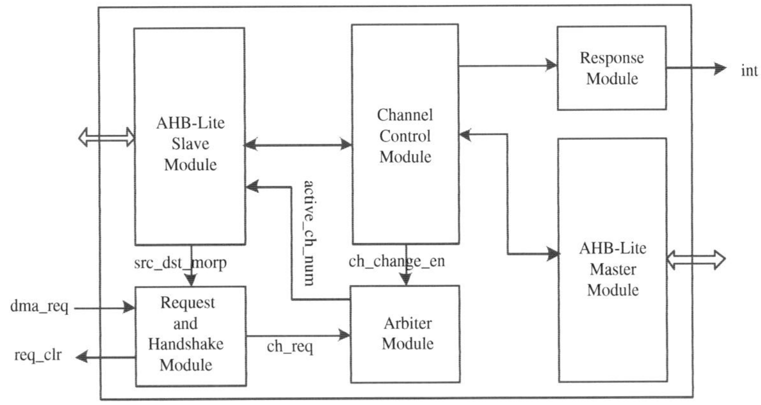

Fig.1 overall architecture of DMA controller

AHB-Lite slave module is applied in the function configuration from CPU. It transmits the configuration information of the current transfer channel to the transfer control and the AHB-Lite master module. In the meantime, relevant registers will be updated in accordance with the feedback of these two modules. Channel request module generate or process the transfer request signal according to the source and destination devices and send it to the arbitration module, which arbitrates the highest priority request based on the transfer request signal of each channel to obtain the transfer control right. And the transfer control module controls the data transfer process from the source device to the destination device. The function of the handshake module is to generate clear signal to clear request signal and sends it to peripheral unit when completing a burst transfer or all data of the transferring channel is completed. The AHB-Lite master module transmits the address, control information to the bus according to the requirements of channel transfer and the AHB-Lite protocol specification and performs source data reception and destination data transmission.

3. The Concrete Realization of DMA Controller

3.1 Design of Request Module and Handshake Module

In the design, the source and destination transfer both support two type devices: memory and peripherals. When the transfer device is a peripheral, it will input a transfer request signal to the controller to request data transfer, and most of the peripherals are integrated in the lower frequency APB bus, so the controller has the demand for the process of crossing clock domain when receiving a peripheral request to minimize effect of metastability. In this design, a two-stage synchronization method is deployed to synchronize the request signal to the controller clock domain. The memory cannot generate a transfer request signal, if the transfer device is a memory, the request signal should be generated in this module, which means when the source or destination device is configured as the memory the software, the source or destination request defaults to be valid. In the meantime, in order to reduce the response time of the DMA to the request, the signal and the DMA don't need synchronization. The source request and the destination request are combined to get the request signal to the arbitration module in order to reduce the long-term occupation of the DMA controller on the bus, and the request module structure is shown in the figure 2.

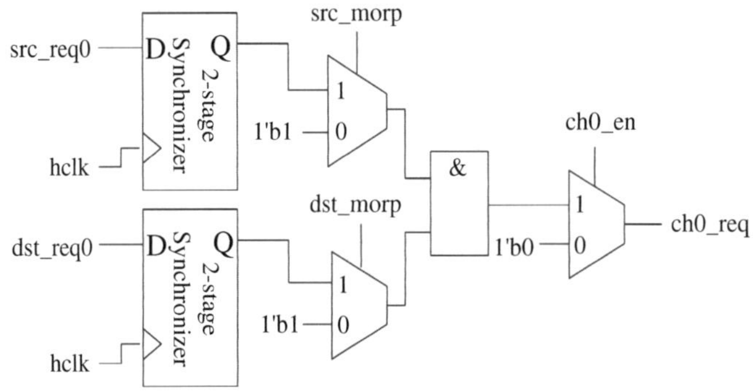

Fig.2 Request module structure diagram

The function of handshake module is to generate a request clear signal and make output to peripherals, clearing peripheral request signal when the transfer is completed. After the corresponding peripheral request signal is cleared, the request signal is fed back to the handshake module, which clears the request clear signal. And the structure of the handshake module is shown in Figure 3.

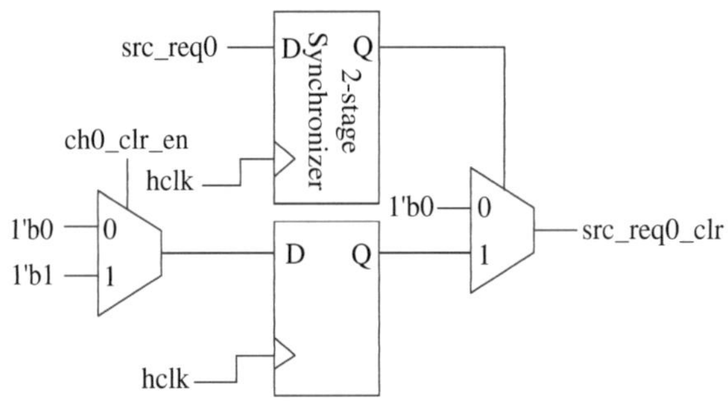

Fig.3 Handshake module structure diagram

3.2 Design of Arbitration Module and Channel Transmission Control Module

In the arbiter module, the arbitration algorithm adopts a fixed priority algorithm. The priority of each channel decreases as the channel number descends, which means channel 7 has the highest priority, and channel 0 has the lowest priority. When multiple channel request signals are valid simultaneously, this module will select the channel whose request signal is valid and which has the highest channel number to obtain the transfer control right. The higher-priority channel can preempt the transfer control after a DMA transfer is completed. For example, when channel 3 is transmitting data, the request signal of channel 5 is valid. Channel 5 will obtain the transfer control right after the channel 3 completed of the current burst transfer in the system. After the data transfer of channel 5 is completed, it then returns the transfer control unit to the channel 3 to continue the transfer that has not been finished before channel 3 is preempted.

The transfer control module is the core module in this design. Containing 3 sub-modules, and they are the transfer state machine module, the control information recalculation and the FIFO module respectively.

The main function of the transfer state machine module is to control and coordinate other modules to complete the entire transfer process. In this module, a whole DMA transfer includes three stages: the recalculation, the source transfer and the destination transfer stage, and then the source and destination transfer stages are subdivided into several states, which are shown in the figure 4. In the LOOP_RELOAD state, the control signal is recalculated on the basis of the data width, burst type and transfer address of the current transfer channel's source and destination transfer. All these generate the control signal conforming to the AHB-Lite protocol and configuration information and some of them are discussed specifically in next parts.

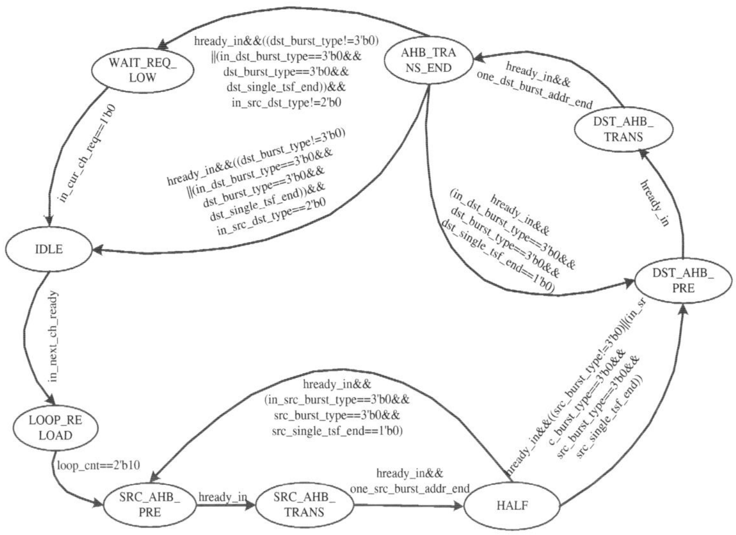

Fig.4 State transition diagram of transfer state machine

The control information calculation module is the key design of this controller. Its main feature is to recalculate based on the information transmitted into this module from the AHB-Lite slave module. And it attempts to calculate burst types of the source and destination transfer. The type of burst will get changes in two situations: one is that next multi-beat burst transfer will cross the 1K address boundary. The other is that the number of remaining un-transferred data is less than the length of the configured burst type. While burst type is single-beat or transfer address is fixed, it is unnecessary to care the condition of crossing the 1K address boundary. When multi-beat burst transfer is performed and the transfer address is incremented, it is crucial to calculate the address after a burst transfer is completed according to the current transfer data width and the type of burst. If the lower 10bits of the address is greater than the lower 10bits of the address before transfer, it means that the burst transfer will not cross the 1K address boundary, and there is no need to change the burst type. Otherwise, it implies that the 1K address boundary will be crossed. And in this case, multi-beat burst transfer is required to renew to several single-beat burst transfer to cross the 1K address boundary, besides, the number of single-beat burst transfer is up to the configured burst type. If the number of remaining un-transmitted data is less than the configured one burst type and does not cross the 1K address boundary, then the burst type is refreshed to a burst transfer with indefinite length, otherwise the burst transfer type is updated to several single-beat burst to cross the 1K address boundary. When the 1K address boundary is crossed and number of remaining un-transmitted data is greater than the beats of burst transfer type, the data is transferred according to the configured burst type.

The FIFO module is applied to store the source data in transfer process. In the source transfer stage, data is transferred from the source device to the FIFO, and then the data in the FIFO is transferred to the destination device in the destination transfer stage. In this design, the source data width and the destination data width and transfer size are configurable, so the data will be packed and unpacked in the FIFO module. The data in the FIFO is stored in the little-endian mode. That means if the source device data is reserved in big-endian mode, which must be converted to little-endian mode when stored in FIFO.

3.3 AHB-Lite Master Module Design

This module mainly generates address and control information that comply with the AHB-Lite protocol based on the current configuration of the transfer channel and the update situation across the

1K boundary. At the end of the source transfer or destination transfer, the corresponding address is output to the AHB-Lite slave module, renewing the relevant register information of the current transfer channel.

4. Verification of DMA Controller

This design adopts a SoC system based on ARM Cortex-M3 to verify the function of DMA controller. The system architecture is shown in Figure 5. The simulation verification was carried out with Modelsim, which mainly verifies the functions of DMA controller's reading or writing transfer, transferring based on crossing 1K address boundary, software and hardware triggering transfer, and transfer channel switching.

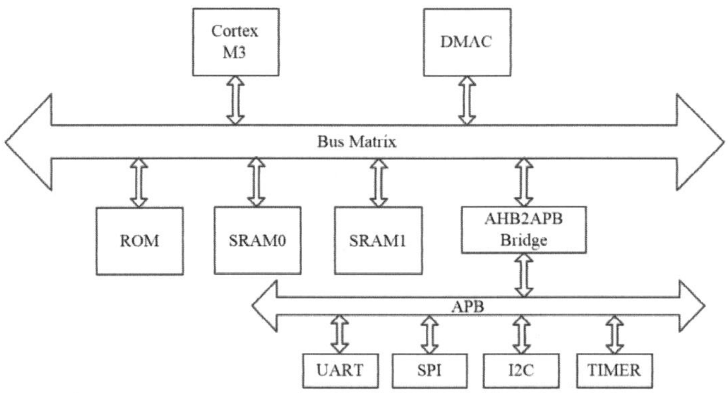

Fig.5 environment diagram of DMA controller verification

Figure 6 depicts the higher-priority channel is preempting the channel control module to transfer data. It can be confirmed when channel 0 is in the transmission process, channel 1 is sending a transfer request. After the completion of the current DMA transfer of channel 0, channel 1 is starting to transfer data. After the data transfer of channel 1 is finished, channel 0 obtains the transfer control right to transfer the remaining data again.

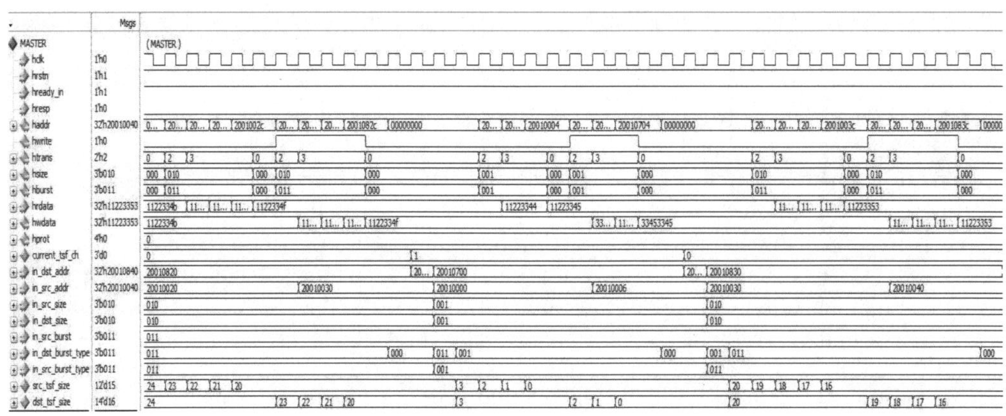

Fig.6 simulation diagram of channel switching

Figure 7 is the functional simulation when crossing the 1K address boundary during the destination transfer. The results indicate that when a transfer crosses 1K address boundary, the multi-beat burst transfer will be converted into several single-beat burst transfer to cross the 1K address boundary.

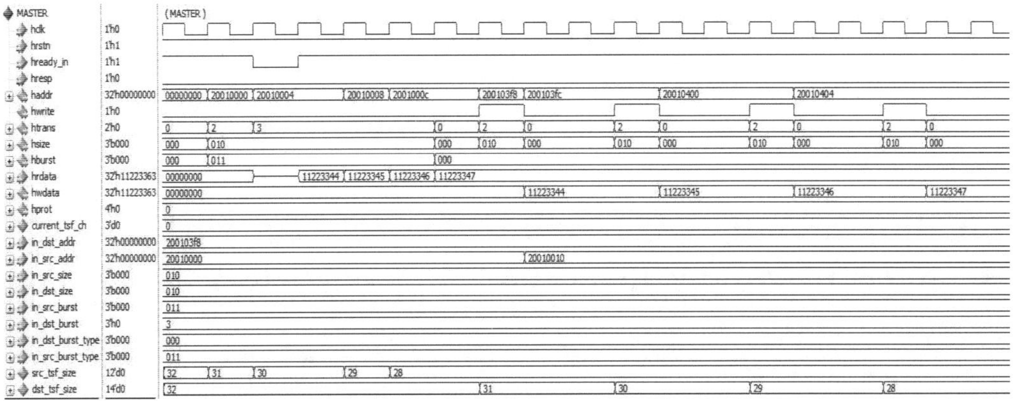

Fig.7 Simulation diagram of crossing 1K address boundary

5. Conclusion

This paper proposes a design method of an 8-channel DMA controller based on AHB-Lite protocol in response to the MCU's requirements for data transfer performance improvement, which attributes the completion of RTL design and function verification of Controller. The results signifies that the controller makes improvement of the efficiency of data transmission in MCU system, reducing the time in the process of transporting data in the CPU and enhancing system performance. Presently, the design has been successfully applied to the motor drive chip based on Cortex-M3.

References

[1] Wang, Y, Wang, T, Zhou, P. (2014) Design and Implementation of a Flexible DMA Controller in Video Codec System. In: Proceedings of the 19th International Conference on Digital Signal Processing, Hong Kong. pp. 78-82.

[2] Ahmed, M.A., Aljumah, A., Ahmad, M.G. (2019) Design and Implementation of a Direct Memory Access Controller for Embedded Applications. J. International Journal of Technology. 10(2): 309-319.

[3] Shirur, Y.J.M., Sharma, K.M., A, A.(2018) Design and implementation of Efficient Direct Memory Access (DMA) Controller in Multiprocessor SoC. In: International Conference on Networking, Embedded and Wireless Systems (ICNEWS). Bangalore. pp. 1-6.

[4] Gao, Y. X. The Design and Implementation of DMA Controller based on AHB Bus[D]. Shenyang: Liaoning University,2020.

[5] Zhang, D.Y. The Analysis and Verification of Direct Memory Access Controller based on AHB[D]. Xi'an: Xidian University, 2015.

Strain distribution and electrical characteristics of flexible low-temperature polysilicon thin film transistors under biaxial bending

ZhiXuan Fang[1,a*], Zeng Huang[1,b], Na Xie[1,c], Zheng Liu[2,d], Dong Li[2,e], YiLin Li[1,f], YeRong Sun[1,g], ZhiWen Yao[1,h]

[1]Beijing University of Technology, Faculty of Information Technology, School of Microelectronics, Beijing, China.

[2]BOE Technology Group Co., Ltd. Beijing, China.

[a]email: fzxkyx@163.com , [b]email:2549048799@qq.com, [c]email:997698827@qq.com, [d]email: liuzheng@boe.com.cn, [e]email: ld@boe.com.cn, [f]email: lyljs1998@163.com, [g]email: Sunyr@emails.bjut.edu.cn, [h]email: 1415440316@qq.com

Abstract. This paper investigates the strain distribution of flexible p-channel low temperature polycrystalline silicon thin film transistors under biaxial bending. The strain distribution of LTPS TFTs under different bending conditions is analysed by finite element analysis. εr and εθ are essentially equal and constant within the load ring. The finite element analysis shows that when the biaxial bending strain condition is 2.5%, the strain distribution is mainly concentrated at the contact angle between the electrode and the oxide layer. This strain is significantly greater than the overall strain of the device, which is the main reason for the bending performance of the device. We found that under biaxial bending, the threshold voltage shifts towards the negative axis as the biaxial strain increases. Within this strain range, these electrical properties are evident under biaxial strain conditions. Comparing theoretical threshold voltage shifts based on energy band changes with experimental results under biaxial strain, it was found that changes in trap density should also contribute to the threshold voltage shift.

1. Introduction

In the past few decades, technology continues to develop at a rapid pace, the development of flat-panel displays is equally rapid, the following main types of flat-panel displays, field emission flat panel displays (FED), plasma flat panel displays (PDP), thin film transistor liquid crystal flat panel displays (TFT-LCD), active matrix organic light-emitting diode Displays (AMOLED) [1-3]. To meet the requirements of rimless, thin and light, bendable and flexible information technology, flexible thin-film transistors (TFTs) are receiving increasing attention for flexible applications such as wearable electronics, high-resolution foldable displays and e-paper. Low temperature polysilicon (LTPS) TFTs are the most promising foldable electronics for small size and high-speed operation because they combine good electrical properties such as carrier mobility (40-100 cm2/Vs), good stability, and the possibility of developing CMOS architectures [4-5]. As an active device commonly used in flexible systems, the reliability of the thin-film transistor is a key issue. It must operate under the combined stress of electronic and mechanical loads. In addition, actual LTPS TFT products are considered to have a biaxial bending behaviour when applied. Previous studies [6-7] have reported strain distributions for uniaxial bending, where the inverted co-face with SiOx is least stressed for mechanical rolling and

Content from this work may be used under the terms of the Creative Commons Attribution 3.0 licence. Any further distribution of this work must maintain attribution to the author(s) and the title of the work, journal citation and DOI.
Published under licence by IOP Publishing Ltd

extreme local stresses can also induce film cracking or peeling at the corners. More significant degradation occurs when the TFT is bent biaxially and becomes greater with increasing bending. Many studies [8-12] have analysed the failure of TFTs under mechanical stress and found that the threshold voltage drops and the stress increases the trapping of the TFT, affecting the degradation of its electrical properties. In addition, carrier transport is also affected by changes in the effective mass of the carriers, which is caused by strain. As a result, the mobility of electrons is enhanced under uniaxial tensile strain and reduced under uniaxial compressive strain parallel to the channel. Biaxial bending occurs more extensively than uniaxial bending, particularly in the wearable sector, where biaxial deformation is often seen in flexible display devices, so the reliability of biaxial bending of flexible TFTs is worth investigating. Therefore, in the research work in this paper, we investigate the effect of strain distribution on LTPS TFTs during biaxial bending under tensile conditions. The stresses on the samples and the strain distribution of LTPS TFTs in the biaxial bending case were analysed using finite element analysis tools and the effect of performance in the biaxial bending case is given. It is hoped that this will provide some assistance to the research work on improving the reliability of thin film transistors.

2. Structure modeling

The structure of the LTPS TFT studied in this paper is shown in Fig.1. The device is a normal planar type with a top gate structure.

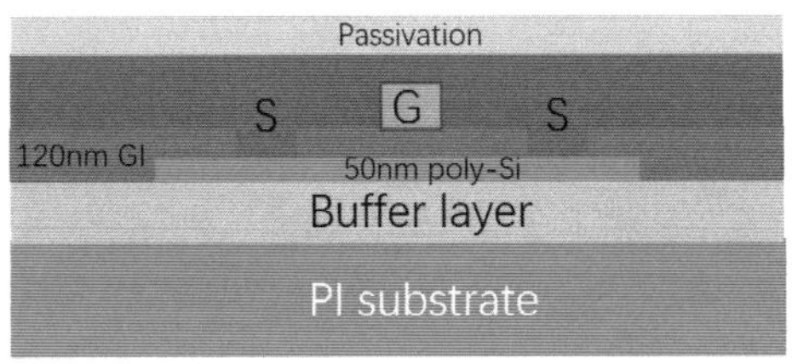

Fig. 1. Schematic structure of flexible p-channel LTPS TFTs

In the preparation process, an excimer laser annealing process was used. The substrate material is a flexible polyimide, which is then on a substrate. The buffer layer is made of 300nm SiO_2 and 100nm SiN_x, the active layer is a 50nm polycrystalline silicon film, the gate insulating layer is made of 120nm SiO_2 and the device surface is passivated with 2um polyimide to protect the device properties from surface oxidation and degradation. The gate material is Mo and the source and drain materials are Ti/Al/Ti. the polysilicon films prepared by the ELA process have a low defect density and a high degree of crystallinity and smoothness on the polysilicon surface.

The biaxial bending case for flexible LTPS TFTs was operated using a designed ring-on-ring device. The stress analysis for the biaxial bending test is very complex. The relationship between the transverse displacement of the disc and the load needs to be solved using the biharmonic equation and the entire boundary conditions must be satisfied [13]. The analytical equations for widely used single-layer discs have now been greatly simplified and can be easily applied to the engineering and analysis of thin discs. All simulations in this paper have been carried out by the finite element software ANSYS.

3. Results & Discussion

The schematic diagram of the ring-on-ring test is shown in fig.2 is the experimental device, the LTPS TFTs is polyimide material with a radius of 12.5mm, 0.12mm thickness. During the experiment, the sample is loaded by the smaller coaxial inner ring and supported by the outer ring. For a single-layer disc, the central area subjected to biaxial stress is decomposed into two directions, tangential θ and radial r.

A top view schematic of the ring-on-ring test is shown in Fig. fig. 2, where F is the anchoring force between the loading device and the support surface. where l_2, l_1 and R are the radii of the outer ring, inner ring and disc respectively.

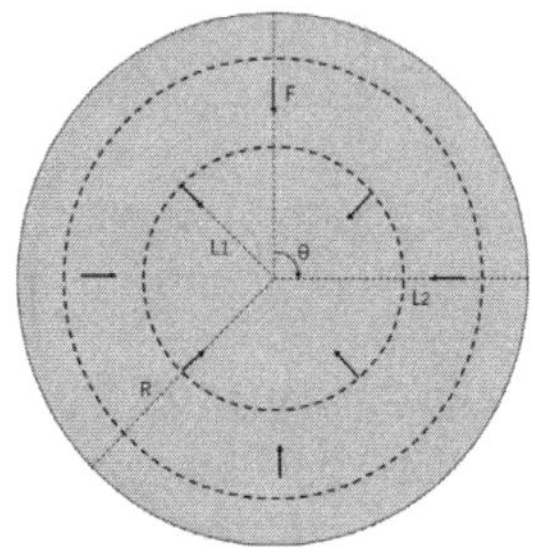

Fig. 2. Biaxial bending diagram.

The sample is a polyimide material with a radius R of 12.5 mm, L1 of 5 mm and L2 of 7.5 mm. In the loading ring position, the pressure is distributed over the L1 position to apply the test load. In the support ring position, the element nodes are held in the vertical direction only to restrain the sample.

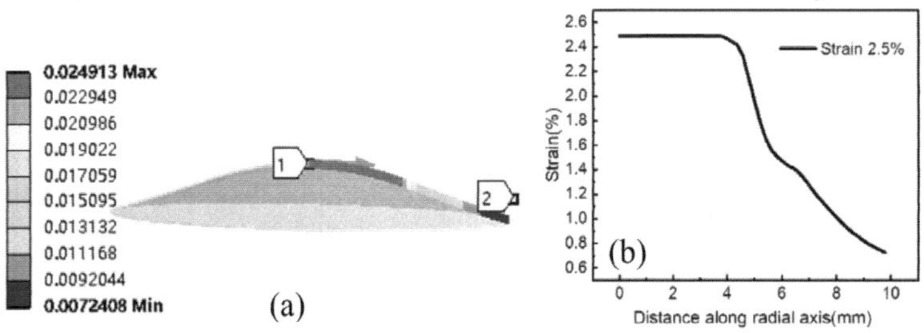

Fig. 3. (a) Strain simulation diagram. (b) Simulation results of strain variation with radius.

The finite element analysis was carried out at a theoretical condition of 2.5% strain. As shown in the simulation results in Figure 3(a), the strain is greatest in the red region in the middle of the sample. The simulation cloud shows that the maximum tensile biaxial stresses occur within the loading ring. Within the region enclosed by the loading ring, the centres εr and εθ of the tensile surface of the disc specimen are equal [14]. In the on-ring simulations, the maximum principal stresses vary radially along the tensile surface and reach an inflection point in the region enclosed by the loading ring, indicating that the stresses are no longer uniform. The same behaviour was found in the simulations for the finite element analysis in this paper. The finite element analysis of the LTPS TFT in the load ring shows that the strain control within the re-load ring is reliable. Controlling the LTPS TFT in the load ring allows for more reliable strain control and enables biaxial experimental operation.

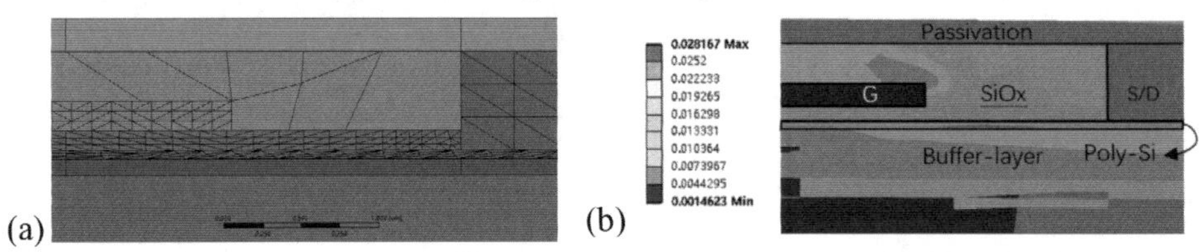

Fig. 4. (a) LTPS TFTs meshing. (b) LTPS TFTs biaxial bending strain distribution.

The finite element grid is shown in fig.4(a), where the active layer, gate oxide layer and electrode of the device are refined, and the grid subdivision can ensure the accuracy of the simulation results.

The biaxial stress is also distributed in the gate insulation layer near the source and drain. The Young's modulus of polycrystalline silicon is greater than that of sio2, which absorbs more of the stress. the strain in the active region of the LTPS TFT also undergoes abrupt changes near the electrodes. Secondly, the distribution of the different material layers is also very different, and the strain concentration in the buffer layer is also more pronounced. During biaxial bending, these areas of stress concentration may be critical in causing changes in the electrical properties of the device.

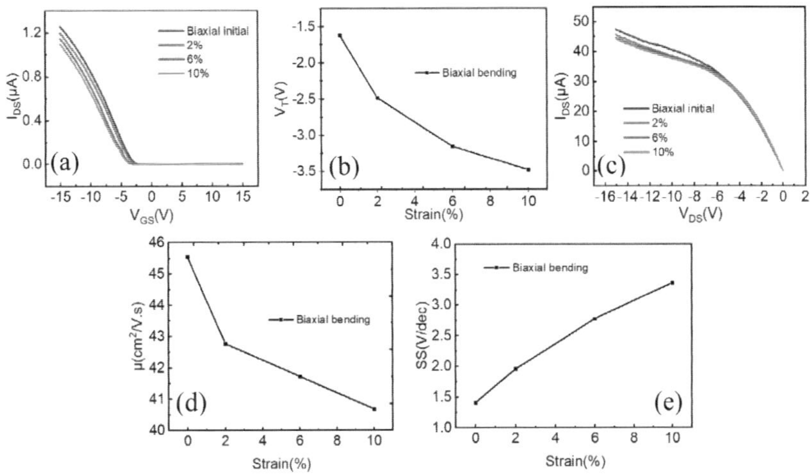

Fig. 5. (a) Transfer characteristic curve under different strain conditions in biaxial bending. (b) VT variation under biaxial bending. (c) Output characteristic curves of different strains at VGS=-10V. (d) Mobility μ variation under biaxial bending. (e) Variation of sub-threshold pendulum SS under biaxial bending.

Applying biaxial strain will cause the VT of the device to shift to the left and the performance of the device will degrade. As can be seen from the strain distribution in the devices studied above, the stresses are on the gate insulator (GI) and the electrodes. There are sharp peaks at the corners of the junctions and mechanical bending creates more defects at the grain boundaries, the polysilicon/silicon dioxide interface and the gate insulator. During biaxial bending, stress concentration leads to the introduction of traps close to the GI. Trap trapping of carriers is the main degradation mechanism for the leftward shift of their VT. The biaxial strain case is complicated by the concentration of areas of stress in various parts of the device, with significant performance changes and degradation. Similarly, its mobility drops from 45.55 cm2/V.s to 40.68 cm2/V.s in Fig.5(d). Combined with the strain distribution of the TFT, the active layer is subjected to stress, leading to a decrease in mobility and degradation of device performance. A comparison of their switching characteristics is shown in Fig.5(e). Due to the stress on the TFT, traps appear within it and this leads to weaker switching characteristics. The performance of the device degrades. A comprehensive comparison of the results of the effects of the biaxial bending approach on the TFT shows that the degradation of VT, SS and μ in biaxial bending is associated with the stress concentration introducing traps.

4. Conclusion

In conclusion, this paper gives a finite element model for the case of biaxial bending and performs a strain analysis for biaxial bending. The biaxial strain distribution of the TFT is combined with experimental results to obtain a relationship between the stress on the device and the degradation of the electrical characteristics. Stress concentration leads to the introduction of traps. Considering the effect of stress-induced traps on device performance degradation, this paper provides a closely linked experimental scenario to investigate the effect of device stress distribution and electrical property degradation on real TFTs under biaxial bending conditions. A viable approach is provided for the continued study of the reliability of flexible LTPS TFTs under biaxial strain.

Acknowledgments

This work was supported in part by the Beijing Natural Science Foundation of China under Grant 4202009 and 4162013, and in part by the National Natural Science Foundation of China under Grant 61201046.

References

[1] Brotherton, S D. (1995) Polycrystalline silicon thin film transistors. Semiconductor Science & Technology, 10: 721-738.

[2] Meng Z.,Wong M.. (2002) Application of metal-induced unilaterally crystallized polycrystalline silicon thin-film transistor technology to active-matrix organic light-emitting diode displays. International Electron Devices Meeting, 49: 991-996.

[3] Meng Z.,Wong M.. (2002) Active-matrix organic light-emitting diode displays realized using metal-induced unilaterally crystallized polycrystalline silicon thin-film transistors. IEEE Transactions on Electron Devices, 49: 991-996..

[4] Long, K. , Kattamis, A. Z. , Cheng, I. C. , Gleskova, H. , Wagner, S. , Sturm, J. C. , et al. (2006) Active-matrix amorphous-silicon tfts arrays at 180$^circhboxc$on clear plastic and glass substrates for organic light-emitting displays. IEEE Transactions on Electron Devices, 53: 1789-1796.

[5] Kim, D. H. , Lu, N. , Ma, R. , Kim, Y. S. , Kim, R. H. , Wang, S. , et al. (2011) Epidermal electronics. World Neurosurgery, 76: 485-490.

[6] Lee, M. H. , Hsu, S. M. , Shen, J. D. , Liu, C. . (2015) Stress distribution of igzo tfts under mechanical rolling using finite element method for flexible applications. Microelectronic Engineering, 138: 77-80.

[7] Lee, S. , Jeong, D. , Mativenga, M. , Jang, J. . (2017) Highly robust bendable oxide thin - film transistors on polyimide substrates via mesh and strip patterning of device layers. Advanced Functional Materials, 27: 1700437.

[8] L Petti, N Münzenrieder , et al. (2014) Influence of Mechanical Bending on Flexible InGaZnO-Based Ferroelectric Memory TFTs. IEEE Transactions on Electron Devices, 61:1085-1092.

[9] Kuo, P. C. , Jamshidi-Rou D Bari, A. , Hatalis, M. . (2009) Electrical characteristics and mechanical limitation of polycrystalline silicon thin film transistor on steel foil under strain. Journal of Applied Physics, 106: 6224-6229.

[10] Liao, P. Y. , Chang, T. C. , Su, W. C. , Chen, B. W. , Chen, L. H. , Hsieh, T. Y. , et al. (2017) Impact of repeated uniaxial mechanical strain on flexible a-igzo thin film transistors with symmetric and asymmetric structures. Applied Physics Letters, 110: 608-612.

[11] Huang, C. F. , Yang, Y. J. , Peng, C. Y. , Yuan, F. , Liu, C. W. . (2006) Mechanical strain effect of n-channel polycrystalline silicon thin-film transistors. Applied Physics Letters, 89: 221-225.

[12] Peng, I. H. , Liu, P. T. , Wu, T. B. . (2009) Effect of bias stress on mechanically strained low temperature polycrystalline silicon thin film transistor on stainless steel substrate. Applied Physics Letters, 95: 2661-2668.

[13] Peng, C. Y. , Yang, Y. J. , Fu, Y. C. , Huang, C. F. , Chang, S. T. , Liu, C. W. . (2009) Effects of applied mechanical uniaxial and biaxial tensile strain on the flatband voltage of (001), (110), and (111) metal–oxide–silicon capacitors. IEEE Transactions on Electron Devices, 56: 1736-1745.

[14] Vepakomma, K.H, Westbrook, J, Carley, S, et al. (2013) Finite element analysis of ring-on-ring test on lcd panels. Journal of Display Technology, 9: 673-677.

Design of Continuous Class-F Mode Power Amplifier with High Gain Flatness

Qinmeng Ji[1,a], Du Chen[2,b], Zehui Chen[3,c], Yongjiu Zhao[4,d*]

[1] College of Electronic and Information Engineering, Nanjing University of Aeronautics and Astronautics

[2] College of Electronic and Information Engineering, Nanjing University of Aeronautics and Astronautics

[3] College of Electronic and Information Engineering, Nanjing University of Aeronautics and Astronautics

[4] College of Electronic and Information Engineering, Nanjing University of Aeronautics and Astronautics

[a]email: 275427222@qq.com, [b]email: 1757803295@qq.com,

[c]email: chenzehui@nuaa.edu.cn,

* [d]email: 931703946@qq.com

Abstract. In this letter, a novel RF power amplifier (PA) with high gain flatness and high efficiency operating in continuous Class-F mode is introduced. Since most continuous Class-F PAs can't meet the requirement of high gain flatness, a new structure which can achieve high gain flatness is presented. A 1.7-2.5GHz PA is designed, which uses CGH40010F GaN device. Simulation results show that the drain efficiency (DE) of the continuous Class-F PA is between 69% and 79%, output power is more than 41dBm across the whole bandwidth. Gain is between 11dB and 11.8dB, and gain flatness is 0.8dB.

1. Introduction

With the rapid development of wireless communication technology, there has been much more requirements on the communication system. For instance, high efficiency, broadband, high flatness, high data rates, etc. As a key part of the communication system, power amplifiers consume the most power. What's more, the most expensive component in the communication system is also RFPAs. Therefore, the research of high efficient broadband RFPAs has been attracting more and more attention. Although traditional harmony-tuned amplifiers can achieve high efficiency, such as Class-F. They can only achieve high efficiency in narrowband conditions. In order to overcome the problem of bandwidth limitation, the concept of continuous Class-F was proposed and proved [1] [2], which can deliver Class-F efficiency across broadband. Therefore, the continuous Class-F theory is the most promising for realizing broadband and high efficiency. Recently, PAs designed based on continuous Class-F theory have achieved good results. However, most continuous Class-F PAs pay little attention to the gain flatness. The gain flatness of these continuous Class-F power amplifiers is poor, so it is important to solve this problem. Although the structure of negative feedback can improve gain flatness, it increases the complexity of the circuit and reduces the gain of the PAs.

Content from this work may be used under the terms of the Creative Commons Attribution 3.0 licence. Any further distribution of this work must maintain attribution to the author(s) and the title of the work, journal citation and DOI.
Published under licence by IOP Publishing Ltd

In this paper, a continuous Class-F PA with high gain flatness is designed, which uses CGH40010F GaN device. A new structure composed of inductance and capacitor which cascade multi-section impedance converter is proposed. It can achieve high gain flatness. Simulation results show that the drain efficiency of the continuous Class-F is between 69% and 79%. Gain is between 11dB and 11.8dB, and gain flatness is 0.8dB.

This paper is organized as follows. Section 2 introduces simple theory of continuous Class-F PAs. Section 3 introduces the design of the continuous Class-F PA with high gain flatness. Simulation results is presented in Section 4. Section 5 gives conclusions.

2. Continuous Class-F

2.1. Traditional Class-F Theory

This theory can generate a voltage square wave and a half rectified current waveform. If the voltage and current waveform do not have any overlap, the theoretical efficiency will be 100%. Realizing short-circuit termination at even harmonics and open-circuit termination at odd harmonics can realize traditional Class-F power amplifiers. Strict harmonic impedance requirements make Class-F PAs only suitable for narrow-band PA designs. Harmonics above the third order have little effect on efficiency. Therefore, the harmonics are generally matched to the third. Square wave of the drain voltage V_D normalized by V_{dc}, can be given by [3]

$$V_D = 1 - \frac{2}{\sqrt{3}}\cos\theta + \frac{1}{3\sqrt{3}}\cos 3\theta + \cdots \tag{1}$$

Half rectified sine wave of the drain current i_D normalized by Imax, can be given by [4]

$$i_D = \frac{1}{\pi} + \frac{1}{2}\cos\theta + \frac{3}{2\pi}\cos 2\theta - \frac{2}{15\pi}\cos 4\theta + \cdots \tag{2}$$

The voltage of Class-F mode can only be realized at a single frequency, so the realization of Class-F PAs is limited to a single frequency. This greatly limits the development of Class-F PAs.

2.2. Continuous Class-F (CCF) Mode

Recent research on CCF mode have shown that open-circuit termination at odd harmonics and short-circuit termination at even harmonics are not the only way to achieve maximum efficiency and output power. A parameter factor γ is introduced to the voltage waveform of the Class-F theory in continuous Class-F mode, which makes it possible to describe a cluster of high-efficiency waveforms, all of which can achieve high efficiency of 90.7%. This provides the possibility for the realization of broadband. The CCF drain waveform equation is described by [5]

$$V_{ds} = (1 - \frac{2}{\sqrt{3}}\cos\theta + \frac{1}{3\sqrt{3}}\cos 3\theta)(1 - \gamma\sin\theta), \quad |\gamma| \leq 1. \tag{3}$$

Where γ is a parameter obtained through experience. The voltage needs to remain positive throughout the frequency band, and the parameter γ must be between -1 and 1 [6]. The CCF theory extends the theory of Class-F power amplifiers to a wide bandwidth by introducing a parameter factor γ. As can be seen in Fig.1. When γ=0, it is the waveform of traditional Class-F PAs.

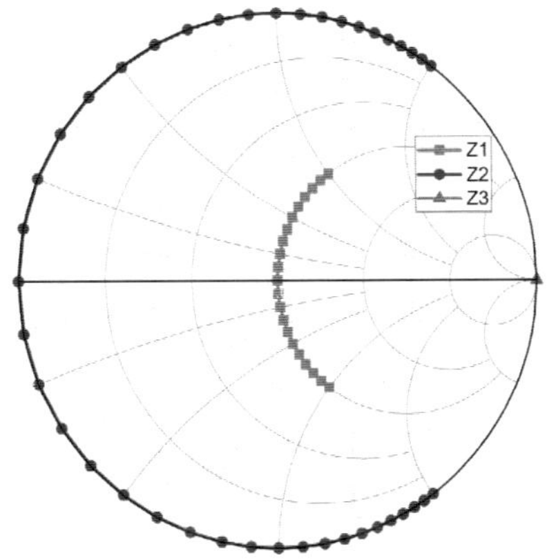

Fig. 1. Theoretical harmonics impedance for CCF

In order to use the CCF mode over a broadband, determining the impedance as γ varies is very necessary. At each harmonic frequency, the equation determining the load impedance is as follows:

$$Z_{n,f} = -\frac{V_{D,n}}{i_{D,n}} \tag{4}$$

Where n represents the nth harmonic. The harmonic impedances of the CCF PA are shown as follows [6]:

$$Z_F = R_{opt}\frac{2}{\sqrt{3}} + jR_{opt}\gamma \tag{5}$$

$$Z_{2,F} = -jR_{opt}\frac{7\sqrt{3}}{24}\gamma \tag{6}$$

$$Z_{3,F} = \infty \tag{7}$$

If all harmonics are short-circuited, Ropt can be achieved when the transistor works in the traditional Class-B mode. Based on (5) and (6), the resistance value of the fundamental wave remains unchanged, and the reactance value changes in a circular trajectory with the change of empirical parameter γ. The second harmonic resistance value remains short-circuited, and the reactance value changes with γ at the edge of the smith chart.

Based on CCF theory, broadband and high efficient RFPA can be designed. However, the gain flatness of CCF attracts little attention. Because bandwidth and efficiency are the primary indicators of research.

3. Design of high gain flatness CCF

3.1. Design philosophy

As mentioned above, it is not necessary to constrain harmonic impedances above the third order. However, the third harmonic has only a few percents on efficiency. For the simplicity of the circuit structure, it can be matched to the second harmonic impedance. The introduction of inductors and capacitors next to the transistor adds flexibility to the design. Suitable inductor and capacitor value can be chosen to improve gain flatness. What's more, this structure can also reduce the difficulty of broadband matching. The use of multi-section impedance converters can easily achieve the matching of output impedance. This may come at the cost of efficiency, but such a loss is acceptable.

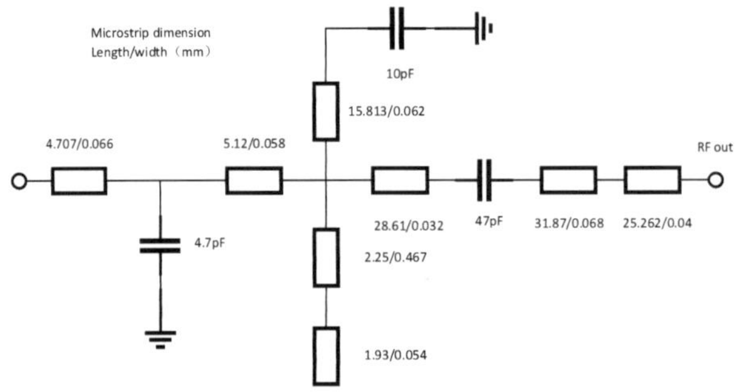

Fig. 2. Output matching network of the designed CCF PA

3.2. Output matching network

This section introduces the realization of the harmonic control circuit and the basic design principles of the output matching network, as shown in Fig. 2. Inductance and capacitance close to the transistor add flexibility to design. Appropriate value of inductor and capacitor can improve PA performance.

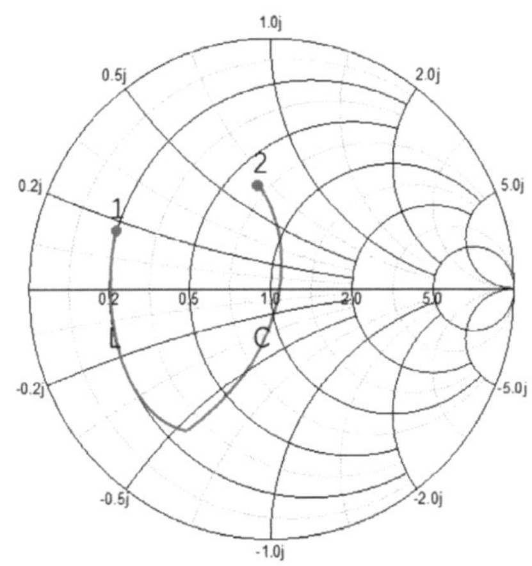

Fig. 3. Inductance and capacitance in Smith Chart

Add series inductance and parallel capacitance to transistor output can make broadband matching easier to achieve. As shown in Fig. 3. Inductance and capacitance change the impedance from position 1 to position 2. It is obvious that position 2 has a larger impedance value which makes it easier to achieve broadband. What's more, proper inductance and capacitance can improve gain flatness. Keysight advanced design system (ADS) is used to achieve the device value of the corresponding frequency, the capacitance value obtained by ADS is 4.7pF, the inductor is replaced by a microstrip line, Harmonic control network uses double stubs to control harmonics in broadband. and the final parameter value is shown in Fig. 2.

3.3. Input matching network

The design of the input matching network is similar to the design of output matching network. The optimal impedances for the input matching network can be obtained based on the output matching network implemented. Its configuration and circuit are given in Fig. 4.

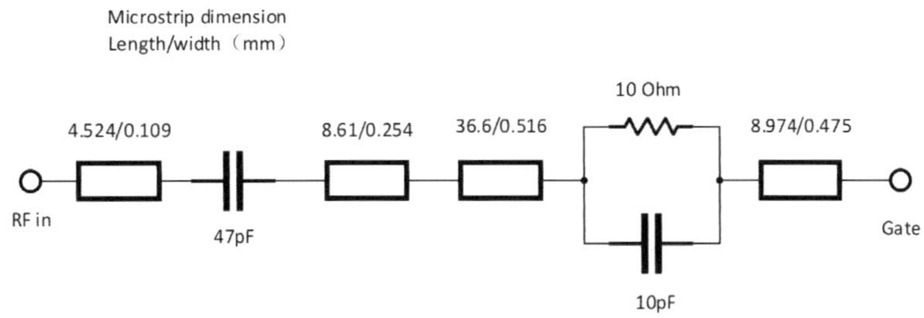

Fig. 4. Input matching network of the designed CCF PA

4. Results

The design The CCF PA with high gain flatness is simulated in Keysight advanced design system (ADS), with the GaAs substrate whose thickness is 100um and relative dielectric constant is 12.9. Input power is 30dBm. For simulation, the transistor gate is biased at -2.7V, drain bias is 28V.

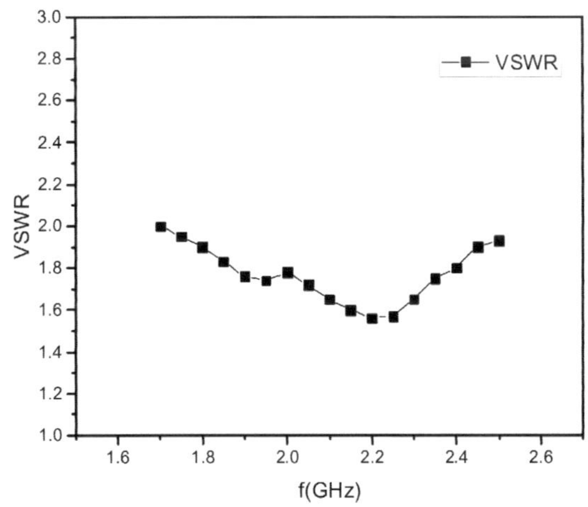

Fig. 5. VSWR of designed PA over the entire bandwidth

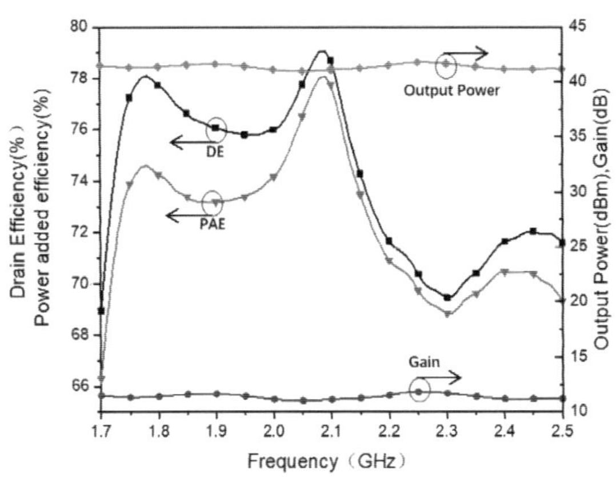

Fig. 6. Simulated results of the designed PA over the entire bandwidth

Figure. 5 shows the VSWR of the designed PA, which is between 1.56 and 2 over the bandwidth of 1.7-2.5GHz. Figure. 6 shows the output power, drain efficiency (DE), power added efficiency (PAE), and gain simulations of the designed CCF PA. As can be seen in the figure, PAE is between 66% and 78%, which achieves high-efficiency requirements within the frequency band. DE is between 69% and 79%, which has a higher value than PAE due to the difference in definition. DE and PAE both achieve the maximum value at 2.1GHz, this is because 2.1GHz is the center frequency of the design, so the performance is the best. The output power is more than 41dBm. The gain is between 11dB and 11.8dB with the gain flatness of 0.8dB.

5. Conclusion

In this paper, a CCF PA with high gain flatness is presented, its structure is shown above. Comparing to CCF PA previously designed, series inductance and parallel capacitance are added to the transistor output which can improve the gain flatness. What's more, they can also increase design flexibility and make broadband matching easier. After applying series inductance and parallel capacitance in designing the CCF PA, a CCF PA with gain flatness of 0.8dB is achieved in the design method.

References

[1] S. C. Cripps, P. J. Tasker, and A. L. Clarke, "On the continuity of high efficiency modes in linear RF power amplifiers," IEEE Microw. Wireless Compon. Lett., vol. 19, no. 10, pp. 665-667, Oct. 2009.

[2] Carrubba, J. Lees, and J. Benedikt, "A novel highly efficient broadband continuous class-F RFPA delivering 74% average efficiency for an octave bandwidth," IEEE MTT-S Int. Microw. Symp. Dig., pp.1-4, Jun. 2011.

[3] S. C. Cripps, RF Power Amplifiers for Wireless Communications. Artech House, 2006.

[4] F. H. Raab, "Class-F power amplifiers with maximally flat waveforms," IEEE Trans. Microwave Theory Tech., vol. 45, no. 11, pp.2007–2012, Nov. 1997.

[5] V. Carrubba et al., "The Continuous Class-F Mode Power Amplifier,"5th Eur. Microw. Integr. Circuits Conf., no. September, pp. 432–435,

[6] V. Carrubba et al., "The continuous inverse Class-F mode with resistive second-harmonic impedance," IEEE Trans. Microwave Theory Tech., vol. 60, no. 6, pp. 1928–1936, Jun. 2012.

Transformer evaluation strategy based on improved machine learning

Tianyi Wu[1*], Xiangyi Xu[1], Lu Chen[1], Weijie Tang[2]

[1]Shanghai Electric Power Research Institute State Grid Shanghai, China

[2]State Grid Shanghai Municipal Electric Power Company

The corresponding author's e-mail address: TYW_sh@sh. sgcc.com.cn

Abstract. A transformer state assessment method based on an improved extreme learning machine is proposed in this paper, which introduces an adaptive evolution algorithm into the extreme learning machine. This method is used in the evaluation model to evaluate the state of the transformer. The algorithm is trained and tested through sample sets. Furthermore, the test results are analyzed to prove the feasibility of the proposed control strategy.

1. Introduction

As the "transition station" of the power transmission and distribution system, power transformers undertake the conversion function of power from high voltage to low voltage. Power transformers affect important power customers, such as government agencies, large steel companies, scientific research institutions, military industries, etc.. If the power transformer cannot guarantee reliable operation, it will inevitably bring about huge hidden dangers in production. Therefore, power transformers are very important in the production field [1-2]. According to the relevant operation and maintenance experience of the operation and maintenance department, if the maintenance strategy can be adjusted in time according to the actual operation of the power transformer, the failure of the power transformer can be effectively avoided. Furthermore, the economic, political, and social risks caused by transformer aging or failure are reduced.

A large amount of research data has shown that many reasons cause power transformers to malfunction or abnormal working conditions, such as severe internal insulation aging of power transformers, long-term overload operation under extreme weather, and excessive short-circuit current leading to excessive short-circuit electromotive force [3]. With the continuous changes of working conditions and environments, the working status of power transformers is also constantly changing. It is impossible to determine the working status of transformers only by relying on a single index or very few operating parameters. Therefore, it is necessary to integrate a series of data to make a comprehensive judgment for the judgment of the operating state of the transformer.

Accurate evaluation of the operation status of power transformers is conducive to the maintenance of transformer status, and only then can a reasonable maintenance strategy be formulated. With the in-depth development of research work and the initial maturity of condition maintenance theory, transformer condition assessment technology has also received key attention [4]. At present, many researchers have proposed many research methods in the evaluation of power transformer status, and there have been mathematical methods such as Bayesian network, fuzzy theory and information fusion.

The intelligent method described above establishes a good evaluation model between the state of the transformer and the actual operating state of the transformer. This is of great significance to the

Content from this work may be used under the terms of the Creative Commons Attribution 3.0 licence. Any further distribution of this work must maintain attribution to the author(s) and the title of the work, journal citation and DOI.
Published under licence by IOP Publishing Ltd

evaluation of the state of the transformer. However, in practice, it is difficult to collect or collect data from many indicators [5-6]. The actual operating state is very different. The problems of the classification of transformer grades, the selection of membership functions, and the weighting of evaluation indexes proposed by these methods can only be solved theoretically, and there is inevitable subjectivity. The above evaluation methods lack the judgment of transformer equipment with operating years, regions, manufacturers, voltage levels, geographic locations, and operating conditions. Therefore, it is urgent to explore a transformer condition evaluation method that is closer to the actual operating condition of the transformer, which can comprehensively consider the factors that may affect the operating condition of the transformer, so that it can play an important role in the transformer condition maintenance. In this paper, a comprehensive state assessment strategy based on machine learning of temperature, load, current, and three-phase unbalance data is proposed, which can accurately estimate the transformer.

2. Machine learning theory

The extreme learning machine is a single hidden layer feedforward neural network (SLFN) learning algorithm proposed by G.B. Huang et al. Compared with the traditional neural network model, the extreme learning machine effectively overcomes the main shortcomings that the network model is difficult to determine and easy to cause over-learning. Its advantages are good generalization performance, small training error, fast learning speed, etc. As shown in Figure 1, the structure is composed of input layer, hidden layer, and output layer. The neurons in the hidden layer are connected with the neurons in the input layer and the output layer.

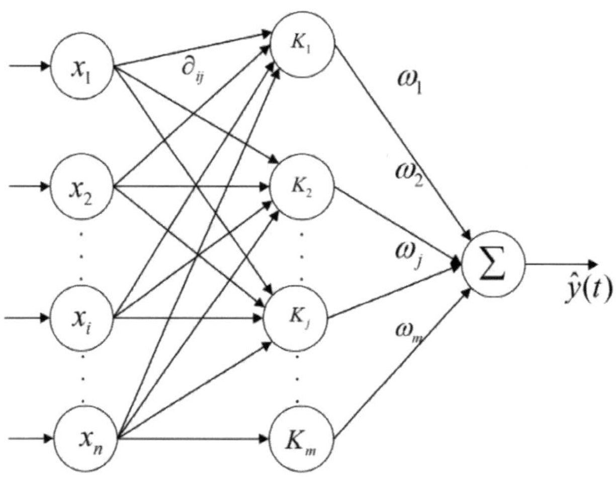

Figure 1 Single hidden layer feedforward network structure

Figure 1 shows the structure of the neural network, its principle is

$$K_j = \chi_j\left(\sum_{i=1}^{n}\partial_{ij}x_i\right), \tag{1}$$

$$\chi_j(l) = \frac{2}{1+e^{-2l}} - 1, \tag{2}$$

$$\hat{y}(t) = \sum_{j=1}^{m}\omega_j K_m \tag{3}$$

where $i=1,2,3\cdots n$, $j=1,2,3\cdots m$. x_i represents the element of the input layer neuron. K_j represents the activation function, which is the Sigmoid function. The network output $\hat{y}(t)$ is an estimate of the disturbance. The weights of and are randomly selected at [-1, 1].

The weight vector is updated through the Back-Propagation Algorithm (BPA). $y(t)$ is the error between the actual value and the observed value.

$$D(t)=y(t)-\hat{y}(t) \tag{4}$$

The goal of BPA is to ensure the smallest $e(t)$, that is

$$e(t)=\frac{1}{2}D^2(t). \tag{5}$$

where $e(t)<a$, a is a normal number. According to the chain rule, we can update the weight based on $e(t)$.

$$\dot{\partial}_{ij}(t)=-\eta\frac{\partial e(t)}{\partial\hat{y}(t)}\frac{\partial\hat{y}(t)}{\partial\chi_j(t)}\frac{\partial\chi_j(t)}{\partial_{ij}(t)} \tag{6}$$

$$\dot{\omega}_j(t)=-\eta\frac{\partial e(t)}{\partial\hat{y}(t)}\frac{\partial\hat{y}(t)}{\partial\omega_j(t)}=\eta D(t)K_j \tag{7}$$

where η is the learning rate. Combining equations (4) and (5), the neural network output is calculated as follows

$$\hat{y}(t)=\sum_{j=1}^{m}\omega_j(t)\beta_j\left(\sum_{i=1}^{n}\partial_{ij}x_i\right). \tag{8}$$

As mentioned above, the two parameters of the hidden layer input weight and hidden layer node bias in the traditional neural network are randomly selected, and they remain unchanged during the training phase, so there are non-optimized hidden layer nodes. The parameter appears in the matrix, which has a certain influence on the output function. In view of this, in the improved extreme learning, the adaptive evolution algorithm optimizes the hidden layer parameter input weights and node offsets, and calculates the hidden layer output weights through the extreme learning machine.

Given a set of hidden layer nodes of training data as x_i and novel activation function as $G(a_i,b_i,x)$. The improved algorithm is as follows:

Step 1: initialization

Initialize the first-generation parameters including hidden layer input weights and node offsets, and generate vector groups of corresponding dimensions (NP) as follows:

$$\theta_{k,g}=\left[a^T_{1,(k,g)},\cdots,a^T_{\mathrm{m},(k,g)},b_{1,(k,g)},\cdots,b_{\mathrm{m},(k,g)}\right] \tag{9}$$

where a_i and $b_i(i=1,2,3,\cdots,\mathrm{m})$ are randomly generated; $k=1,2,3,\cdots,\mathrm{NP}$; g represents evolutionary algebra.

Step 2 Calculation of output weight and root mean square error

Calculate the output weight matrix of hidden layer nodes. The formula is as follows

$$H_{k,g}=\begin{bmatrix} G\left(a_{1,(k,g)},b_{1,(k,g)},x_1\right) & \cdots & G\left(a_{m,(k,g)},b_{m,(k,k)},x_1\right) \\ \vdots & \ddots & \vdots \\ G\left(a_{1,(k,g)},b_{1,(k,g)},x_n\right) & \cdots & G\left(a_{m,(k,g)},b_{m,(k,g)},x_n\right) \end{bmatrix}_{n\times m} \tag{10}$$

$$\beta_{k,g}=H_{k,g}^{+}\hat{y}(t) \tag{11}$$

where $h(x_i)=\left[G\left(a_1,b_1,x\right),\cdots G\left(a_m,b_m,x\right)\right]$, $H=\left[h\left(x_1\right),\cdots,h\left(x_n\right)\right]^T$. $H_{k,g}^{+}$ is called the Moore-Penrose generalized matrix of the hidden layer output matrix $H_{k,g}$.

Step 3 Crossover and mutation

To obtain the contemporary target vector, according to the adaptive possibility, the test vector can be obtained by the following mutation strategy. The crossover probability CR and the control parameter F are randomly selected and generated and obey the normal distributions N (0.5, 0.1) and N (0.5, 0.3) respectively.

Step 4 Evaluation

The output weight standard is added as a standard for the selection of test vectors. Repeat steps 3 and 4 until the desired effect is achieved or the maximum number of iterations is reached.

3. Experiment analysis

In this paper, a total of 300 sets of data are obtained by screening the historical data of the transformer. The data is normalized and divided into levels. These sample sets are randomly divided based on 2:1. Among them, 200 sets of samples are generated as the training sample set, and the remaining 100 sets of samples are used as the test sample set. Perform multiple tests as required and analyze the test results.

3.1. Data set classification

It is necessary to combine these data with the operating status of the transformer into a sample set. The state evaluation model is trained through the sample set, and then the best mathematical model is obtained. Based on the characteristic parameters that can be used for comprehensive evaluation of transformers, the basic parameters, operating parameters and parameters of bad conditions are combined for evaluation.

Table 1 Data classification

Health index	Transformer operating status	Number of samples	Percentage(%)
P1	Excellent	50	16.67%
P2	Good	123	41.00%
P3	General	100	33.33%
P4	Fault	27	9.00%

Table 1 shows the transformer health status classification based on temperature, load, current, and three-phase unbalanced data. These sample sets are used for the test set, training set, and validation set of the state evaluation algorithm. Next, these data are used to evaluate the operating state of the transformer.

3.2. Test result analysis

To better improve the power transformer operating state evaluation index system, this paper considers the temperature, load, current, three-phase unbalanced data and other indicators. The effects of these factors on the accuracy of the transformer operating state evaluation are as follows:

Table 2 Evaluation result

Health index	Number of test samples	The exact number of tests	Correct rate (%)
P1	40	34	85%
P2	100	95	95%
P3	80	74	92.5%
P4	20	17	85%

It can be seen from the data in Table 2 that the temperature, load, current, and three-phase unbalanced data have an impact on the operating state of the transformer. Using it as an evaluation index for the operating state of the transformer can improve the accuracy of the evaluation of the operating state of the transformer.

4. Conclusion

This paper proposes a transformer operating state evaluation model based on an improved machine learning algorithm. To improve the accuracy of transformer state evaluation, temperature, load, current, and three-phase unbalance data are considered as evaluation indicators for transformer operating state. The test results show that considering these factors can improve the accuracy of the transformer condition assessment.

Acknowledgments
This work was supported by Science and Technology Project of State Grid Shanghai Municipal Electric Power Company (520940200061).

References

[1] Fan R, et al. "Comparison of transformer legacy protective functions and a dynamic state estimation-based approach." Electric Power Systems Research 184(2020):106301.

[2] Ippolito, L., A. Vaccaro, and D. Villacci. "The use of affine arithmetic for thermal state estimation of substation distribution transformers." COMPEL International Journal of Computations and Mathematics in Electrical 23.1(2004):237-249.

[3] Korres, G. N., P. J. Katsikas, and G. C. Contaxis. "Transformer Tap Setting Observability in State Estimation." IEEE Transactions on Power Systems 19.2(2004):699-706.

[4] Rui F, et al. "Transformer inter-turn faults detection by dynamic state estimation method." 2016 North American Power Symposium (NAPS) IEEE, 2016.

[5] Chen, Q. , D. Kaleshi , and F. Zhong . "Inferring low voltage transformer state using only Smart Metering data." Innovative Smart Grid Technologies Europe IEEE, 2013.

[6] Rui F, et al. "Dynamic state estimation-based protection of power transformers." 2015 IEEE Power & Energy Society General Meeting IEEE, 2015.

A high-range reference voltage low-dropout regulation design for voltage-controlled oscillator circuit

Shengyu Song*, Hengzhou Yuan, Jianjun Chen

Collage of Computer, National University of Defence Technology, Changsha, 410073, China

Corresponding author's e-mail: yhfang@nudt.edu.cn

Abstract: The stability of VCO as a clock recovery module in high-speed transmission circuits cannot be ignored. The LDO circuit as its power module not only needs to ensure the stability of the VCO under high-speed working conditions, but also needs to satisfy the demand for changes in its supply voltage swing. To meet the above requirements, a high-range reference voltage LDO (Low-dropout Regulator) was designed under the 28nm process. The input of the error amplifier of the LDO adopts a rail-to-rail structure, the reference voltage can be changed in a larger range, and the output voltage can also fluctuate in the range of 0.3V-0.9V. After simulation, the power supply rejection ratio (PSRR) of this LDO is 1.53MHz, the line regulation at full load is 8.29mV/V, the internal noise of the circuit is low, the transient response time from on load to full load is 42ns, and the phase margin is close to 80deg.

1. Introduction

Low-dropout regulator (LDO) is the most commonly used power management circuit in integrated circuit systems, providing accurate, stable, and low-noise power supply voltage for high-performance digital circuits or digital-analog hybrid circuits. Under normal circumstances, the LDO outputs a fixed voltage value for use in subsequent circuits, but in some special cases, the output voltage of the LDO needs to change as the circuit function changes.

Voltage Controlled Oscillator (VCO) is a frequency-controllable oscillator that coordinates the output frequency according to the frequency of the control signal during operation. The oscillation condition of the VCO satisfies the Barkhausen oscillation condition[1]. Behind the circuit oscillation is the charging and discharging of the capacitor at the corresponding frequency. When the frequency increases to a certain level, the charging and discharging of the capacitor will be incomplete, and finally When the frequency of the oscillation is different, the swing amplitude of the output signal will also be different[2]. As shown in Figure 1, for output signals with different swing amplitudes, the power supply voltage in the signal path should be consistent with the swing amplitude. This requires the LDO to be able to quickly adjust the value of the output voltage according to changes in the reference voltage.

Content from this work may be used under the terms of the Creative Commons Attribution 3.0 licence. Any further distribution of this work must maintain attribution to the author(s) and the title of the work, journal citation and DOI.
Published under licence by IOP Publishing Ltd

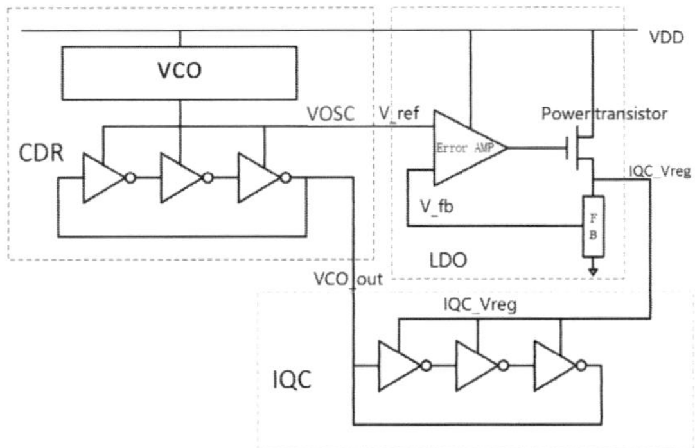

Figure 1. LDO in VCO circuit

Aiming at the above working characteristics, this paper designs an LDO based on 28nm bulk silicon CMOS process. The error amplifier of this LDO adopts rail-to-rail input design. The reference voltage can be changed in a large range, and the control output voltage is 0.3V-0.9 V varies within the range. Not only that, the LDO has a higher bandwidth, lower linear adjustment rate and fast transient response capability.

2. Design of the LDO Circuit

Under the 28nm process conditions, the core voltage is 0.9V. In the actual working process, the swing range of the clock signal output by the VCO is 0.3V-0.9V. Therefore, considering the large swing at the differential input of the op amp, the op amp needs to adopt a rail-to-rail structure[3]. Figure 2(a) shows three different rail-to-rail operating modes: rail-to-rail input and output, rail-to-rail output, and rail-to-rail input design method.

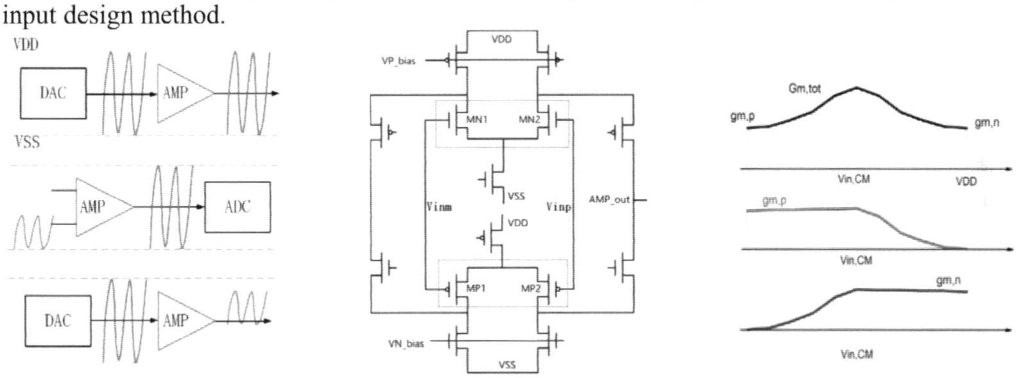

Figure 2. (a)Rail-to-rail operating modes (b)Rail-to-rail input op amp (c)Transconductance changing with the common mode point

The differential signal input circuit of the error amplifier is shown in Figure 2(b). Unlike ordinary differential amplifiers, a mixed NMOS differential pair and PMOS differential pair are used here. This differential pair symmetrical folding method pays more attention to the two differences. The total transconductance of the moving pair changes with the CM level of the input common mode point. The bottom two curves in Figure 2(c) are the changes in transconductance of a single PMOS differential pair and NMOS differential pair with the input common-mode CM level. The top curve reflects the equivalent overall span of a symmetrical hybrid differential pair. It can be clearly seen from Figure 2(c) that the symmetrical hybrid differential pair design method can effectively increase the common-mode input range of the operational amplifier and can increase the transconductance of the first stage of the operational amplifier[4].

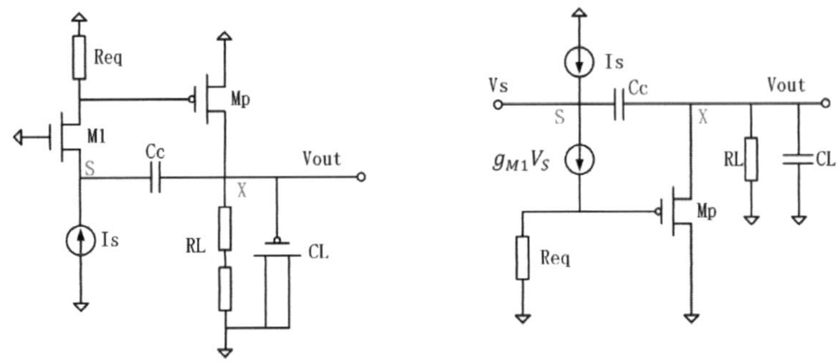

Figure 3. (a)Frequency compensation circuit and (b)the small-signal equivalent model

According to the above analysis, it can be found that the output voltage of the LDO in this design changes with the reference voltage, so for the closed-loop feedback system here, the feedback coefficient β is slightly less than 1. The larger feedback coefficient causes the longitudinal shift of the gain Bode plot in the multi-pole system, the phase margin becomes smaller, and the circuit stability is more difficult to guarantee. Here, the common-gate frequency compensation method shown in Figure 3(a) is used, and one end of the compensation capacitor Cc is connected to the output node and the other end is connected to the node S. This method can use the compensation capacitor to isolate the DC level of the active feedback stage and the DC level of the output stage. The capacitor Cc and the common gate M1 convert the swing of the output voltage into a current, and the current is input to the gate of Mp. The small-signal equivalent model of this circuit is shown in Figure3(b). Calculate the circuit transfer function:

$$\frac{V_{out}}{I_S} = \frac{RL(Ccs + g_{M1})}{RLg_{M1}Ccs\left(1 + g_{Mp}Req\right) + (1 + RLCLs)(Ccs + g_{M1})} \tag{1}$$

When the compensation capacitor is placed between the source and output node of the cascode circuit, the main pole is located at $(\mathrm{Req}g_{Mp}\mathrm{RLCL})^{-1}$, which is the same position as when Cc is directly connected to the gate of the power transistor. However, this method can effectively avoid the current conduction from the output node to the gate of the power transistor during the traditional Miller compensation and prevent the zero from moving to a very high frequency[5].

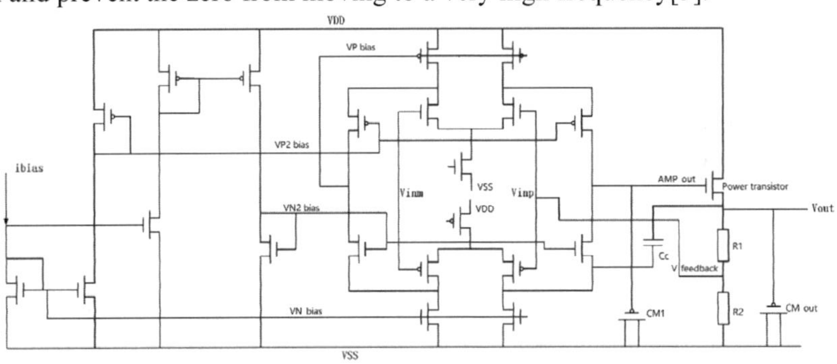

Figure 4. LDO overall circuit

3. Simulation

3.1 PSRR and Frequency Response

Figure 5(a) is a Bode plot of the power supply rejection ratio(PSRR), gain and phase of the LDO at a temperature of 25°C, a power supply voltage of 1.8V, and a process angle of tt. The reference voltage value is set to ideal 0.6V during simulation. This design ensures that the loop has a gain of 54.7dB at low frequencies, and the phase margin is close to 80deg to ensure that the loop is sufficiently stable. The PSRR has a gain of 39.9dB at low frequencies, with an inflection point at 1.53MHz, and has a minimum gain of 19.2dB when the frequency reaches 100MHz.

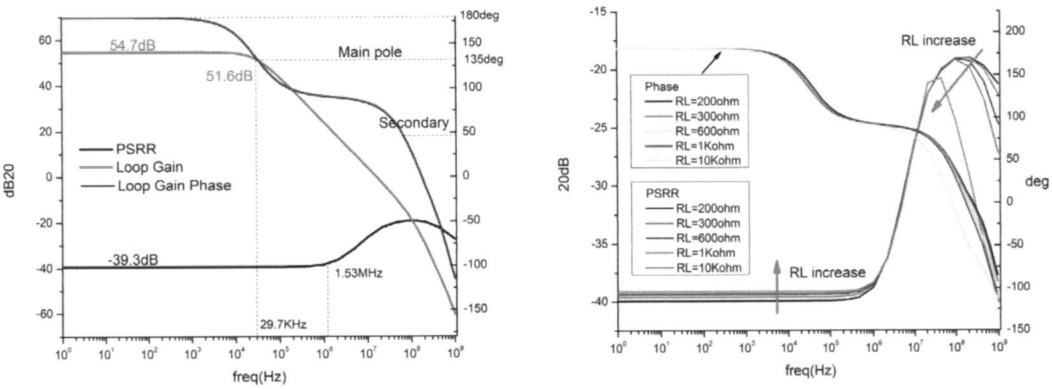

Figure 5. (a)PSRR, gain and phase Bode plot (b)PSRR and phase under different loads

Figure 5(b) reflects the phase Bode plot and PSRR of the circuit when the LDO goes from no load to full load. For mid-to-high frequency bands above 10MHz, a larger load resistance will form a larger RC filter circuit, and the better the suppression of high-frequency noise. Therefore, in the high frequency band above 100MHz, the PSRR curve of the circuit has more divergence.

3.2 Line Regulation and Load Regulation

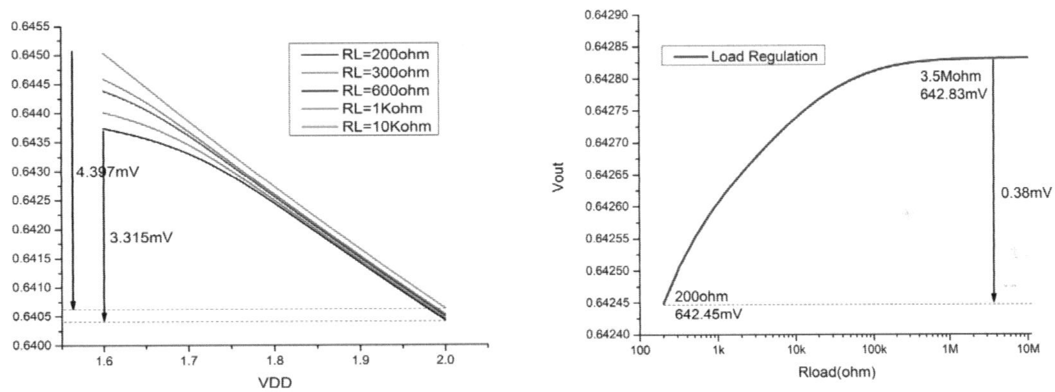

Figure 6. (a)Line regulation under different loads (b)Load regulation

For the simulation of the line regulation, the power supply voltage variation range is ±10% of the standard voltage under the DC condition. The reference voltage is the ideal reference voltage 0.6V, the load changes from no-load to full-load, and the change curve of the LDO output voltage is drawn as figure 6(a). When the circuit is no-load, the low-frequency gain is the smallest, and the greater the output voltage change caused by the power supply change, the worse the line regulation performance. On the contrary, when the circuit is under high load, the low frequency gain is large, and the line regulation is better. After calculation, the LDO circuit has a line regulation of 10.99mV/V at no load and of 8.29mV/V at full load.

The load regulation reflects the change of the output voltage when the load current of the LDO circuit changes. Under the condition that other simulation conditions remain unchanged, set the LDO load to change from no load to full load, and draw the output voltage change curve as figure 6(b). When the load current decreases from 3mA to close to 0mA (considering the leakage current of the load circuit, the current is slightly greater than 0mA under no-load conditions), the output voltage increases by about 0.38mV, and the calculated load regulation is 0.19mV/mA.

3.3 Transient Response and Noise

The simulation results of transient response are shown in Figure 7(a) below. The black curve is the size of the load current. The leakage current at no load is set to 10uA, and the load current at full load is 2.5mA. The load current is switched at 3us and 8us respectively, and a load switch of 2.5mA to 1mA is added at 6-7us to simulate the load switch when the LDO circuit is working normally. The switching time of each load current above is 1ns. The blue curve is the change of the output voltage. It takes 42ns for the output voltage to recover after the first load switching, and 156ns for the output voltage to recover after the last load switching.

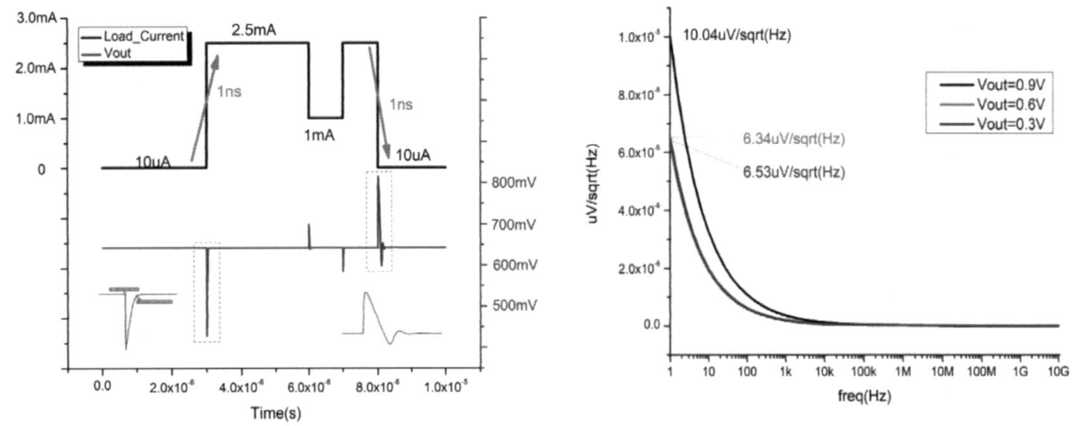

Figure 7. (a)Transient response curve (b)Noise density spectrum under different reference voltage

Simulate the output noise of the circuit when the output voltage is 0.3V, 0.6V and 0.9V. The output noise frequency domain diagram is as figure 7(b). At low frequencies, the maximum noise amplitude is 6.34uV/sqrt(Hz) when the output voltage is 0.6V, the maximum noise amplitude is 6.53uV/sqrt(Hz) when the output voltage is 0.3V, and the maximum noise amplitude is 10.04uV when the output voltage is 0.9V /sqrt(Hz), at this time 1/f noise dominates, and the thermal noise of transistors and resistors dominates at high frequency.

Table 1. LDO performance

Load current (mA)	PSRR (dB)	PSRR inflection point (KHz)	Phase margin (deg)	Load regulation (mV/V)	Transient reponse time (ns)	Maximum low frequency noise (uV/sqrt(Hz))
2.5	-39.3	1530.0	80.0	0.19	42/156	6.34

4. Conclusion

In the high-speed interface circuit, the VCO can generate a clock signal of the corresponding frequency through the received data. A VCO circuit in a high-speed working state has higher requirements for power supply stability and noise, so it is necessary to design a corresponding LDO circuit to provide it with a stable power supply voltage. After briefly introducing the working principle of the VCO, in view of the feature that the clock voltage swing follows the frequency change, an LDO with a high variation range of the reference voltage is designed. The differential signal input of the error amplifier of the LDO adopts a rail-to-rail symmetrical design, which increases the range of the common-mode input voltage. When performing frequency compensation, a common gate compensation method is used to effectively avoid current conduction from the output node to the gate of the power transistor comparing with traditional Miller compensation, and to prevent the zero from moving to a very high frequency. The performance indicators of the LDO circuit are simulated. The low-frequency gain of the loop is 54.7dB, the PSRR inflection point is 1.53MHz, the line regulation at full load is 8.29mV/V, and the transient response time of the load from no load to full load is 42ns, and the phase margin is close to 80deg. The LDO designed in this paper can meet the requirements of the actual working conditions of the VCO, and the next research direction can face the adaptability of analog LDOs of other structures and digital LDOs in high-speed circuits.

References

[1] Kao, H. L., Chang, L. C., & Fu, J. S. (2011). A wide tuning range and low phase noise 20 Ghz 0.18 μm cmos voltage controlled oscillator. AEUE - International Journal of Electronics and Communications, 65(9), 763-766.

[2] Setoudeh, F., & Nouri, N.. (2021). A novel approach for low phase noise voltage controlled oscillator design based on tsmc 0.18 um technology.

[3] Dessouky, M., & Kaiser, A.. (2015). Input switch configuration suitable for rail-to-rail operation of switched-op amp circuits. Electronics Letters, 35(1), 8-10.

[4] Ojha, Amit, Chaturvedi, Pradyumn, Mittal, & Arvind, et al. (2016). Carrier based common mode voltage reduction techniques in neutral point clamped inverter based ac-dc-ac drive system. Journal of power electronics: A publications of the Korean Institute of Power Electronics.

[5] [1] Leung, K. N., & Mok, P.. (2001). Analysis of multistage amplifier-frequency compensation. IEEE Transactions on Circuits and Systems I Fundamental Theory and Applications, 48(9), 1041-1056.

A comparative research of innovative comparators

Bangyu Song[1†], Yi Zhao[2†]

[1]School of Integrated Circuit Science and Engineering, Beihang University, Beijing, Beijing, 100191, China

[2]School of Software, Academy of Computer and Information Science, Southwest University, Chongqing, Chongqing, 400715, China

Corresponding author email: 18373216@buaa.edu.cn, zy3456412717@email.swu.edu.cn

[†]These authors contributed equally.

Abstract. This paper studies four novel design comparators and gives a detailed analysis and summary of them. edge-pursuit comparator (EPC) improved energy efficiency and noise over conventional comparators by a circuit loop consisting of numbers of delay units. The triple-tail fully dynamic comparator minimizes the comparator's total delay time and enhances the sample rate. The dynamic bias architecture of the double-tail latch-type comparator can provide a relatively high voltage gain while ensuring a low power consumption by stabilizing the static operating point. It also has advantages over conventional comparators in noise and delay. A triple-latch feedforward (TLFF) comparator improves on the triple-tail fully dynamic comparator. The triple-latch feedforward (TLFF) dynamic comparator consists of three-stage latches and a parallel feedforward path. It has a smaller delay time than other circuit designs, especially for large differential input signals.

1. Introduction

Comparator is a crucial part in lots of integrated circuits, especially of analog-to-digital converters (ADCs). Nearly 50% to 60% of ADC's power consumption comes from comparators. Furthermore, the noise and the delay of comparators also have a critical effect on ADC's performance. However, conventional comparators cannot satisfy these requirements because integrated circuits have larger and larger scale. Thus, improvements of existing structures are made and new topologies are invented to achieve lower power consumption, noise, and delay.

EPC, edge-pursuit comparator is famous for its self-adjustment of power consumption. It uses the difference between the velocity of signals caused by input differential voltage to control comparison time and power [1]. The next one is dynamic bias latch-type comparator. It achieves lower power consumption by reducing the amount of discharge and saving them in tail capacitor. Besides, this method also offers a lower input referred noise. Furthermore, triple-tail fully dynamic comparator is invented by separating three stages and realizing maximum optimization in each stage independently. In addition, feedforward transistors are added to triple-tail fully dynamic comparator to charge the output latch directly. Consequently, a lower delay can be realized for large differential voltage.

This paper is organized as follows. Section 2 studies the distinctive structure of the edge-pursuit comparator. Section 3 introduces the dynamic bias latch-type comparator. The triple-tail fully dynamic

Content from this work may be used under the terms of the Creative Commons Attribution 3.0 licence. Any further distribution of this work must maintain attribution to the author(s) and the title of the work, journal citation and DOI.
Published under licence by IOP Publishing Ltd

comparator and an improvement type of it is discussed in Section 4. Section 5 shows the comparative results of each comparator mentioned above.

2. Edge-pursuit comparator

Conventional clocked comparators usually consume nearly constant energy for each comparison since they are designed according to the most accurate and power-hungry comparison as shown in figure 1(a). If it can be adjusted the energy for comparison according to the input difference level can greatly help in reducing the total comparison energy [figure 1(b)] as well as the overall energy consumption. So this paper presents a new energy-efficient ring oscillator collapse-based comparator, named edge-pursuit comparator (EPC) [1].

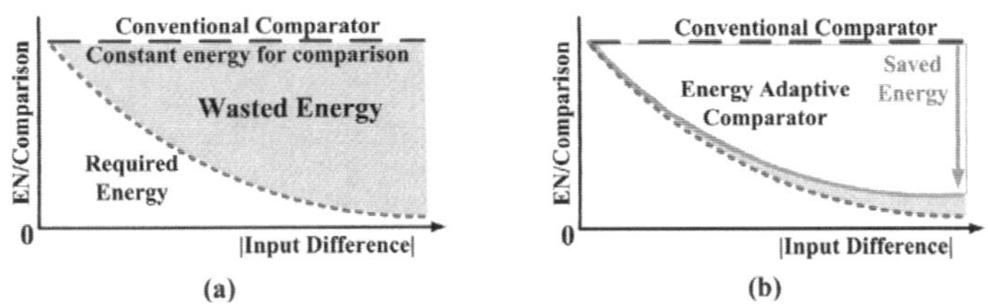

Figure 1. Required energy for comparison versus input difference.

As shown in the figure 2 below, EPC consists of two NAND gates and eight inverter delay cells. There are two inputs called VINP and VINM, and it can be divide two lines which is composed of one NAND gate and four inverters. One of the input nodes of two NAND gates and node COMP is the output on the intersection of the two parts of the circuit.

There are groups of two as inverters delay calls. when one of the inputs VINP be chosen, the signal will go through the input VINM of next inverter delay cells and out of the next group. Through this path, the signal will travel faster and faster. On the other hand, when the inputs VINM be chosen, the signal will go through the input VINP of next inverter delay cells and out of the next group.so the velocity of signals will be difference [2]. After one propagating edge overtakes the other edge, the oscillation collapses and the stage outputs settle to either VDD or GND, dictated by which edge was slower and hence overtaken [figure 3(a)]. The comparator output COMP is sampled from an internal stage that goes high when VINP > VINM and low otherwise. When the voltage difference between VINM and VINP is small, the two injected edges have similar propagation delays and the number of cycles required to make a decision automatically increases [figure 3(b)]. So the comparator naturally adjusts its energy dissipation without external control, and realizes both high accuracy and low power operation.

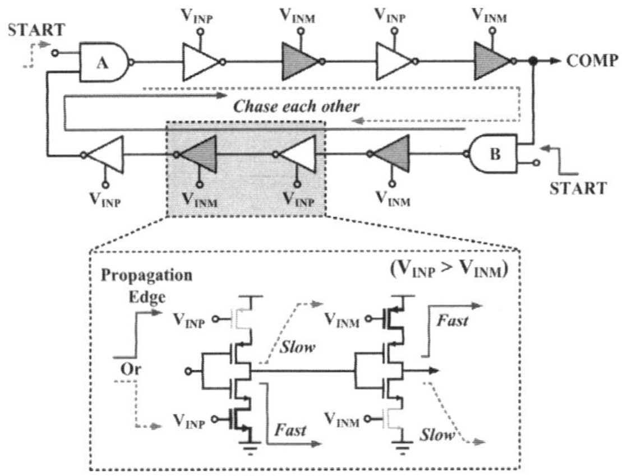

Figure 2. Structure of the EPC.

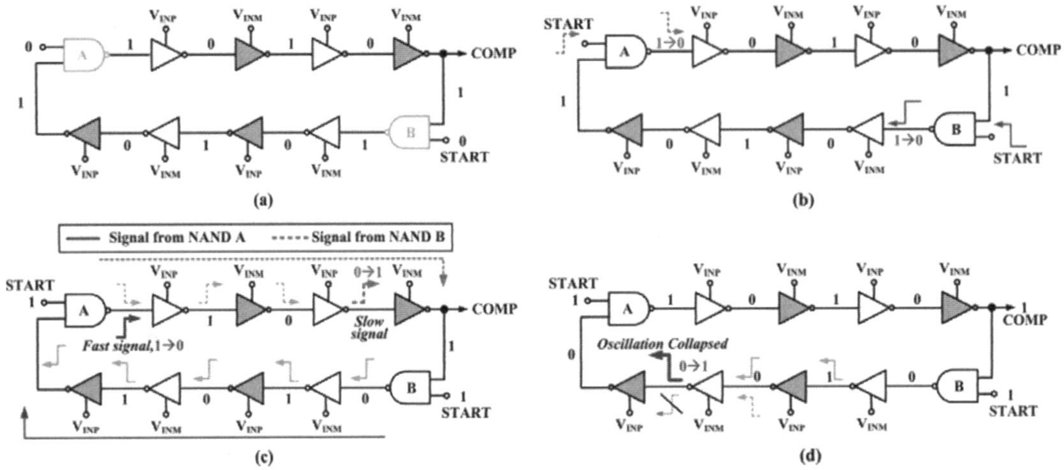

Figure 3. Operation of the EPC.

2.1. Input Referred Noise:

The comparator's input-referred noise power σ_{v_n} is obtained as

$$\sigma_{v_n} = \sqrt{\sigma_{v_n}^2} = \frac{\pi}{\sqrt{3}} f_0 \frac{kT}{I}\left(2(\gamma_N + \gamma_P) + \frac{2V_{ov}}{V_{DD}}\right) \tag{1}$$

Where I is the average current that flows from the supply to ground for the inverters, f_0 is the oscillation frequency, V_{ov} is the overdrive voltage of the current-limiting transistors, γ_N and γ_P are body-effect parameter of NMOS and PMOS.

2.2. Comparison Time and Energy:

They define M and $\sum$, the "drift" and "diffusion" coefficients of this edge-pursuit random process as

$$M \equiv \frac{\mu_{\Delta_\varphi}}{\tau} = 4\pi f_0 \frac{v_{in}}{V_{ov}} \tag{2}$$

$$\Sigma \equiv \frac{\sigma_{\Delta_\varphi}^2}{\tau} = 8\pi^2 f_0^2 \frac{kT}{I}\left(\frac{2}{V_{ov}}(\gamma_N + \gamma_P) + \frac{2}{V_{DD}}\right) \tag{3}$$

where τ is the oscillation period, $\mu_{\Delta\varphi}$ is a defined period, $\sigma_{\Delta\varphi}^2$ is variance of the phase difference shift at this period.

The average comparison time $E\,[T]$ is

$$E[T] = \left(\frac{M}{\Sigma}\right)\frac{\pi^2}{\Sigma} S \tag{4}$$

where the scaling factor S, which is dependent on the ratio M/Σ, is defined as

$$S(K) \equiv \frac{\tanh(k\pi)}{k\pi} \tag{5}$$

To evaluate the EPC's energy efficiency and compare it with other comparators, we shall define a norm for comparison

$$N \equiv E \times \frac{\sigma_{\Delta_\varphi}^2}{V_{DD}^2} \tag{6}$$

According to this, we get the norm value of the EPC

$$N_{EPC} = \frac{\pi^2}{6} S\left(\frac{M}{\Sigma}\right) kT \frac{\left(V_{DD}(\gamma_N + \gamma_P) + V_{ov}\right)V_{ov}}{V_{DD}^2} \tag{7}$$

Assume $\gamma_N = \gamma_P = \gamma$ and $V_{DD}(\gamma_N + \gamma_P) >> V_{ov}$, we can get the norm value, simplifies to

$$N_{\mathrm{EPC}} \cong \frac{\pi^2}{3} S\left(\frac{M}{\Sigma}\right) kT\gamma \frac{V_{\mathrm{ov}}}{V_{\mathrm{DD}}} \tag{8}$$

per comparison. Using (6), Nuzzo's performance norm is

$$N_{\mathrm{SS}} \cong 7kT\gamma \frac{V_{\mathrm{ov}}}{V_{\mathrm{th}}} \tag{9}$$

Elzakker's performance norm is

$$N_{\mathrm{TS}} \cong 4kT\gamma \frac{V_{\mathrm{ov}}}{V_{\mathrm{th}}} \tag{10}$$

All comparators' performance norms estimated in (2.8), (2.9), and (2.10) share the same dimension and similar form factored by γ . Ratios among those norms are

$$N_{\mathrm{EPC}} : N_{\mathrm{SS}} : N_{\mathrm{TS}} = S\left(\frac{A}{B}\right)\frac{\pi^2}{3}(V_{\mathrm{DD}})^{-1} : \frac{7}{V_{\mathrm{th}}} : \frac{4}{V_{\mathrm{th}}} \tag{11}$$

According to the measured results, this efficiency gain of EPC is decreasing consumed energy for output regeneration (versus a single-stage comparator only), by 4/7;

Figure. 4 shows the simulation results of the three comparators compared. With similar levels of input-referred noise [figure 4(a)], the EPC shows large energy savings from automatic energy scaling, while the other two comparators show nearly constant energy consumption. Figure 4(b) shows the simulation results when the V_{OV} is similar among all comparator types considered.

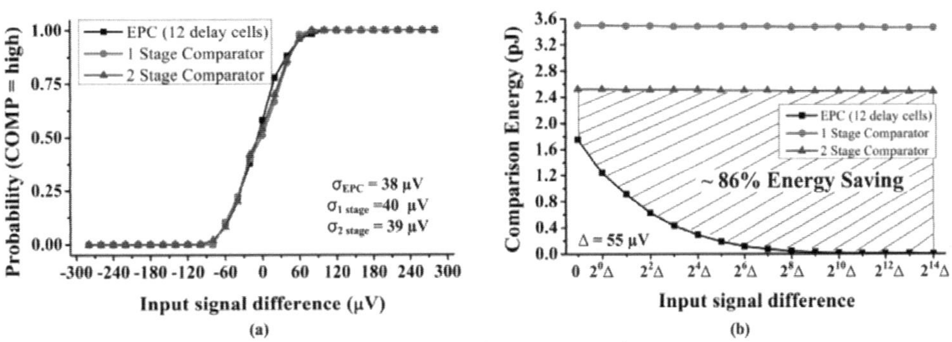

Figure 4. performances comparison among three comparators.

In the meanwhile, the number of delay cells can also affect the noise. Figure 5 shows that a 12.5x noise level change only by changing the number of stages from 8 to 14, while conventional comparators require more than 100x design parameter tuning for a similar noise level change.

All in all, the energy efficiency and noise of EPC are greatly improved compared to conventional comparators. The shortcoming of EPC is the improvement of delay compared with conventional comparators.

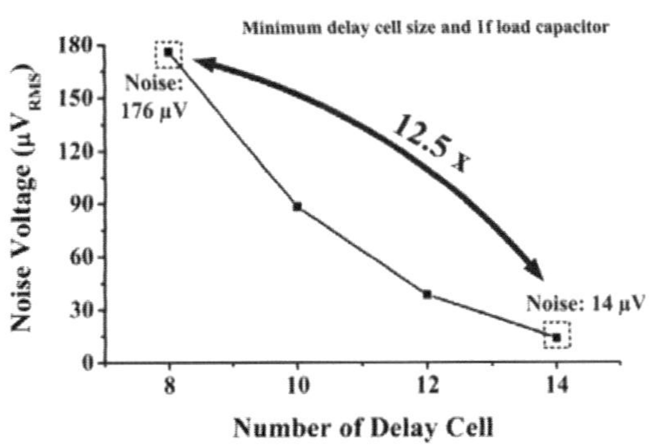

Figure 5. Simulated input-referred noise.

3. Dynamic bias latch-type comparator

The dynamic bias architecture is invented to solve the problem of power consumption of the double-tail latch-type comparator. And its performance need to be compared with another solution, the Elzakker comparator.

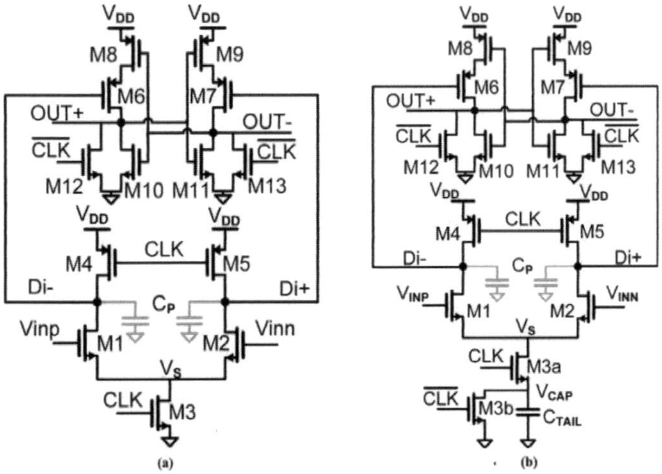

Figure 6. Elzakker comparator and dynamic bias comparator

The Elzakker comparator uses PMOS transistors M6/M7 to accept the output of pre-amplifier stage in the latch stage and remove the tail transistor [3]. Before the comparison starts, Di-/Di+ equals to Vdd, so M6/M7 cut off. When the comparison starts, CLK is raised to VDD. And M3 turns on. As a result, Di-/Di+ starts to fall. At the same time, differential voltage is added to Vinp/Vinn. Thus when Di-/Di+ falls low enough, M6/M7 will turn on and receive enough Di-/Di+ from pre-amplifier stage. Figure 6(b) shows another structure called dynamic bias comparator. Through reducing the amount of discharge on Di-/Di+, dynamic bias comparator can provide a relatively high voltage gain while ensure a low power consumption. When the comparison starts, CLK equals to VDD, M3a turns on. This result in discharging of Di-/Di+ and charging of CTAIL, thus the voltage of VCAP is raised. This phenomenon will stop until Vgs of M1/M2 reaches the threshold voltage.

Compared with the Elzakker comparator, Di-/Di+ in dynamic bias comparator will end up at VD1/VD2, which is not zero as them in the Elzakker comparator. This means the power consumption of the pre-amplifier stage in dynamic bias comparator is

$$2 \cdot C_P V_{DD}^2 - C_P V_{DD}\left(V_{D1} + V_{D2}\right) \tag{12}$$

This is lower than

$$2 \cdot C_P V_{DD}^2 \tag{13}$$

in the Elzakker comparator. Simulation results confirm this conclusion.

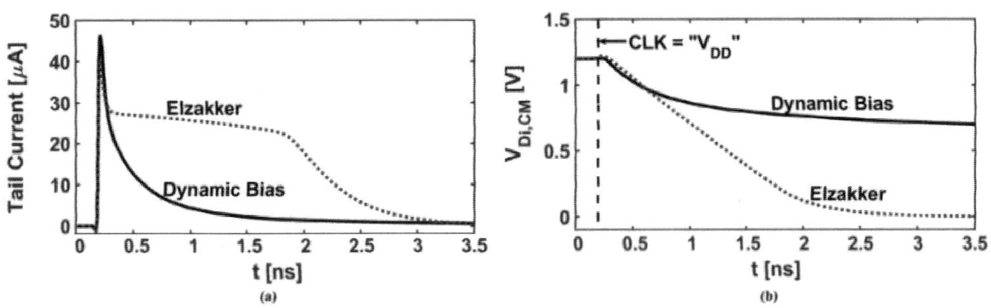

Figure 7. Tail current and Common-mode voltage of two comparators.

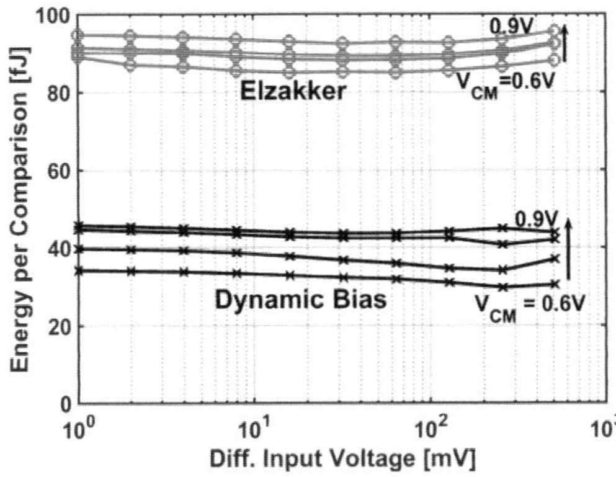

Figure 8. Energy per comparison of two comparators.

Figure 7 shows the tail current of two comparators. Clearly, the Elzakker comparator has a larger tail current than the dynamic bias comparator from 0.25ns to 3ns. This means differential pair in the Elzakker comparator is discharged completely than the dynamic bias comparator. Thus, as figure 8 shows, it has a zero common-mode voltage while the dynamic bias comparator has a 700mV one. Therefore, according to (12) and (13), the dynamic bias comparator has a lower power consumption than the Elzakker comparator. At the same time, according to the simulations which applying different VCM and the differential voltage, the dynamic bias comparator shows a fabulous power consumption about half of the Elzakker comparator in figure 8.

Besides, these to comparators also have great noise performance. For noise performance, only the thermal noise of differential pair is considered [4]. This is because flicker noise is fair small compared with thermal noise, which is verified through experiments. The input referred noise of the Elzakker comparator is shown as below,

$$E\left[v_{\mathrm{n,SI,in}}^2\left(T_{\mathrm{INT}}\right)\right]=\frac{4kT\gamma}{C_P\cdot\Delta V_{\mathrm{Di,CM}}\left(T_{\mathrm{INT}}\right)}\cdot\left(\frac{g_{\mathrm{m}}}{I_{CM}}\right)_{\mathrm{SI},T_{\mathrm{INT}}}^{-1} \tag{14}$$

where SI means the differential pair operates in strong inversion region.

The input refer noise of the dynamic bias comparator is shown as below,

$$E\left[v_{\mathrm{n,WI,in}}^2\left(T_{\mathrm{INT}}\right)\right]=\frac{2nkT}{C_P\cdot\Delta V_{\mathrm{Di,CM}}\left(T_{\mathrm{INT}}\right)}\cdot\left(\frac{g_{\mathrm{m}}}{I_{CM}}\right)_{\mathrm{WI},T_{\mathrm{INT}}}^{-1} \tag{15}$$

where WI means the differential pair operates in weak inversion region. Simulations about input referred noise are conducted and results are shown below.

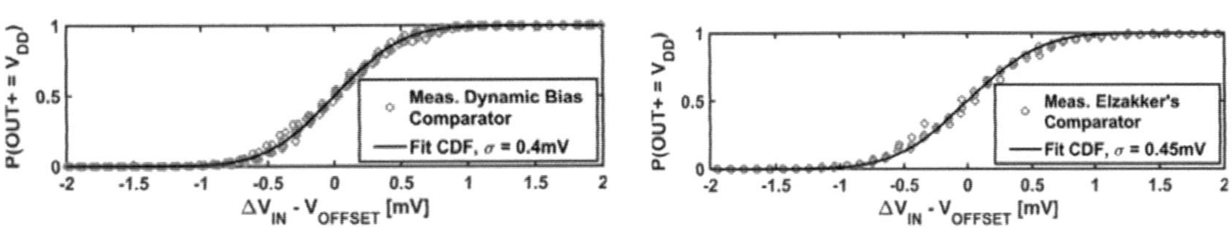

Figure 9. Input referred noise of two comparators versus VIN.

Figure 9 shows the input referred noise of two comparators by applying small increments on VIN. Through fitting the measurements to a Gaussian CDF, the simulation shows an rms input referred noise voltage of 0.4 mV for the dynamic bias comparator.This is smaller than 0.45 mV for the Elzakker comparator.

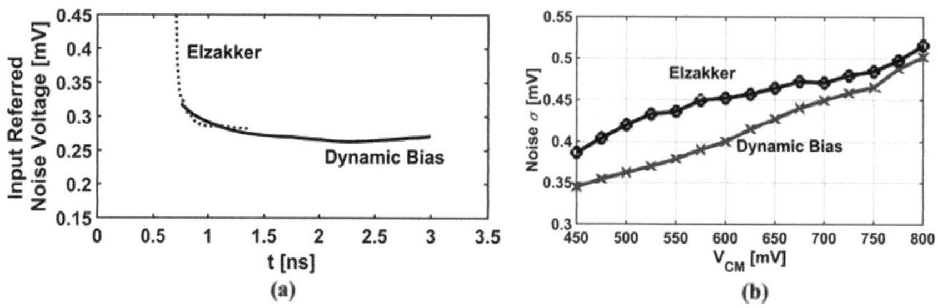

Figure 10. Noise of two comparators versus time and VCM.

Figure 10(a) shows the input referred noise of two comparators changing through time. Apparently, the Elzakker comparator has a larger input referred noise at the beginning of the comparison than the dynamic bias comparator. In the whole range of VCM, the dynamic bias comparator has smaller noise than the Elzakker comparator. However, noise of the dynamic bias comparator increases near 50% while the Elzakker comparator increases about 30%.

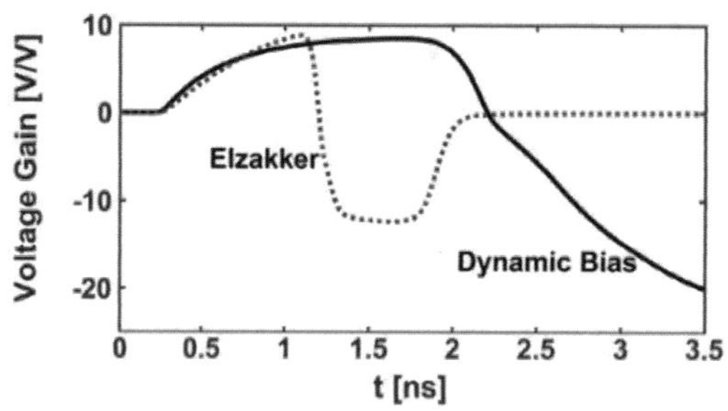

Figure 11. Voltage gain of two comparators.

Furthermore, the dynamic bias comparator has a more stable voltage gain than the Elzakker comparator as the figure above shows. The Elzakker comparator has a positive gain in only 1ns and then becomes negative and finally zero while the dynamic bias comparator has a positive gain about 2ns. Nevertheless, the dynamic bias comparator has a tradeoff on delay.

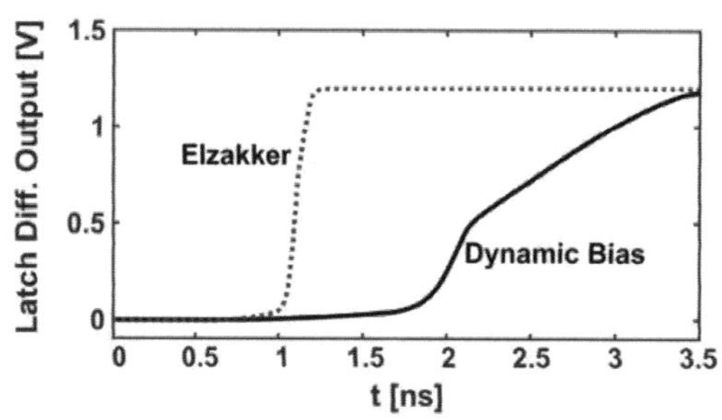

Figure 12. Output voltage of two comparators.

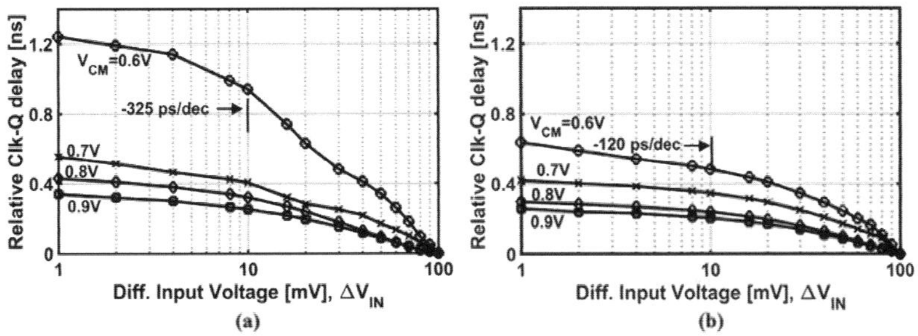

Figure 13. Clk-Q delay of two comparators.

As figure 12 shows, the output voltage of the Elzakker comparator changes rapidly to 1.2V in 1ns, but the dynamic bias comparator uses 3.5ns to reach it. Additionally, the CLK-Q delay of variant VCM and the differential voltage for the dynamic bias comparator is larger than the Elzakker comparator according to figure 13.

4. Triple-tail fully dynamic comparator and TLFF

Triple-Tail Dynamic Comparator is a key component in every high-speed ADC. Its noise, kickback, and common-mode sensitivity determine to a large extent the total accuracy [5]. The two main design parameters, speed in the form of resolving time and noise, which is our special attention.

Figure 14 shows the schematic of the proposed comparator. It comprises a cascoded integrator as the first pre-amplifier followed by a second pre-amplifier, which acts as both integrator and latch, driving finally the latching stage in a triple-tail fully dynamic arrangement.

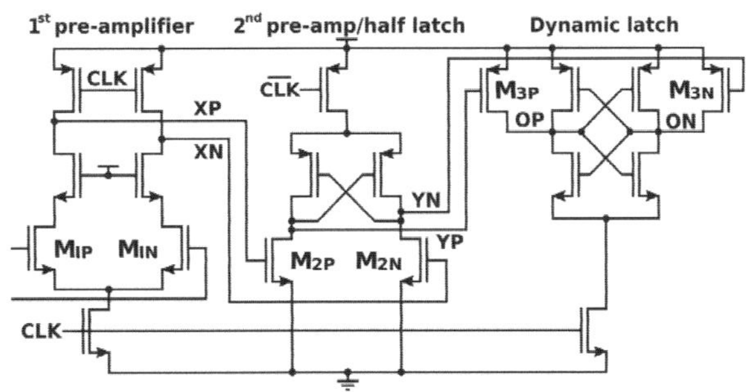

Figure 14. Triple-tail fully dynamic comparator.

The first pre-amplifier is designed to provide low noise/offset and high gain to attenuate the noise of the following stages. An NMOS cascode on the top of M_{IP}/M_{IN} isolates nodes XP/XN from the parasitic capacitance of the input pair during integration. It is optimized to have a very low time constant τ_{comp}. The second pre-amplifier suppresses the output noise and provides further signal gain, enhanced by the cross-coupling, prior to the latch, thus minimizing its regeneration time. The intermediate devices M2P/M2N and M3P/M3N act both as gain stages to provide further shielding from latch output noise, as well as reset devices for nodes YP/YN and OP/ON, respectively, obviating the need for additional reset transistors, which reduces the capacitance at those nodes, further minimizing the latch regeneration time.

Figure 15 shows the comparator clock and its differential outputs for input voltages as small as 1pV. The shortcoming of Triple-tail comparator schematic is that it cannot provide valid digital levels to the CDAC before the end of its allocated period [6]. The ways of improving this situation include allocating more time in the loop and/or creating more gain in the comparator, which come at the expense of speed and/or power.

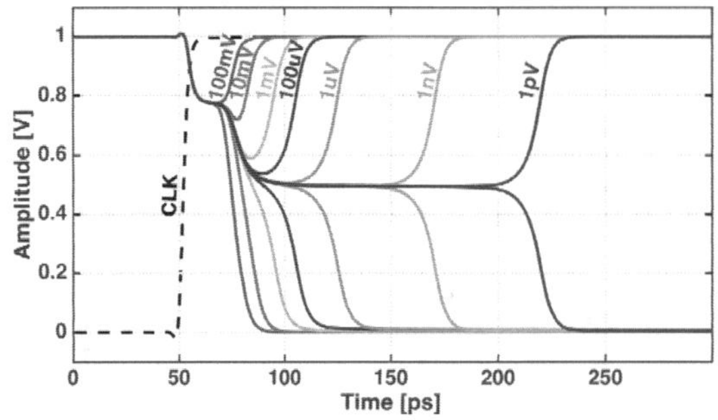

Figure 15. Output for different input voltages.

From figure 15, when its differential outputs for input voltages as small as 1uV, the time[ps] is about 35 unit. Then when its differential outputs for input voltages as small as 1nV, the time[ps] is about 85 unit. when its differential outputs for input voltages as small as 1pV, the time[ps] is about 135 unit. As its differential outputs for input voltages time increases, the times are creeping up.

Input common-mode voltage $V_{CM,IN}$ is an important aspect of comparator design, which affects both its resolving time and noise. The resolving time [see figure 16(a)] and input referred noise [see figure 16(b)] versus $V_{CM,IN}$ for the triple-tail comparator are shown. The triple-tail comparator shows a faster resolving time with a lower common-mode dependence for a wide range of voltages compared with others. An optimum exists between 500 and 600mV, so $V_{CM,IN}$ of 500mV was chosen for a near optimum resolving time and a small enough input referred noise.

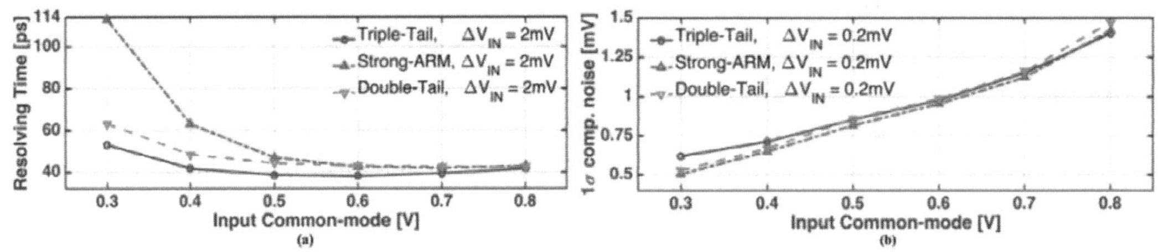

Figure 16. Resolving time and input referred noise versus $V_{CM,IN}$.

The resolving time [see figure 17(a)] and energy/comparison [see figure 17(b)] versus input signal for the three different comparators are also shown. The proposed comparator achieves similar energy/comparison with the double-tail and about 35% higher than the strong-ARM latch.

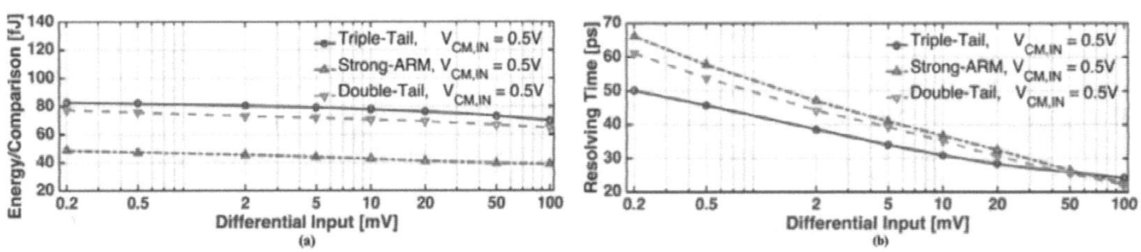

Figure 17. Resolving time and Energy/comparison versus differential input.

The triple-tail fully dynamic comparator enhances the sampling rate. It minimizes the comparator's total resolution time by using two preamplifiers that provide more input-output isolation and signal amplification before the latch [7]. The disadvantage is that the triple-tail fully dynamic comparator does not improve the power, noise and delay.

Three-stage fully dynamic triple-latch comparator separates the stacking latches into three stages and add the parallel feedforward path to stage 3 [8]. As a result, this structure can achieve an extremely short CLK–OUT delay.

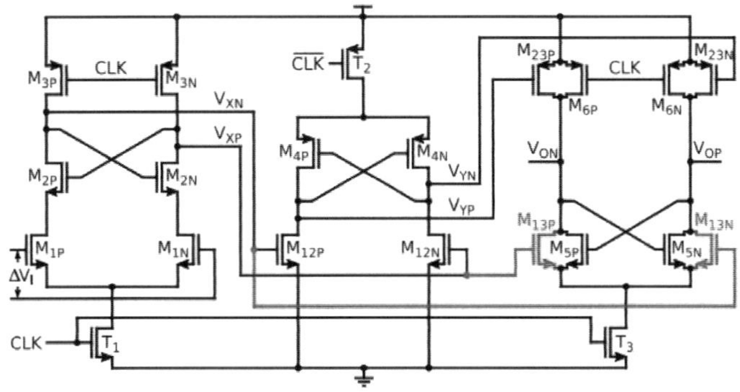

Figure 18. Triple-latch feedforward comparator.

Figure 18 shows the detailed circuit schematic of the proposed triple-latch feedforward (TLFF) comparator. The first stage is designed to provide a high gain and a certain bias voltage for the following stages [9]. The second stage is designed to minimize the delay at small ΔV_1. In the last stage, M5P/M5N is latch driven by either M13P/M13N or M23P/M23N to achieve minimum regeneration time. This is the most essential part of this circuit. The working condition of the feedforward path and direct path is depending on the input signal ΔV_1. Furthermore, the size of transistors in each path can control which path is activated to each ΔV_1. Thus, minimum delay time can be realized in full ΔV_1 range.

For all ΔV_1 range, stage 1 will always be working, therefore the delay time equals to the time constant of stage 1 plus the time constant of the feedforward path. For small ΔV_1, VYP/VYN amplified by stage 2 is high enough to drive M23P/M23N and M5P/M5N latch. For large ΔV_1, VXP/VXN is high enough to drive M13P/M13N and M5P/M5N latch directly. As a result, the feedforward time constant and the delay time can be described as (16), (17) and (18)

$$\tau_{\mathrm{comp}} \approx \tau_1 + \tau_{\mathrm{FF}} \tag{16}$$

$$\tau_{\mathrm{FF}} \approx \tau_2\left(1+\frac{g_{\mathrm{m12}}}{g_{\mathrm{m4}}}\right)^{-1} + \tau_3\left(1+\frac{g_{\mathrm{m23}}}{g_{\mathrm{m5}}}\right)^{-1} \tag{17}$$

$$\tau_{\mathrm{FF}} \approx +\tau_3\left(1+\frac{g_{\mathrm{m13}}}{g_{\mathrm{m5}}}\right)^{-1} \tag{18}$$

gm12, gm4, gm13, gm23, and gm5 are the transconductances of M12P/M12N, M4P/M4N, M13P/M13N, M23P/M23N, and M5P/M5N, and $\tau 2$, $\tau 3$ are the time constants of stages 2 and 3.

In order to prove the TLFF has a smaller delay time than other system in the whole input range, a simulation is conducted to measure the delay time of all the system mentioned in the article and a three-stage comparator without feedforward. The input referred noise (~1m Vrms), offset (~11m Vrms), and latching drive strength, while the same output load is provided to all of them to ensure the comparison is fair. The result is shown in figure below.

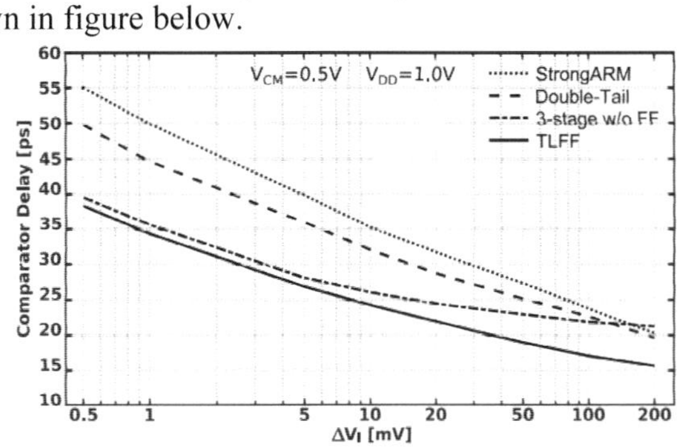

Figure 19. Delay of four comparators.

According to figure 19, for smallest $\Delta V_1 = 0.5\text{mV}$, both three-stage comparator without feedforward and TLFF shows approximately 30% shorter delay time than other structures. However, for the largest $\Delta V_1 = 200\text{mV}$, three-stage comparator without feedforward has almost the same delay time as Double-Tail and StrongARM [10]. This is because the three stage is always on during comparison, thus the delay time equals to the sum of all three stages' time constant. But for TLFF, the feedforward path will be activated under large ΔV_1. As a result, it can still have a 25% shorter delay time than other structures. Nevertheless, a shorter delay time also have tradeoff on power consumption. TLFF shows about 32% and 15% larger power than the StrongARM and the Double-Tail according to the authors. This is acceptable because TLFF enhance the overall performance of the whole system, therefore the benefit overcomes the power consumption.

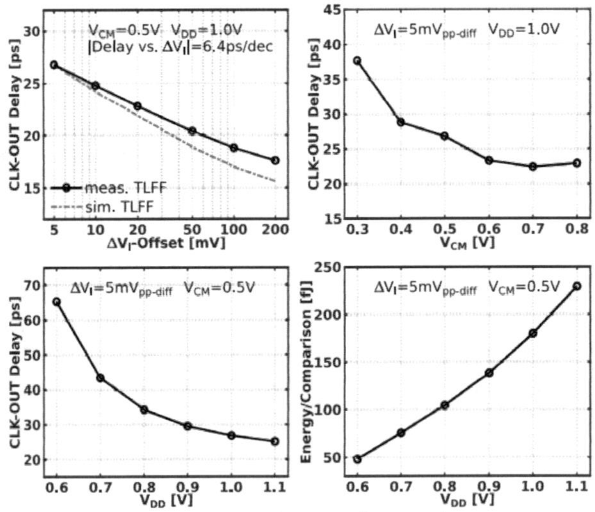

Figure 20. Clk-out delay and energy/comparison ratio versus VDD and VCM

According to figure 20, TLFF has a CLK-OUT delay under 70ps for VCM from 0.3V to 0.8V and VDD from 0.6V to 1.1V. This means TLFF can operate stably under disturbances which could affect VCM and VDD. Besides, energy/comparison increases almost linearly when VDD increases. However, CLK-OUT delay only decreases a little when VCM changes from 1V to 1.1V. At the same time, CLK-OUT becomes almost stable for VCM larger than 0.5V. Thus, a 1V VDD and 0.5V VCM will be a acceptable choice for both delay and power consumption.

5. Comparative Result

Table 1. Performance comparison of the comparators.

	Edge-pursuit comparator (EPC)	Dynamic bias comparator	Triple-tail fully dynamic comparator	Triple-latch feedforward (TLFF)
Technology	40nm	65nm	28nm	28nm
Supply Voltage	-	1.2V	1.0V	1.0V
Data Rate [Gb/s]	-	0.05	-	13.5
Sampling Rate	20KS/s	-	1.25GS/s	-
Delay Slope [ps/decade]	-	120	-	6.4

Input Offset	-	1.0mV	-	5.0mV
Power Consumption [mW]	0.00117	-	3.56	2.2
FOM[fJ/conv-step]	173.4	34~45	34.4	163.0
Noise (V)	-	400	-	-

Table 1 shows the performance of the four comparators, each of these comparators has its own strengths. The edge-pursuit comparator automatically adjusts the performance by changing the comparison energy according to its input difference without any control, eliminating unnecessary energy spent on coarse comparisons, so EPC has the highest energy efficiency per comparison [11]. The triple-tail fully dynamic comparator minimizes the total resolving time of the comparator by employing two pre-amplifiers, more input-output isolation, and signal amplification before the latch is provided, so that its sampling rate is much higher than the other comparators. With the dynamic bias architecture of the double-tail latch-type comparator, there is a big performance improvement, which compared with the Elzakker comparator [12]. The triple-latch feedforward comparator has a high transferred data rate and low delay with its unique proposed TLFF architecture. In the meanwhile, TLFF is also a re-creation of the triple-tail fully dynamic comparator, which be improved in smaller delay time, but it has larger power than others.

6. Conclusion

These comparators have their own advantages and disadvantages and are improved from each other to achieve the performance requirements of different ADC. There is a very high sampling rate in the triple-tail fully dynamic comparator. Nevertheless, the energy efficiency and noise are not focused on improving. The triple-tail fully dynamic comparator reduces delay by increasing the gain through the cascode of the tier 3 amplifiers. Triple-latch feedforward adds a new path, which optimizes the delay under the whole size signal input based on the triple-tail fully dynamic comparator. The dynamic bias architecture's constant tail current is optimized to reduce power consumption, and whose performance of power consumption, noise, and delay is greater than Elzakker. The edge-pursuit comparator is an innovative circuit. It improves energy efficiency, signal-to-noise, and distortion ratio by a new circuit structure, which contains a lot of delay units. However, it does not improve the delay of the comparators. The EPC shows 67 times automatic energy scaling between the MSB and LSB bit decisions without any external control, saving a significant portion of the energy for the MSB decision. It also has a 184dB FOMC, which is the best number among the designs compared. Overall EPC is a valuable innovation.

References

[1] Shim, M., Jeong, S., Myers, P D., Bang, S., Shen, J., Kim, C., Sylvester, D., Blaauw, D., Jung, W. (2017) Edge-Pursuit Comparator: An Energy-Scalable Oscillator Collapse-Based Comparator With Application in a 74.1 dB SNDR and 20 kS/s 15 b SAR ADC. IEEE Journal of Solid-State Circuits., vol. 52, no. 4: 1077-1090.

[2] Kobayashi, T., Nogami, K., Shirotori, T., Fujimoto, Y and Watanabe, O. (1992) A current-mode latch sense amplifier and a static power saving input buffer for low-power architecture. In: Proc. VLSI Circuits Symp. Dig. Tech. pp. 28–29.

[3] Elzakker, M., Tuijl, E., Geraedts, P., Schinkel, D., Klumperink, E A M and Nauta, B. (2010) A 10-bit charge-redistribution ADC consuming 1.9 μW at 1 MS/s. IEEE Journal of Solid-State Circuits., vol. 45, no. 5: 1007-1015.

[4] Bindra, H S., Lokin, C E., Schinkel, D., Annema, A and Nauta, B. (2018) A 1.2-V Dynamic Bias Latch-Type Comparator in 65-nm CMOS With 0.4-mV Input Noise. IEEE Journal of Solid-State Circuits., vol. 53, no. 7: 1902-1912.

[5] Ramkaj, A T., Strackx, M., Steyaert, M S J., and Tavernier, F. (2018) A 1.25-GS/s 7-b SAR ADC With 36.4-dB SNDR at 5 GHz Using Switch-Bootstrapping, USPC DAC and Triple-Tail Comparator in 28-nm CMOS. IEEE Journal of Solid-State Circuits., vol. 53, no. 7: 1889-1901.
[6] Harpe, P., Cantatore, E and Roermund,A V. (2013) A 2.2/2.7fJ/conversionstep 10/12b 40kS/s SAR ADC with data-driven noise reduction. In: IEEE Int. Solid-State Circuits Conf. San Francisco, CA, USA. pp. 270-271.
[7] Yang, J., Naing, T L and Brodersen, R W. (2010) A 1 GS/s 6 bit 6.7 mW successive approximation ADC using asynchronous processing. IEEE Journal of Solid-State Circuits., vol. 45, no. 8: 1469–1478.
[8] Ramkaj, A T., Steyaert, M S J., and Tavernier, F. (2019) A 13.5-Gb/s 5-mV-Sensitivity 26.8-ps-CLK–OUT Delay Triple-Latch Feedforward Dynamic Comparator in 28-nm CMOS. IEEE Solid-State Circuits Letters., vol. 2, no. 9: 167-170.
[9] Stepanovic, D and Nikolic, B. (2013) A 2.8 GS/s 44.6 mW time-interleaved ADC achieving 50.9 dB SNDR and 3 dB effective resolution bandwidth of 1.5 GHz in 65 nm CMOS. IEEE Journal of Solid-State Circuits., vol. 48, no. 4: 971–982.
[10] Wicht, B., Nirschl, T and Schmitt-Landsiedel, D. (2004) Yield and speed optimization of a latch-type voltage sense amplifier. IEEE Journal of Solid-State Circuits., vol. 39, no. 7: 1148–1158.
[11] Razavi, B. (2015) The strongARM latch [a circuit for all seasons]. IEEE Solid-State Circuits Mag., vol. 7, no. 2: 12–17.
[12] Wang, Y and Razavi, B. (2000) An 8-bit 150-MHz CMOS A/D converter. IEEE Journal of Solid-State Circuits., vol. 35, no. 3: 308–317.

Innovative Techniques in Comparator Designs

Yiming Chen[1]

[1] Department of Electrical & Computer Engineering, McMaster University, Hamilton, Ontario, L8S 4L8, Canada

*cheny466@mcmaster.ca

Abstract. This paper presents four innovative designs of comparators proposed these years. A latch-type dynamic bias adds a tail capacitor to prevent fully discharging at the pre-amplifier output nodes to reduce energy consumption. The comparator is analysed and then compared with floating inverter amplifier (FIA) type. The pre-amplifier of the FIA type adopts an inverter-based input pair by a floating reservoir capacitor, greatly boosting gm/Id and improving the energy efficiency. The edge-pursuit comparator (EPC) provides a new perspective when designing comparators. According to the input difference, it will automatically adjust the comparison energy, avoiding wasting unnecessary energy spent on coarse comparison. Finally, for the edge-race comparator (ERC), it further improves the performance of saving energy and addressing the relatively long comparison time in the EPC.

1. Introduction

Comparators play a core role in bridging physical and digital worlds nowadays. As the development of highly integrated circuits such as biomedical implants and sensing devices in the daily life, the demand for low-power and lower noise analog-to-digital converter (ADC) is growing rapidly. Normally, 40% to 50% of energy consumed by ADCs can trace back to the power distributed to the comparators [1-3]. Meanwhile, low-voltage operation requires strict conditions on quantization noise. As a result, improving the energy efficiency becomes both critical and challenging to pursue high-performance digital integrated circuits.

Many reformative structures of comparators are proposed in the recent years. The innovations of the improved designs mainly focus on reducing energy consumption, lowering the decision delay, minimizing the inferred input noise, and optimizing the voltage gain. Based on the Elzakker's comparator, H. S. Bindra proposed the dynamic-bias latch-type comparator in 2018 [1, 4]. This new architecture greatly improved the energy efficiency by adding a tail capacitor to prevent fully discharging for the transistors. The noise performance of the dynamic-bias comparator was also ameliorated. In 2020, X. Tang proposed a comparator with the dynamic floating inverter amplifier (FIA) under 180-mm process [5]. It greatly reduced the influence of the process corner and the input common-mode voltage on the comparator performance. M. Shim brought an innovative structural design named edge-pursuit comparator (EPC) in 2017 [6]. By adjusting the comparison energy automatically with different input voltage, it greatly saved energy and optimizes the tunability with the input-referred noise. In 2020, H. Zhuang proposed the edge-race comparator (ERC) [7]. It compared the differential input voltage by generating two propagating edges in two inverter loops and measured the distance between them to determine the comparison result. This new architecture further reduced the energy consumption and resolves relatively long comparison time in previous designs.

Content from this work may be used under the terms of the Creative Commons Attribution 3.0 licence. Any further distribution of this work must maintain attribution to the author(s) and the title of the work, journal citation and DOI.
Published under licence by IOP Publishing Ltd

This paper is organized as follow: in Section 2, compared with the conventional double tail latch type comparator, the advantages of the dynamic bias comparator proposed by H. S. Bindra are discussed [4]. Due to the existing trade-offs in dynamic bias comparator, further amended architecture presented by X. Tang will be indicted in Section 3 [5]. Section 4 provides a brand-new perspective with the edge-pursuit comparator proposed by M. Shim [6]. In Section 5, the performance of H. Zhuang's edge-race comparator will be analysed quantificationally [7]. Comparisons for all the four innovative designs are mentioned in Table I in Section 6. Finally in Section 7 concludes this paper.

2. Dynamic bias latch-type comparator

Based on the conventional double-tail latch-type comparator, M. Elzakker proposed a new as shown in figure 1 [1]. Only when the output common-mode voltage is lower than the threshold voltage of (M6/M7), the latch stage starts working. By this mechanism, pre-charge energy of the pre-amplifier stage is minimized, resulting in decrease in the energy consumption.

The proposed dynamic bias comparator not only further improves the energy efficiency but also optimizes the overhead issues compared to the previously proposed design "bi-directional dynamic comparator [2]. The dynamic bias comparator adds a tail capacitor M3b and a tail transistor M3a (as shown in figure 2) to control the discharge from nodes Di+ and Di-. When CLK=0, CTAIL is discharged to ground through M3b. When CLK=VDD, all reset transistors are turned OFF while M3a turns ON and capacitances on the drain nodes Di+ and Di- starts to discharge. This leads to the tail current charging the capacitor M3b and increase in the voltage VCAP, thus contributing to diminishing VGS of the differential part (M1/M2) and finally provides a dynamic bias for the differential pair.

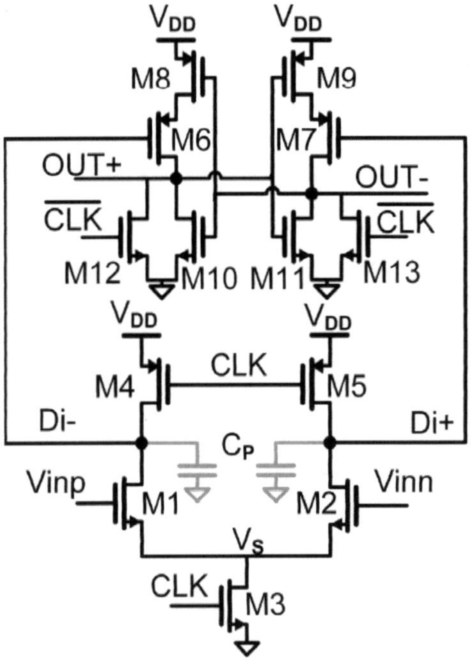

Figure 1. Circuit of Elzakker's comparator.

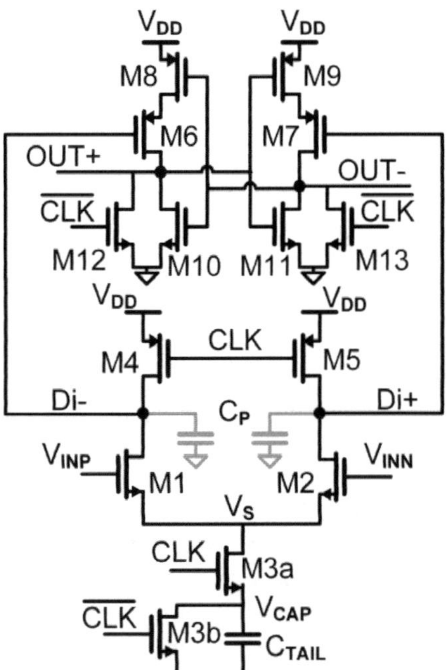

Figure 2. Circuit of the proposed dynamic bias comparator

The required energy for the preamplifier in the dynamic bias comparator to pre-charge the drain nodes is 2*Cp*VDD2-Cp*VDD*(VD1+VD2), which is lower than the conventional pre-charge energy for the 2*Cp*VDD2 in the case of Elzakker's comparator. Figure 3 indicates the lower power consumption of dynamic bias compared with Elzakker's latch.

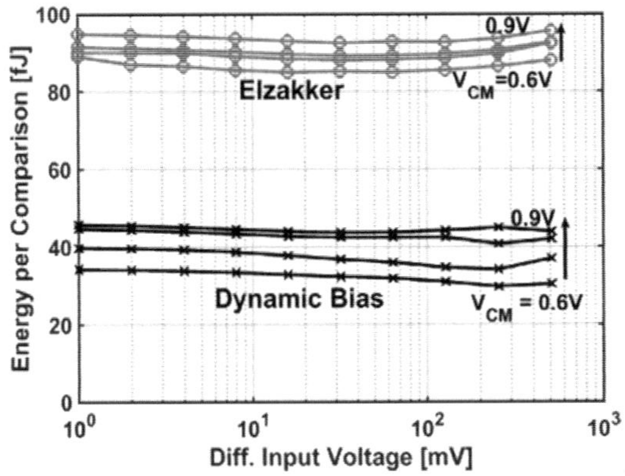

Figure 3. Energy consumption of dynamic bias and Elzakker's latch-type comparator across differential input voltage.

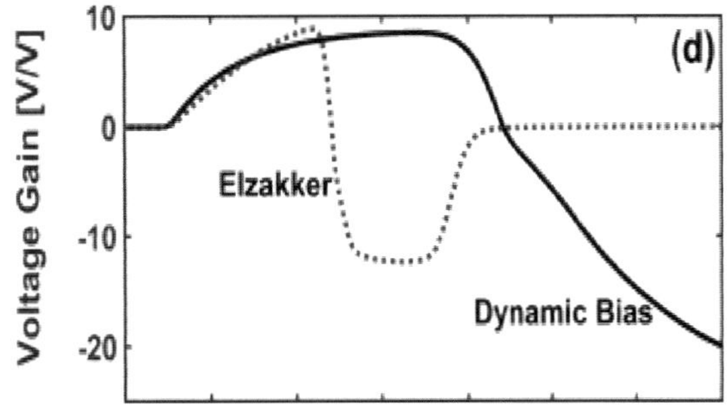

Figure 4. Simulated pre-amplifier differential voltage gain of the two comparators.

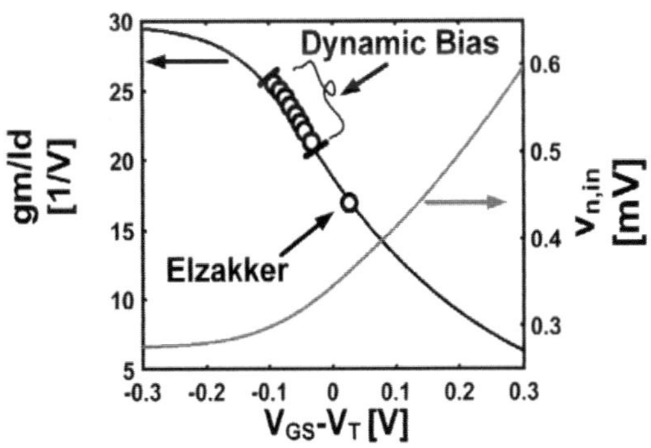

Figure 5. gm/Id and input noise voltage (calculated) versus Vov.

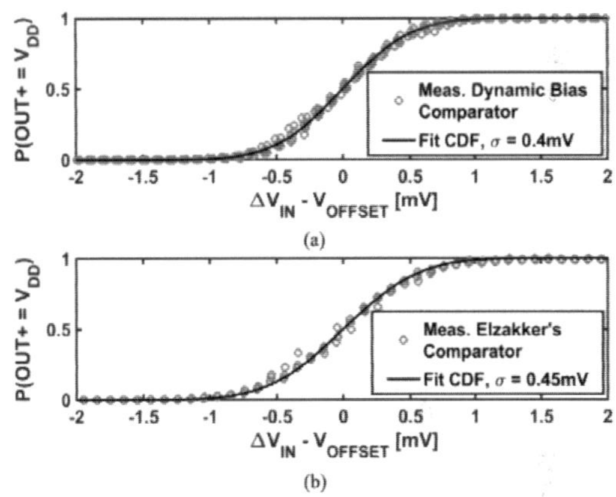

Figure 6. Measured cumulative probability density distribution and fit to Gaussian distribution for (a) dynamic bias comparator and (b) Elzakker's comparator.

As for the voltage gain of the dynamic bias comparator, the dynamic bias comparator performs more stable than Elzakker comparator, the result can be derived from figure 4. The detailed expression of voltage gain of dynamic bias can be written as

$$A_V(T_{\mathrm{INT}}) = \frac{C_{\mathrm{TAIL}}}{2n \cdot C_P} \frac{V_s(T_{\mathrm{INT}})}{\frac{kT}{q}} \tag{1}$$

Where $V_s(T_{\mathrm{INT}})$ is

$$V_s(T_{\mathrm{INT}}) = \frac{2 \cdot \Delta V_{\mathrm{Di,CM}}(T_{\mathrm{INT}}) \cdot C_P}{C_{\mathrm{TAIL}}} \tag{2}$$

The maximum drop in $\Delta V_{\mathrm{Di,CM}}(T_{\mathrm{INT}})$ is

$$\Delta V_{\mathrm{Di,CM},max} = \frac{C_{\mathrm{TAIL}}}{C_{\mathrm{TAIL}} + 2 \cdot C_P} V_{\mathrm{DD}} \tag{3}$$

From the expressions, it can be observed that the Di nodes do not fully discharge to ground even under the extreme conditions. Therefore, the conclusion that dynamic bias comparator is more energy efficient than Elzakker's comparator can also be proved.

In addition to the optimization in energy consumption, the proposed dynamic bias comparator also performs better than Elzakker's comparator in thermal noise contribution of the differential pair. The input-referred noise of the pre-amplifier in Elzakker's comparator in the strong inversion operation can be calculated for

$$E\left[v_{\mathrm{n,SI,in}}^{2}(T_{\mathrm{INT}})\right]=\frac{4kT\Upsilon}{C_P\cdot\Delta V_{\mathrm{Di,CM}}(T_{\mathrm{INT}})\cdot\left(\frac{g_m}{I_{\mathrm{CM}}}\right)_{\mathrm{SI},T_{\mathrm{INT}}}} \tag{4}$$

Where $\Delta V_{\mathrm{Di,CM}}(T_{\mathrm{INT}})$ is

$$\Delta V_{\mathrm{Di,CM}}(T_{\mathrm{INT}})=\frac{I_{\mathrm{CM}}\cdot T_{\mathrm{INT}}}{C_P} \tag{5}$$

Meanwhile, under the weak inversion operation the noise of the dynamic bias pre-amplifier can be derived as

$$E\left[v_{\mathrm{n,WI,in}}^{2}(T_{\mathrm{INT}})\right]=\frac{2nkT}{C_P\cdot\Delta V_{\mathrm{Di,CM}}(T_{\mathrm{INT}})\cdot\left(\frac{g_m}{I_{\mathrm{CM}}}\right)_{WI,T_{\mathrm{INT}}}} \tag{6}$$

Where Cp is the capacitor at the pre-amplifier's output nodes, gm is the transconductance of M1 and M2 and ICM is the current through M1 and M2.

From formula (4) and (6), it can be observed the input-referred noise is generally inversely proportional to gm. So larger gm or smaller VGS can ameliorate the performance of both comparators. Figure 5 shows the relationship between gm/Id with the overdrive voltage V_{OV} of the two comparators. When V_{OV} decreases, the input-referred noise also reduces. The increasing Vs ensures the gm/Id of dynamic bias comparator at the integration time is higher than that of Elzakker's comparator. In addition, it can also be observed that for the dynamic bias, the overdrive voltage lies in the weak inversion region where Vov is lower than 0V, so gm/Id for the dynamic bias comparator raises for about 25% compared to Elzakker's comparator. Moreover, figure 6 shows the rms input referred noise of the dynamic bias comparator is 0.4 mV and the Elzakker's comparator is 0.45mV when fitting the measurements to a Gaussian CDF

However, the performance of CLK-Q delay of the dynamic comparator is not that satisfactory. The results for both comparators are shown in figure 7, and it can be concluded that the dynamic bias comparator is more sensitive to the change of the input common-mode voltage than Elzakker's comparator. This problem is improved by the architecture discussed in Section 3.

3. Floating inverter pre-amplifier comparator

X. Tang proposed an energy-efficient dynamic comparator with a floating inverter amplifier (FIA)-based pre-amplifier, as shown in figure 8. It comes from the further modification of CMOS DB integration pre-amplifier shown in figure 9 by connecting two tail capacitors and replacing with C_{RES}. The presence of the FIA is because CMOS DB integration is infeasible in the extreme corner like SF, where nMOS is in the slow corner while pMOS is in the fast corner. FIA comparator fixes this problem and to ensure the input-output current for C_{RES} to be the same. Take the SF corner as an example, the comparison result is shown in figure 10.

The integration gain of the FIA can be expressed as

$$A_V(T_{\mathrm{INT}})=\frac{2\cdot C_{\mathrm{RES}}\cdot\Delta V_S(T_{\mathrm{INT}})}{n\cdot C_X\cdot U_T} \tag{7}$$

Where $\Delta V_S(T_{\mathrm{INT}})$ is the source voltage change at the integration time, also the input referred noise at the integration time can be calculated

$$\sigma_{in,\mathrm{FIA}}^{2}(T_{\mathrm{INT}}) = \frac{2nkT}{C_{\mathrm{RES}} \cdot \Delta V_{S}(T_{\mathrm{INT}})} \cdot \frac{I_{D}}{G_{m}} \tag{8}$$

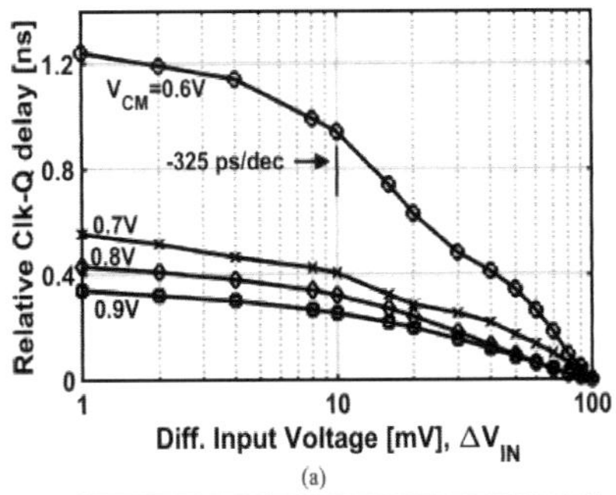

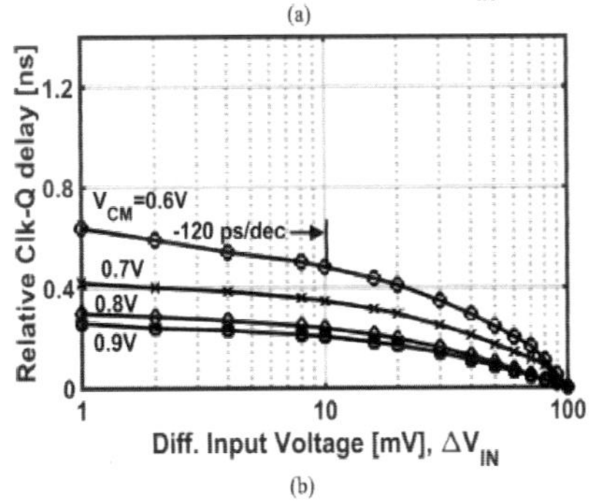

Figure 7. Relative CLK-Q delay measured for the (a) dynamic bias comparator and (b) Elzakker's comparator for various V_{CM}.

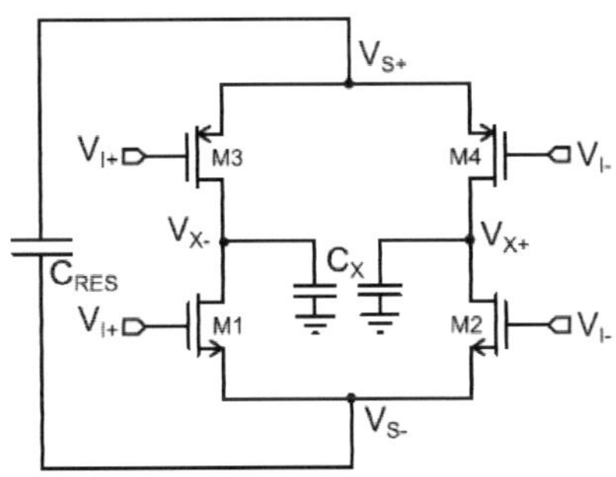

Figure 8. Proposed CMOS DB integration model.

Journal of Physics: Conference Series **2221** (2022) 012022 doi:10.1088/1742-6596/2221/1/012022

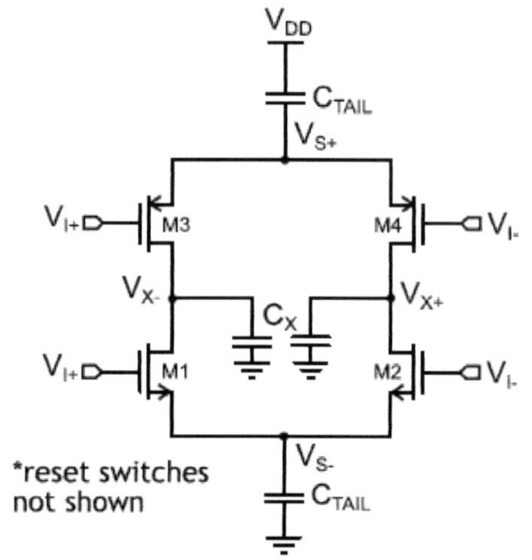

Figure 9. Proposed CMOS DB integration model.

The fundamental trade-off between energy consumption and thermal noise is described as the parameter FoM (figure of merit). It is defined as the product of energy consumption and the input-referred noise power. The energy efficiency improvement of FIA can be calculated as

$$\frac{FoM_{SA}}{FoM_{FIA}} = \frac{V_{\mathrm{DD}}}{V_{\mathrm{THN}}} \cdot \frac{(G_m/I_D)_{\mathrm{FIA}}}{(g_m/I_D)_{\mathrm{SA}}} \tag{9}$$

The equation indicates two main advantages of the proposed FIA. First, the coefficient (VDD/VTHN) comes from the circumvent of fully discharging the integration capacitor. Besides, ratio of gm/Id for the two kinds of pre-amplifiers is about 2.5, thus predicts the higher energy-efficiency and lower input-referred noise in the FIA structure. Figure 11 and figure 12 are the simulation results and proves the previous conclusion.

As described before, the design mainly solves the relatively long CLK-Q delay time from the nMOS DB integration pre-amplifier. Figure 13 shows the simulation of CLK-Q delay with 1-mV differential input and energy efficiency of the proposed comparators with different choices of CRES values. Figure 13(a) indicates that when the choice of CRES lies between 1pF to 3.5pF, the absolute value of the rate of change in simulated CLK-Q delay is large and it becomes mild when coming to 3.5pF to 5pF region. Figure 13(b) shows when the CRES value is chosen between 2pF to 3.5pF the simulated FoM is enough to suppress the latch stage noise for avoiding the degradation of comparator precision. Besides large CRES will lead to the reduction in the dynamic bias effect and diminish gm/Id. By reason of the foregoing, X. Tang chooses 2-pF CRES to reach the balance between energy efficiency, comparison speed and area consumption.

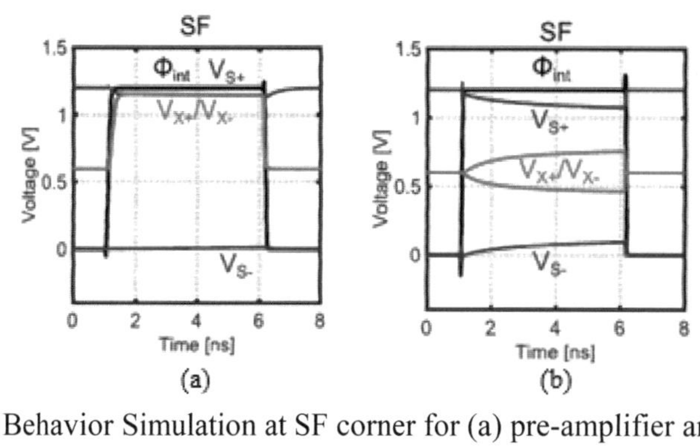

Figure 10. Behavior Simulation at SF corner for (a) pre-amplifier and (b) FIA.

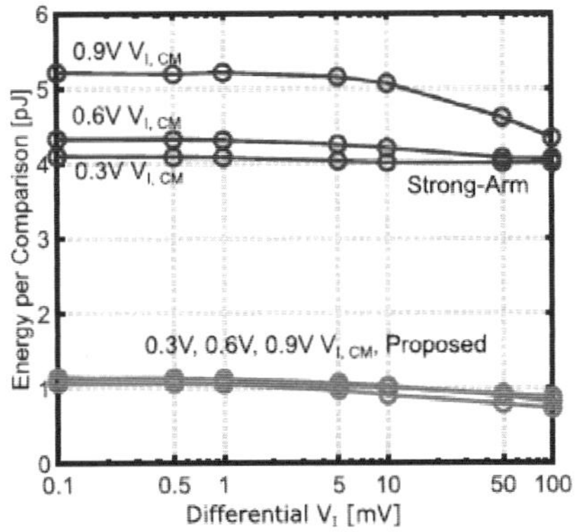

Figure 11. Measured energy consumption versus input voltage for the SA latch and proposed comparator.

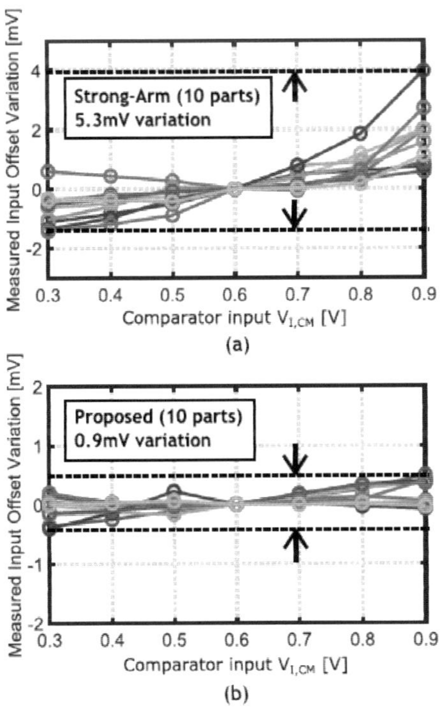

Figure 12. Measured ten-part input-referred offset versus input CM voltage for (a) SA latch and (b) proposed comparator.

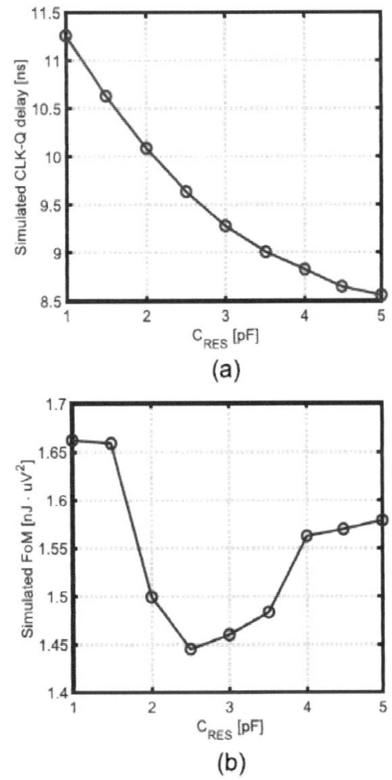

Figure 13. Simulated (a) CLK-Q delay with 1-mV differential input and (b) energy efficiency of the proposed comparator.

After choosing the applicable CRES, figure 14 presents that the simulated CLK-Q delay of the proposed comparator is much more stable than the previous strong-arm design. Also, when using higher input common-mode voltage, CLK-Q delay is even lower.

Figure 15 describes the measured cumulative probability density distribution and fit to Gaussian distribution for both proposed comparator and SA latch. The result reveals that FIA reduces the noise variation by four times due to its input common-mode-insensitive operation compared with the traditional architecture. The comparison of strong-arm and proposed type is shown in figure 16. It indicates that with the same common-mode input voltage, the input noise of the proposed type is about

25% lower when $V_{I,CM}$ is 0.3-0.6V. Apart from that, with the input voltage increasing, the FIA type comparator shows more stable performance for the changes in the measured input noise is much lower than that of the strong-arm type.

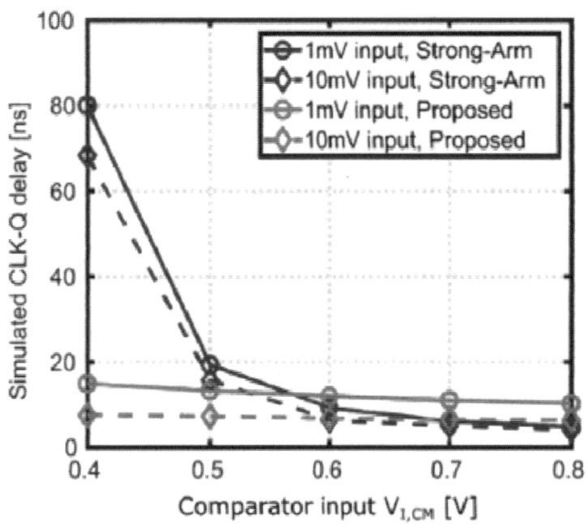

Figure 14. Simulated CLK-Q delay versus input common-mode voltage.

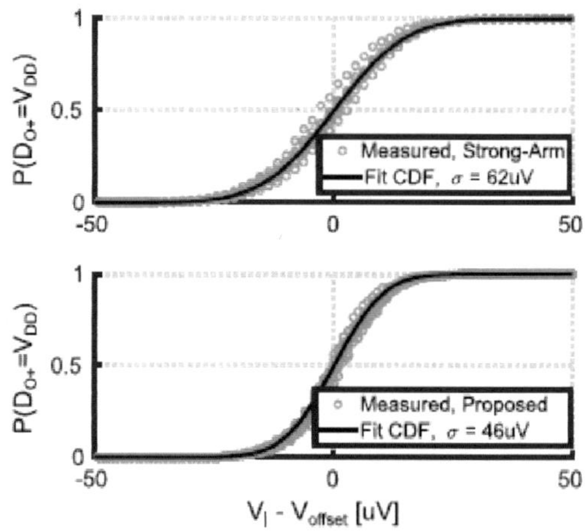

Figure 15. Measured cumulative probability density distribution and fit to Gaussian distribution for proposed comparator and SA latch with 1.2-V supply and 0.6-V input common-mode voltage.

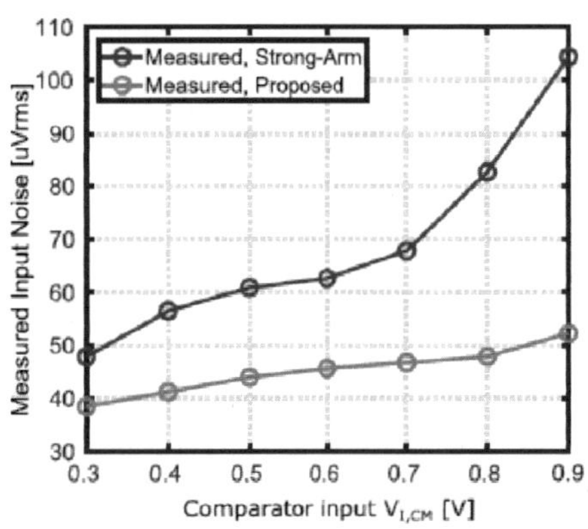

Figure 16. Measured input-referred noise versus input common-mode voltage for the SA latch and proposed comparator.

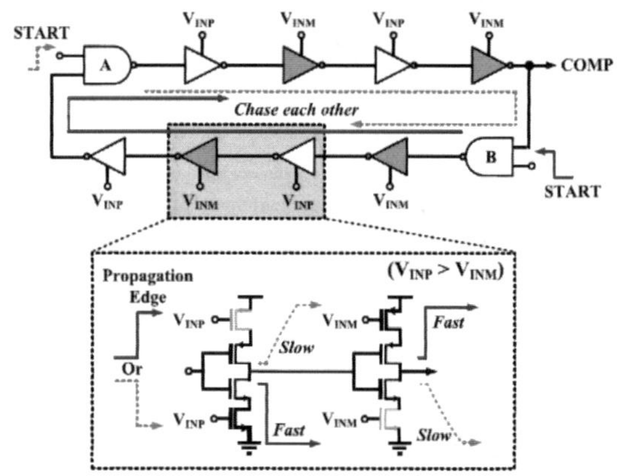

Figure 17. Structure of the EPC.

4. Edge-pursuit comparator

By changing the comparison energy automatically according to the input difference to save energy, M. Shim proposed an innovative comparator named edge-pursuit comparator (EPC) in 2017 [6]. This new structure is composed of two NAND gates and several inverter delay cells as shown in figure 17. When the comparator initiates with the signal START goes high at both NAND gates, two propagating edges will be injected into the oscillator and then travel around the comparator. By increasing V_{INP}, one edge will propagate faster (and vice versa for V_{INM}), resulting in the oscillation collapses and indicating the slower edge is overtaken. Therefore, the logic level is set to either V_{DD} or GND. For large voltage differences, the oscillation collapses quickly, reducing the energy consumption for coarse comparisons. By this methodology, the comparator not only optimizes the energy dissipation but also realizes high accuracy.

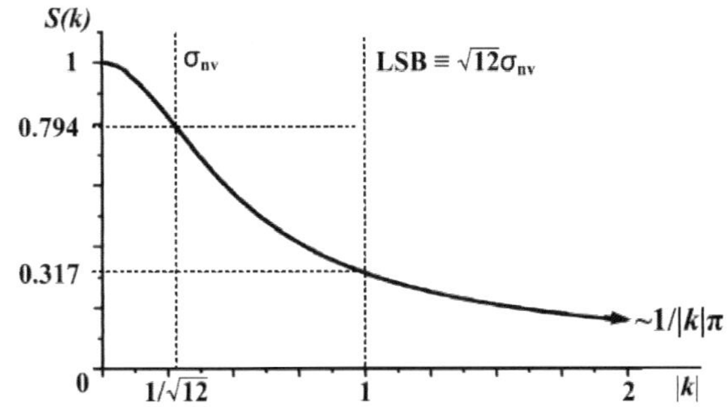

Figure 18. EPC's scaling factor S(k).

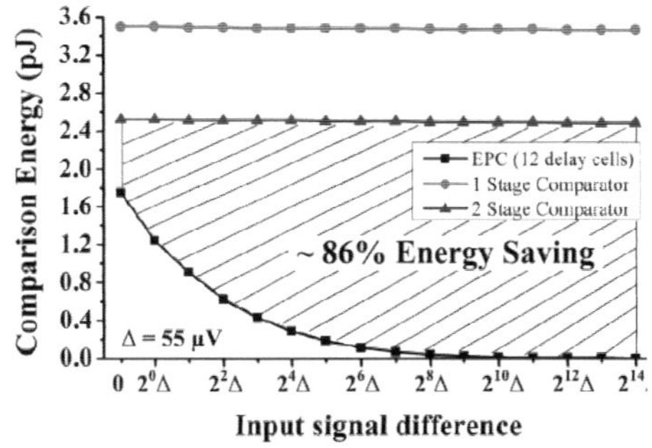

Figure 19. Comparison of simulated comparator performance among EPC and conventional comparators for energy.

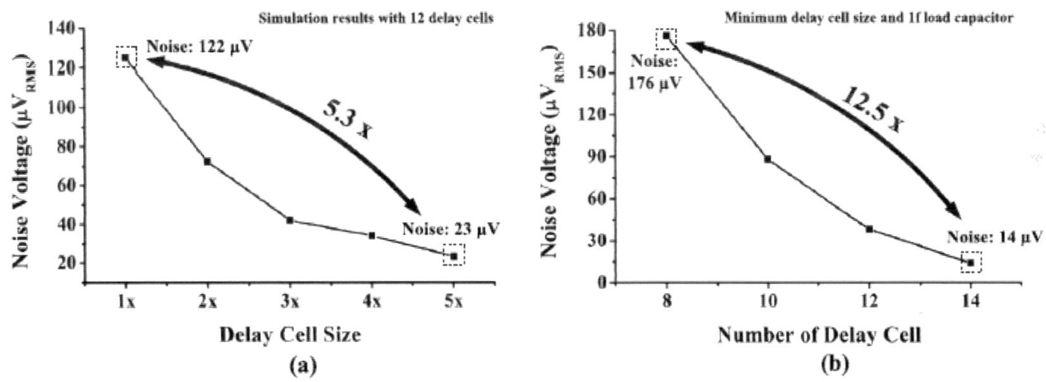

Figure 20. Simulated input-referred noise versus (a) delay cell size and (b) number of delay cells.

By analysing quantitatively, the average energy consumption per comparison for EPC is

$$E = 2IV_{\mathrm{DD}}\frac{\pi^2}{\Sigma}S\left(\frac{M}{\Sigma}\right) \tag{10}$$

Where M and Σ are the "drift" and "diffusion" coefficients of this random process defined as

$$M \equiv \frac{\mu\Delta\phi}{\tau} = 4\pi f_0\frac{v_{\mathrm{in}}}{V_{\mathrm{ov}}} \tag{11}$$

$$\Sigma \equiv \frac{\sigma_{\Delta\phi}^2}{\tau} = 8\pi^2 f_0^2\frac{kT}{I}\left(\frac{2}{V_{\mathrm{ov}}}(\gamma_N+\gamma_P)+\frac{2}{V_{\mathrm{DD}}}\right) \tag{12}$$

Also, the scaling factor, which is dependent on the ratio M/Σ, is defined as

$$S(k) \equiv \frac{\tan h\,(k\pi)}{k\pi} \tag{13}$$

The function S(k) decreases as |k| becomes larger, reflecting the automatic energy scaling behaviour of EPC comparator. Figure 18 shows the scaling factor S is about 1 when Vin is at the noisy region. In contrast, it will decrease rapidly towards 0 when it is out of the noisy region. As a result, the comparator can save approximately all the energy in most of the voltage regions except for the small noisy region.

To evaluate the EPC's energy efficiency compared with other comparators, a norm is defined as

$$N \equiv E \times \frac{\sigma_{v_n}^2}{V_{\mathrm{DD}}^2} \tag{14}$$

The formula illustrates the energy consumption per SNR assuming maximum signal power is VDD2. After deriving the expression for EPC comparator, single-stage comparator and two-stage comparator [8-9]. The ratios among those norms are

$$N_{\mathrm{EPC}}: N_{\mathrm{SS}}: N_{\mathrm{TS}} = S\left(\frac{A}{B}\right)\frac{\pi^2/3}{V_{\mathrm{DD}}}:\frac{7}{V_{\mathrm{th}}}:\frac{4}{V_{\mathrm{th}}} \tag{15}$$

According to (15), the EPC has relatively smaller norms than the other two kinds of comparators even when the input voltage and scaling factor do not benefit the EPC at Vin = 0 and S = 1. The simulation result for the three comparators is shown in figure 19. It demonstrates that with similar input signal difference and overdrive voltage V_{OV}, the EPC shows much better performance by saving energy for 86% compared to the other two types. Apart from that, distinguishing from the other two types with constant energy consumption. Figure 19 also illustrates the EPC saves energy through automatic energy scaling.

In addition to energy saving, the design also shows great performance in input noise tunability. The noise rms level σ_{V_n} can be expressed as

$$\sigma_{v_n} = \sqrt{\sigma_{v_n}^2} = \frac{\pi}{\sqrt{3}} f_0 \frac{kT}{I}\left(\frac{2}{V_{\mathrm{ov}}}(\gamma_N + \gamma_P) + \frac{2}{V_{\mathrm{DD}}}\right) V_{\mathrm{ov}} \tag{16}$$

Figure 20(a) shows that with the increase in delay cell size, the noise rms level approximately follows the inverse of the total capacitance. As shown in figure 20(b), due to the positive feedback the noise change is more sensitive than expected. This positive feedback will affect more with less stages and this is alike to the regeneration of the output signal in traditional comparators. Nevertheless, it does not consume much energy while the conventional comparators consume a fixed amount of dynamic energy. Therefore, the positive feedback further increases the tuneable noise range and even with few stages, the EPC comparators can still exhibit better energy efficiency.

In summary, the proposed edge-pursuit comparator is an innovative structural design which dramatically improves the energy efficiency and tunability with input-referred noise compared with the conventional comparators with an automatic energy scaling capability according to the input difference. This new design is different from the conventional SA latch + pre-amplifier structure, it measures the phase difference between the two input edges and then determine the comparison result. It provides a new perspective for the designers when improving the performance of comparators.

5. Edge-race comparator

In Section 4, the proposed EPC improves the energy efficiency and input-referred noise greatly compared with the conventional comparators [6]. However, it has limitation that in fine comparisons, the large delay is not negligible. In 2020, H. Zhuang proposed a novel voltage comparator named edge-race comparator (ERC) which overcomes this limitation, saving significant energy ad time in comparisons. As can be seen in figure 21, ERC consists of inverter delay units and NAND gates.

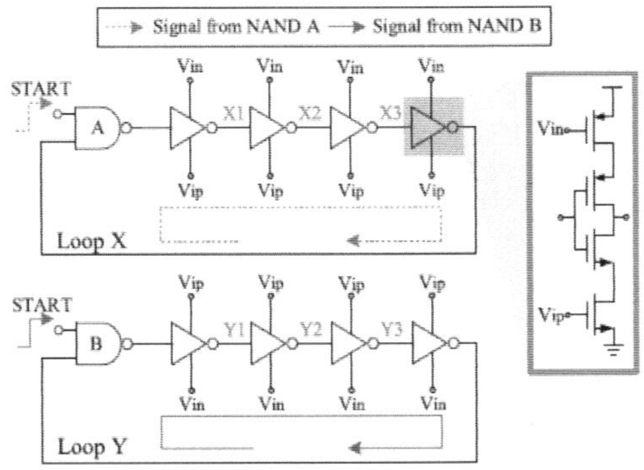

Figure 21. The ciurcuit of proposed ERC.

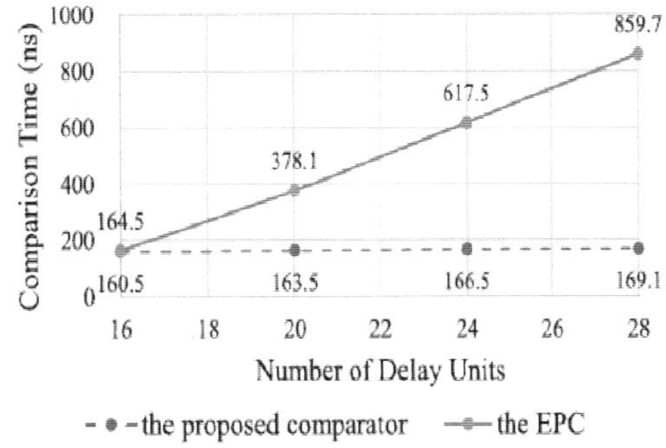

Figure 22. Number of delay units versus comparison time.

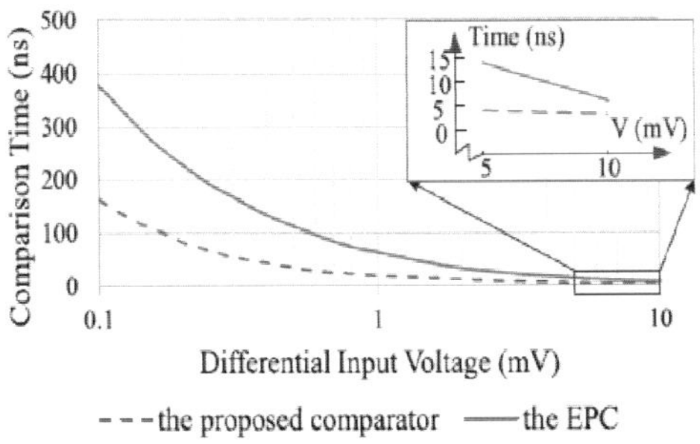

Figure 23. Comparison time versus differential input voltage.

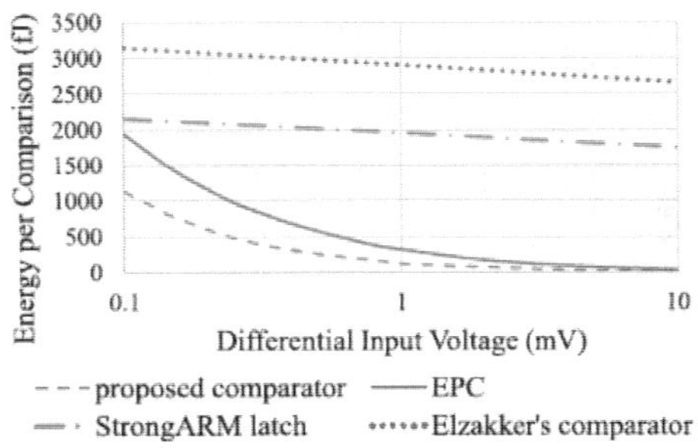

Figure 24. Energy per comparison at the same 50-µV input-referred noise.

When START rises to high, two propagating edges are generated by the two NAND gates, and they then start to race with each other. Due to the different propagating speed, the distance between them will gradually increases and finally reaches a pre-set value d0. Afterwards the comparison stops, and the comparison result is determined by the winner. The pre-set value d0 is 2 inverter delays which is 2.5 time smaller than that of the EPC, which is 5 inverter delays, thus the proposed type will have much faster comparison speed. The comparison time for both EPC and ERC can be calculated as follows:

$$t_{\mathrm{comp}} \approx \frac{C_{\mathrm{L}} V_{\mathrm{DD}} d0}{g_{\mathrm{m}}\left(V_{\mathrm{ip}} - V_{\mathrm{in}}\right)} \tag{17}$$

Where C_L is the load capacitance of each delay unit and gm is the transconductance of the transistor connected to V_{ip} or V_{in}. Figure 22 describes the delay time performance for ERC and EPC with different number of delay units. It can be observed that the ERC comparator is both shorter and more stable under different amounts of delay units than EPC type at V_{ip}-V_{in} = 0.1 mV. In figure 23, with the differential input voltage increasing, the result also reveals the better time delay performance of the proposed comparator. The comparison time is inversely proportional to the differential input and the time of the ERC comparator is 2 times smaller than the EPC type.

Apart from the faster comparison speed, the proposed ERC comparator also greatly optimizes the energy efficiency. Figure 24 demonstrates the energy consumption of 4 different types of comparators at the same 50-µV input-referred noise. As can be found that the StrongARM latch and the Elzakker's comparator consume relatively constant high energy. Meanwhile, the ERC and EPC can effectively adjust the energy with various differential input voltage and significantly save the energy consumption.

To further compare the performance between the ERC and EPC type, a norm for comparison is defined:

$$\mathrm{NM} \equiv t_{\mathrm{comp}} \times E_{\mathrm{comp}} \times \sigma_{\mathrm{n\_comp}}^{2} \tag{18}$$

In the norm ratio, the comparison between the EPC and ERC is given as follows:

$$\mathrm{NM}_{\mathrm{EPC}}{:}\mathrm{NM}_{\mathrm{ERC}} = d0_{\mathrm{EPC}}{:}\,(1.42 \cdot d0_{\mathrm{ERC}}) \tag{19}$$

When there are 20 inverter delay units, $d0_{EPC}$ is 11 inverter delays while $d0_{ERC}$ is 2 inverter delays. Under this circumstance, the norm ratio is 11:2.84 and shows that the ERC comparator is approximately 4 times better than the EPC type.

Moreover, the input-referred noise analysis should also be noticed. Assuming each loop is equivalent to a voltage-controlled inverter chain, the input-referred root mean square (rms) can be expressed as:

$$\sigma_{\mathrm{n}} = \frac{1}{\sqrt{N \cdot C_{\mathrm{L}}}} \frac{2I_{\mathrm{SS}}\sqrt{\alpha kT}}{V_{\mathrm{DD}} g_{\mathrm{m}}} \tag{20}$$

Because the EPC comparator has a longer comparison time, it performs noise averaging, for which reason the EPC should have a smaller input-referred noise than the ERC. However, as shown in the simulation result, the EPC has much larger noise than the ERC under 16 and 20 delay units. This comes from the coupling of parasitic capacitance, degrading the noise performance.

Table 1. Comparison of the four comparators.

	[4]	[5]	[6]	[7]
Process [nm]	65	180	40	40
Architecture	Dynamic Bias	Proposed FIA+SA	SAR (EPC)	SAR (ERC)
Energy [pJ]	0.034	0.98	0.2	0.1
Noise [μV]	400	46	65	74
Area [μm^2]	125	9800	315000	-
Comparison Time [ns]	-	-	63.1	19.9

6. Comparison result

Table 1 summarizes the measured performance of the four reviewed comparators, and they all have outstanding performance in a certain field. The comparator with dynamic floating inverter amplifier has the smallest input referred noise. The Dynamic Bias Latch-Type Comparator has the lowest energy consumption per comparison. These two different comparators are designed based on the conventional latch-type comparators. Due to the process of the FIA, the size of it is inevitably large. However, it resolves the relatively high noise lies in the dynamic bias comparator also stays in low level of energy consumption compared to the conventional comparators [9, 10]. Thus, they are all outstanding reformative designs. Besides, the idea of EPC is innovative, distinguishing from the traditional designs, it does comparisons by measuring the phase difference between two input edges. The performance of EPC is very impressive, it reaches both high energy efficiency and low input referred noise. ERC can be treated as the modification for the EPC comparators. It comprehensively improves the performance compared with the EPC especially for the comparison time. That is because it only needs two inverter delays during comparison thus greatly optimizing time efficiency. Also, as the development of manufacturing process, the overall performance in energy efficiency and input referred noise is getting better.

7. Conclusion

This paper presents four innovative designs dynamic-bias comparator, comparator with dynamic floating inverter amplifier, edge-pursuit comparator, and edge-race comparator. All these new designs improve the energy efficiency and input-referred noise performance. There exist trade-offs when designing, so these four innovative designs can meet various requirements. The modifications methodology of the dynamic bias and FIA comparators are based on gm/Id, which uses the voltage overdrive V_{OV} as the key parameter and it is strongly related to the performances of analog circuits, giving an indication of device operating region and tools for calculating the transistors dimensions. For the EPC and ERC comparators, they provide a brand dimension when designing comparators, measuring the phase difference between input edges and scale automatically to various input voltages. The new perspective of these kinds of innovations will inspire a lot for designers and is valuable when exploring new ways to optimize the devices' performance.

References

[1] M. Elzakker, E. vanTuijl, P. Geraedts, D. Schinkel, E. A. M. Klumperink, and B. Nauta, (2010) A 10-bit charge-redistribution ADC consuming 1.9uWat1MS/s, IEEE JSSC, vol. 45, no. 5, pp. 1007–1015.

[2] M. Liu, K. Pelzers, R. van Dommele, A. van Roermund, and P. Harpe, (2016) A 106nW 10b 80kS/s SAR ADC with duty-cycled reference generation 65 nm CMOS, IEEE JSSC, vol. 51, no. 10, pp. 2435–2445.

[3] P. Harpe, E. Cantatore, and A. V. Roermund, (2013) A 2.2/2.7fJ/conversionstep 10/12b 40kS/s SAR ADC with data-driven noise reduction, IEEE Int. Solid-State Circuits Conf. (ISSCC) Dig. Tech. Papers, San Francisco, CA, USA, pp. 270–271.

[4] H. S. Bindra, C. E. Lokin, D. Schinkel, A. Annema and B. Nauta, (2018) A 1.2-V Dynamic Bias Latch-Type Comparator in 65-nm CMOS With 0.4-mV Input Noise, in IEEE Journal of Solid-State Circuits, vol. 53, no. 7, pp. 1902-1912.

[5] X. Tang et al., (2020) An Energy-Efficient Comparator with Dynamic Floating Inverter Amplifier, in IEEE Journal of Solid-State Circuits, vol. 55, no. 4, pp. 1011-1022.

[6] Shim, M., Jeong, S., Myers, P., Bang, S., Shen, J., Kim, C., Sylvester, D., Blaauw, D. and Jung, W., (2017) Edge-Pursuit Comparator: An Energy-Scalable Oscillator Collapse-Based Comparator with Application in a 74.1 dB SNDR and 20 kS/s 15 b SAR ADC. IEEE Journal of Solid-State Circuits, 52(4), pp.1077-1090.

[7] H. Zhuang, C. Tong, X. Peng, and H. Tang, (2020) Low-Power, Low-Noise Edge-Race Comparator for SAR ADCs, in IEEE Transactions on Very Large-Scale Integration (VLSI) Systems, vol. 28, no. 12, pp. 2699-2707.

[8] Yun-Ti Wang and B. Razavi, (2000) An 8-bit 150-MHz CMOS A/D converter, in IEEE Journal of Solid-State Circuits, vol. 35, no. 3, pp. 308-317.

[9] B. Razavi, (2015) The StrongARM Latch [A Circuit for All Seasons], in IEEE Solid-State Circuits Magazine, vol. 7, no. 2, pp. 12-17.

[10] D. Schinkel, E. Mensink, E. Klumperink, E. van Tuijl, and B. Nauta, (2007) A double-tail latch-type voltage sense amplifier with 18ps setup hold time, in IEEE Int. Solid-State Circuits Conf. Dig. Tech. Papers, San Francisco, CA, USA, pp. 314–605.

A Study on Design and Optimization methods in Latch Type Comparator

Ziyan Chen[1, *, †] and Zun Yang[2, †]

[1]Department of Electrical and Computer Engineering, University of Illinois, Urbana-Champaign, Illinois, 61820, United States

[2]Department of Electrical and Computer Engineering, Ohio State University, Ohio, 43210, United States

*Email: ziyanc3@illinois.edu, yang.4998@osu.edu

[†]These authors contributed equally.

Abstract The current state-of-the-art communication systems require analog-to-digital converters (ADCs) to sample the analog signals in digital representation at high speed and low power consumption with very high accuracy. Performance demand for ADCs directly promote the advancement of their basic component – the comparator. The traditional Strong-Arm dynamic comparator has been widely accepted for its fast decision, but it is limited by its significant voltage headroom and kickback noise. The double-tail latched dynamic comparator mitigates the aforementioned problems by allowing separate stages for the pre-amplifier and the latch so that it can have near- operation. However, energy consumption and propagation delay are still problems to be solved. In this article, we present a review of the recent improvements for the dynamic comparator architecture. The advantages and disadvantages of these techniques are also illustrated with simulation and measurement results.

1. Introduction

Comparators play an essential role in the design of Analog-to-digital converters (ADC). The first, and most important, is the comparator working at a high operation speed. However, circuit performance like power consumption, input offset voltage, and propagation delay are significant factors that need to be considered in a comparator design. In addition, problems like channel doping and current leakage should not be overlooked either. This review focus on the design method from the recent dynamic CMOS comparator designs. The compression result between the simulation and measurement results serves as support points to verify the conclusion for each design.

In recent years, various comparators have been proposed to improve energy consumption, reduce input-referred noise, and enhance comparison speed. In 2018, H. S. Bindra designed a latch-type comparator with a dynamic bias pre-amplifier in a 65-nm CMOS process [1]. The comparator stops the discharge of capacitors to reduce the power consumption by adding one capacitor in the tail of the pre-amplifier. The stimulation result has shown a comparison between the proposed comparator and Elzakker's comparator to prove the dynamic bias can lower the average power consumption. Then, in 2019, a novel design technique for the dynamic comparator was proposed by Y. Wang [2]. Due to the rising demand for high-speed dynamic comparators in high-speed analog-to-digital converters and digital I/O circuits, the proposed comparator enhances the effective total transconductance at the start of the comparison stage. By replacing the cross-coupled inverters with separated gate-biasing cross-

Content from this work may be used under the terms of the Creative Commons Attribution 3.0 licence. Any further distribution of this work must maintain attribution to the author(s) and the title of the work, journal citation and DOI.
Published under licence by IOP Publishing Ltd

coupled transistors, the proposed comparator has a faster comparison speed and lower energy consumption compared to the traditional dynamic comparator. Moreover, X. Tang presented an energy-efficient comparator with a novel dynamic pre-amplifier in 2019 [3]. It reduces noise and improves the comparator performance by using an inverter-based input pair. The stimulating result shows the efficiency of the proposed comparator is seven times larger than the conventional Strong-Arm latch. Furthermore, S. Chevella presents a low voltage supply dynamic comparator in a 65 nm CMOS against a double tail latch comparator in 2020 [4]. The proposed comparator drops the power consumption by adding cross-coupled devices while keeping a similar noise level.

This paper is organized as follows. Section II reviews a dynamic floating pre-amplifier-based energy-efficient comparator by X. Tang. A low power supply dynamic comparator proposed by S. Chevella is discussed in Section III. Section IV studies a dynamic bias latch-type comparator presented by H. S Bindra. Furthermore, Section V studies a high-speed comparator applies in the analog-to-digital converters proposed by Y. Wang. Section VI concludes comparative results of all four reviewed comparators.

2. Dynamic floating inverter amplifier comparator

Xiyuan Tang et al presented an energy-efficient comparator with a floating inverter amplifier (FIA)-based pre-amplifier in 2020. The floating reservoir capacitor powered an inverter-based input pair in a pre-amplifier. The amplifier implements current reuse and dynamic bias, thus outstandingly increasing the gm/ID and reducing noise simultaneously. In addition, the effects of input common-mode voltage and process corner on the comparator performance, including delay, noise and offset have also been reduced by the pre-amplifier simultaneously.

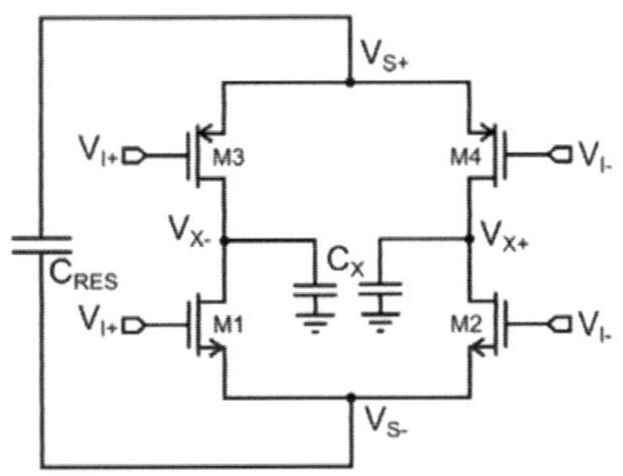

Figure 1. The floating reservoir capacitor in the proposed comparator

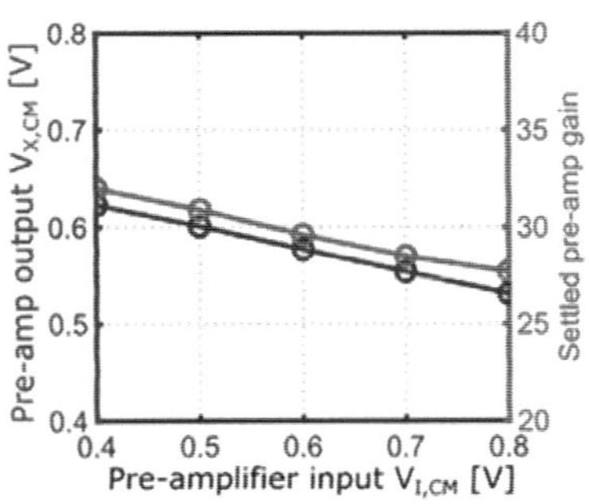

Figure 2. Stimulating result of output common-mode voltage and pre-amplifier input voltage

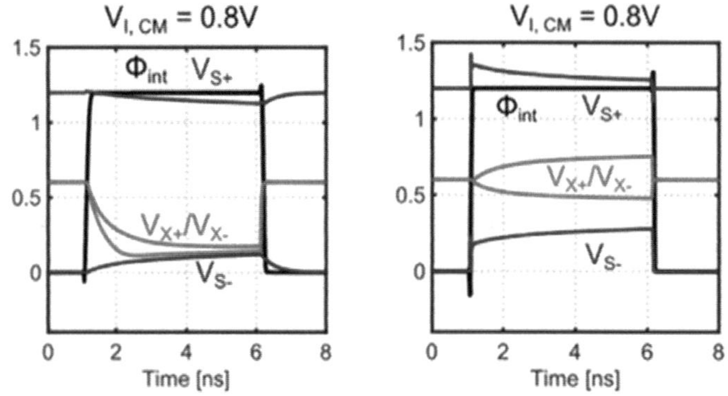

(a) (b)

Figure 3. The common-mode voltage variation vs. time in conventional (a) FIA (b) structure

An energy-efficient dynamic comparator with a floating inverter amplifier (FIA)-based pre-amplifier has been selected to show the efficiency improvement compared to the Strong-Arm latch, as shown in Fig. 1. Current reuse has been achieved by using the inverter-based input stage in the FIA [8]. Powering the FIA with a floating reservoir capacitor furnishes an isolated power domain, thereby making the amplification unrelated to the input common-mode voltage. The output common-mode voltage in the FIA stays constant, as shown in Fig.2. The pre-amplifier output voltage drops from 0.62 to 0.53 when the pre-amplifier input voltage increases from 400 mV to 800 mV. The output voltage nearly has no change while the input voltage doubled. Fig. 3 (a) and (b) compares the input common-mode voltage variation vs. time for conventional and FIA structures. The voltage for the conventional one drops as time increases. However, the voltage for the FIA stays unchanging along with time increment. As a result, the constant output common-mode voltage increases the gain and extends the integration time.

Furthermore, the dynamic source degeneration provided by the reservoir capacitor enhances the gm/ID and avoids the integration capacitors getting fully discharged. Equation (1) shows the input-referred noise of the FIA can be calculated using voltage gain of the pre-amplifier. In contrast, equation (2) shows the calculation of input-referred noise in the conventional structure. The noise decreases by increasing the CRES and reducing the ID/gm.

$$\sigma_{in,FIA}^2(T_{INT}) = \frac{2nkT}{C_{RES} \cdot \Delta V_S(T_{INT})} \cdot \frac{I_D}{G_m} v \quad (1)$$

$$\sigma_{in,SA}^2(T_{INT}) = \frac{2nkT}{C_{THN} \cdot C_X} \cdot \frac{I_D}{g_m} \quad (2)$$

The Gm/ID of the FIA is twice larger than the one in the SA latch due to the current reuse at the start of the dynamic integration phase. During the integration phase, the Vs- increases while Vs+ decreases, thus decreasing the overdrive voltage and increasing the transconductance of both nMOS and pMOS input pairs. The average Gm/ID of the FIA increases due to the integration process, leading to an improvement in power efficiency.

Comparison of CLK-Q Delay between Strong-Arm and proposed FIA is shown in Fig. 4. The delay of FIA is much smaller compared to the conventional structure at the low input common mode. Under a low input common mode, the current reduces and increases the delay in the SA latch. In contrast, the delay is about four times smaller for the proposed FIA compared to the Strong-Arm at the same input voltage.

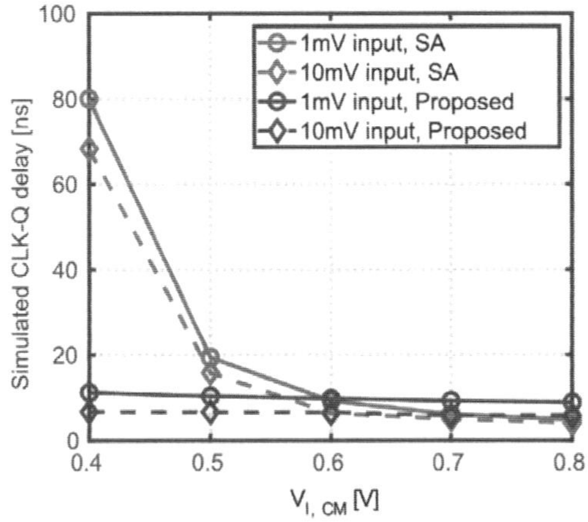

Figure 4. CLK-Q delay vs. input common-mode voltage

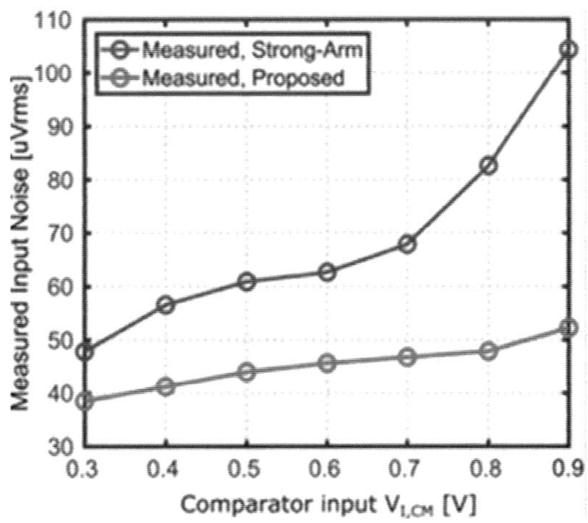

Figure 5. Stimulating result of measured input-referred noise vs. comparator input voltage

The integration of the differential charge happens only on the loading capacitor while the common-mode voltage keeps steady. It eliminates the bounded common-mode drop limitation for the pre-amplifier gain to prevent Cx's full discharge. The FIA structure merges two-tail capacitors into one floating reservoir capacitor CRES. It lowers the total capacitance size and provides an isolated voltage domain for the pre-amplifier. Moreover, the zero-output common-mode change has been guaranteed by the isolated voltage domain. The input common-mode-insensitive operation from the FIA drops the noise variation by approximately twice compared to the conventional structure at high comparator input voltage, as shown in Fig. 5.

The proposed FIA improves the energy efficiency by realizing current reuse, enhancing Gm/ID, and preventing full discharge of the integration capacitor during the integration process. Overall, the proposed design achieves the best energy efficiency compared to prior designs.

Based on the stimulating result shown in Fig. 6 and Fig. 7, CRES changes affect the delay and theoretical energy efficiency (FoM) in an FIA. It is necessary to separate each graph in half since the value of CRES may be slightly different from the stimulation process due to aged and overheating issues in the actual world. A small variation of delay and FoM change simultaneously is preferred. Thus, the range of CRES is between 2 to 3.5 pF. In this article, the author chooses 2 pF to balance the trade-off between area consumption, comparison speed, and energy efficiency.

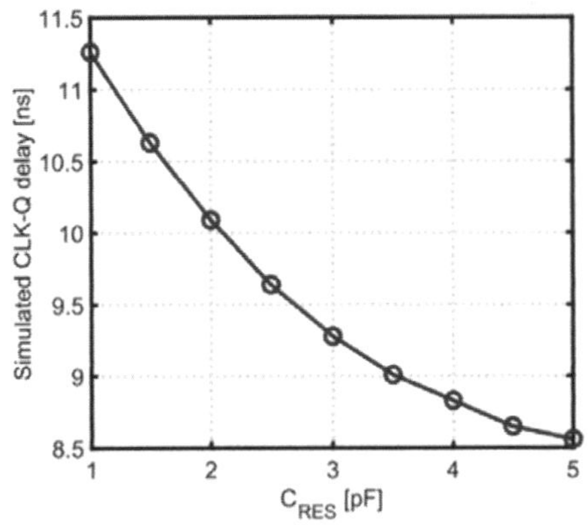

Figure 6. Measured CLK-Q delay vs. CRES of proposed comparator

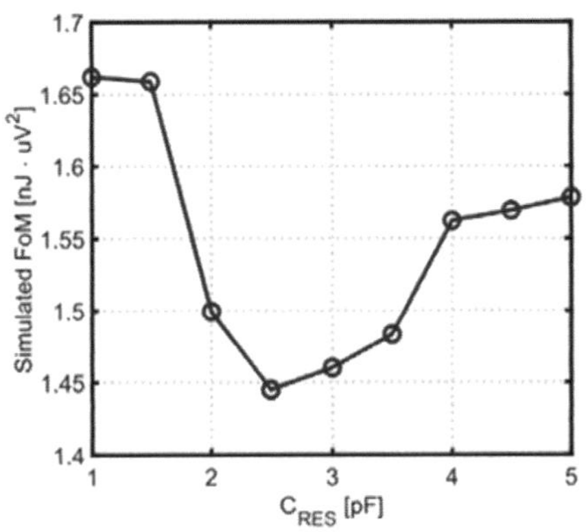

Figure 7. Stimulated FoM vs. CRES of proposed comparator

3. Low power dynamic comparator

Dynamic comparators play an important role in mixed-signal conditioning circuits by amplifying the input differential signal and comparing the latch. Due to the extensive use of battery-powered applications, the low-power signal processing circuits have become indispensable. As a result, dynamic comparators become the first choice in these application designs. Subhash Chevella et al. present a low-power dynamic comparator for ultralow-power applications in 2020. The designed comparator based on a 65-nm CMOS process was compared to a double tail latch comparator for power efficiency and input-referred rms noise.

The circuit topology of a double tail latch comparator is shown in Fig. 8. In each comparison process, the capacitors (Cp) first discharge to the ground then recharge to the supply voltage (VDD). To trigger the latch and turn on M11 and M12, Vlatch needs to stay low. As a result of this, the node voltage, VinN and VinP will cost more energy for raising back to VDD. Moreover, the raising process does not affect noise or conversion time reduction. This process can prevent energy waste by stopping the discharge of these nodes once the latch triggers. The node voltage drops to 0 and recharges back to the VDD in a double tail latch comparator.

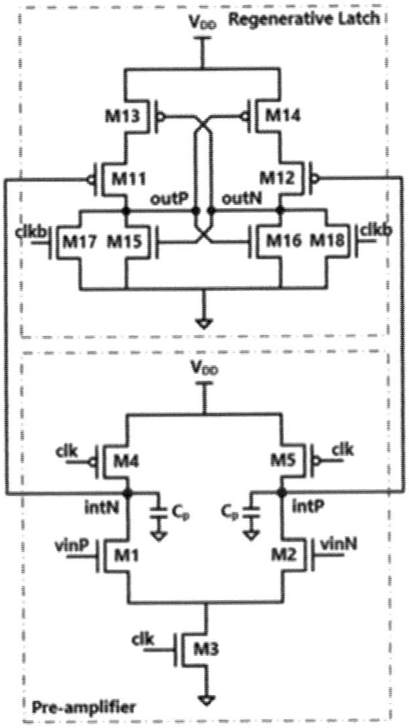

Figure 8. Proposed comparator circuit layout

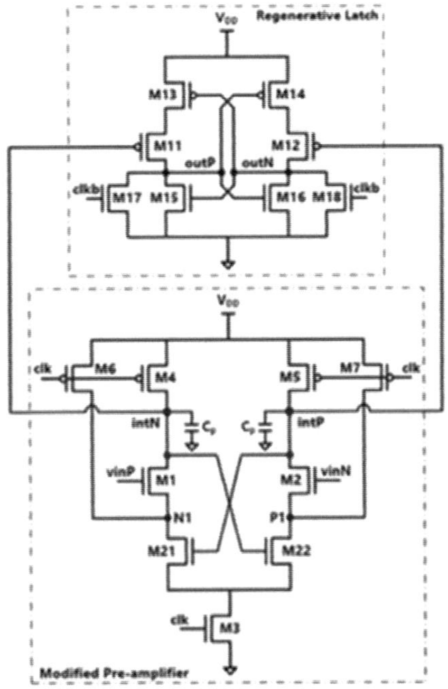

Figure 9. Double tail latch comparator circuit layout

In contrast, the circuit topology of the proposed comparator is shown in Fig. 9 [7]. The proposed comparator saves energy from stopping the discharge and recharge procedure by adding two cross-coupled devices, M21 and M22, and a regenerative latch to a double tail latch comparator. Transistors M1 and M21 act like a cascode current source since M22 is turned off. At the same time, the gate potential of M21 is fixed while the node intP stays steady. As a result of this, the node intN discharge at a lower rate. The voltage drops from 1 to 0.5 in 20 ns. The reduced rate for the proposed comparator is twice compared to rate of the double tail latch comparator.

The stimulating result of the relationship between the differential input voltage and comparator delay for the proposed comparator and a double tail latch comparator is shown in Fig. 10 (a). The proposed comparator has a higher delay compared to the other one at small differential input voltage. This was due to the additional time spent in the weak-inversion region. The less effective Vgs and reduction of starting gm value could also be a reason.

As Fig. 10 (b) shows the power consumption at differential input voltage, the proposed comparator has a smaller value than the double tail latch comparator. Since the proposed comparator will be widely used in low-power signal processing circuits, the differential input voltage region should focus from 0 to 10 mV. The proposed comparator has about a 30% reduction, resulting in lower power consumption than the other.

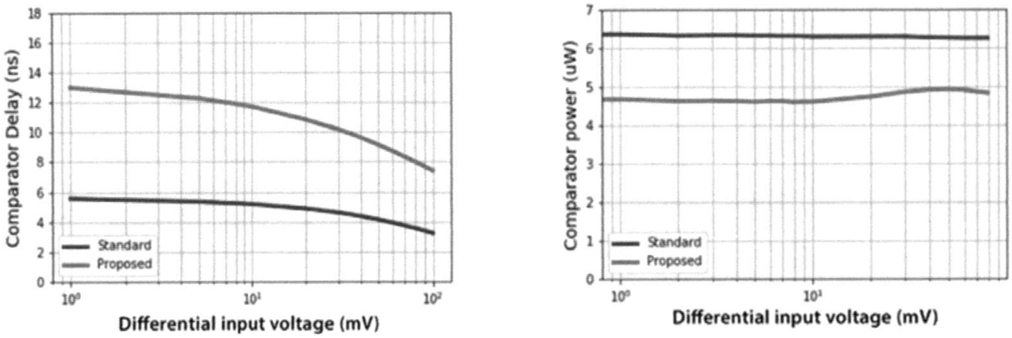

(a) (b)

Figure 10. Measured comparator delay(a) and power(b) as a function of differential input voltage

4. Dynamic bias latch-type comparator

H. S. Bindra, et al. (2018) analyzed the performance and energy consumption of the double-tail latch comparator and the Elzakker's comparator, and they proposed a modification to Elzakker's comparator by adding a capacitor in the tail of pre-amplifier to reduce the pre-charge energy consumption. They demonstrate that the dynamic bias comparator can be a simple, low overhead solution to reduce the energy consumption of the pre-amplifier.

Double-tail latch-type comparator [5], along with its variant type - the Elzakker's comparator [6], is suited for low-voltage operation since it achieves near-VDD operation by allowing greater input common-mode voltage range. The separate tail transistors also enable independent control of the common-mode current and the regeneration time. However, the sizing of charge package capacitance is a main problem for the double-tail comparators to design the input referred noise and the energy consumption, of which the pre-amplifier consumes up to 80%. The Elzakker's comparator has the same pre-amplifier architecture, but the output is fed to the P-transistors in the latch state instead of the N-transistors. The Elzakker's comparator can reduce the energy consumption by partially charging the pre-amplifier nodes to half supply and hence discharging the output nodes of the pre-amplifier partially. However, the energy-saving expectation is only limited to 33% instead of supposed 50% due to logic overhead. The proposed dynamic bias comparator reduces the amount of discharge from the Di+/Di- nodes with less overhead using a capacitor in the tail of the pre-amplifier. The tail transistor M3a is sized to ensure the differential pair to be biased near weak inversion, which further allows M1 and M2 to operate near or in the weak inversion region at the beginning of the comparator operation. Such a design can obtain a minimum noise at a given current. However, the dynamic bias comparator has a higher CLK-Q delay when small differential input voltages is applied.

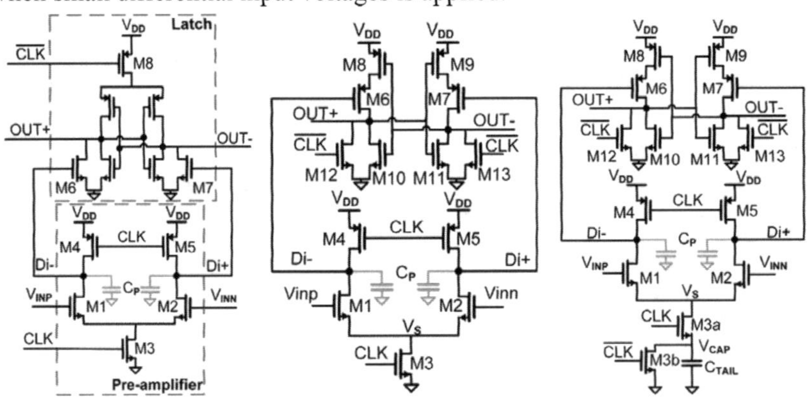

Figure 11. (a) Double-tail latch-type comparator (b) the Elzakker's comparator (c) proposed dynamic bias comparator

The energy consumption for the pre-amplifier in pre-charging in the Elzakker's comparator and the dynamic bias comparator is given below:

$$P_{Elzakker} = 2 \cdot C_p \cdot V_{DD}^2 \quad (3)$$

$$P_{dynamic\ bias} = 2 \cdot C_p \cdot V_{DD}^2 - C_p \cdot V_{DD} \cdot (V_{Di+} + V_{Di-}) \quad (4)$$

A comparison of voltage gain between constant tail current pre-amplifier and the proposed dynamic bias pre-amplifier is shown below:

$$A_V(T_{INT})_{ctc} = \frac{\Delta V_{Di}(T_{INT})}{\Delta V_{IN}(T_{INT})} = \frac{g_m \Delta V_{Di,CM}(T_{INT})}{I_{CM}} \quad (5)$$

$$A_V(T_{INT})_{dynamic\ bias} = \frac{C_{tail}}{2n \cdot C_p} \cdot \frac{V_s(T_{INT})}{kT/q} \quad (6)$$

The thermal noise of the Elzakker's comparator and the proposed dynamic bias comparator are:

$$E\left[V_{n,SI,in}^2(T_{INT})\right]_{Elzakker} = \frac{4kT\gamma}{C_p \cdot \Delta V_{Di,CM}(T_{INT}) \cdot \left(\frac{g_m}{I_{CM}}\right)_{SI,T_{INT}}} \quad (7)$$

$$E\left[V_{n,WI,in}^2(T_{INT})\right]_{dynamic\ bias} = \frac{2nkT}{C_p \cdot \Delta V_{Di,CM}(T_{INT}) \cdot \left(\frac{g_m}{I_{CM}}\right)_{WI,T_{INT}}} \quad (8)$$

In the simulation, the Elzakker's comparator and the dynamic comparator are designed under 65-nm CMOS process. The results show the gm/Id between 20/V to 25/V for dynamic bias comparator compared to that at 16/V for the Elzakker's comparator. This means that the C_p of the dynamic bias comparator can be 40-50% smaller to achieve the same input referred noise, and thus less power consumption is expected.

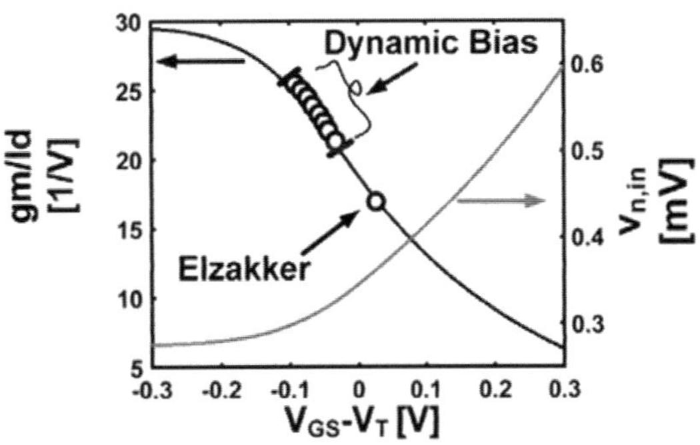

Figure 12. Gm/Id and input referred noise voltage vs. V_GS-V_T

The measurement results further confirm the simulation results. The energy consumption for a dynamic bias comparator is 34 fJ/comparison with V_CM at 0.6 V and 45 fJ/comparison with V_CM at 0.9 V. The corresponding results for the Elzakker's comparator are 88 fJ/comparison for V_CM at 0.6 V and around 95 fJ/comparison for V_CM at 0.9 V. With a given V_CM, The Elzakker's comparator consumes more than two times than the dynamic bias comparator does. However, the CLK-Q delay with the same V_CM of a dynamic bias comparator is higher as expected than that for the Elzakker comparator.

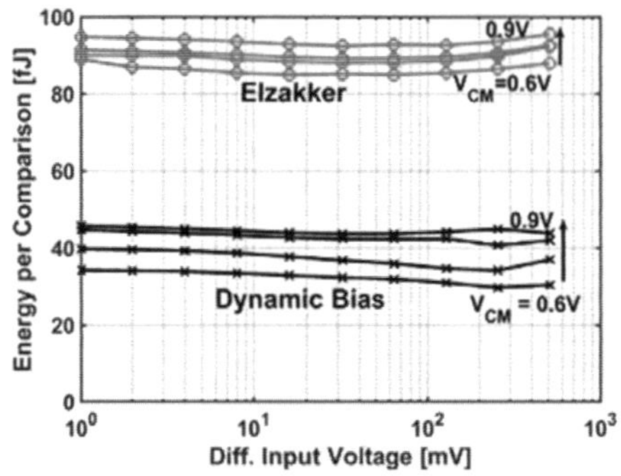

Figure 13. Energy consumption of the Elzakker's comparator and the proposed comparator

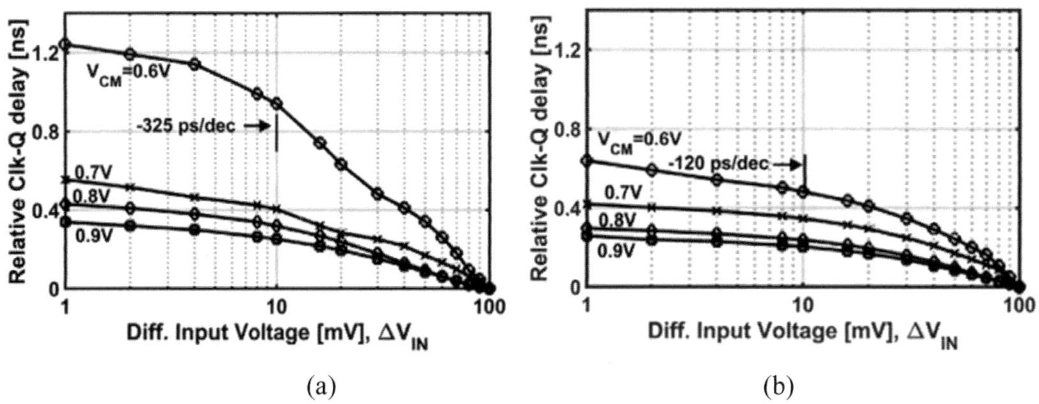

Figure 14. CLK-Q delay for the (a) dynamic bias comparator and (b) Elzakker's comparator

5. Separated gate-biasing dynamic comparator

Power consumptions and processing speed are the two most important metrics for a comparator in Analog-to-Digital Converters (ADC). Many designs in conventional latched comparators have limits in the maximum operating frequency and power consumption due to the degradation of the total effective transconductance of the latching stage. Y. Wang, et al. (2019) proposed a novel design technique for the dynamic comparator by employing separated gate-biasing cross-coupled transistors instead of the cross-coupled inverters in the traditional dynamic comparator design. The separated gate-biasing enhances the effective total transconductance at the early phase of the latching stage.

In the conventional double-tail latched comparator, the pre-amplifier stage is used to reduce the input offset voltage, while the latch stage is designed to operate at a higher current that increases the speed. It has advantages in lower kickback noise, low-voltage operation, and a flexible trade-off between speed and offset. However, as mentioned in [2], the power consumption of the conventional dynamic comparator is still a problem. Several works [9][10] on modified two-stage comparators or creative latching stage are reported, but many suffer from problems including higher kickback noise, larger device area, and higher delay.

The delay of the conventional double-tail latch-type comparator is calculated by the time that the output difference reaches half V_DD. In the formula for the total delay, a clear relationship between delay and effective transconductance is shown.

$$t_0 = \frac{V_{THN}C_{OUT}}{I_p} \approx \frac{V_{THN}C_{OUT} \cdot 2\mu_p C_{OX} \cdot \dfrac{W_p}{L_p}}{g_{m,eff}^2} \tag{9}$$

$$t_{latch} = \frac{C_{OUT}}{g_{m,eff}} \cdot \ln \frac{V_{DD}^2}{\Delta V_0} \tag{10}$$

$$t_{total} = \frac{V_{THN} C_{OUT} \cdot 2\mu_p C_{OX} \cdot W_p / L_p}{g_{m,eff}^2} + \frac{C_{OUT}}{g_{m,eff}} \cdot \ln \frac{V_{DD}^2}{\Delta V_0} \tag{11}$$

$$g_{m,eff} = g_{mn} + g_{mp} \approx g_{mp} \tag{12}$$

where V_{THN} is the threshold voltage of transistor M2 and M3. C_{OUT} is the load capacitance at the output node, and C_{OX} is the gate oxide capacitance. μp, W_p, and L_p are the hole mobility channel width and length of PMOS transistor M1. $g_{mp}{}^2$ and $g_{mp,eff}{}^2$ are the transconductance and effective transconductance of M1. g_{mn} can be neglected since NMOS M3 is off when measuring the delay. ΔV_0 is the initial voltage difference between the two output nodes.

The proposed dynamic comparator with transconductance-enhanced latching state includes cross-coupled transistors that are biased in the strong inversion region during the reset stage. This will improve the total effective transconductance at the beginning of the comparison and thus reduce the delay and energy consumption. The delay and transconductance of the proposed comparator are shown below. The effective total transconductance of the proposed latch stage increases much faster than the conventional cross-coupled latch at the early stage of the comparison therefore reducing the delay.

$$\boldsymbol{t_{total} = t_{opm} + t_0 + t_{delay}} \tag{13}$$

$$g_{m,eff} = g_{mn} + g_{mp} = \mu_n C_{OX} \left(\frac{W}{L}\right)_n V_{dsn} + \mu_p C_{OX} \left(\frac{W}{L}\right)_p |V_{dsp}| \tag{14}$$

where t_{opm} is the delay between CLK and CLK1. V_{dsn} and V_{dsp} are the drain-to-source voltages of the NMOS and PMOS transistors.

The simulation results shows that the proposed comparator's effective transconductance increases faster and reaches its peak about three times than the conventional comparator. Fig. 15 (b) illustrates that increasing width may help reduce delay until around 1μm. In Fig. 16, circuit is simulated under 2GHz operating frequency and 0.1V input differential voltage. With input common-mode voltage ranges from 0.2-0.7V, the circuit achieves a minimum delay of around 210ps and very low power consumption at around 225 μW.

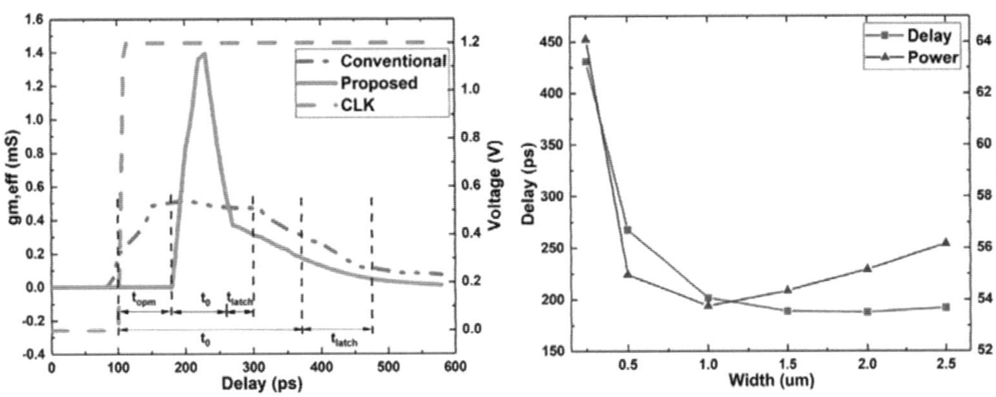

(a) (b)

Figure 15. Simulation results of (a) $g_{m,eff}$ of the conventional and proposed comparator related to delay (b) delay and power consumption of proposed comparator vs. width of M2 & M3

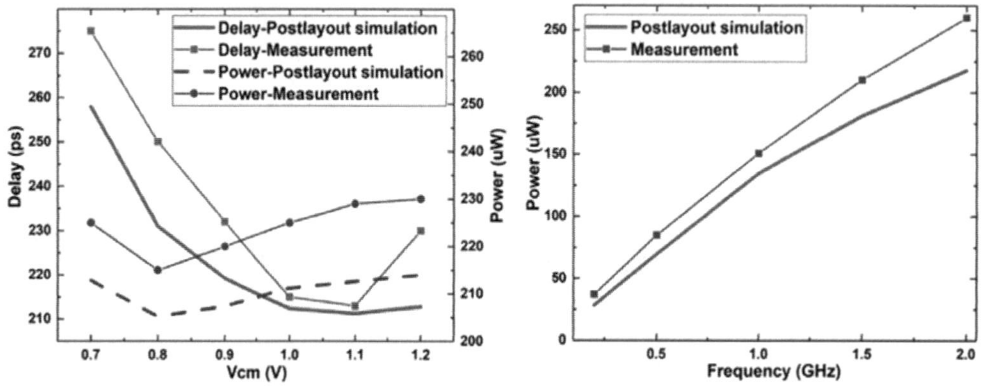

Figure 16. Comparison between post layout simulation and measurements

6. Conclusion

In this paper, we presented a review on current dynamic comparators. We compared and analyzed conventional comparators with several novel designs that improve the metrics, including power consumption, propagation delay, thermal noise, effective transconductance, and input offset voltage. The novel energy-efficient comparator with dynamic floating inverter amplifier demonstrates enhancement in transconductance, energy efficiency, input offset variation and noise reduction. A partially discharging type dynamic comparator using cross-coupled technique shows a good reduction in power consumption. The dynamic bias comparator that modified the pre-amplifier stage shows improvements in reducing energy consumption but also increases the delay due to longer operation time in the weak inversion regime. The dynamic comparator using separated gate-biasing cross-coupled transistors illustrate enhanced transconductance and shorter delay with even smaller size. Even though the improvements in these designs all come with newer tradeoffs, it is still promising for these devices in the current industry since some specialized requirements for certain parameters are preferred. Further research is needed to better tune the existing design and may achieve better performance for these devices.

References

[1] H. S. Bindra, C. E. Lokin, D. Schinkel, A. Annema and B. Nauta, "A 1.2-V Dynamic Bias Latch-Type Comparator in 65-nm CMOS With 0.4-mV Input Noise," in IEEE Journal of Solid-State Circuits, vol. 53, no. 7, pp. 1902-1912, July 2018.

[2] Y. Wang, M. Yao, B. Guo, Z. Wu, W. Fan and J. J. Liou, "A Low-Power High-Speed Dynamic Comparator With a Transconductance-Enhanced Latching Stage," in IEEE Access, vol. 7, pp. 93396-93403, 2019.

[3] X. Tang, B. Kasap, L. Shen, X. Yang, W. Shi and N. Sun, "An Energy-Efficient Comparator with Dynamic Floating Inverter Pre-Amplifier," 2019 Symposium on VLSI Circuits, 2019, pp. C140-C141, doi: 10.23919/VLSIC.2019.8777942.K. Elissa, "Title of paper if known," unpublished.

[4] S. Chevella, D. O'Hare and I. O'Connell, "A Low-Power 1-V Supply Dynamic Comparator," in IEEE Solid-State Circuits Letters, vol. 3, pp. 154-157, 2020.

[5] D. Schinkel, E. Mensink, E. Klumperink, E. van Tuijl and B. Nauta, "A Double-Tail Latch-Type Voltage Sense Amplifier with 18ps Setup+Hold Time," 2007 IEEE International Solid-State Circuits Conference. Digest of Technical Papers, 2007, pp. 314-605.

[6] M. van Elzakker, E. van Tuijl, P. Geraedts, D. Schinkel, E. A. M. Klumperink and B. Nauta, "A 10-bit Charge-Redistribution ADC Consuming 1.9 μW at 1 MS/s," in IEEE Journal of Solid-State Circuits, vol. 45, no. 5, pp. 1007-1015, May 2010.

[7] M. Elzakker, E. vanTuijl, P. Geraedts, D. Schinkel, E. A. M. Klumperink and B. Nauta, " A 10-bit charge-redistribution ADC consuming 1.9 μ W 1MS/s ", IEEE J. Solid-State Circuits, vol. 45, no. 5, pp. 1007-1015, May 2010.
[8] M. Liu, K. Pelzers, R. van Dommele, A. van Roermund and P. Harpe, "A 106 nW 10 b 80 kS/s SAR ADC with duty-cycled reference generation in 65 nm CMOS", IEEE J. Solid-State Circuits, vol. 51, no. 10, pp. 2435-2445, Oct. 2016.
[9] A. Khorami and M. Sharifkhani, "A Low-Power High-Speed Comparator for Precise Applications," in IEEE Transactions on Very Large Scale Integration (VLSI) Systems, vol. 26, no. 10, pp. 2038-2049, Oct. 2018, doi: 10.1109/TVLSI.2018.2833037.
[10] J. Gao, G. Li, and Q. Li, ‘‘High-speed low-power common-mode insensitive dynamic comparator,’’ Electron. Lett., vol. 51, no. 2, pp. 134–136, Jan. 2015.

Theoretical study of the discharge decomposition mechanism of environment-friendly insulation gas HFO-1234ze(E) in the presence of trace water

Hui Gan [1,a], Feng Wang [1,b], Lipeng Zhong [1,c*], Jie Liu [1,d], Bin Hai [1,e], Nian Tang [2,3,f], Li Li [2,3,g]

[1]College of Electrical and Information Engineering, Hunan University, Changsha, Hunan, China

[2]Key Laboratory of Sulfur Hexafluoride of China Southern Power Grid Co., Ltd., Guangzhou, Guangdong, China

[3]Electric Power Research Institute of Guangdong Power Grid Co., Ltd., Guangzhou, Guangdong, China

[a]email: ganhuisx@163.com, [b]email: fengwang@hnu.edu.cn, [d]email: liujie18@hnu.edu.cn, [e]email: haibin@hnu.edu.cn, [f]email: tangnian_2005@163.com, [g]mail: liligz@163.com

*corresponding author: [c]email: zhonglipeng@hnu.edu.cn

Abstract: HFO-1234ze(E) has excellent insulating and environmental properties, but its discharge decomposition mechanism in the presence of trace water is still unclear. Based on the density functional theory (DFT), the discharge decomposition characteristics of HFO-1234ze(E) in the presence of trace water were analyzed at the B3LYP/6-311+G(d,p) level. The results show that in the presence of trace water, HFO-1234ze(E) will decompose and generate H_2, HF, CH≡CF, CH≡CH, CF_3OH, CF_3H, CF_4, C_2F_6, CF_2O, CH_2=CHF and CHF_2CH_3. By analyzing the ionization parameters of the main decomposition products, it can be found that the dielectric strengths of most of the products were lower than that of HFO-1234ze(E). Among the products, CF_3H, CF_4 and C_2F_6 are hazardous to the environment, while CH_2=CHF, CHF_2CH_3 and CF_2O are flammable or toxic, which may endanger the safety of staff. Therefore, water should be avoided when HFO-1234ze(E) is used as the insulating medium.

1. Introduction

Sulfur hexafluoride (SF_6) has excellent insulation and arc extinguishing performance, so it has been widely used in the power system [1]. However, SF_6 has a global warming potential (GWP) of 23,500 times than that of CO_2 and an atmospheric lifetime of 3,200 years, which makes it have a strong greenhouse effect [2]. According to statistics, SF_6 emissions mainly originate from the electrical equipment sector [3], so it is urgent to find ways to limit the use of SF_6 in power equipment. In order to reduce SF_6 emissions as much as possible, researchers have devoted themselves to finding new insulating gases, and have successively reported several kinds of gases with low GWP and high dielectric strength, such as CF_3I, $C_5F_{10}O$ and C_4F_7N.

Recently, the environment-friendly insulating gas trans-1,3,3,3-tetrafluoropropene (HFO-1234ze(E), chemical formula: $C_3H_2F_4$) has received a lot of attention. As a non-toxic, low

Content from this work may be used under the terms of the Creative Commons Attribution 3.0 licence. Any further distribution of this work must maintain attribution to the author(s) and the title of the work, journal citation and DOI.
Published under licence by IOP Publishing Ltd

flammability refrigerant, HFO-1234ze(E) with a quite low GWP of 6 has a great potential to replace SF_6 [4][5]. Many studies have proved that HFO-1234ze(E) has good dielectric strength [6][7]. As for the discharge decomposition characteristics, although the decomposition characteristics of HFO-1234ze(E) have been studied extensively, few studies have focused on its discharge decomposition characteristics in the presence of trace water. In gas-insulated power equipment such as gas-insulated transmission line (GIL) and gas-insulated switchgear (GIS), the insulation gas may be mixed with a small amount of water, which may cause HFO-1234ze(E) to decompose and produce products with poor insulation characteristics, thus reducing the dielectric strength of the original gas. In this paper, the discharge decomposition products of HFO-1234ze(E) in the presence of trace water were studied based on the DFT, and the insulation performance and safety properties of the discharge decomposition products were discussed. The calculation results will provide a basis for the application of HFO-1234ze(E) in power equipment.

2. Computational methods

DFT has been widely used in the field of physical and chemical calculation. Its core idea is to replace the many-body electronic wave function with the electronic density [8], which has been widely used to study the decomposition of compounds. The calculation method is B3LYP/6-311+G(d,p). B3LYP which has the advantages of high calculation accuracy and low calculation cost and is especially suitable for energy calculation of systems composed of C and F [9]. Figure 1 shows the optimized structure of HFO-1234ze(E) at the B3LYP/6-311+G(d,p) level. For the sake of description, the three C atoms from left to right are named C(α), C(β) and C(γ), respectively.

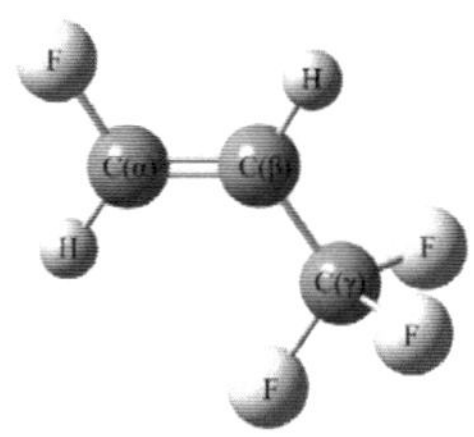

Figure 1. The optimized structure of HFO-1234ze(E)

In this paper, the structures of all reactants and products were optimized, and their exact energies were calculated. The intrinsic reaction coordinate (IRC) of each transition state structure was calculated to ensure that all the transition states were located at the saddle point of the potential energy surface and the two ends of IRC were connected with the reactants and products respectively. In addition, the frequencies of the transition states, reactants and products were calculated to ensure that all the transition states had only one imaginary frequency while the reactants and products had no imaginary frequency. All calculations were carried out in the Gaussian09 program package at a temperature of 298.15 K and a pressure of 0.1 MPa. In addition, the ionization energy (IE), electron affinity (EA) and molecular orbital energy gap (E_G) of the main decomposition products were calculated to evaluate the dielectric strength of the HFO-1234ze(E)/H_2O system.

3. Results and discussion

3.1. Dissociation of HFO-1234ze(E)

Table 1 presents the energy changes of the dissociation of HFO-1234ze(E). The dissociation of HFO-1234ze(E) are barrierless reactions without transition states, and the order of enthalpies of the reactions R1-R6 is R5 < R4 < R1 < R2< R3 < R6. It can be seen that CFH and $CHCF_3$ radicals are the most difficult to generate, which proves that C(α)=C(β) bond is the most stable in HFO-1234ze(E).

Table 1 Enthalpy of the main dissociation pathways of HFO-1234ze(E)

number	chemical equation	Enthalpy (kcal/mol)
R1	$CFH{=}CHCF_3 \rightarrow CF{=}CHCF_3\bullet + H\bullet$	110.21
R2	$CFH{=}CHCF_3 \rightarrow CFH{=}CCF_3\bullet + H\bullet$	112.91
R3	$CFH{=}CHCF_3 \rightarrow CH{=}CHCF_3\bullet + F\bullet$	118.11
R4	$CFH{=}CHCF_3 \rightarrow CFH{=}CHCF_2\bullet + F\bullet$	105.21
R5	$CFH{=}CHCF_3 \rightarrow CFH{=}CH\bullet + CF_3\bullet$	102.52
R6	$CFH{=}CHCF_3 \rightarrow CFH\bullet + CHCF_3\bullet$	169.27

3.2. *Primary decomposition products of HFO-1234ze(E) in the presence of trace water*

3.2.1. *Reactions between HFO-1234ze(E) and H radical*

Table 2 details the energy barrier and enthalpy of reactions R7-R13. The optimized structures and key parameters of the transition states are shown in figure 2. In reactions R7 and R8, the H atoms connected to C(α) and C(β) in HFO-1234ze(E) combine with H radicals to generate H_2, $CF{=}CHCF_3$ and $CFH{=}CCF_3$ radicals via TS1 and TS2. The energy barriers of these two reactions are 13.10 and 13.77 kcal/mol, respectively. In reactions R9 and R10, H radicals attack the F atoms connected to C(α) and C(γ) in HFO-1234ze(E) through TS3 and TS4 to form HF, $CH{=}CHCF_3$ and $CFH{=}CHCF_2$ radicals. These two reactions need to cross 29.29 and 24.33 kcal/mol energy barriers and then release 13.85 and 26.75 kcal/mol, respectively. Obviously, reactions R9 and R10 are not as competitive as reactions R7 and R8.

3.2.2. *Reactions between HFO-1234ze(E) and OH radical*

In reactions R11 and R12, the H atoms connected to C(α) and C(β) combine with OH radicals to form $CF{=}CHCF_3$, $CFH{=}CCF_3$ and H_2O. The energy barriers of these two processes are 3.96 and 5.02 kcal/mol, respectively. When the OH radical attacks the C(γ)-F bond, F radical and $CFH{=}CHCF_2(OH)$ will be formed via TS7 with an energy barrier of 58.14 kcal/mol. Therefore, from an energy perspective, reaction R13 is unlikely to occur.

Table 2 Reactions between HFO-1234ze(E) and H and OH radicals

number	chemical equation	Imaginary frequency (cm^{-1})	Energy barrier (kcal/mol)	Enthalpy (kcal/mol)
R7	$CFH{=}CHCF_3 + H\bullet \rightarrow CF{=}CHCF_3\bullet + H_2$	TS1 (-906.67)	13.10	6.55
R8	$CFH{=}CHCF_3 + H\bullet \rightarrow CFH{=}CCF_3\bullet + H_2$	TS2 (-671.38)	13.77	9.25
R9	$CFH{=}CHCF_3 + H\bullet \rightarrow CH{=}CHCF_3\bullet + HF$	TS3 (-1525.56)	29.29	-13.85
R10	$CFH{=}CHCF_3 + H\bullet \rightarrow CFH{=}CHCF_2\bullet + HF$	TS4 (-1475.07)	24.33	-26.75
R11	$CFH{=}CHCF_3 + OH\bullet \rightarrow CF{=}CHCF_3\bullet + H_2O$	TS5 (-1562.68)	3.96	-3.45
R12	$CFH{=}CHCF_3 + OH\bullet \rightarrow CFH{=}CCF_3\bullet + H_2O$	TS6 (-1509.07)	5.02	-0.75
R13	$CFH{=}CHCF_3 + OH\bullet \rightarrow CFH{=}CHCF_2(OH) + F\bullet$	TS7 (-348.55)	58.14	16.02

3.3. *Consumption of primary decomposition products*

Table 3 shows the energy barriers and enthalpies of the consumption pathways of the primary decomposition products. The absorbed energies of reactions R14, R15, R16 and R17 are 42.38, 55.03, 52.32 and 29.10 kcal/mol, respectively. It can be inferred that $CH{\equiv}CF$ and $CH{\equiv}CH$ are the main products in the dissociation of $CF{=}CHCF_3$, $CFH{=}CCF_3$ and $CH{=}CHCF_3$ radicals.

Table 3 The consumption pathways of the primary particles

number	chemical equation	Imaginary frequency (cm^{-1})	Energy barrier (kcal/mol)	Enthalpy (kcal/mol)
R14	$CF{=}CHCF_3\bullet \rightarrow CH{\equiv}CF + CF_3\bullet$	-	-	42.38
R15	$CF{=}CHCF_3\bullet \rightarrow CF{\equiv}CCF_3 + H\bullet$	-	-	55.03
R16	$CFH{=}CCF_3\bullet \rightarrow CF{\equiv}CCF_3 + H\bullet$	-	-	52.32

R17	$CH{=}CHCF_3\bullet \rightarrow CH{\equiv}CH+CF_3\bullet$	-	-	29.10
R18	$CF_3\bullet+H\bullet \rightarrow CF_3H$	-	-	-102.43
R19	$CF_3\bullet+OH\bullet \rightarrow CF_3OH$	-	-	-104.12
R20	$CF_3\bullet+F\bullet \rightarrow CF_4$	-	-	-119.45
R21	$CF_3\bullet+CF_3\bullet \rightarrow C_2F_6$	-	-	-86.03
R22	$C_2F_6 \rightarrow C_2F_5\bullet+F\bullet$	-	-	114.56
R23	$C_2F_5\bullet+CF_3\bullet \rightarrow C_3F_8$	-	-	-80.23
R24	$CF_3OH \rightarrow CF_2O+HF$	TS8 (-1701.92)	40.57	1.79
R25	$CH{\equiv}CH+HF \rightarrow CH_2{=}CHF$	TS9 (-1817.67)	45.35	-23.97
R26	$CH_2{=}CHF+HF \rightarrow CHF_2CH_3$	TS10 (-1689.42)	40.65	-18.19
R27	$CH_2{=}CHF+HF \rightarrow CH_2FCH_2F$	TS11 (-1822.91)	51.60	-6.52

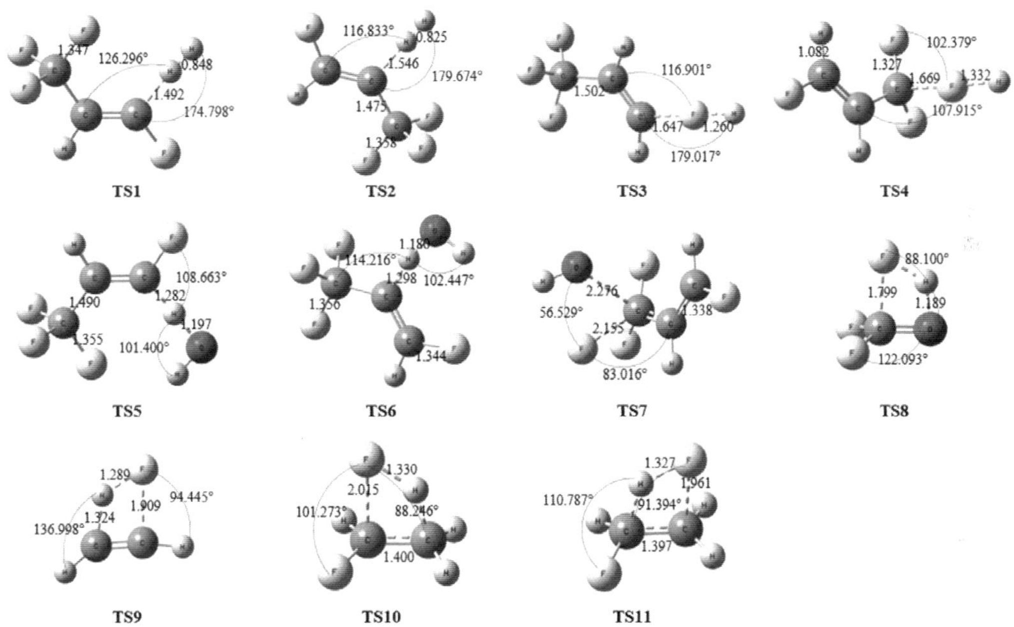

Figure 2 Optimized molecular structures of the transition states

Reactions R18-R27 are the consumption pathways of particles such as H, OH, F, CF_3, CH≡CH and HF. Reactions R18-R23 have no energy barrier, among which reactions R22 need to absorb 114.56 kcal/mol, so C_3F_8 is difficult to generate.

Reactions R24-R27 need to cross the energy barriers. To produce CF_2O and HF, the reaction R24 needs to overcome the energy barrier of 40.57 kcal/mol via TS8. CH≡CH and HF are important products, and the reaction between CH≡CH and HF to form CH_2=CHF requires an energy barrier of 45.35 kcal/mol. Reactions R26 and R27 give two possible pathways for the addition of HF to CH_2=CHF. Compared with the generation process of CHF_2CH_3, the energy barrier for the generation process of CH_2FCH_2F is higher, so the reaction R27 is difficult to occur. The optimized structures and key parameters of the transition states are shown in figure 2. In summary, CH≡CF, CH≡CH, CF_3OH, CF_3H, CF_4, C_2F_6, CF_2O, CH_2=CHF and CHF_2CH_3 are the main products of reactions R14-R27.

3.4. Dielectric strength, environmental characteristics and safety properties of the main products

Ionization energy reflects how difficult it is for a molecule to lose electrons. The higher the ionization energy is, the more difficult the molecule is to ionize in the electric field. Electron affinity reflects the ability of the molecule to absorb free electrons. The higher the electron affinity is, the stronger the ability of the molecule to adsorb free electrons is, and the higher the electronegativity of the molecule is. The formation of negative ions can inhibit the discharge. The molecular orbital energy gap value reflects the stability of the molecule. The larger the molecular orbital gap value is, the higher the chemical stability is [1]. The calculations of ionization energy (IE), electron affinity (EA) and molecular orbital energy gap (E_G) are expressed as the following expressions [10].

$$IE = E^{+}(M) - E(M) \quad (1)$$

$$EA = E(M) - E^{-}(M) \quad (2)$$

$$E_G = E(LUMO) - E(HOMO) \quad (3)$$

Where $E^{+}(M)$, $E^{-}(M)$ and E(M) are the energy values of cation, anion and neutral molecule of M, respectively, E(LUMO) is the energy of the lowest unoccupied molecular orbital of M, E(HOMO) is the energy of the highest occupied molecular orbital of M.

Figure 3 contains the values of ionization energy, electron affinity and molecular orbital energy gap of SF_6, HFO-1234ze(E) and the main decomposition products. The ionization energy of SF_6 is 15.18 eV, which is only smaller than that of CF_4, and its electron affinity is 1.23 eV, which is the largest, indicating that SF_6 has high dielectric strength. The electron affinity of HFO-1234ze(E) is only smaller than that of SF_6, so it has a strong electronegativity.

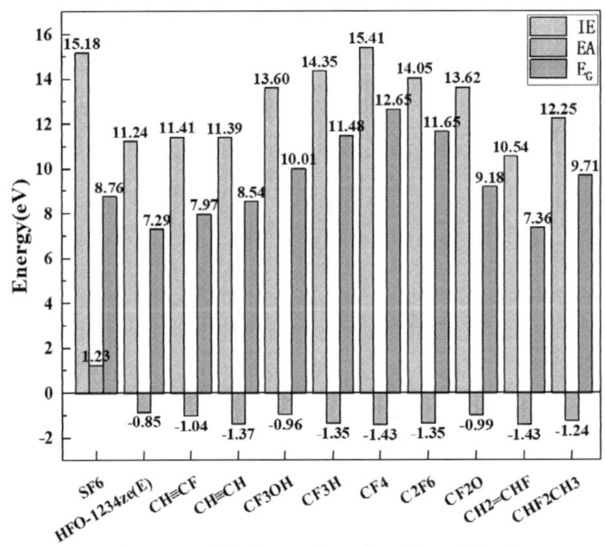

Figure 3 Ionization energy, electron affinity and molecular orbital energy gap of the products

The ionization energies of CH≡CF and CH≡CH are both small, but the electron affinity of CH≡CF is greater than that of CH≡CH. Therefore, CH≡CH is not a good insulating medium. The molecular orbital energy gap values of CF_3H and CF_3OH are 11.48 eV and 10.01 eV, respectively, so CF_3H has better stability than CF_3OH. Although their ionization energies are relatively high, it can be found that both CF_3H and CF_3OH are less electronegative than HFO-1234ze(E) by comparing the electron affinity. CF_3H is a greenhouse gas with a GWP value of 11,700 times that of CO_2 [10].

The electron affinity value of C_2F_6 is larger than that of CF_4. Therefore, with the increase of the number of carbon atoms, the electronegativity and dielectric strength of CF_4 and C_2F_6 gradually increase. In terms of environmental hazards, the GWP values of CF_4 and C_2F_6 is 6500 and 9200 times that of CO_2, respectively, which is extremely harmful to the environment [11]. CF_4 and C_2F_6 are the most stable among all products.

The ionization energy and electron affinity of CF_2O are 13.62 eV and -0.99 eV respectively, which are higher than most of the products. Therefore, CF_2O may have good insulation strength. However, CF_2O is extremely toxic, with a 10-minute lethal concentration 50 (LC50) of 1083 ppm in rats, which is 45 times as lethal as HF. Exposure to CF_2O may lead to tissue necrosis and airways nerve stimulation [12].

The electron affinity values of CH_2=CHF and CHF_2CH_3 are small, so their electronegativity is poor. Especially for CH_2=CHF, because its ionization energy and electron affinity are the smallest among all products, the dielectric strength of CH_2=CHF may be the worst. The molecular orbital energy gap value of CH_2=CHF is 7.36 eV, so its stability is relatively poor. CH_2=CHF is a flammable gas [13]. CHF_2CH_3 is an extremely flammable gas, which can be used as a propellant [14]. Avella et al pointed out that inhalation of CHF_2CH_3 may lead to poisoning and loss of muscular coordination, and their experimental results showed that rats exposed to 20 L/min CHF_2CH_3 had toxic symptoms within 20 s [15].

4. Conclusions

In this paper, the discharge decomposition characteristics of HFO-1234ze(E) in the presence of trace water were calculated and analyzed at the level of B3LYP/6-311+G(d,p). The main decomposition products were analyzed according to the energy barrier and enthalpy, and their insulation performance and safety properties were evaluated.

(1) The dissociation of HFO-1234ze(E) requires high energy and is not easy to occur. However, H_2O will accelerate the decomposition of HFO-1234ze(E) and produce H_2, HF, CH≡CF, CH≡CH, CF_3OH, CF_3H, CF_4, C_2F_6, CF_2O, CH_2=CHF, CHF_2CH_3 and other gases.

(2) Among the main products, most of them have low electronegativity and good stability, so the dielectric strength of the gas mixtures may be poor. H_2, CH≡CH, CH_2=CHF and CHF_2CH_3 are flammable, and HF, CF_2O and CHF_2CH_3 are toxic, which may pose a threat to the safety of operators. CF_3H, CF_4 and C_2F_6 are harmful to the environment. Therefore, when HFO-1234ze(E) is used as an insulating medium, water must be avoided.

This study reveals the decomposition mechanism of HFO-1234ze(E) in the presence of trace water, and the results will help to formulate relevant standards for the application of HFO-1234ze(E) in power equipment.

References

[1] Zhang, X.X., Li, Y., Xiao, S., Tang, J., Tian, S.S., Deng, Z.T. (2017) Decomposition mechanism of $C_5F_{10}O$: an environmentally friendly insulation medium. Environ. Sci. Technol., 51: 10127-10136.

[2] Liu, J., Wang, F., Zhong, L.P., Gan, H., Hai, B., Tang, N., Li, L., Zhou, Y.Y. (2022) Theoretical study of the decomposition mechanism of a novel eco-friendly insulation medium HFO-1336mzz(E) considering the effect of trace humidity. J. Phys. D: Appl. Phys., 55: 045201.

[3] Fang, X.K., Hu, X., Janssens-Maenhout, G., Wu, J., Han, J.R., Su, S.S., Zhang, J.B., Hu, J.X. (2013) Sulfur hexafluoride (SF_6) emission estimates for China: an inventory for 1990–2010 and a projection to 2020. Environ. Sci. Technol., 47: 3848-3855.

[4] Irriyanto, M.Z., Lim, H.S., Choi, B.S., Myint, A.A., Kim, J. (2019) Thermal stability and decomposition behavior of HFO-1234ze(E) as a working fluid in the supercritical organic Rankine cycle. J. Supercrit. Fluids, 154: 104602.

[5] Liang, Y.D., Wang, F., Sun, Q.Q., Zhai, Y.J. (2017) Molecular structural and electrical properties of trans-1, 3, 3, 3-tetrafluoropropene and 2, 3, 3, 3-tetrafluoropropene under external electric fields. Comput. Theor. Chem., 1120: 79-83.

[6] Jia, Z.K., Chen, Q.G., Lin, L., Wang, X.Y. (2020) Study of partial discharge characteristics in HFO-1234ze(E)/N_2 mixtures. Plasma Sci. Technol., 22: 115403.

[7] Preve, C., Piccoz, D., Maladen, R. (2017) Application of HFO1234zeE in MV switchgear as SF_6 alternative gas. CIRED-Open Access Proceedings Journal, 2017: 42-45.

[8] Fu, Y.W., Rong, M.Z., Wang, X.H., Yang, A.J. (2019) Rate constants of $C_5F_{10}O$ decomposition reactions at temperatures of 300–3500 K. J. Phys. D: Appl. Phys., 52: 035202.

[9] Fu, Y.W., Yang, A.J., Wang, X.H., Rong, M.Z. (2019) Theoretical study of the decomposition mechanism of C_4F_7N. J. Phys. D: Appl. Phys., 52: 245203.

[10] Zhong, L.P., Deng, Y.C., Liu, J., Wang, F., Chen, S., Sun, Q.Q., Duan, X.L., Huang, H.B. (2020)

Theoretical study by density functional theory calculations of decomposition processes and primary products of $C_5F_{10}O$ with moisture content. J. Phys. D: Appl. Phys., 53: 485204.
[11] Cooke, M. (2005) Tackling the plasma etch villains. III-Vs Review, 18: 40-42.
[12] Januszkiewicz, A.J., Bazar, M.A., Crouse, L.C.B., Chapman, M.A., Hodges, S.E., McCormick, S.J., O'Neill, A.J. (2018) Morbidity and mortality resulting from acute inhalation exposures to hydrogen fluoride and carbonyl fluoride in rats. Inhalation Toxicol., 30: 114-123.
[13] Chattopadhyay, A., Podder, S. (2014) Vinyl Fluoride. Encyclopedia of Toxicology, 939-941.
[14] Hirata, K., Sasao, A., Ohtsu, Y., Tsutsumi, H., Furukawa, S., Yonemitsu, K., Nishitani, Y. (2020) A sudden death related to 1, 1-difluoroethane inhalation—A case report and brief review of the literature. Forensic Science International: Reports, 2: 100062.
[15] Avella, J., Kunaparaju, N., Kumar, S., Lehrer, M., Zito, S.W., Barletta, M. (2010) Uptake and distribution of the abused inhalant 1, 1-difluoroethane in the rat. J. Anal. Toxicol., 34: 381-388.

Evolution of Silicon Wafer Dimensions

Jialiang Peng[1*], Anwei Liu[2], Muyang Wang[3]

[1]Faculty of Information Technology, Beijing University of Technology, Beijing 100124, China

[2]School of Electrical Information, Hunan Institute of Engineering, Xiangtan, Hunan 411104, China

[3]School of Electrical Engineering, Zhejiang University, Hangzhou, Zhejiang 310058, China

*Corresponding author email: pjl18027407@emails.bjut.edu.cn

Abstract: Solar energy is a renewable energy that the world is endeavouring to utilize. The solar cell is the essential invention that makes energy transfer from light to electricity possible. Solar cells are normally made from silicon wafers due to silicon's semi-conductivity. The thesis will focus on the development of silicon wafers' size and fabrication technology. Through analysis of resent (in 5 years) statistics from reports released by government or incorporations, the link between the trend and the reasons behind is interpreted. ITRPV has pointed out that silicon wafers are switching to the mono-crystalline ones, the larger ones, and the PERC processed ones. The mono-crystalline ones excel in efficiency due to uniform lattice structure, but is more costly because of the Czochralski process. The larger cells can absorb more sunlight. The PERC cell technique reduces recombination and reflects unabsorbed sunlight, therefore gradually eliminates the traditional BSF technique.

1. Introduction

Solar energy is an inexhaustible energy that the world is seeking to replace fossil fuels, so people designed solar cells to capture the sunlight and transform the absorbed photo-energy into electronic energy.

1.1 ITRPV Results

Crystalline Silicon is by far the major material used to produce solar cells, making up 95% of solar panels sold today due to silicon's semi-conductivity and abundance in earth crust, and the lifetime of solar panels can last for about 25 years [1]. The statistics of wafers' production reflects the trend in solar cell fabrication. Dimensions such as silicon type, silicon wafer size and cell technology are all worth considering.

The tenth edition of International Technology Roadmap for Photovoltaic (ITRPV) indicates the followings:

1. Mono-crystalline silicon are going to take over multi-crystalline silicon.
2. Larger silicon wafers will become the mainstream.
3. N-type silicon will surpass p-type silicon.
4. Traditional BSF cell technology will be replaced with PER technologies and hetero-junctions.

Content from this work may be used under the terms of the Creative Commons Attribution 3.0 licence. Any further distribution of this work must maintain attribution to the author(s) and the title of the work, journal citation and DOI.
Published under licence by IOP Publishing Ltd

1.2 The Objective of this Thesis

The ITRPV report states the predicted development, but is lack of concrete evidence of what made the predictions plausible. Further research need to be done to answer the following questions:

1. Why mono-crystalline silicon outdoes multi-crystalline silicon.
2. What makes vast-production of mono-crystalline silicon possible.
3. What makes solar cells grow larger.
4. Why BSF technology is predicted to be eliminated
5. What makes the extensive use of PER technologies and hetero-junctions available.

1.3 The Innovation of this Thesis

This thesis will integrate the trend in the following three aspects: silicon type, wafer size, wafer technology. The analysis of all three aspects will provide evidence corresponding to the original ITRPV report.

2. Mechanism

This thesis is a meta-analysis of statistics from various sources. The analysis mainly focuses on silicon type, silicon wafer size, and cell technology, relevant literatures are reviewed and referred to, in order to integrate into a general conclusion about how and why the silicon wafer dimensions are changing.

As mentioned in the introduction, our research will explain the reasons behind the figures shown in the ITRPV 2018 report, which means we need to find supporting evidence and materials showing that mono-crystalline silicon, n-type silicon, PER or hetero-junction technology deserve promising prospects. We will browse for scientific explanations that show how mono-crystalline silicon surpasses multi-crystalline silicon in energy efficiency, what specific difference lies between p-type and n-type another than the type of majority carriers, the definition of PER, or hetero-junction, and what makes those technologies possible. This thesis will endeavor to provide answers to the above questions.

We first browsed through manufacture or education website to get a general idea of several conceptions, such as recombination, passivation. Then, we gather reports or theses written by specific authors. Afterwards, we work on academic papers from ResearchGate to deepen our understandings.

3. Results

3.1 Analysis of silicon type

According to ITRPV 2018 [2], the type of silicon wafer is predicted to switch from multi-crystalline silicon wafers to mono-crystalline silicon wafers. It is predicted that the 50% world market share (2018) of high performance multi-crystalline silicon wafers will dramatically drop to 10% in 2029. The p-type mono-crystalline silicon wafers will remain 45% world market share until 2029. By contrast, the world market share of n-type mono-crystalline silicon wafers will increase from 5% (2018) to 43% (2029).

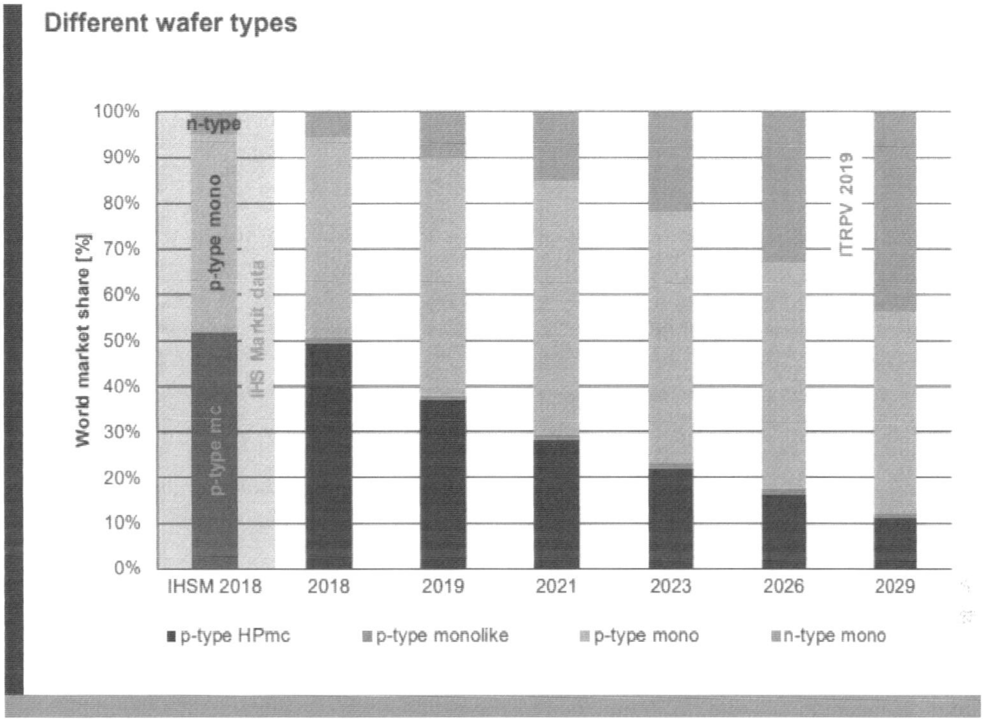

Figure 1. World market shares for different wafer types (ITRPV 2019) [2].

3.1.1 Difference Between Mono- and Multi-crystalline Silicon. The fundamental difference between mono-crystalline silicon and multi-crystalline silicon is their atomic structures. Mono-crystalline silicon is made up of a single slice of silicon, the lattice is uniform, therefore electrons have plenty of room to flow. While multi-crystalline silicon is made up of multiple pieces of silicon, the lattice is not uniform, so the electron's motion is hindered by irregular lattice structures. This explains why mono-crystalline silicon is more efficient than multi-crystalline silicon. Table 1 provides precise data showing that the efficiency of mono-crystalline silicon is slightly higher than multi-crystalline silicon.

Table 1. The features of various photovoltaic cells [3]

Photovoltaic Cell	**Efficiency**	**Characteristic**
Mono-crystalline silicon	15% ~ 17%	Stable, high efficiency, 300μm thick, hard, non-crimp-able, black
Multi-crystalline silicon	12% ~ 14%	Stable, high efficiency, 300μm thick, hard, non-crimp-able, dark blue
Non-crystalline silicon	6% ~ 10%	Unstable during the first stage of crystallization, low efficiency, then stabilizes, crimp-able, dark red
Thin-film	10%~18%	Appropriate for large-scale production, adaptive to different solar spectrum, black

Though less in efficiency, multi-crystalline silicon is more affordable than mono-crystalline silicon [4].

Generally, producing multi-crystalline silicon requires two essential chemical reactions [5]:

$SiO_2 + C = Si$ (metallurgical grade silicon) + $2CO$

Si (metallurgical grade silicon) + $3HCl = SiHCl_3 + H_2$

$SiHCl_3 + H_2 = Si$ (high purity multi-crystalline silicon) + $3HCl$

While mono-crystalline silicon requires an additional Czochralski process, during which high-purity silicon is melted in crucible, then a seed crystal is dipped into the molten silicon and slowly extracted with rotation [6], during the cutting process 40%-50% of the extracted ingot will be wasted [7].

3.1.2 Difference Between P-type and N-type. The distinction between p-type and n-type is not just the dopants phosphorous (or arsenic) and boron (or gallium), but also that n-type is less iron-sensitive than p-type [8].

On the other hand, boron doped p-type silicon may form boron-oxygen complexes, causing boron-oxygen light-induced degradation (B-O LID) [9], which can be avoided if selecting n-type silicon or choosing gallium to be the p-type dopant, this advantage of n-type will be further discussed in 3.3.3. Nevertheless, p-type silicon are less sensitive to radiation damages and require lower temperature to diffuse [10].

As for silicon ingots, p-type boron doped silicon has segregation factor of 0.7, resistivity variation of 3 varies from bottom to top, while n-type phosphorous doped silicon has segregation factor of 0.35, resistivity variation of 10 varies from bottom to top [10].

As for the cost, though p-type and n-type share similar ingot manufacturing process, n-type involves more expensive fabrication process afterwards [11].

3.1.3 Potential Alteratives. The types of solar cells are not restricted to crystalline silicon. Amorphous silicon, cadmium telluride (CdTe), copper indium gallium selenide (CIGS) are also potential options for solar cell materials [12].

Because the ITRPV focuses on crystalline silicon products, the above alternatives are hardly mentioned in the report. These three solar cells are categorized as thin-films. Amorphous silicon is fabricated by coating doped silicon on the back of the substrate, therefore less energy is consumed when processing. Cadmium telluride is efficient in light absorption and is recyclable. CIGS is also efficient in light absorption, and requires a much thinner film [12].

3.2 Analysis of silicon wafer size

According to ITRPV 2019 [2], both multi-crystalline silicon wafers and mono-crystalline silicon wafers are predicted to feature prevailing growth in wafer size.

It is predicted that the 90% world market share (2018) of M2 (156.75*156.75 mm^2) multi-crystalline silicon wafers will slightly shrink to 83% in 2029, while the upcoming M4 (161.75*161.75 mm^2) and M6 (166*166 mm^2) will gradually increase from 5% (predicted 2019) to 15% (2029).

It is predicted that the 90% world market share (2018) of M2 (156.75*156.75 mm^2) mono-crystalline silicon wafers will sharply decrease to 30% in 2029, while the emerging M4 (161.75*161.75 mm^2) mono-crystalline wafer's world market share will increase from 3% (2018) to 30% (2029).

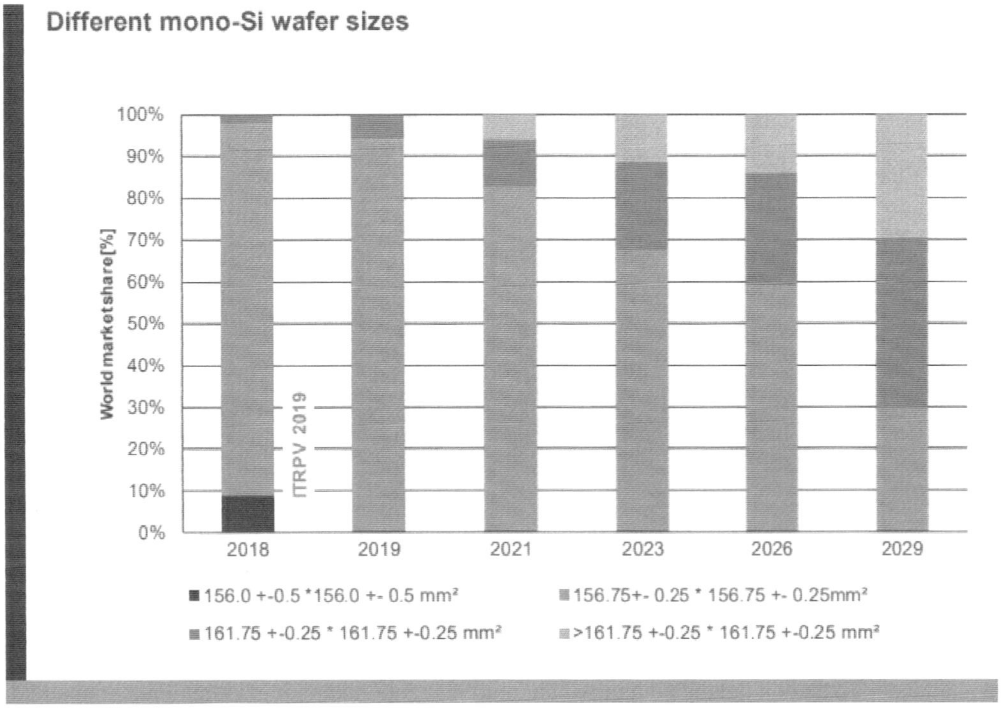

Figure 2. Expected trend of mono-Si wafer size in mass production (ITRPV 2019) [2].

It seems that larger wafers are occupying the global market share. It can be implied that mono-crystalline silicon wafers are more likely to have larger size than multi-crystalline silicon wafers. It goes without saying that larger silicon wafer captures more sunlight, but PV-manufacturing.org indicates that larger size lengthens the distance the electrons need to travel, which may cause loss in current intensity [13]. PV-manufacturing.org also warns that the handling tools such as holders and carriers need to be updated in order to adapt to larger wafer sizes [13].

3.3 Analysis of cell technology. ITRPV [2] also analyzes the trend in cell technology. The world market share of traditional back-surface-field (BSF) technology will steeply drop from 56% (2018) to 15% (2023) and get eliminated in 2026. On the contrary, the silicon hetero-junction technology will have its world market share expanded from 3% (2018) to 15% (2029). The mainstream passivated emitter and rear (PER) technologies will grow from 40% (2018) to 70% (2029).

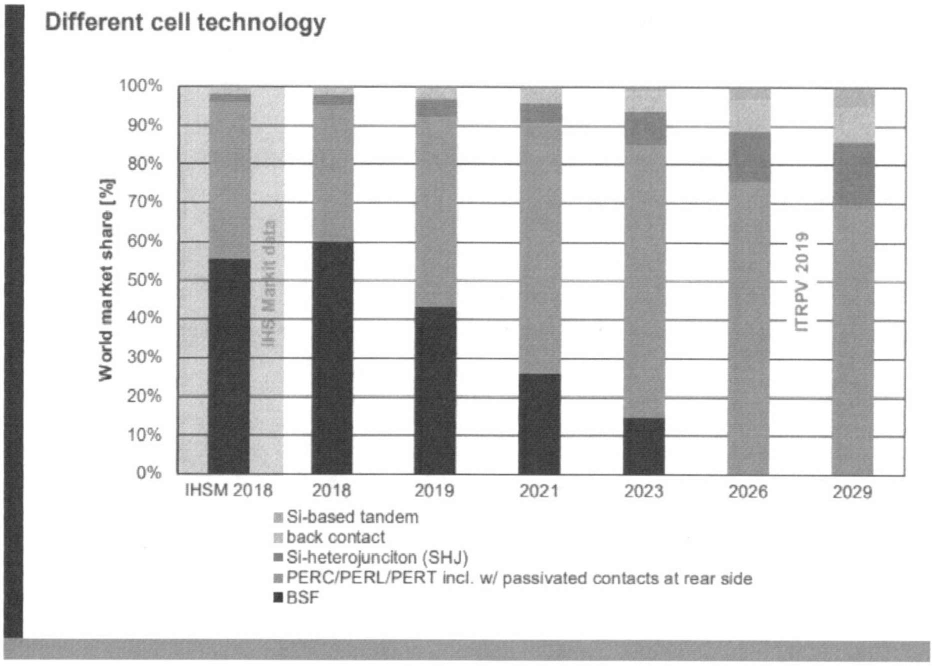

Figure 3. Worldwide market shares for different cell technologies (ITRPV 2019) [2].

As for the stabilized efficiency, the BSF technologies (taking the M2 wafers as an example) shows poor efficiency below 21% and growth rate no more than 1 percentage point. By contrast, the PER technologies shows promising prospective of 21% to 24%, and the hetero-junction technology has slightly higher expectation of 25%.

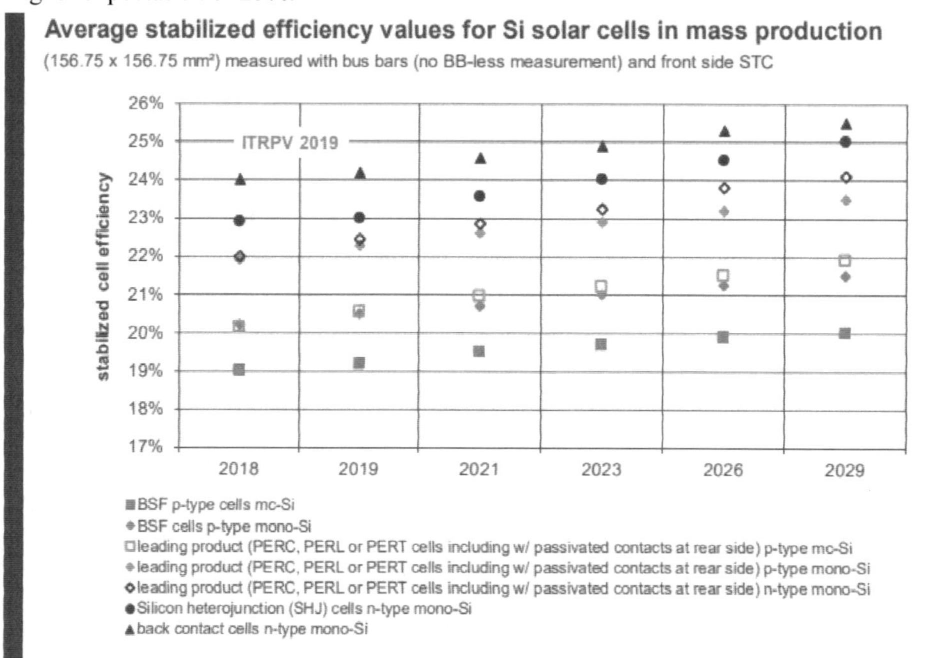

Figure 4. Average stabilized efficiency values of crystalline-silicon solar cells in mass production (M2)(ITRPV 2019) [2].

Undoubtedly, the BSF technology will be eliminated because of low product efficiency, while the PER and hetero-junction technologies shall be adapted to enhance product efficiency.

3.3.1 Back-Surface-Field. Back-surface-field (BSF) involves highly doping the rear surface of the wafer with high concentration of major carriers to prevent the minor carriers from recombination [14]. An experiment has been done on n-type and p-type mono-crystalline silicon wafers. Both wafers have bulk doping level of $1.5*10^{16}/cm^2$ and emitter doping level of $7.5*10^{18}/cm^2$, the varying BSF concentration varies from $1*10^{17}$ to $1*10^{20}/cm^2$. The result shows that the highest short circuit current density and efficiency can be produced when the BSF concentration is $1*10^{17}/cm^2$, slightly higher than the bulk doping level. But if the BSF concentration exceeds the optimal level, the efficiency decreases [15].

3.3.2 Passivated Emitter and Rear. The PERC (passivated emitter and rear cell) technology adds an additional passivation layer (to prevent unwanted premature recombination, usually made of aluminum oxide [16]) between the BSF, and a silicon-nitrate capping layer beneath [17]. The passivation layer also reflected the unabsorbed light back to the solar cell for an extra absorption attempt, therefore this technique reduces electron recombination, improves light absorption, enhances internal reflectivity [18]. Improved efficiency of more than 2% and prevalent efficiency above 22% [19] makes the PERC silicon wafers a promising choice to make do with non-ideal conditions, such as lack of area and weak illumination [18].

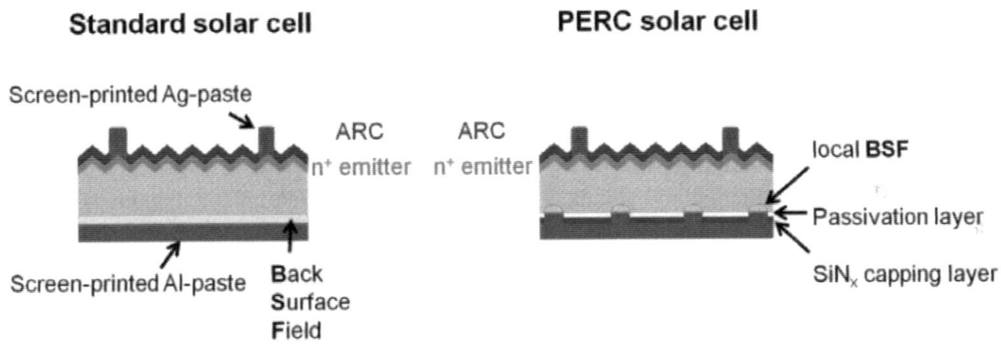

Figure 5. Cross Sections of Standard and PERC Solar Cell (ISFH) [17].

The rise in efficiency of PERC is also contingent with high quality wafer, improved optical performance, optimized emitter profile, optimized rear BSF [20]. By optimizing the firing condition and aluminum paste, the void ratio will be reduced by more than half [20], the advancement of PERC relies on the basic BSF technology.

On the other hand, few reforms should be done on the fabrication line to adapt to the PERC technique [18]. The slight difference between BSF cells and PERC cells is that the PERC requires rear-side deposition of silicon oxide or aluminum oxide layer (Figure 7 Step 5), PECVD (plasma-enhanced chemical vapor deposition) of SiNx for backside reflection and surface passivation [21].

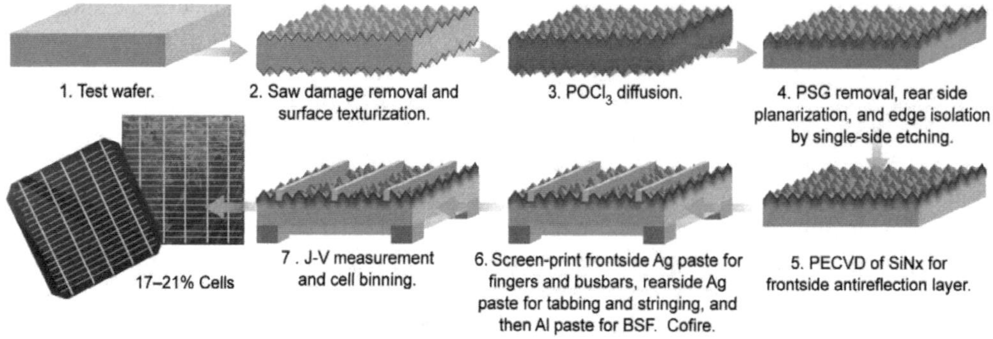

Figure 6. Process flow for manufacturing standard full-area Al_BSF cells [21]

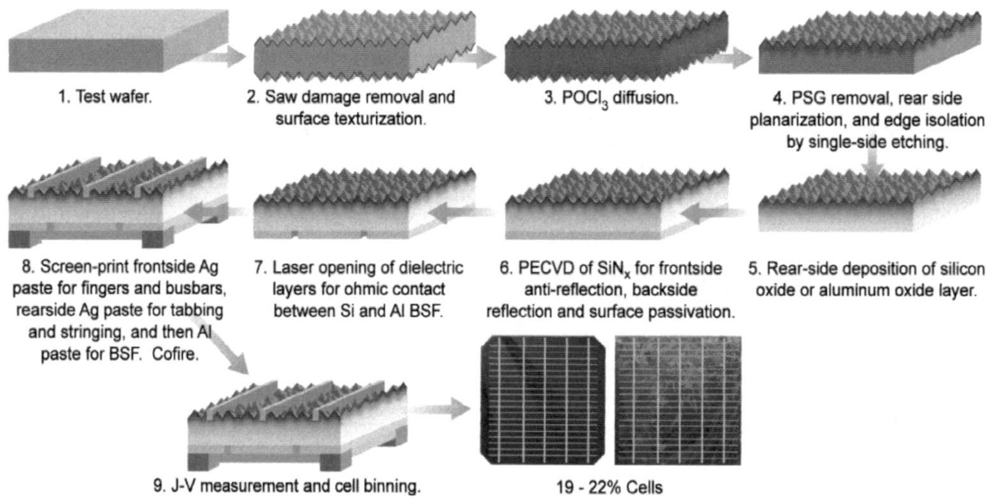

Figure 7. Process flow for manufacturing standard full-area PERC cells [21]

The theoretical principle and fabrication of PERC is derivative from the traditional BSF technology, PERC's future prevalence is rooted in the current PERC-fashioned production. PERT and PERL are also similar technologies that make up the PER-tech series.

PERC uses oxidized metals (such as aluminum oxide), so n-type will be more suitable than p-type since n-type is more resistant to iron contaminants (as mentioned in 2.1) [8].

PERT (passivated emitter with rear totally diffused) is another specific PER technology, which reduces the cell series resistance from higher resistivity substrates [22].

PERL (passivated emitter with rear locally diffused) is also another specific PER technology, the rear is locally diffused only at the metal contacts, to reduce rear recombination while keeping proper electrical contact [23].

According to Figure 8, the PERC technique on multi-crystalline silicon is $0.02/Watt (DC) cheaper and 0.5 percent points more efficient than the BSF, which is a supporting evidence that PERC will take over BSF to become the mainstream cell technology.

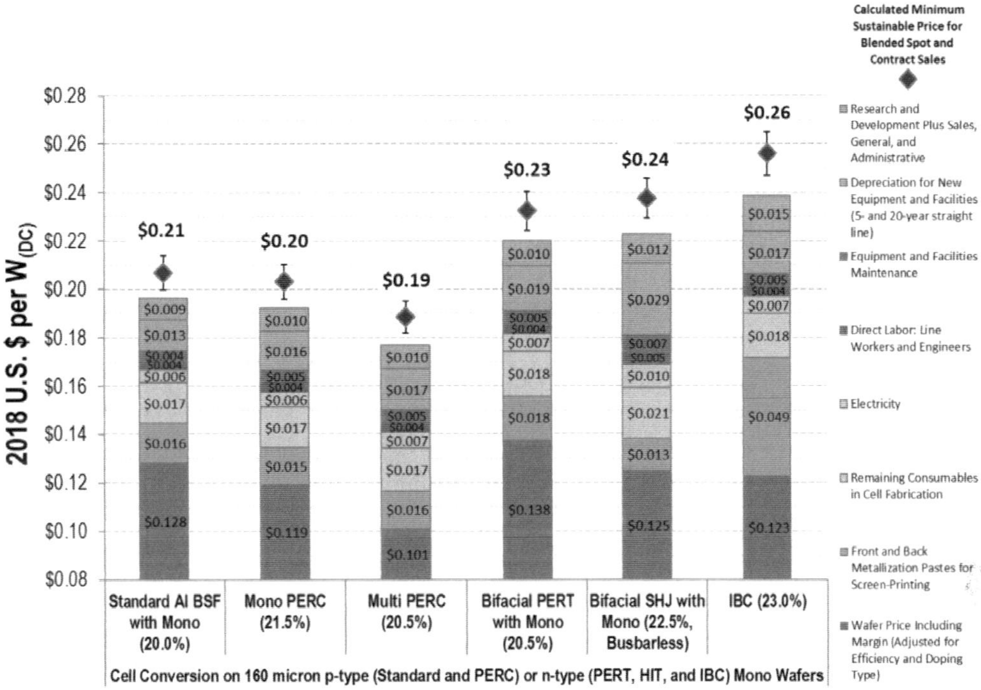

Figure 8. Modeled costs and MSPs for Al-BSF, PERC, PERT, SHJ, and IBC cell technologies [21]

3.3.3 Hetero-junction. Hetero-junction means using two or more different materials to create a junction. The kind of materials, the mixing rate of materials etc. are contingent factors of the creation of an effective hetero-junction. The typical hetero-junction solar cell is made of a piece of mono-crystalline silicon layer sandwiched between two ultra-thin amorphous silicon layer [24].

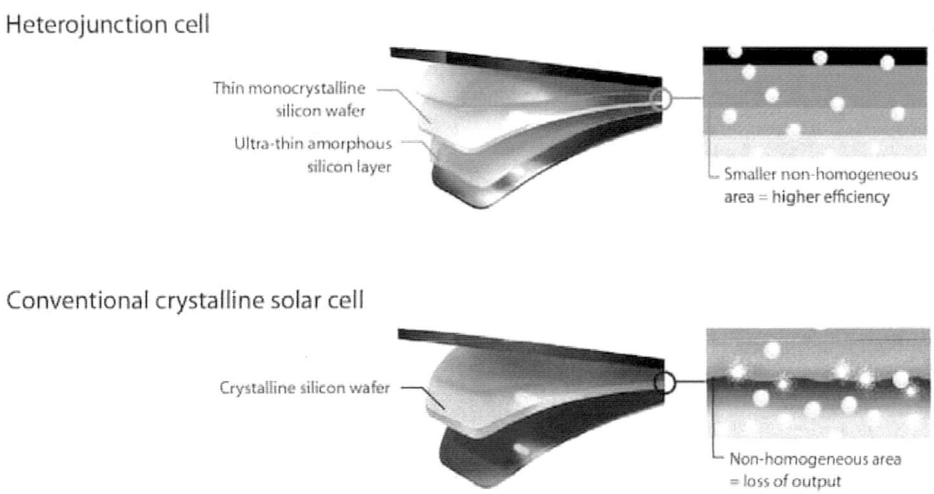

Figure 9. Structure of hetero-junction silicon wafers [24].

The top silicon layer of a hetero-junction silicon wafer absorbs sunlight and retrieves the sunlight reflected from the two layers below. The bottom silicon layer of a hetero-junction silicon wafer captures the sunlight that have passed through the two layers above, and also collects sunlight that bounced off the ground [24].

The hetero-junction silicon wafers mainly applies phosphorous-doped n-type silicon, since boron-doped p-type silicon is susceptible to boron-oxygen light-induced degradation (B-O LID) [9]. However, gallium is a proper substitution of boron in p-type doping, and is estimated that gallium's world market share in multi-crystalline silicon will increase from less than 2% (2018) to 34% (2029). According to an experiment done by a Russian-Australian group, the boron-doped wafers show an efficiency of 20.5% while the gallium-doped wafers have an efficiency of 22.6% [9].

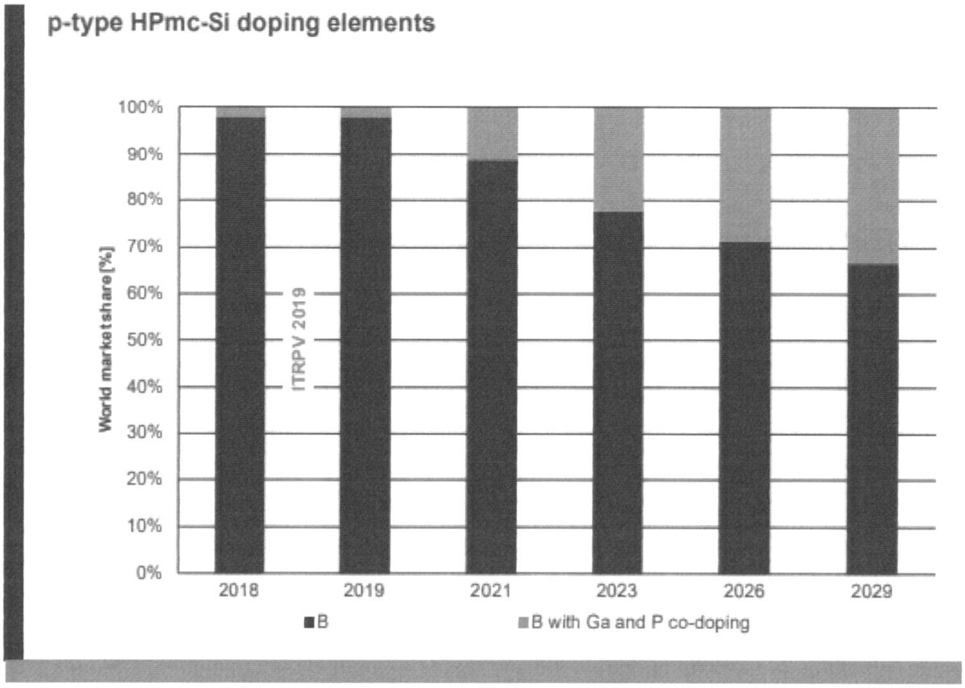

Figure 10: Expected market share of dopants used for doping of p-type multi-crystalline silicon material (ITRPV 2019) [2].

Moreover, with an advanced hydrogenation technique, B-O LID can be solved so that the function of boron-doped p-type hetero-junction silicon wafers get stabilized, the efficiency of gallium-doped p-type and phosphorous doped n-type silicon wafers also get improved by 0.4-0.7% [9]. However, fill factor is still a limitation of p-type hetero-junction silicon wafers [9].

4. Conclusion

Through the ITRPV report, we have learned the following:

1.Mono-crystalline silicon is surpassing multi-crystalline silicon.

2.The size of silicon cells is growing larger.

3.N-type is emerging, while the prevailing p-type fades.

4.PER-tech, hetero-junction solar cells are predicted to become the mainstream.

Referring to various literatures, we have gradually discovered the supporting evidence:

1.Mono-crystalline silicon ensures sufficient drift current because of its neat lattice structure.

2. A larger silicon wafer absorbs more sunlight and is not difficult to let the production tools adapt to larger wafers.

3.N-type silicon prevents B-O LID, which is a subtle advantage that excels from p-type silicon.

4. PER-tech combats recombination through a layer of passivation.

5. Hetero-junction improves sunlight reflection and absorption.

Though the above silicon wafers excel in efficiency, most of them are higher in cost, the question of how the high cost of the above silicon wafers can be handled is still to study. In other words, we still need to answer how the above silicon wafers can prevail in solar cell production so that the ITRPV predictions can come true.

Our study is of value to suppliers and consumers of solar cells, informing them what kind of silicon wafers are promising and the science principles behind. Reference to our idea and further research into current or new silicon types may lead to new solar cell designs and inventions . The science and technology of silicon solar cells is an ongoing subject, more answers lead to further questions. We hope that the future solar cell industry will make all ITRPV predictions come true.

References:

[1] Solar Energy Technologies Office. Solar Photovoltaic Cell Basics. Retrieved from https://www.energy.gov/eere/solar/solar-photovoltaic-cell-basics

[2] International Technology Roadmap for Photovoltaic (ITRPV) 2018 Results. Tenth Edition. March 2019. P40-53. Retrieved from https://pv-manufacturing.org/wp-content/uploads/2019/03/ITRPV-2019.pdf

[3] Zhang, M. (2015). The Current Development and Application of Photovoltaic Cells.

[4] DIY Solar Panels - Monocrystalline vs. Polycrystalline Design. Retrieved from https://www.mysolargoods.com/blogs/news/diy-solar-panels-monocrystalline-vs-polycrystalline-design

[5] Polysilicon - from Sand to Solar Cells, it starts here. Retrieved from https://www.solarmango.com/scp/polysilicon-from-sand-to-solar-cells-it-starts-here/

[6] Czochralski Process and Silicon Wafers. (2016). Retrieved from https://www.waferworld.com/post/czochralski-process-and-silicon-wafers

[7] Bayod-Rujula, A. A. (2019). Solar Hydrogen Production. Retrieved from https://www.sciencedirect.com/topics/engineering/monocrystalline-silicon-cell

[8] Kenning, T. (2018). Mono vs Multi, N-type vs P-type: Outlooks from PV CellTech 2018. Retrieved from https://www.pv-tech.org/mono-vs-multi-n-type-vs-p-type-outlooks-from-pv-celltech-2018/

[9] Bellini, E. (2021, July 20). New tech may bring heterojunction solar cells based on p-type wafers closer to mass production. Retrieved from https://www.pv-magazine.com/2021/07/20/new-tech-may-bring-heterojunction-solar-cells-based-on-p-type-wafers-closer-to-mass-production/

[10] MacDonald, D. The emergence of n-type silicon for solar cell manufacture. The Australian National University. Retrieved from http://pv.cecs.anu.edu.au/files/AuSES2012Macdonaldfullpaper.pdf

[11] The difference between p-type and n-type solar cells. Retrieved from https://www.aleo-solar.com/difference-n-type-p-type-solar-cells/

[12] Mughees, A. (2020). Solar Cell Evolution: Past, Present, Future. Retrieved from https://electronics360.globalspec.com/article/15210/solar-cell-evolution-past-present-future

[13] Wafer Sizes. Retrieved from https://pv-manufacturing.org/wafer-sizes/

[14] Surface Recombination. Retrieved from https://www.pveducation.org/es/fotovoltaica/design-of-silicon-cells/surface-recombination

[15] Ferdiansjah, Faridah, Tirakusuma, M., Kelvian (2018). Analysis of Back Surface Field (BSF) Preformance in P-type and N-type Monocrystalline Silicon Wafer. E3S Web of Conferences. Retrieved from https://ui.adsabs.harvard.edu/abs/2018E3SWC..4301006F/abstract

[16] Thin aluminum oxide passivation layers for solar cells. Application note A088-EN04. Retrieved from https://www.bronkhorst.com/en-us/markets-en/energy-research-production/application-note-a088-en04-thin-aluminium-oxide-passivation-layers-for-solar-cells/

[17] PERC Cell Technology Explained. Retrieved from

https://www.aleo-solar.com/perc-cell-technology-explained/
[18] Pickerel, K. (2016). What is PERC? Why Should You Care? Retrieved from https://www.solarpowerworldonline.com/2016/07/what-is-perc-why-should-you-care/
[19] Li, Y. (**2021**). The Research on the Current Status and Trend of Photovoltaic Industry
[20] Deng, W., W., Zhen, X., Chen, D., M., Verlinden, P., J., et al., (2015) 20.8% PERC Solar Cell on 156 mm x 156 mm P-Type Multi-crystalline Silicon Substrate, IEEE Journal of Photovoltaics. Retrieved from https://www.researchgate.net/publication/304286548_208_efficient_PERC_solar_cell_on_156_mm156_mm_p-type_multi-crystalline_silicon_substrate
[21] Woodhouse, M., Smith, B., Ramdas, A., Margolis, R. Crystalline Silicon Photovoltaic Module Manufacturing Costs and Sustainable Pricing: 1H 2018 Benchmark and Cost Reduction Road Map, National Renewable Energy Laboratory.P21-29. Retrieved from https://www.nrel.gov/docs/fy19osti/72134.pdf
[22] Zhao, J. H., Wang, A. H., & Green, A. M. (2000, January 18). 24.5% Efficiency silicon PERT cells on MCZ substrates and 24.7% efficiency PERL cells on FZ substrates. Retrieved from https://onlinelibrary.wiley.com/doi/10.1002/%28SICI%291099-159X%28199911/12%297%3A6%3C471%3A%3AAID-PIP298%3E3.0.CO%3B2-7#:~:text=A%20PERT%20%28passivated%20emitter%2C%20rear%20totally-diffused%29%20cell%20structure,passivation%20quality%20of%20MCZ%20and%20some%20FZ%20substrates.
[23] High Efficiency Solar Cells. Retrieved from https://www.pveducation.org/pvcdrom/manufacturing-si-cells/high-efficiency-solar-cells
[24] What is a heterojunction solar cell?. Retrieved from https://www.thesolarnerd.com/blog/what-is-heterojunction-solar-panel/

The Influence of Novel Power System Characteristics on 330kV Power Grid Short-circuit Current

Xingzhong Yuan[1a], Zhe Wang[1b], Huan Yan[1c], Na Yan[1d], Yi Sun[2a*], Yiqing Xu[2b]

[1] Economic and Technological Research Institute of State Grid Shaanxi Electric Power Co., Ltd., Shanxi 710000, China

[2] East China Electric Power Design Institute Co., Ltd., Shanghai 210000, China

[2a*]3130@ecepdi.com

Abstract—Under the development goal of "Carbon Peak, Carbon Neutralization", the concept of novel power system is proposed. Its representative features include AC/DC interconnection of power grid and large-scale access of new energy power stations. Under these characteristics, higher requirements are put forward for the short-circuit current breaking capacity of the circuit breaker. The DC component of short-circuit current and the injection of new energy into the system short-circuit current cannot be ignored. Based on the 330kV local power grid in Northwest China, this paper studies the power grid operation mode, DC component of short-circuit current and the influence of photovoltaic power station on short-circuit current. The influence law of DC component of short-circuit current on the actual breaking capacity of circuit breaker and access of photovoltaic power station on the short-circuit current are summarized. It can provide a theoretical basis for the selection of circuit breaker in novel power system.

1. Introduction

Under the development goal of "Carbon Peak, Carbon Neutralization", a novel power system based on new energy is proposed [1~2]. Wind power, photovoltaic and other new energy power stations will develop on a large scale. With the construction of novel power system, the form of power grid has gradually changed from AC backbone grid and DC long-distance transmission to the interconnection of AC power grid and DC network [3~4].

Under the characteristics of novel power system, the factors affecting the short-circuit current of power grid are increasing, such as the DC component of short-circuit current (SCC) and the access of new energy power stations. The breaking capacity of circuit breaker is determined by AC component and DC component. In East and Northwest China, the problem of DC component of power grid is more prominent, and the impact of DC component on the breaking capacity of circuit breaker cannot be ignored [5~6].

In the traditional short-circuit calculation methods, such as using BPA software to calculate the short-circuit current, the wind and solar power station is still modeled in the form of negative load, that is, the contribution to the short-circuit current is not considered [7~8]. However, with the large-scale development of new energy, the impact on short-circuit current cannot be ignored.

In summary, based on a 330kV power grid in Northwest China, this paper focuses on the impact of novel power system characteristics on power grid short-circuit current.

Content from this work may be used under the terms of the Creative Commons Attribution 3.0 licence. Any further distribution of this work must maintain attribution to the author(s) and the title of the work, journal citation and DOI.
Published under licence by IOP Publishing Ltd

2. Influence of novel power system characteristics on SCC

According to the above analysis, the impact of AC/DC interconnection and large-scale access of new energy on power grid short-circuit current is mainly considered. The influence of AC/DC interconnection is mainly reflected in the DC component of short-circuit current. The large-scale access of new energy provides more short-circuit current injection for the power grid. This paper focuses on photovoltaic as an example.

2.1. Influence of DC component

The total short-circuit current is composed of AC component and DC component [9]. The DC component part focuses on calculating its attenuation time constant T_{dc}. The calculation formula is shown in formula (1).

$$T_{dc} = X/(R\omega) \quad (1)$$

X and R are the equivalent reactance and resistance of the system respectively, ω is the angular frequency.

The calculation of attenuation time constant of DC component of short-circuit current is mainly used to check the breaking capacity of AC circuit breaker. According to relevant standards [10], the standard time constant of DC component attenuation is 45ms.

After considering the influence of DC component, the relationship between the actual AC breaking capacity of the circuit breaker ($I_{ac.R}$) and the rated breaking current ($I_{ac.N}$) is shown in the formulas (2) and (3).

$$I_{ac.R} = K \cdot I_{ac.N} \quad (2)$$

$$K = \sqrt{\frac{1+2e^{-2t_{min}/T_{dc.N}}}{1+2e^{-2t_{min}/T_{dc.R}}}} \quad (3)$$

t_{min} is the minimum breaking time of the circuit breaker, which is equal to 40ms [10]; $T_{dc.N}$ is the rated standard time constant of the circuit breaker, which is equal to 45ms; $T_{dc.R}$ is the attenuation time constant of DC component of short-circuit current calculated in practice. K is the conversion coefficient after considering the influence of DC component.

According to the formula (2) and (3), when the $T_{dc.R}$ is larger than $T_{dc.N}$, the AC component breaking capacity of the circuit breaker is lower than the rated value.

2.2. Influence of photovoltaic power station

According to the principle and typical design of photovoltaic power generation, the photovoltaic element generates DC power through sunlight and inverts it into AC output. It is assumed that the DC power generated by the power generation system is constant at the moment of short circuit, and the AC output after inverter is also approximately constant power. Therefore, after a short circuit occurs, due to the sharp drop of bus voltage, the output current of the inverter also increases until the inverter protection acts and the output is closed [10].

For the centralized photovoltaic power station, since the photovoltaic power generation part is separated from the main network system through the frequency converter, the output power and current are controlled by the vector control strategy of the frequency converter. When the photovoltaic system provides dynamic reactive power, it will increase the short-circuit current of the AC system [9~10].

3. A 330kV local power system grid

The local 330kV power grid is powered by a transmission network composed of two 750kV substations and ten 330kV substations. The schematic diagram of grid structure is shown in Fig.1. The main transformer scale of 750kV substation S1 is 3×2100MVA. The main transformer scale of 750kV substation S2 is 2×2100MVA. The scale of main transformers in 330kV substation is 2×360MVA. In S1 power supply area, two thermal power plants are connected to S1-1 substation and S1-3 substation respectively. One DC line is connected to 330kV side of 750kV substation S1. In S2 power supply area, one thermal power plant is connected to S2-5 substation.

Two operation modes are considered in this study. One is that two 750kV substations operate independently, i.e. S1-2~S2-1 channels are disconnected. The other is the joint operation of two 750kV substations through S1-2~S2-1. The conductor section of each line in the grid is 4×400mm^2. The line length information is shown in Figure 1. The scale of S1-G1 is 4×300MW. The scale of S1-G2 is 2×330MW. The scale of S2-G1 is 2×300MW. The DC transmission power flow is 1500MW.

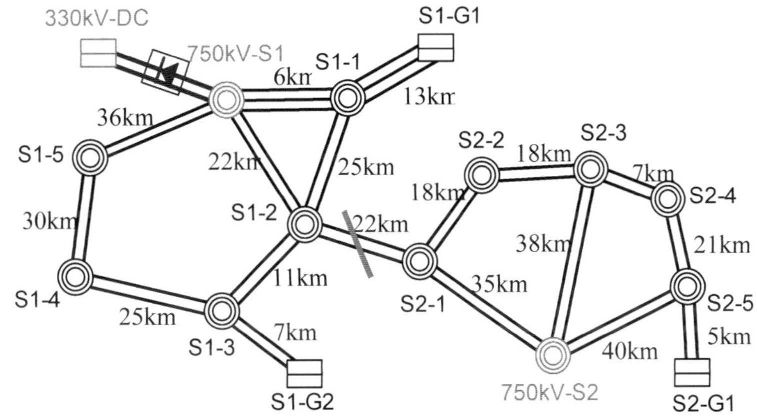

Fig.1 Diagram of grid structure

4. Calculation of SCC under various influencing factors

4.1 Grid operation mode

The short-circuit currents of two different modes of the power grid are calculated. As shown in Table 1, when the power grid operates jointly, the short-circuit current of each bus is greater than that of independent operation.

The operation mode of 330kV power grid has relatively little influence on 750kV voltage level, and the influence amplitude is no more than 6%. The short-circuit current of two regional interconnection substations is greatly affected. After the two areas are disconnected, the short-circuit current of S1-2 substation is reduced by about 24%, and that of S2-1 substation is reduced by about 44%.

Table 1 Calculation results of short circuit current under two operation modes

		Single phase /kA	Three phase /kA
750kV	S1	39.5/39.3	43.5/43.2
	S2	29.1/27.9	31.4/29.6
330kV	S1	58.6/51.2	55.0/47.0
	S2	46.9/33.7	42.5/28.2
	S1-1	51.5/45.2	50.4/43.1
	S1-2	48.8/37.2	48.2/36.8
	S1-3	38.0/31.9	38.9/31.8
	S1-4	24.8/22.6	27.6/24.3
	S1-5	25.5/23.6	27.9/25.1
	S2-1	40.7/22.7	41.6/21.8
	S2-2	33.2/22.9	34.7/21.7
	S2-3	35.9/25.6	34.8/22.5
	S2-4	33.5/25.0	31.9/21.8
	S2-5	35.5/27.3	33.7/23.8

Note: in "A/B", A is the result of joint mode and B is the result of independent mode.

4.2 DC component of short circuit current

Calculate the attenuation time constant of DC component of short-circuit current at each station. Taking three-phase short-circuit current as an example, the main substation results are shown in Table 2 and Table 3. The rated breaking capacity of 750kV and 330kV side circuit breakers is 63kA.

According to the calculation results, the attenuation time constant of DC component in this area is larger than the standard time constant of 45ms. According to formula (2) and (3), the actual AC breaking capacity of circuit breakers in each station is about 76%~89% of the rated breaking capacity. The actual AC breaking capacity of the circuit breaker is generally reduced.

Comparing the attenuation time constants at the buses of two voltage levels in 750kV substation, the DC component of short-circuit current in this area has a more serious impact on 330kV side. Comparing the two power grid operation modes, the attenuation time constant at 750kV side of S1 and S2 substations has little change. The breaking capacity of 750kV side circuit breaker is basically not affected.

Table 2 Actual breaking capacity under independent operation mode

Substation	S1		S2		S1-2	S2-1
Voltage/kV	750	330	750	330	330	330
SCC/kA	43.2	47.0	29.6	28.2	36.8	21.8
$T_{dc.R}$/ms	80.3	113.6	77.9	97.4	77.1	74.4
K	0.877	0.820	0.883	0.844	0.885	0.892
$I_{ac.R}$/kA	55.3	51.7	55.6	53.2	55.8	56.2

Table 3 Actual breaking capacity under joint operation mode

Substation	S1		S2		S1-2	S2-1
Voltage/kV	750	330	750	330	330	330
SCC/kA	43.5	55.0	31.4	42.5	48.2	41.6
$T_{dc.R}$/ms	80.2	151.7	76.7	185.1	77.3	107.8
K	0.878	0.783	0.886	0.763	0.884	0.828
$I_{ac.R}$/kA	55.3	49.4	55.8	48.1	55.7	52.2

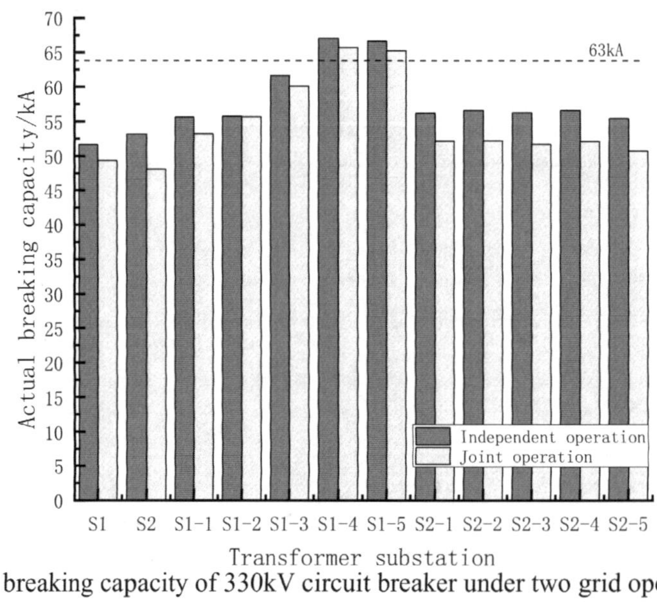

Fig.2 Actual breaking capacity of 330kV circuit breaker under two grid operation modes

The actual breaking capacity of the bus circuit breaker of each 330kV substation is shown in Fig.2. When the actual breaking capacity is greater than 63kA, it shall be considered as 63kA. Under the joint operation mode, the 330kV short-circuit current of 750kV substation S1 is 55kA, which exceeds the actual breaking capacity of the circuit breaker by 49.4kA.

For the area S1 where the DC is located, the actual breaking capacity of the circuit breaker of each 330kV substation is 55.6kA~63kA for independent mode. The breaking capacity is reduced by about 0.1kA~2.4kA when joint operation is adopted. For the area S2, the actual breaking capacity of the circuit breaker of each 330kV substation is 55.4kA~56.6kA for independent mode. The breaking capacity is reduced by about 4.0kA~4.7kA when joint operation is adopted. When the power grid operates jointly, the relative electrical distance of each substation is shortened, so the DC component of short-circuit current has a greater impact on S2 area under this mode.

4.3 Photovoltaic power station

In order to develop green power, it is considered to build 200MW photovoltaic power plant in this area. This paper studies the impact of different access schemes on short-circuit current.

Two photovoltaic access schemes are proposed. For Scheme I: 200MW photovoltaic power station is centralized construction which connected to S1-4 substation. For Scheme II: 100MW will be built in each of the two areas, respectively connected to S1-4 substation and S2-4 substation.

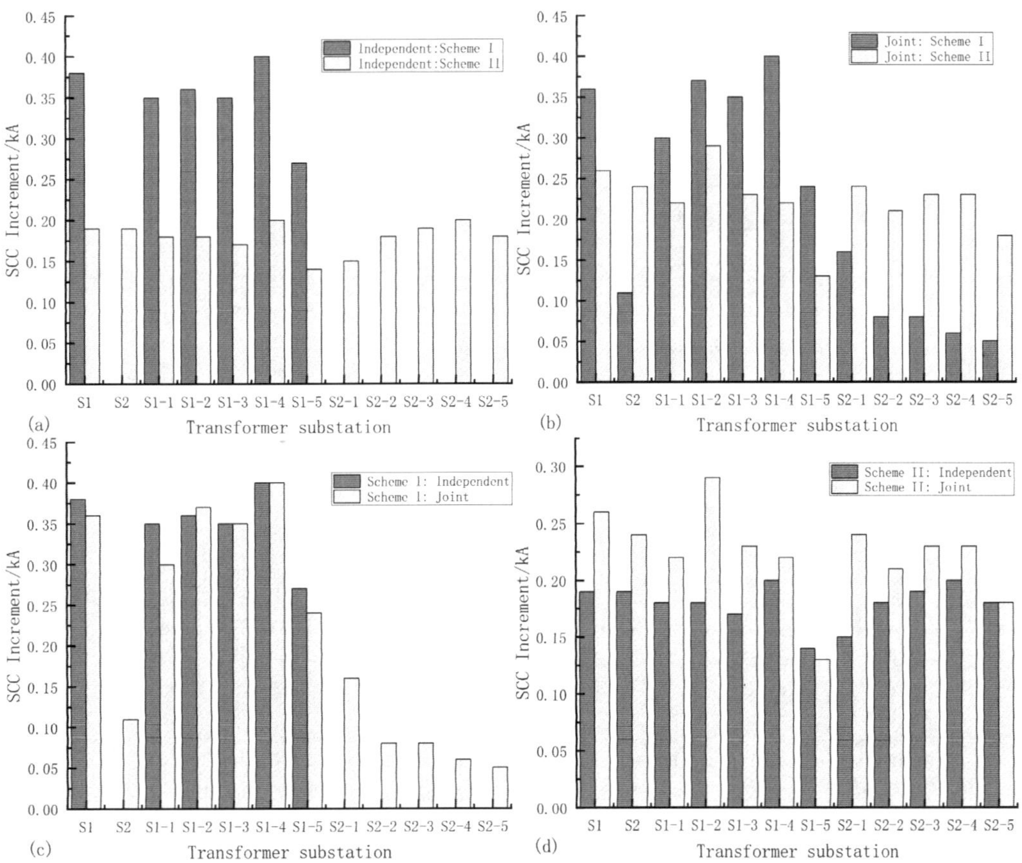

Fig.3 Influence amplitude of photovoltaic power station on short-circuit current

In independent mode and joint mode, the short-circuit current increase amplitude after photovoltaic power station accessed to each substation is shown in Fig.3. The access of photovoltaic power station has limited short-circuit current increase to the system. The access of 100MW and 200MW photovoltaic power stations increases the short-circuit current at the access point by about 0.2kA and 0.4kA respectively.

According to Fig.3 (a), under the independent mode, the photovoltaic power station will not affect the 330kV short-circuit current in other areas. According to Fig.3 (c), when 200MW photovoltaic power station is connected to s1-4 substation, the grid operation mode does not affect the variation range of short-circuit current at the access point. In the combined mode of power grid, the increase of short-circuit current of photovoltaic power station is inversely proportional to the electrical distance. According to Fig.3 (d), when two 100MW photovoltaic power stations are connected to S1-4 and S2-4 substations respectively, the short-circuit current increment of each substation under the joint mode is greater.

5. Conclusion

This paper focuses on the influence of DC short-circuit component and photovoltaic power station on the short-circuit current of power grid. The conclusions are obtained as below:

(1) The attenuation time constant of DC component of short-circuit current in DC near area power network is large. The influence of DC component should be comprehensively considered when checking the breaking capacity of circuit breaker. In this example, considering the influence of DC component, the actual breaking capacity of the circuit breaker is about 76%~89% of the rated breaking capacity.

(2) The access of photovoltaic power station will slightly increase the short-circuit current of regional power grid. When the 100MW~200MW photovoltaic power station is connected to the power grid, the short-circuit current at the access point increases by about 0.2kA~0.4kA.

(3) When the power grid adopts the joint mode, the equivalent electrical distance between each substation decreases. The influence of short-circuit DC component and photovoltaic power station on short-circuit current of power grid has increased.

References

[1] Jin Haosheng, et al. Improved short-circuit current calculation for distribution network considering distributed photovoltaic with capability low voltage ride-through[J]. Electrical Measurement and Instrumentation, 2020, v.57;No.735(10):35-40.

[2] PENG Shengjiang, et al. Research on the influence of photovoltaic access on a power grid[J]. Power System Protection and Control, 49(5):8.

[3] XI Wei, ZHANG Zhe, et al. Research on fault characteristic of the PV inverter[J]. Power System Protection and Control, 2015, 43(7):70-75.

[4] WU Yan, ZHUANG Jufeng, et al. Analysis of short circuit calculation model of photovoltaic power station[J]. Electronic Component and Information Technology, 2020, v.4; No.31(01):130-132.

[5] ZHANG Zengqiang, et al. Analysis and suppression measures of short circuit current in 750 kV Fenghuang area[J]. Sichuan Electric Power Technology, 2015, 000(004):20-23.

[6] CHEN Wei, et al. Transmission Expansion Planning Considering Adaptability of Grid to Short — circuit Current[J]. Sichuan Electric Power Technology, 2018, 041(001):74-77,82.

[7] Fang Buxun, et al. 220 kV network expansion planning considering short-circuit current constraint [J]. Electrical Measurement and Instrumentation,2018,55(21):25-30.

[8] CAO Wei, et al. Study on calculation methods for DC component and decay time constant of branch short circuit current[J]. JOURNAL OF EIECTRIC POWER SCIENCE AND TECHNOLOGY ,2021,36(2):31-39.

[9] PAN Wei, et al. Research on Operating Modes of 750 kV Northwest China Power Grid Electro-Magnetically Coupled with 330 kV Power Grid[J]. POWER SYSTEM TECHNOLOGY, 2007,31(15):33-38.

[10] CUI Jingchun. High voltage AC circuit breaker[M]. China Electric Power Press, 2016.

Study on Short Circuit Current Calculation of EHV Overhead Line-GIL Hybrid Line

Honggang Jia[1a], Na Yan[1b], Wei Wang[1c], Zhe Wang[1d], Yi Sun[2a*], Yiqing Xu[2b], Shan Lu[2c]

[1] Economic and Technological Research Institute of State Grid Shaanxi Electric Power Co., Ltd., Shanxi 710000, China

[2] East China Electric Power Design Institute Co., Ltd., Shanghai 210000, China

[2a*]3130@ecepdi.com

Abstract—GIL electrical parameters are different from overhead lines. With the increasing application of GIL in actual projects, the influence of GIL needs to be considered when calculating the short-circuit current of the power grid. In this paper, the calculation of three-phase short-circuit current is carried out for the EHV double-circuit overhead line on the same tower with GIL. First, the electrical parameters of EHV GIL are calculated. The positive sequence resistance, reactance and capacitance values of a typical GIL are calculated. Secondly, EMTP-ATP software is used to modelling and simulation. The influence of GIL length on the short-circuit current and the position of short-circuit fault is studied respectively. Finally, the influence of the GIL unit length modelling accuracy and the grounding mode of the metal shell on the short-circuit current is studied. The research results of this paper can provide a theoretical reference for GIL simulation modelling and the evaluation of the impact of system short-circuit current.

1. Introduction

In recent years, GIL has been used more and more in transmission lines. The commissioning of the UHV Sutong GIL in 2019 represents that China has mastered the world's highest level of GIL design and laying technology. GIL generally forms a transmission line together with overhead lines. This type of hybrid line will cause a new impact in the power system. This article focuses on the calculation of the short-circuit current of the overhead line-GIL hybrid line.

At present, domestic and foreign scholars have carried out a large number of short-circuit current calculation studies for the power grid composed of overhead lines. There are few studies on the influence of GIL on short-circuit current. Literature [1] pointed out that each phase of the GIL in the tunnel is arranged separately, and the probability of a three-phase short-circuit failure is extremely low. There is a possibility of a three-phase short-circuit fault at the terminals on both sides of the GIL. Therefore, only three-phase short-circuit faults on both sides of the GIL are considered in the study. The GIL grounding system is studied in the literature [2]. The impedance parameters of the equivalent circuit between the GIL shell and the ground are analysed in the article. Literature [3~4] studied the structural design and electrical parameters of the GIL, which showed that the phases of the GIL are completely decoupled.

At this stage, there is little research on short-circuit current calculation of overhead line-GIL hybrid line. In the future, there will be more GIL applications in power transmission lines. In this paper, simulation modelling and calculation of three-phase short-circuit current are carried out for the EHV double-circuit overhead line on the same tower and the double-loop GIL hybrid line. The influence of GIL length, fault

Content from this work may be used under the terms of the Creative Commons Attribution 3.0 licence. Any further distribution of this work must maintain attribution to the author(s) and the title of the work, journal citation and DOI.
Published under licence by IOP Publishing Ltd

location, GIL modelling method and metal sheath grounding method on the short-circuit current is studied in detail.

2. GIL electrical parameter calculation

The cross-sectional view of the EHV dual-circuit GIL layout is shown as Fig.1. The GIL loop spacing is 5400mm, and the GIL phase spacing is 1000mm. The phase sequence of the two circuit of GIL conductor is the same. From top to bottom are A-phase, B-phase, and C-phase. The inner diameter and outer diameter of the single-phase GIL conductor is 62.5mm and 75mm. The inner diameter and outer diameter of the metal shell is 260mm and 266mm.

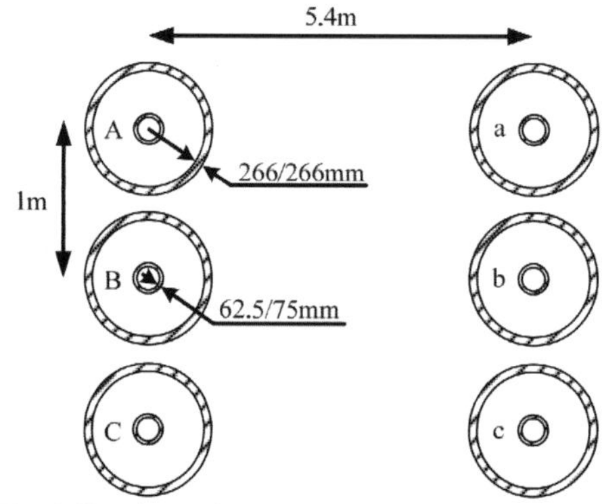

Fig.1 Section of EHV double-circuit GIL

The conductor shell of the GIL transmission line is made of aluminum material, which has good metal conductivity and electric field shielding effect. The conductor shell is well grounded to form a return flow, which cancels the power frequency magnetic field generated by the center conductor current on the outside of the shell. The trans-impedance parameter of the GIL conductor per unit length is much smaller than the self-impedance parameter. Therefore, the trans-impedance parameter can be ignored in the approximate calculation. GIL self-resistance R, self-inductance L and self-capacitance C calculation formulas are shown in formulas (1) ~ (3) [5~8].

$$R = \frac{\sqrt{f\mu}}{2r_1\sqrt{\sigma\pi}} \quad \Omega/\mathrm{m} \tag{1}$$

$$L = \frac{\mu}{2\pi}\ln(\frac{r_2}{r_1}) \quad \mathrm{H/m} \tag{2}$$

$$C = \frac{2\pi\varepsilon}{\ln(\frac{r_2}{r_1})} \quad \mathrm{F/m} \tag{3}$$

σ is the conductivity of the conductor. f is the operating frequency. μ is the permeability of the conductor. ε is the permittivity value. r_1 is the outer radius of the conductor. r_2 is the inner radius of the shell.

According to R, L, C, impedance matrix Z and admittance matrix Y can be obtained. The order parameter of GIL can be obtained by matrix change of Z and Y. Taking the order parameter Z_s of impedance as an example, its change formula is shown in formulas (4) ~ (6). The order parameter Y_s can be calculated in the same way [9~12].

$$Z_s = S^{-1} \cdot Z \cdot S \tag{4}$$

$$S = \begin{bmatrix} 1 & 1 & 1 \\ 1 & a^2 & a \\ 1 & a & a^2 \end{bmatrix} \tag{5}$$

$$a = e^{\frac{2}{3}\pi \cdot i} \tag{6}$$

3. EHV overhead line-GIL hybrid line model

Taking the 750kV voltage level as an example to carry out the research, the double-circuit overhead line on the same tower and the double-circuit GIL hybrid line model is shown in Fig.2. GIL is on the bus side of substation B. The cross section of the overhead line conductor is $4\times630\text{mm}^2$. The positive sequence resistance, reactance and capacitance per unit length of this line are 0.0136Ω/km, 0.2478Ω/km and 0.0144μf/km respectively [13~14]. The parameters of the line poles and towers are shown in Table 1. Both ends of the GIL metal shell are grounded. According to formulas (1) ~ (6), the positive sequence resistance, reactance, and capacitance per unit length of GIL are calculated as 0.0108Ω/km, 0.0845Ω/km, and 0.0141μf/km, respectively.

This paper ATP-EMTP is used to build the overhead line-GIL hybrid line model. The equivalent system impedances on both sides are set as: positive sequence resistance 15Ω, positive sequence reactance 15Ω, negative sequence resistance 7.5Ω, and negative sequence reactance 7.5Ω.

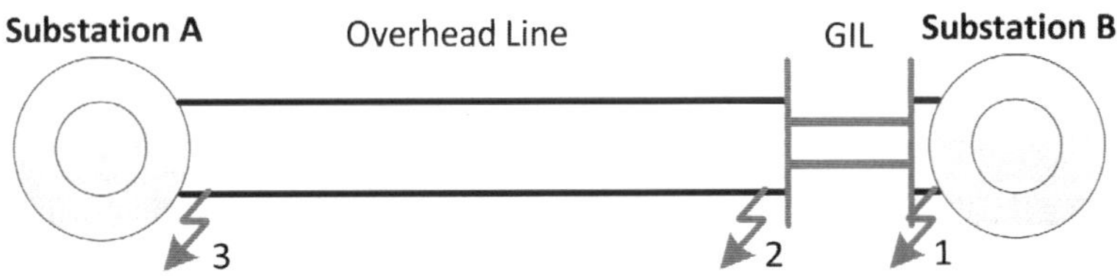

Fig.2 Schematic diagram of overhead line-GIL hybrid line

Table 1 Overhead line pole tower hanging line parameters

Phase sequence	Horizontal distance/m	Vertical distance/m	Sag/m
A	-8.0	58.0	8.0
B	-10.7	47.0	
C	-8.9	36.4	
a	8.0	58.0	
b	10.7	47.0	
c	8.9	36.4	
Ground wire	-10.7	68.1	6.0
Ground wire	10.7	68.1	

4. Calculation of short-circuit current of hybrid line

4.1 The influence of GIL length

The total line length is 30km. Calculate the three-phase short-circuit current when the GIL accounts for different lengths in the line. The three-phase short-circuit fault is set on the bus side of substation B. The short-circuit current calculation results are effective values as shown in Table2 and Fig.3. According to the calculation results, the short-circuit current is relatively low when all the lines are overhead lines, and the short-circuit current is relatively maximum when all the lines are GIL. When the entire section of the 30km overhead line is replaced with GIL, the short-circuit current increases by approximately 6.3%. According to the calculation of electrical parameters, this law can be verified. The positive sequence resistance and reactance of GIL are all smaller than the electrical parameters of overhead lines. Therefore, the equivalent impedance of the line system formed by GIL is relatively small.

As shown in Fig.3, because the overhead lines are not completely transposed, there is a certain degree of imbalance in the short-circuit current of each phase. The GIL of each phase is completely decoupled. In theory, the operating conditions of each phase are completely balanced. Therefore, when a symmetrical short-circuit fault occurs, the short-circuit current amplitude of each phase is the same.

Table 2 The influence of GIL length on three-phase short-circuit current

Overhead wire/km	GIL/km	Proportion of GIL	Maximum three-phase short-circuit current/kA
30	0	0%	50.4
25	5	17%	51.5
15	15	50%	52.5
5	25	83%	53.3
0	30	100%	53.6

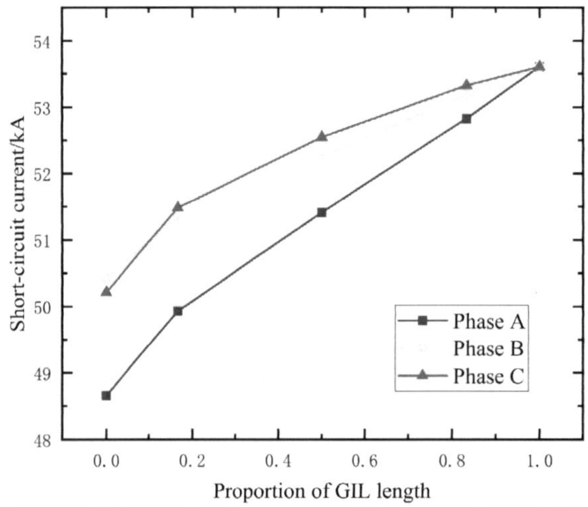

Fig.3 Change of short-circuit current with proportion of GIL length

4.2 Effect of fault location

According to different fault locations in the line, the short-circuit current calculation study of the overhead line-GIL hybrid line is carried out. The line conditions are 25km for overhead lines and 5km for GIL lines. The three-phase short-circuit faults are respectively set at the bus bar on the substation B (Position 1), the line side of the GIL (Position 2) and the bus bar on the substation A (Position 3), which are shown in Fig.1 respectively.

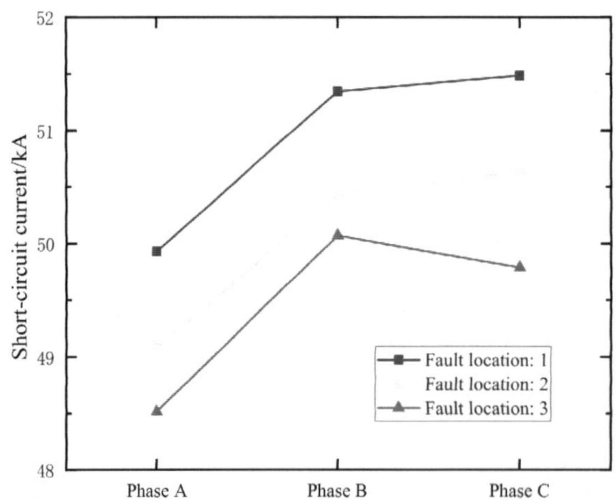

Fig.4 The effect of fault location on short-circuit current

According to the calculation results as shown in Figure 4, when the fault occurs at the bus bar on the substation B, the three-phase short-circuit current is the largest. When the fault occurs at the bus bar of substation A, the three-phase short-circuit current is the smallest. The difference between the short-circuit currents on both sides is about 1.5kA, and the amplitude is about 3%.

4.3 The influence of metal shell grounding method

In the actual project, the GIL metal shell is grounded every 50m. In order to save the calculation cost, in the simulation calculation, the unit length of GIL modelling is generally not 50m. This section focuses on the influence of GIL modelling length and grounding method on the calculation results of short-circuit current in the simulation calculation. The line conditions are 25km for overhead lines and 5km for GIL lines. For the grounding method, set the metal shell to be grounded on both sides, grounded on one side and not grounded on both sides in the simulation calculation. For the GIL unit length modelling accuracy, 5km GIL is divided into 1, 3 and 5 sections respectively. The three-phase short-circuit fault occurred on the bus bar substation B, and the calculation results are shown in Table 3.

In the simulation calculation, the GIL unit length modelling accuracy has almost no effect on the short-circuit current calculation. The grounding method on both sides of the GIL metal shell has an impact of about 0.2~0.6kA on the short-circuit current. The short-circuit current is relatively large when both sides of the GIL metal shell are grounded. According to the results, in order to save calculation cost, GIL modelling in simulation can simplify the number of segments, but it is necessary to ensure that the metal shells on both sides are grounded.

Table 3 The influence of grounding method and GIL modelling on short-circuit current

Modelling mode	Metal shell grounding method	Phase A/kA	Phase B/kA	Phase C/kA
One section	Grounded on both sides	49.9	51.3	51.5
	Grounded on one side	49.5	51.2	51.3
	Not grounded on both sides	49.5	51.2	51.3
Three sections	Ground on both sides of each section	49.9	51.4	51.5
Five sections	Ground on both sides of each section	49.9	51.4	51.5

5. Conclusion

This paper calculates and compares the electrical parameters of $4\times630mm^2$ overhead lines and EHV GIL. GIL positive sequence resistance and reactance are smaller than overhead lines, and the equivalent impedance of the line system formed by GIL is relatively small. According to the simulation calculation, the short-circuit current of the overhead line in the equivalent system is relatively low compared to the GIL.

For overhead line-GIL hybrid line, the three-phase short-circuit current is relatively large when the fault occurs on the GIL side. The short-circuit current is relatively small when the fault occurs on the opposite side of the GIL.

In the simulation calculation, the model length accuracy of GIL does not affect the short-circuit current calculation results. The grounding method of GIL metal shell affects the calculation result of short-circuit current. The short-circuit current is relatively large when both sides of the shell are grounded.

References

[1] LIU Z H, WANG C Y, LU S J, et al. Key Technical Parameters of UHV GIL in Sutong Utility Tunnel Project[J]. Power System Technology, 2020,44(6):2377-2385.
[2] TANG L L, CAI W, WANG Q, et al. Circulation Characteristics and Influencing Factors of GIL Grounding System[J].High Voltage Engineering,2010, 29(3):49-53.
[3] WANG Y N, DING W D, GOU Y, et al. Discussion of the Grounding Method of Gas Insulated

Transmission Line [J]. High Voltage Apparatus, 2016,52(4):98-102.
[4] WANG T D, WANG X Q, WAN F H, et al. Design and Application of High Voltage AC GIL [J]. High Voltage Apparatus,2014,50(10):107-111.
[5] YE Shengyong, et al. Research on calculation mode of short-circuit current based on BPA[J]. Power System Protection and Control,2015,43(18):138-143.
[6] YANG Shan, et al. Short-Circuit Current Calculation of Distribution Network With Distributed Generation[J]. Power System Technology,2015,39(7):1977-1982.
[7] CHE Renfei, et al. SHORT CIRCUIT CALCULATION OF DISTRIBUTION NETWORK BASED ON SUPERPOSITION THEOREM[J]. AUTOMATION OF ELECTRIC POWER SYSTEMS,2001,25(24):22-25,44.
[8] ZHANG Zhi, et al. An Improved Short-Circuit Current Calculation Method for Multi-Bus Ship Power System[J]. Power System Technology,2013(8):2212-2217.
[9] TIAN Hua, et al. Comparative study of short-circuit current calculation results based on PSASP software[J]. POWER SYSTEM PROTECTION AND CONTROL,2010,38(1):56-60,65.
[10] LUO Qinyue, et al. Structure Transform Calculation of Power Network Short-circuit Current[J]. ELECTRIC POWER CONSTRUCTION,2004,25(6):21-23.
[11] ZHANG Yunpeng. Analysis and Calculation of Fault Current for Optical-fiber Composite Overhead Ground Wire in Power System Fault of Complex[J]. Automation of Electric Power Systems,2015(16):132-137.
[12] ZOU Jun, et al. COMPUTATION OF FAULT CURRENT DISTRIBUTION AND SHIELDING COEFFICIENT OF GROUND WIRES FOR OVERHEAD TRANSMISSION LINES[J]. POWER SYSTEM TECHNOLOGY,2000,24(10):27-30.
[13] JIANG Taoning, et al. Calculation of the Single Phase Short-Circuit Current in Complex Ground System of UHV AC Transmission Lines[J]. Power System Technology,2011,35(10):16-21.
[14] ZHAO Jun, et al. Distribution Characteristics of Spatial Magnetic Field in UHV GIL Gallery in Short-circuit Fault Condition[J]. High Voltage Engineering,2021,47(5):1866-1871.

Analysis of power quality problems in DC distribution network

Shangshu Guan[a], Yidan Song, Shun Tao

North China Electric Power University, Changping, Beijing, China

[a]email: 18910852457@163.com

Abstract: With the application and popularization of distributed power supply and power electronic equipment in power grid, AC distribution is faced with problems such as low efficiency and low power quality. At the same time, due to the advantages of large power supply capacity, long power supply radius, high efficiency, good controllability and adaptability to renewable energy access, DC power distribution has become a crucial development direction of the future distribution system. Reasonable and systematic analysis of power quality problems of DC distribution network plays an important role in its construction and application. Firstly, this paper introduces the background, significance and status of the research on DC distribution network and its power quality. Then introduces the three main topology of DC distribution network structure and connection mode. After selecting pseudo bipolar structure to build the simulation circuit, the power quality phenomena such as voltage sag are simulated and analyzed by PSCAD. The results obtained are consistent with theoretical analysis.

1. Introduction

With the popularization of distributed power supply and power electronic equipment in the power grid, DC devices connected to the power grid are increasing, and the power consumption forms of users begin to change. Moreover, the introduction of a large number of power electronic devices will inevitably lead to power quality problems. Therefore, a new distribution mode is urgently needed to improve this series of power quality problems. In order to adapt to this change, the concept of DC distribution network is put forward. DC distribution network, with its advantages of large power supply capacity, long power supply radius, high operation efficiency, good controllability and adaptation to renewable energy access, has become a crucial development direction of future distribution system [1], and has become a focus of academic attention and research in recent years. In 1997, experts from the Dutch Energy Research Center proposed a plan to use DC distribution technology in residential buildings [2]. In 2004, Tokyo Institute of Technology proposed the use of DC power distribution network to build microgrid, and built a 10kW DC power distribution system model based on this concept. In 2010, Virginia Tech University proposed a sustainable DC power supply structure from the perspective of building construction.

DC distribution technology is popular in numerous fields due to its absence of phase, frequency, reactive power and other problems, but it has not been widely used in practical projects [3]. At the beginning of this century, plenty of DC distribution demonstration projects have been carried out by a wide variety of countries including China [4]. Due to the short time of research, most current studies on DC distribution network focus on topological structure and fault protection [5-6], and few studies on power quality.[7] analyzes the formation mechanism of voltage deviation. [8] analyzes the generation mechanism of ripple in DC distribution network. It is essential to deeply discuss the mechanism of each representative power quality problem and formulate their evaluation indexes.

Content from this work may be used under the terms of the Creative Commons Attribution 3.0 licence. Any further distribution of this work must maintain attribution to the author(s) and the title of the work, journal citation and DOI.
Published under licence by IOP Publishing Ltd

2. DC distribution network topology and PSCAD model

2.1 The main topology of DC distribution network

2.1.1 DC distribution network topology
Radial, also called tree, is the most basic topology of distribution network. Because the structure is simple, the control protection requirements are low, so the operation is convenient. But in this kind of structure, if the fault of higher voltage grade line occurs, its influence will spread to the whole distribution network, so the reliability of radial structure is low. Compared with the radial topology, the two-end power distribution topology has better reliability, more complex structure and higher requirements on protection devices. In the ring topology, when a fault occurs at any point in the loop, the protection device can move quickly. After fault location and isolation, the rest of the loop can still work normally. Its main disadvantages are complex structure, high cost and requirement for protection device. However, it has many advantages, such as high power supply reliability and easy to find fault points. It is still the trend of DC distribution network.

2.1.2 DC distribution network connection mode
The wiring modes of DC distribution network system mainly include unipolar asymmetric wiring mode, bipolar wiring mode consisting of a single converter (pseudo bipolar), and bipolar wiring mode consisting of series and parallel converters (true bipolar). Unipolar asymmetry only serves as a transition from the early stage of engineering. At present, the DC side connection mode of the projects that have been put into use is mostly pseudo-bipolar. Compared with the pseudo-bipolar connection, the true bipolar connection has a larger transmission capacity, but its significant disadvantage is higher cost.

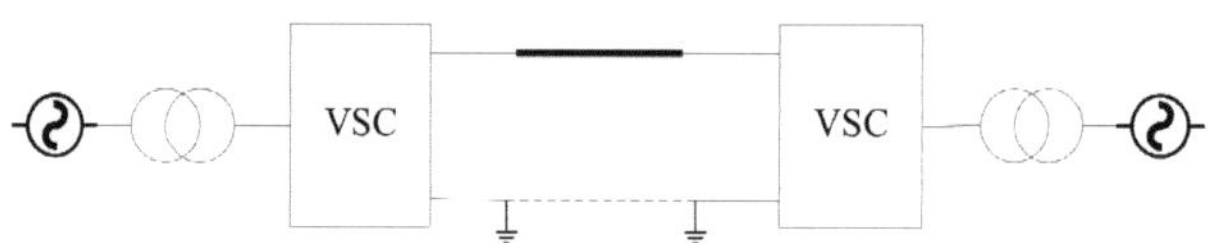

Figure 1. Unipolar asymmetric connection mode

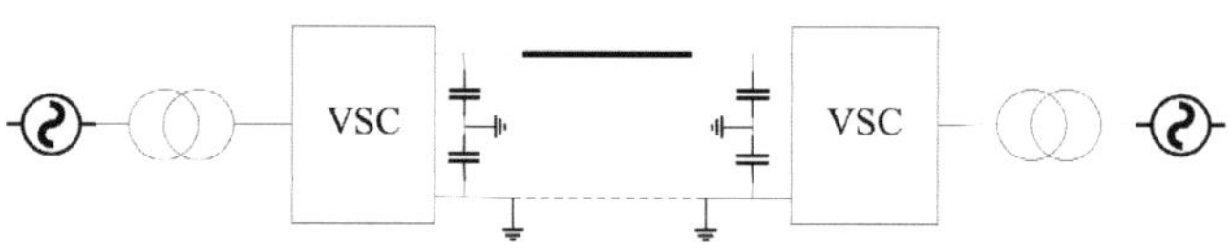

Figure 2. Pseudo bipolar connection mode

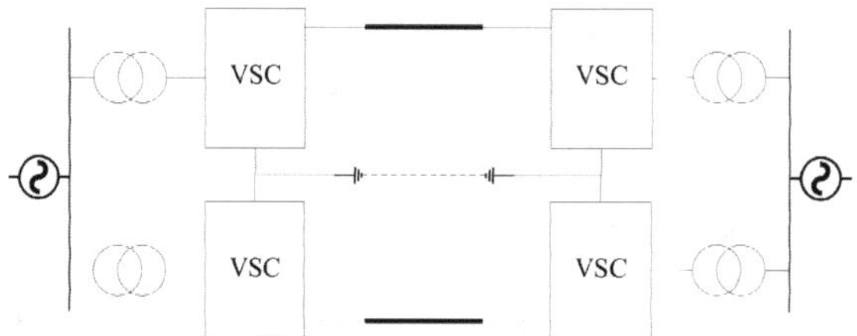

Figure 3. True bipolar connection mode

Considering the actual application scenarios of DC power distribution network, the capacity of a single converter need not be very large. In addition, after the DC circuit breaker is configured, when a converter or a section of line fault, can be quickly cut off, the withdrawal of a single converter will not constitute a great impact on the distribution system. In combination with economic factors, this paper finally chooses pseudo bipolar as the main connection mode of DC side, and builds a simulation circuit on PSCAD based on abovementioned theory.

2.2 PSCAD simulation design

Voltage source converter is often used in AC-DC conversion, which can be used not only as the grid-connection interface between AC distributed power supply and DC distribution network, but also as the grid-connection interface between DC distribution network and large power grid. This paper adopts the topology with voltage source converter as the core as shown in Figure 4.

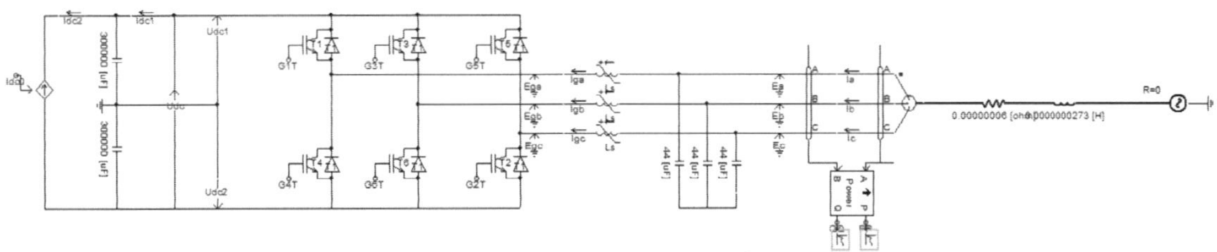

Figure 4. The main circuit used in the simulation

3. Generation mechanism and simulation of voltage sag

3.1 Mechanism of voltage sag

Voltage sag refers to the sudden drop in voltage at a point in a power system that can be restored to normal in a short time. IEEE defines it as a drop to 90-10% of the rating. The time from the onset to the end of the dip was defined as the duration, the ratio of the voltage square root value to the rated voltage square root value during the dip was defined as the temporary drop value, and the ratio of the temporary drop value to the DC bus voltage rating was defined as the dip depth MF%, $MF\% = \frac{u_{sag}}{u_d} \times 100\%$. When the DC power electrical short circuit and cause to happen on any point, voltage sag problems caused short circuit current increases sharply, then cause greatly on the DC power distribution network impedance voltage drop, and then clear the fault protection device, each point of the power system voltage immediately back to close to the previous value, this is a common scenario of voltage sag.

3.2 PSCAD simulation

3.2.1 Influence of fault type on voltage sag

Voltage sag caused by short circuit fault can be divided into: asymmetric voltage sag caused by single-phase grounding short circuit, two-phase short circuit and two-phase ground short circuit and symmetrical voltage sag caused by three-phase short circuit.

Table 1 Influence of different types of faults on voltage sag

fault type		fault duration（s）	drop duration（s）	sag value（kV）	MF%
AC side	single-phase	0.10	0.16	0.6312	78.69
	two-phase	0.10	0.17	0.7170	89.39
	two-phase ground	0.10	0.17	0.7213	89.93
	three-phase	0.10	/	/	/
DC side	pole-to-pole	0.03	0.06	0.5607	69.90
	single-pole ground	0.03	0.05	0.4158	51.84

Analysis of the above data shows that: When a single-phase ground fault or two-phase ground fault or two-phase short circuit fault occurs on the AC side, the bus voltage can be restored to the rated value after a short period of time after the fault is cleared. Similarly, the voltage sag of the DC bus can be caused by the short-circuit fault between the DC side poles and the single-pole grounding fault. After the fault is removed, the voltage of the DC bus can be restored to the rated value.

3.2.2 Influence of fault duration on voltage sag

Table 2 Influence of fault duration on voltage sag

fault duration（s）	drop duration（s）	sag value（kV）	MF%
0.01	0.04	0.4862	60.61
0.02	0.06	0.5212	64.97
0.04	0.09	0.6076	75.75
0.06	0.12	0.6325	78.85
0.10	0.16	0.6413	79.95

According to Table 2, by comprehensive analysis of the above simulation results, it can be seen that the longer the AC side short circuit fault lasts, the larger the DC side voltage sag duration, the greater the temporary drop value and the greater the dip depth, and the greater the harm to the power system operation.

3.2.3 Impact of a DC side fault on the AC side

1） single-pole grounding fault

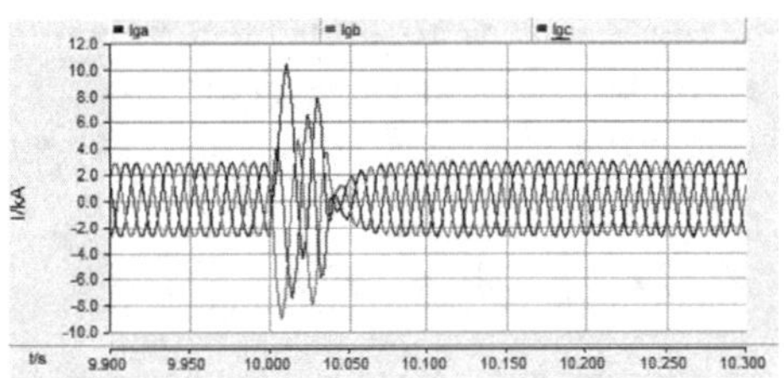

Figure 5. Three-phase current on the AC side

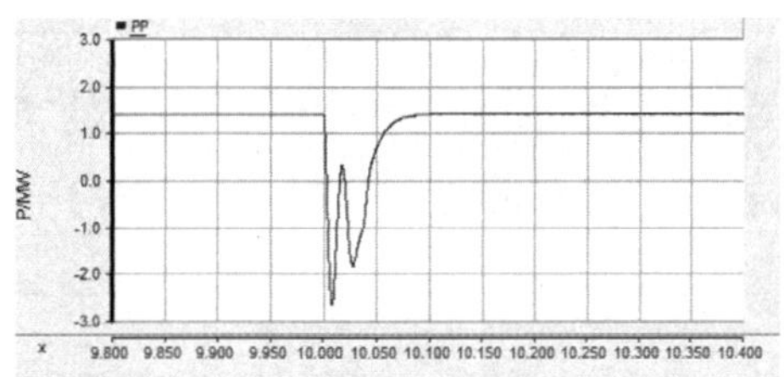

Figure 6. Three-phase active power

2） pole-to-pole fault

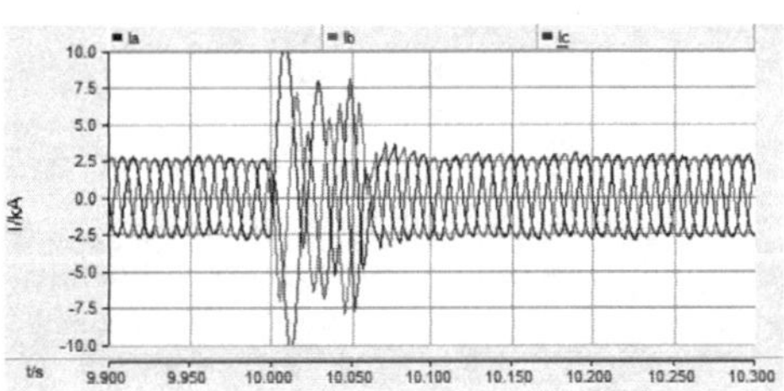

Figure 7. Three-phase current on the AC side

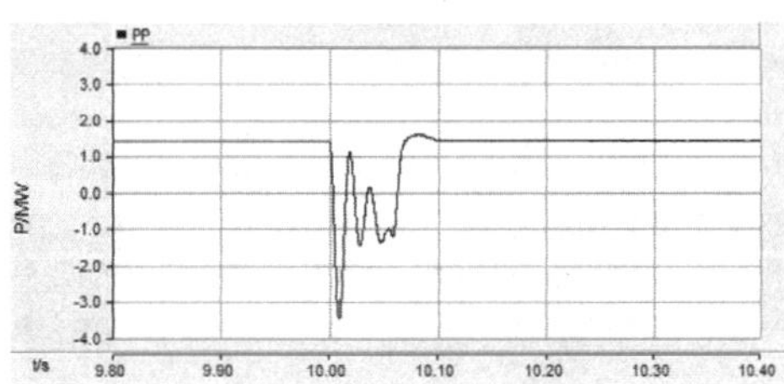

Figure 8. Three-phase active power

According to above waveforms, after the occurrence of the single-pole grounding fault or the pole-to-pole fault, the three-phase current on the AC side has large transient fluctuation and can be restored to the state before the fault within 0.1s, while the three-phase active power on the AC side drops instantly and can be restored to the reference value within 0.1s.Comparing the above two kinds of DC side faults, it can be seen that, under the same conditions, the recovery time of inter-pole short circuit fault is longer than that of single-phase grounding fault, which has a greater impact on the operation of power system.

4. Conclusion

With the application and popularization of distributed power supply and power electronic equipment in power grid, the traditional AC distribution mode faces a series of problems. At the same time, DC distribution network with large power supply capacity, long power supply radius, high efficiency, good controllability and adapt to renewable energy access and other unique advantages, continue to popularize the matter is only a matter of time. This paper takes the power quality related problems of DC distribution network as the research object. The main topology structure and common wiring mode of DC distribution network are introduced. Then the generation mechanism of voltage sag is discussed, and the pseudo-bipolar structure is selected to complete the circuit construction and simulation by PSCAD, and reasonable indicators are used for quantitative evaluation, simulation results are analyzed to verify the power quality phenomenon.

References

[1] Liao J Q, Zhou N C, Wang Q G. (2018) Definition and Correlation Analysis of Power Quality Index of DC Distribution Network. Proceedings of the CSEE, 38(23): 6847-6860+7119.
[2] Zhang Z P. (2014) The development prospects of DC power distribution technology. Distribution & Utilization, 7:16-19.
[3] Chen P W, XIAO X N, TAO S. (2016) Discussion on power quality problems for DC microgrid. Automation of Electric Power Systems, 40(10):148-158.
[4] Song Q, ZHAO B, LIU W H. (2013) An overview of research on smart DC distribution power network. Proceedings of the CSEE, 33(25): 9-19.
[5] Ji Y R, Yuan Z C, Zhao J F. (2016) A suitable voltage control strategy for DC distribution power network. Proceedings of the CSEE, 36(2): 335-341.
[6] Du Y, Jiang D Z, Yin R. (2015) Topological structure and control strategy of DC distribution network. Electric Power Automation Equipment, 35(1): 139-145.
[7] Han M X, Chen Y, Wang H J. (2016) Research on the limitation value of voltage deviation for DC distribution system. High Power Converter Technology, 3: 1-5.
[8] Jiao J R, Jia Q Q, Wang N. (2017) Ripple formation mechanism and superimposition characteristics for flexible DC distribution network. Power System Technology, 41(6):1726-1733.

Automated Quantum Volume Test

Xiaonan Liu[1a], Ming He[1b*], Junchao Wang[1c], Haoshan Xie[1d], Chenyan Zhao[1e]

[1]Information Engineering University, State Key Laboratory of Mathematical Engineering and Advanced Computing, Zhengzhou, China, 450001

[a]email: prof.liu.xn@foxmail.com, [c]email: wangjunchao11@126.com, [d]email: 15237933568@163.com, [e]email: 1098022623@qq.com

[*]Corresponding author's: [b]email: hm_0718@163.com

Abstract: As a benchmark for the overall performance of quantum computers, quantum volume has the advantage of being able to reflect the depth of running quantum circuits. But, the quantum volume test code provided by IBM needs to be executed manually, and the simulation result of the quantum simulator is used as the result of the volume test, so that users cannot quickly and accurately test the quantum volume of the actual quantum computer required. In response to this problem, this paper designs an automated quantum volume test program. The program automatically generates quantum volume sequences, selects the number of executions of quantum circuits, and defines real quantum computers to facilitate users to perform quantum volume tests on quantum computers provided by the IBM Quantum Cloud Platform. Simultaneously, according to the automated test program, the quantum volume of IBM's four small superconducting quantum computers was tested. The test results show that (1) the quantum computer is different, and the qubit layout and execution times *ntrials* are the same, will cause the quantum volume is uncertain; (2) the same quantum computer, whether ntrials is the same, the robustness of qubit coupling will be affected to a certain extent.

1. Introduction

In recent years, with the advent of the quantum age, compared with classical computers, quantum computers have more advantages in dealing with big data search [1], factorization [2], and solving linear equations [3]. Especially in 2019, Google has realized the milestone of "Quantum Supremacy" [4] with the 53-qubit "Sycamore" quantum computer developed by Google, and domestic and foreign teams have increased the development of quantum computers.

As the number of qubits continues to increase, quantum computers will reach the stage of quantum superiority, and their computing power in terms of climate prediction and drug research and development will far exceed that of supercomputers. However, although the advantages of parallel computing of quantum computers are extremely significant, compared with classical computers, quantum computers are still in their infancy, and quantum computers in the true sense have not yet been realized.

In order to design a quantum computer with high operational fidelity and resistance to external environmental noise, it is necessary to evaluate its overall performance. In the classical computing era, based on the test results of the linear system software package LINPACK [5], the peak value of floating-point computing of high-performance computers can be evaluated. In the quantum age, the number of qubits can show the superiority of quantum computers. Therefore, it can be used as one of the indicators to detect the performance of quantum computers. However, because quantum computers

Content from this work may be used under the terms of the Creative Commons Attribution 3.0 licence. Any further distribution of this work must maintain attribution to the author(s) and the title of the work, journal citation and DOI.
Published under licence by IOP Publishing Ltd

are simultaneously affected by many factors, such as connectivity, coherence time, crosstalk between bits, and error rate, it is not comprehensive to measure the superiority of a machine only from the number of qubits.

Considering the influence of many factors, IBM proposed a measurement standard that has nothing to do with the architecture--Quantum Volume (QV) [6-8]. At the same time, IBM released a quantum volume test set [9] on the Quantum Information Software Kit (Qiskit) [10], which aims to facilitate users to test the overall performance of quantum computers. But, the quantum volume test set needs to be executed manually, and the test set only defines the noise model, and does not measure the quantum volume of the real quantum computer. To this end, this article designs an automated quantum volume test program, and at the same time performs volume tests on the small-scale superconducting quantum computers supported by the IBM quantum cloud platform to observe the robustness of qubits.

2. Quantum Volume

How to measure the overall performance of a quantum computer is an urgent problem to be solved in a medium-scale quantum system with noise. Although Google once claimed to achieve "Quantum Supremacy" with the number of qubits, the number of qubits may not be the only measure of the power of a quantum computer. Therefore, IBM proposed an index that has nothing to do with the architecture-quantum volume.

2.1 Quantum volume circuit

The quantum volume test index given by IBM is mainly to measure the size of the state space available to the quantum processor and instantly capture the number of qubits that can generate a reliable random state on the processor [7]. Reference [11] provides the calculation formula of quantum volume:

$$log_2 V_Q = arg \max_m min(m, d(m)) \quad (1)$$

Here, m represents the number of qubits. Ideally, d(m) represents the maximum reachable depth of the quantum circuit, which is equal to the number of qubits m. In the state of noise, the value of d(m) is smaller than m, that is, noise is generated in d(m). And V_Q represents the size of the quantum computer's ability-the quantum volume value.

At the same time, in order to describe the quantum volume model more vividly, this article refers to [11], here is the circuit model of the quantum volume, as shown in Figure 1.

Here, the |0> state indicates that the qubits have been initialized; "π" indicates that in the quantum volume circuit, every two qubits are uniformly and randomly allocated, which can also be called uniform shuffling. If the number of qubits is odd, one bit will be in an idle state each time it is allocated.

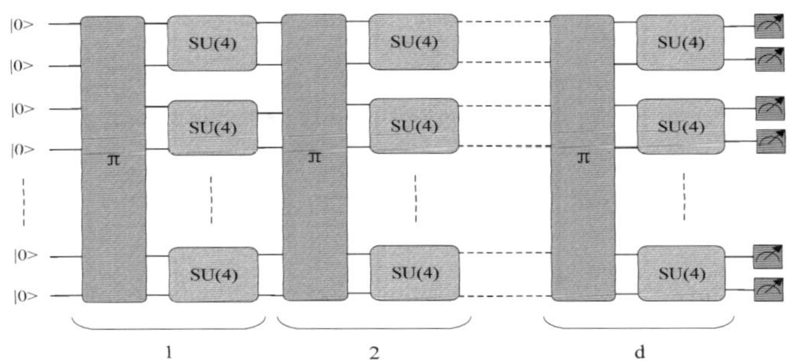

Figure 1. Quantum volume circuit

The SU(4) matrix is a 4 × 4 complex number matrix, and its mathematical expression is [12]:

$$\mathrm{SU}(4) = \{\mathrm{A} = 4 \times 4 \quad \text{comlex matrix} | \mathrm{A}^{\dagger}\mathrm{A} = 1, \det(\mathrm{A}) = 1\} \quad (2)$$

In this formula, "A" represents a 4 × 4 invertible complex matrix. In the quantum circuit, it can be decomposed into a cross form of 3 CNOT gates (Red circle) and 7 single-qubit gates (Pink box) by KAK decomposition [13], the circuit after KAK decomposition [14] is shown in Figure 2.

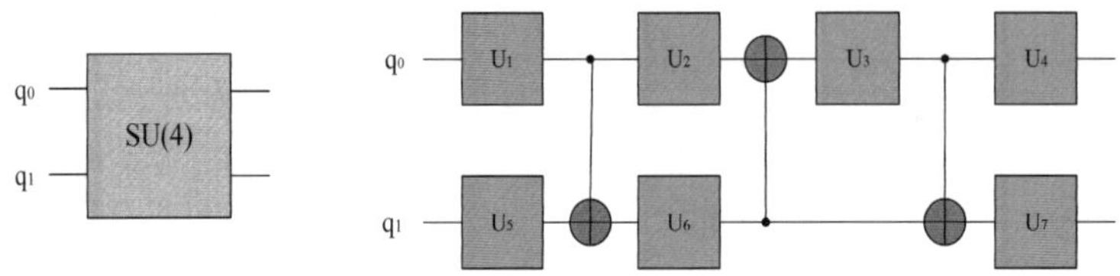

Figure 2. KAK decomposition of SU(4) gate circuit

U(θ,Φ,λ) gate is a general form of single quantum revolving gate, and the mathematical formula can be expressed as [15]:

$$U(\theta, \phi, \lambda) = RZ(\phi - \pi/2)RX(\pi/2)RZ(\pi - \theta)RX(\pi/2)RZ(\lambda - \pi/2) \quad (3)$$

CNOT gate is a kind of two-qubit gate, expressed in matrix form:

$$CNOT = \begin{bmatrix} 1 & 0 & 0 & 0 \\ 0 & 1 & 0 & 0 \\ 0 & 0 & 0 & 1 \\ 0 & 0 & 1 & 0 \end{bmatrix} \quad (4)$$

As shown in Figure 2, the red circle represents the target position, and the black dot represents the control position. According to formula (4), the control bit is the ground state |0>, no flipping occurs; the control bit is the excited state |1>, and the target bit will change from the ground state to the excited state (or the excited state to the ground state).

2.2 Quantum volume test set

In order to facilitate users to test quantum volume, IBM provides a quantum volume test set on Qiskit [9], which mainly includes seven aspects (Figure 3): generate the QV sequence, simulate the ideal QV circuit, calculate the heavy outputs, define the noise model, calculate the average fidelity of the gate, calculate the obtainable depth, and calculate the quantum volume.

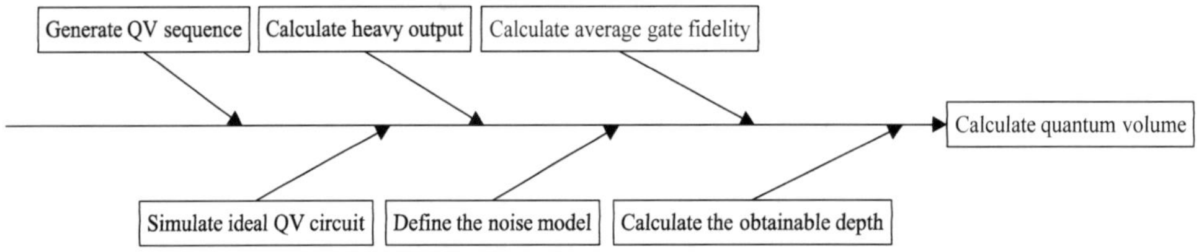

Figure 3. Quantum volume test set process

Here, the heavy output probability (HOP) can be used to describe the average error rate of the gate. Therefore, in the quantum volume test set, calculating the heavy outputs is one of the important links. The output distribution in the ideal state [11] is expressed as:

$$p_U(x) = |\langle x|U|0\rangle|^2 \quad (5)$$

Here, U represents the quantum volume circuit model, and x ∈ {0,1}m is a set of visible binary strings. The problem of heavy output generation refers to outputting a set of binary strings, and the repetition rate of the binary strings exceeds 2/3 [11]. In an ideal state of a quantum computer, the approximate expected value of the heavy output probability can be expressed by (1 + ln (2))/2 ~ 0.85 [16].

3. Automated Program Design Method

Compared with the quantum volume test set provided on Qiskit, this design has been improved in two major aspects: generating QV sequences and defining real quantum computers. The specific improvement methods are as follows.

3.1 Generate QV sequence

The generation of QV sequence includes four aspects: U(θ,Φ,λ) gate error test, CNOT gate error test, construction of quantum volume sequence and construction of execution times *ntrials*. Here, when performing a quantum circuit, there are different degrees of error in three aspects: initial state preparation, quantum gate operation, and final state measurement. Testing U(θ,Φ,λ) gate and CNOT gate operation errors needs to offset the initial state preparation and final state measurement errors.

3.1.1 Single qubit gate--U(θ,Φ,λ) gate error test

Readout error test. Set the qubit to the initial state|0>, and use the direct measurement scheme to measure each qubit of the quantum computer separately to measure the readout error. The error pseudo code is shown in Figure 4.

Add U(θ,Φ,λ) gate operation error test. Set the qubit to the initial state |0>, randomly generate the values of θ,Φ,λ (θ,Φ,λ ∈ [0,2π]), and add the U(θ,Φ,λ) gate operation, measure the operation error of adding U(θ,Φ,λ) gate. Its error pseudo code is shown as in Figure 5.

U(θ,Φ,λ) gate error calculation. Calculate the difference between the error measured by adding the U(θ,Φ,λ) gate and the readout error as the U(θ,Φ,λ) gate error. If a certain qubit U(θ,Φ,λ) gate error is less than 0, the addition of U(θ,Φ,λ) gate error to the bit is re-tested. The error pseudo code is shown in Figure 6.

```
Algorithm: Single-gate readout error
Input: Qubit measurement circuit
Output: Qubit measurement error
import the toolkit
provider ← Visit Quantum Cloud Platform
Quantum Machine, qubit, shots ← Define the name of the IBM quantum computer,
the number of qubits and the number of executions
backend ← Visit the Quantum Machine
qreg_q ← Define quantum register
creg_c ← Define classic registers
circuit ← Define quantum circuits
i ← 0
while i < qubit do
    cq ← Measure qreg_q, creg_c
    job ← Execute quantum circuit
    R_error_results[i] ← Execution circuit result
    i ← i + 1
print R_error_results
```

Figure 4. Single qubit gate readout error test pseudo code

```
Algorithm: Add U(θ,Φ,λ) gate error
Input: Qubit add U(θ,Φ,λ) gate measurement circuit
Output: Qubit add U(θ,Φ,λ) gate measurement error
import the toolkit
provider ← Visit Quantum Cloud Platform
Quantum Machine, qubit, shots ← Define the name of the IBM quantum computer,
the number of qubits and the number of executions
backend ← Visit the Quantum Machine
qreg_q ← Define quantum register
creg_c ← Define classic registers
circuit ← Define quantum circuits
i ← 0
while i < qubit do
    Randomly generate θ, Φ, λ
    Circuit add U(θ,Φ,λ) gate operation
    cq ← Measure qreg_q, creg_c
    job ← Execute quantum circuit
    Add_U_results[i] ← Execution circuit result
    i ← i + 1
print Add_U_results
```

Figure 5. Add U(θ,Φ,λ) gate error test pseudo code

```
Algorithm: U(θ,Φ,λ) gate error
Input: Qubit measurement error, qubit add U(θ,Φ,λ) gate measurement error
Output: U(θ,Φ,λ) gate error result
i ← 0
While i < qubit do
    U_error_results[i] ← Add_U_results[i] - R_error_results[i]
    if U_error_results[i] < 0
        Re-execute Figure 4 and Figure 5
print U_error_results
```

Figure 6. U(θ,Φ,λ) gate error test pseudo code

Construct U(θ,Φ,λ) gate error set. Sort the U(θ,Φ,λ) gate errors from small to large, and construct a single gate error set.

3.1.2 CNOT gate error test

Double-gate readout error test. Similar to the single-qubit gate error test, according to the connection graph of the sub-computer to be measured, the two adjacent qubits with edges are set to the initial state |0>. The direct measurement scheme is adopted to measure the readout error. The error pseudo code is shown in Figure 7.

```
Algorithm: Double-gate readout error
Input: Qubit measurement circuit
Output: Qubit measurement error
import the toolkit
provider ← Visit Quantum Cloud Platform
Quantum Machine, qubit, shots ← Define the name of the IBM quantum computer,
the number of qubits and the number of executions
backend ← Visit the Quantum Machine
qreg_q ← Define quantum register
creg_c ← Define classic registers
circuit ← Define quantum circuits
i ← 0
CL ← Number of connection lines
while i < CL do
    cq ← Measure qreg_q, creg_c
    job ← Execute quantum circuit
    DR_error_results[i] ← Execution circuit result
    i ← i + 1
print DR_error_results
```

Figure 7. CNOT gate readout error test pseudo code

```
Algorithm: Add CNOT gate error
Input: Qubit add CNOT gate measurement circuit
Output: Qubit add CNOT gate measurement error
import the toolkit
provider ← Visit Quantum Cloud Platform
Quantum Machine, qubit, shots ← Define the name of the IBM quantum computer,
the number of qubits and the number of executions
backend ← Visit the Quantum Machine
qreg_q ← Define quantum register
creg_c ← Define classic registers
circuit ← Define quantum circuits
i, j ← 0
CL ← Number of connection lines
while i < CL do
    for j in range(2):
        if j ← 0
            Circuit add CNOT gate operation forward
            cq ← Measure qreg_q, creg_c
            job ← Execute quantum circuit
            Add_CN_results[i][0] ← Execution circuit result
        else j ← 1
            Circuit add CNOT gate operation reverse
            cq ← Measure qreg_q, creg_c
            job ← Execute quantum circuit
            Add_CN_results[i][1] ← Execution circuit result
    Add_CN_results[i] ← (Add_CN_results[i][0] + Add_CN_results[i][1])/2
    i ← i + 1
print Add_CN_results
```

Figure 8. Add CNOT gate error test pseudo code

Add CNOT gate error test. Set the two adjacent qubits with edges to the initial state |0>, add the CNOT gate operation, and measure the corresponding CNOT gate error. Here, the CNOT gate error needs to be tested in both directions. For example, if there are adjacent edges between bit Q0 and bit

Q1, it is necessary to test Q0 as the control bit, Q1 as the target bit Q1(0-1) and Q1 as the control bit, and Q0 as the target bit (1-0) respectively. The average error of bidirectional CNOT gate obtained is the error of adding CNOT gate. Its error pseudocode is shown in Figure 8.

CNOT gate error calculation. Calculate the difference between the error of the added CNOT gate and the readout error as the CNOT gate error. The error pseudo code is shown in Figure 9. If the average value of the CNOT gate error of a certain two qubits is less than 0, the CNOT gate error of the double bit is retested.

Construct CNOT gate error set. Sort the CNOT gate errors from small to large to form a double gate error set.

```
Algorithm: CNOT gate error
Input: Qubit measurement error, qubit add CNOT gate measurement error
Output: CNOT gate error result
i ← 0
While i < qubit do
   CN_error_results[i] ← Add_CN_results[i] - DR_error_results[i]
   if CN_error_results[i] < 0:
      Re-execute Figure 7 and Figure 8
   i ← i + 1
print CN_error_results
```

Figure 9. CNOT gate error test pseudo code

3.1.3 Construct QV sequence

The core of constructing quantum volume sequence is to construct qubit_lists matrix. Because in the qubit_lists matrix, the matrix composed of the same number of qubits cannot be repeated, and for SU(4), the minimum is two qubits to work. So, the subset matrices of the qubit_lists matrix are 2 qubits, 3 qubits, 4 qubits, etc. respectively. At the same time, according to the KAK decomposition diagram of the quantum SU(4) gate and the error of the two-qubit gate is about 10 times [39] the error of the single-qubit gate, it can be seen that the construction of the sequence subset needs to preferentially select the adjacent qubits with the smaller CNOT gate error. Here, take the 5-qubit *ibmq_santiago* device (As shown in Figure 10.1) as an example to construct a sequence set. The process is shown in Figure 11.

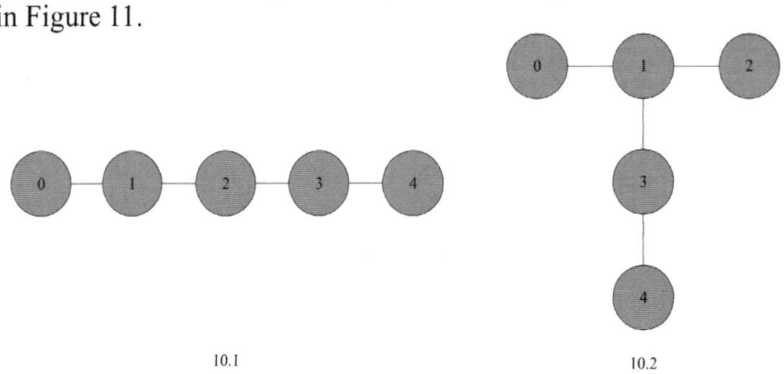

Figure 10. Qubit layout

(1) 2-qubit. The minimum CNOT gate error (such as 0-1, 1-2, 2-3, and 3-4) is selected as the sequence subset.

(2) 3-qubit. Choose the minimum sum of CNOT gate errors of 3 qubits with edges and adjacent (such as 0-1-2, 1-2-3, 2-3-4 and 3-4-5) as sequence subset.

(3) 4-qubit. Similar to the case of 3 qubits, select the minimum sum of CNOT gate errors of 4 qubits with edges and adjacent (such as 0-1-2-3, 1-2-3-4), as sequence subset.

(4) 5-qubit. Directly as a sequence subset (such as 0-1-2-3-4-5).

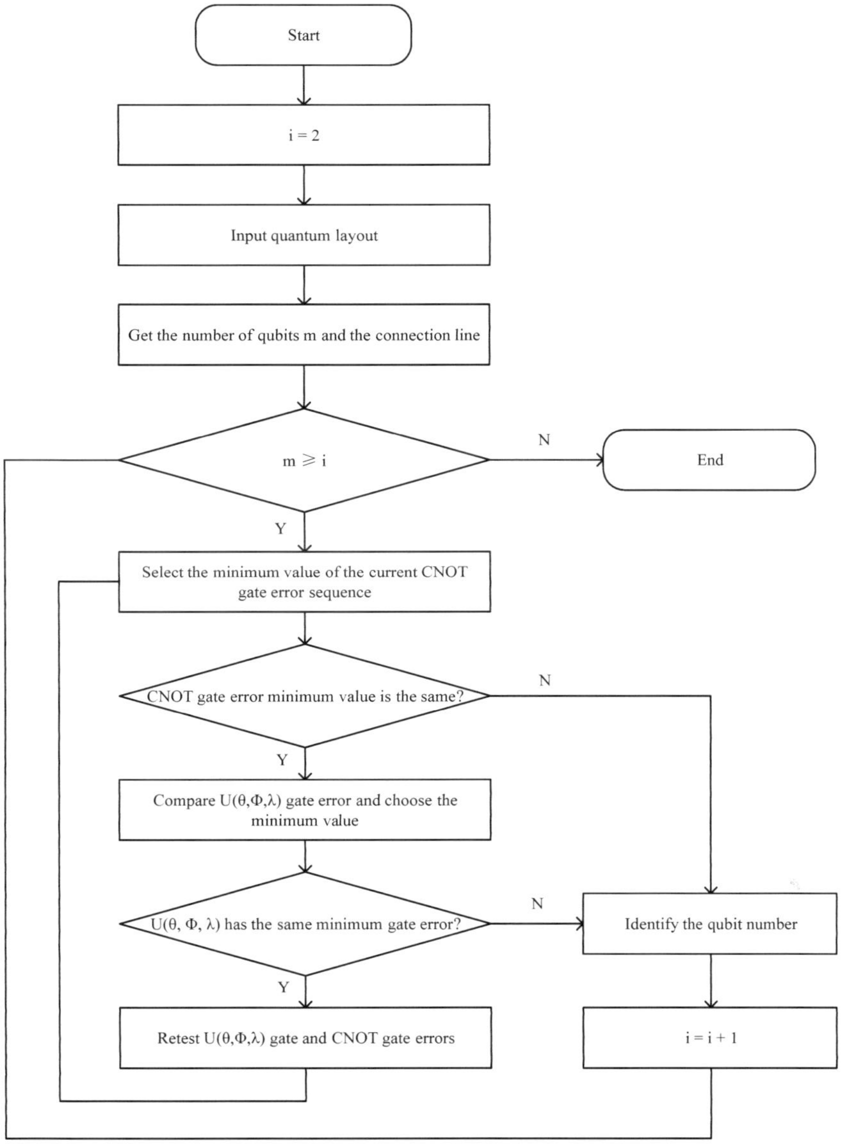

Figure 11. QV matrix selection process

If the sum of the gate errors of the qubit CNOT is equal, compare the sum of the gate errors of the corresponding qubit $U(\theta,\Phi,\lambda)$, and the smallest is regarded as the sequence subset. If the errors of the corresponding bits $U(\theta,\Phi,\lambda)$ gates are equal, the $U(\theta,\Phi,\lambda)$ gates and CNOT gates are tested again and then compared.

3.1.4 Build execution times ntrials

When submitting a job to the IBM quantum cloud platform, due to the impact of the waiting time of the quantum computer and the unstable performance of the quantum computer itself, when constructing execution times *ntrials*, it is not that the larger the value, the more accurate the quantum volume experiment. In this design, when testing $U(\theta,\Phi,\lambda)$ gate and CNOT gate errors, a time function is added. Each time a test is performed, the time is recorded once. The time interval between the two tests is the estimated waiting time in the queue. If the duration is smaller, you can increase ntrials appropriately.

3.2 Defining a real quantum computer

Ideally, by constructing U(θ,Φ,λ) gate and CNOT gate error models, the quantum volume can be quickly calculated. However, a real quantum computer is affected by the inherent errors of the instrument and the noise of the external environment, and its overall performance is constantly changing. The quantum volume calculated by simulation cannot fully demonstrate the overall performance of a quantum computer. Therefore, in order to truly test the quantum volume of a quantum computer, it is necessary to define a quantum computer. This article takes the ibmq_santiago device supported by the IBM Quantum Cloud Platform as an example. The Python code of the device definition is shown in Figure 12.

```
IBMQ.load_account()
provider = IBMQ.get_provider(hub='ibm-q')
backend = provider.get_backend('ibmq_santiago')
```

Figure 12. *ibmq_santiago* device definition code

In order to prevent the influence of noise on the experimental data, the U(θ,Φ,λ) gate and CNOT gate error test are executed in parallel in this design.

4. Quantum Volume Test Results

In this experiment, the quantum volume of four small-scale superconducting quantum computers *ibmq_belem*, *ibmq_lima*, *ibmq_quito* and *ibmq_santiago* connected to the back end of the IBM quantum cloud platform was tested. Here, the qubit layout of the first three quantum devices is the same (As shown in Figure 10.2). The result shows,

(1) The qubit layout and execution times are the same as ntrials, but the quantum computer is different, and the quantum volume is different. As shown in Table 1, the quantum volume comparison of different quantum computers with the same qubit layout.

(2)

Table 1. Quantum volume comparison

	ibmq_belem	ibmq_lima	ibmq_quito
ntrials = 100	1	8	8
ntrials = 200	8	8	8

It can be seen from Table 1 that when the number of executions ntrials = 100, the quantum volume of the *ibmq_belem* device is lower than the other two devices. When ntrials = 200, the quantum volume of the three quantum computers is the same, that is, the number of qubits that can be coupled with high robustness is 3.

(3) In the same quantum computer, ntrials are the same or different, which affects the robustness of qubit coupling to a certain extent. As shown in Table 2-5, the robustness comparison of 3-qubit coupling in *ibmq_belem*, *ibmq_lima*, *ibmq_quito*, and *ibmq_santiago* devices, respectively. Here, "N" means that it has not become a test target.

ibmq_belem device (Table 2). When ntrials = 100, quantum sequence subsets [1,2,3] and [1,3,4] do not meet the robustness requirements for coupling; ntrials = 200, the subset [0,1,3] and [1,3,4] meet the robustness requirements of coupling; the subset [1,3,4] has an impact on the robustness of the coupling when the ntrials are different.

Table 2. Robustness comparison of *ibmq_belem* device 3 quantum coupling

	[0,1,2]	[1,2,3]	[0,1,3]	[1,3,4]
ntrials = 100	N	1	N	1
ntrials = 200	N	N	8	8

ibmq_lima device (Table 3). When ntrials = 100, quantum sequence subset [0,1,2] and [0,1,3] meet the robustness requirements for coupling, the subset [1,3,4] does not meet the requirements; ntrials = 200, the subsets [0,1,2], [1,2,3], [0,1,3] and [1,3,4] all meet the robustness requirements of coupling; the subset [1,3,4] has an impact on the robustness of the coupling when the ntrials are different.

Table 3. Robustness comparison of *ibmq_lima* device 3 quantum coupling

	[0,1,2]	[1,2,3]	[0,1,3]	[1,3,4]
ntrials = 100	8	N	8	1
ntrials = 200	8	8	8	8

ibmq_quito device (Table 4). When ntrials = 100, the quantum sequence subset [0,1,2], [1,2,3] does not meet the robustness requirements of coupling, while [1,3,4] meets the requirements; ntrials = 200, the subsets [1,2,3] and [1,3,4] meet the robustness requirements of coupling; the subset [1,2,3] has an impact on the robustness of the coupling when the ntrials are different.

Table 4. Robustness comparison of *ibmq_quito* device 3 quantum coupling

	[0,1,2]	[1,2,3]	[0,1,3]	[1,3,4]
ntrials = 100	1	1	N	8
ntrials = 200	N	8	N	8

ibmq_santiago device (Table 5). When ntrials = 100, 200, the subsets [0,1,2], [1,2,3] and [2,3,4] all meet the robustness requirements of coupling; and when the same subset is executed at different times, it does not affect the robustness of the coupling.

Table 5. Robustness comparison of *ibmq_santiago* device 3 quantum coupling

	[0,1,2]	[1,2,3]	[2,3,4]
ntrials = 100	8	8	8
ntrials = 200	8	8	8

According to Table 2-4, for three quantum computers with the same qubit layout, the quality of Q1 as the central qubit will affect the robustness of the qubit coupling. As for the *ibmq_santiago* device, the larger the number of coupled qubits, the more SWAP gates it needs, and the higher the performance requirements for qubits.

5. Conclusion And Discussion

This article proposes an automated quantum volume test program. Which improves the original manually executed quantum volume test program, and shows how to generate a quantum volume sequence and select the number of executions ntrials; moreover, transform the quantum simulator into a real quantum computer. Finally, perform quantum volume testing on four small-scale superconducting quantum computers provided to users by IBM for free. Test data shows that different quantum computers with the same qubit layout and execution times *ntrials* have different quantum volumes. Simultaneously, for each quantum computer, the number of 3 qubits is selected as the quantum sequence subset. When the execution times is the same or different ntrials, it will have a certain degree of impact on the robustness of the qubit coupling.

As a measurement method, quantum volume helps to evaluate the overall performance of larger-scale quantum computers. The automated quantum volume test program designed in this paper will provide users with the circuit depth to test the IBM superconducting quantum computer more conveniently and quickly. Thence, this test system has certain practical significance in measuring the capabilities of quantum computers.

Here, in the case of a long waiting time in the queue, how to properly select the number of executions ntrials to ensure that the quantum volume of the quantum computer is accurately measured will be the future work.

References

[1] Lov, K. G. (1996) A fast quantum mechanical algorithm for database search. In: Proceedings of the 28th Annual ACM Symposium on the Theory of Computing., New York. pp. 212-219.

[2] Harry, B. (1996) .A short note on Shor's factoring algorithm. J. Association for Computing Machinery., 27(1): 89-90.

[3] Aram, W. H., Avinatan, H., Seth, L. (2009) Quantum algorithm for solving linear systems of equations. J. Phys. Rev. Lett., 15(103): 150502.

[4] Frank, A., Kunal, A., Ryan, B., Dave, B., Joseph, C. B., Rami, B., Rupak, B., Sergio, B., Fernando, G. S. L. B., David, A. B., Brian, B., Yu, Chen, Zi, J. C., Ben, C., Roberto, C., William, C., Andrew, D., Edward, F., Brooks, F., Austin, F., Craig, G., Marissa, G., Rob, G., Keith, G., Steve, H., Matthew, P. H., Michael, J. H., Alan, H., Markus, H., Trent, H., Travis, S. H., Sergei, V. I., Evan, J., Zhang, J., Dvi,r K., Kostyantyn, K., Julian, K., Paul, V. K., Sergey, K., Alexander, K., Fedor, K., David, L., Mike, L., Erik, L., Dmitry, L., Salvatore, M., Jarrod, R. M., Matthew, M., Anthony, M., Xiao, M, Kristel, M., Masoud, M., Josh, M., Ofer, N., Matthew, N., Charles, N., Murphy, Y. N., Eric, O., Andre, P., John, C. P., Chris, Q., Eleanor, G. R., Pedram, R., Nicholas, C. R., Daniel, S., Kevin, J. S., Vadim, S., Kevin, J. S., Matthew, D. T., Amit, V., Benjamin, V., Theodore, W. Z. J. Y., Ping, Y., Adam, Z., Hartmut, N., & John, M. M. (2019) Quantum supremacy using a programmable superconducting processor. J. Nature., 574(7779): 505-510.

[5] Jack, J. D., Piotr, L., Antoine, P. (2003) The LINPACK Benchmark: past, present and future. J. Concurrency and Computation: Practice and Experience., 15(9): 803-820.

[6] Lev, S. B., Sergey, B., Andrew, C., Jay, M. G., John, S. (2017) Quantum Volume. https://storageconsortium.de/content/sites/default/files/quantum-volumehp08co1vbo0cc8fr.pdf.

[7] Blume-Kohout, R., and Young, K. C. (2020) A volumetric framework for quantum computer benchmarks. J. Quantum., 4: 362.

[8] Petar, J., Ali, J. A., Lev, S. B., Isaac, L., Daniela, F. B., Markus, B., Lauren, C., Oktay, G., Toshinari, I, Naoki, K., Abhinav, K., George, A. K., Kevin, K., William, L., Eric, P. L., Douglas, T. M., Giacomo, N., Adinath, N., Hasan, M. N., Emily, P., Mary, B. R., Srikanth, S., Neereja, S., Cindy, W., Ken, X. W., Christopher, J. W., Jeng-Bang, Y., Eric, J. Z., Oliver, E. D., Jerry, M. C., Jay, M. G. (2021) Demonstration of quantum volume 64 on a superconducting quantum computing system. J. Quantum Science & Technology., 6(2): 025020.

[9] The Jupyter Book Community. (2021) Measuring Quantum Volume. https://qiskit.org/textbook/ch-quantum-hardware/measuring-quantum-volume.html.

[10] IBM. (2021) Qiskit/qiskit. https://github.com/Qiskit/qiskit.

[11] Andrew, W. C., Lev, S. B., Sarah, S., Paul, D. N., and Jay, M. G. (2019) Validating quantum computers using randomized model circuits. J. PHYSICAL REVIEW A., 100(3): 032328.

[12] Mahmoud, A. A. S., Moeen, KH. S., Hamada, M. S., and Fayad, H. M. (2013) Lie Algebra and Representation of SU(4). J. Electronic Journal of Theoretical Physics., 10(28): 9-26 .

[13] Farrokh, V., Colin, W. (2004) Optimal quantum circuits for general two-qubit gates. J. Physical Review A., 69(3): 032315.

[14] Zulehner, A., Wille, R. (2018) Compiling SU(4) quantum circuits to IBM QX architectures. https://arxiv.org/abs/1808.05661.

[15] Qiskit. (2021) U Gate. https://qiskit.org/documentation/stubs/qiskit.circuit.library.UGate.html.

[16] Scott, A, Lijie, C. (2017) Complexity-Theoretic Foundations of Quantum Supremacy Experiments. In: Proceedings of the 32nd Computational Complexity Conference., Dagstuhl, Germany. pp. 1-67 .

Research on path optimization of electric distribution vehicle with hybrid charging strategy

Jing Chang [1,a], Qiqian Zhang [2,b], Honghai Zhang [3,c*]

[1]College of Civil Aviation, Nanjing University of Aeronautics and Astronautics

[1]College of Civil Aviation, Nanjing University of Aeronautics and Astronautics

[1]College of Civil Aviation, Nanjing University of Aeronautics and Astronautics

[a]email: changjing2127@nuaa.edu.cn, [b]email: zhangqq@nuaa.edu.cn

[a]email: zhh0913@163.com

Abstract: At present, the number of new energy vehicles in China is increasing year by year. In the field of logistics and distribution, new energy vehicles are also showing vigorous vitality. This paper presents a two-stage optimization method for electric vehicle routing problem with hybrid charging strategy. In the first stage, a fuzzy transitive closure method is used to group customer orders, in the second stage, the decision variables and constraints related to the charging strategy are introduced, and a mixed integer linear programming model considering the three charging strategies of fast charging, slow charging and fast switching is constructed, finally, the validity of the proposed model is verified by a instance.

1. Introduction

Today, the global climate change is increasingly severe, countries began to seek low-carbon energy-saving new energy as a vehicle fuel.

The traditional vehicle routing problem (VRP) was first proposed by scholars Dantxing and Ramser in 1959. The traditional VRP is generally described as: reasonably organizing the vehicle routing problem under certain constraints, make the vehicles visit a series of delivery points and receiving points in a certain order to achieve a certain goal[1]. Based on the basic VRP (Vehicle Routing Problem) , different extension problems such as VRPTW (Vehicle Routing Problem with Time Window) and multi-Vehicle Vehicle Routing Problem are derived.

For multi-vehicle routing problem, the main research types are divided into two types: vehicle number limit and non-vehicle number limit. The multi-vehicle routing problem with unlimited number of vehicles was first proposed by Golden[2] in 1984. Li Jun[3] et al (2001) became the first scholar in China to study the multi-vehicle routing problem by studying the vehicle load restriction to determine the vehicle type and the vehicle number. Since then, more and more scholars have begun to pay attention to this kind of problem.

The vehicle Routing problem with Time Windows (VRP) is one of the most widely studied VRP related Problems, which was first proposed by Savelsbergh[4] in a paper called "Local Search in Routing Problems with Time Windows". At present, there are two types of vehicle routing problem with hard time windows and vehicle routing problem with soft time windows. Liu Zhishuo[5] (2009) studies the VRP problem with hard time windows, and designs an adaptive hybrid ant colony algorithm based on two-stage strategy of feasible solution. Zhang Kaiqing and Ji Qichun [6](2021) lead the VRP problem

Content from this work may be used under the terms of the Creative Commons Attribution 3.0 licence. Any further distribution of this work must maintain attribution to the author(s) and the title of the work, journal citation and DOI.
Published under licence by IOP Publishing Ltd

with soft time window to the road type and velocity time-varying condition, and design a two-stage algorithm to avoid the influence of traffic jam time on vehicle route.

Compared with the traditional vehicle routing problem, the electric vehicle routing problem is more complex. The Green Vehicle Routing Problem (GVRP) was first proposed by Erdogan and Miller-Hooks[7] (2011) . Frade [8](2011) et al. . Guofang[9](2020) and others put forward the problem of service station location and electric vehicle routing optimization considering the strategy of customer self-pickup, and give a hybrid heuristic based on improved saving algorithm and Tabu search algorithm. Zhang Qiang[10] et al. (2021) proposed an improved moth algorithm to solve the shortest total travel time electric vehicle (EV) path optimization model.

However, in the current urban planning system, the distribution of charging stations is scattered and irregular, the number of charging stations is obviously less than the number of gas stations, the battery charging time is significantly longer than the refueling time of fuel vehicles. Therefore, it is very important to explore a more flexible charging mode and distribution strategy for the limited technical conditions.

In this paper, the research of EV route optimization is divided into two parts: (1) customer grouping based on fuzzy clustering. In order to improve the efficiency of route optimization, this paper determines the key factors of EVRP, and uses the method of transitive closure based on fuzzy equivalent matrix to group customer orders. (2) path optimization of electric vehicle considering hybrid charging strategy. A hybrid integer programming model for EVRP considering three charging strategies is established. Finally, an example is given to demonstrate the effectiveness of the proposed method.

2. Problem Description and mathematical model

2.1. Grouping customers

2.1.1. Selecting decision indicators. In this paper, the following important indicators are selected as decision-making indicators: (1) demand for goods (2) time pressure (3) location (4) external similarity of products (5) product value (6) product safety requirement. The specific definitions of decision indicators are shown in Table 1.

Among the six decision indexes mentioned above, $\varphi^1{}_i$、$\varphi^2{}_i$、$\varphi^3{}_i$ are quantitative decision indexes, $\varphi^4{}_i$、$\varphi^5{}_i$、$\varphi^6{}_i$ are qualitative decision indexes. The qualitative decision indicators are described in five kinds of description: “Very high”, “High”, “Medium”, “Low”and “Very low”.

Table 1 Definition of decision indicator

Notation	Variable definition	Description
$\varphi^1{}_i$	Demand for goods	Expressed in terms of the quality of goods delivered to customer
$\varphi^2{}_i$	Time pressure	measured by the latest delivery time of customer i.
$\varphi^3{}_i$	Geographic location	Refers to the actual geographic location of customer i.
$\varphi^4{}_i$	Product external similarity	Refers to the external similarity of goods and other goods assigned to customer i.
$\varphi^5{}_i$	Product value	The value of the product delivered to customer i.
$\varphi^6{}_i$	Product safety requirements	Refers to the safety requirements on the distribution of products of customer i, determined by customers.

2.1.2. Processing of qualitative decision index. In this paper, the intermediate membership function is used for fuzzy evaluation[11], and the linguistic description of qualitative decision index is transformed into triangular fuzzy number φ, as shown in Table 2:

Table 2 Five kinds of description corresponding triangular fuzzy nembers

Desciption	Triangular fuzzy number	Evaluation scale
Very low	(0,0,0.25)	0
Low	(0,0.25,0.5)	0.25
Medium	(0.25,0.5,0.75)	0.5
High	(0.5,0.75,1)	0.75
Very high	(0.75,1,1)	1

Firstly, the fuzzy numbers make a translation-standard deviation transformation to eliminate the dimensional difference. Then the translation-range transformation is carried out to make the data fall in the range of [0,1] , and then the similarity of fuzzy numbers is calculated by using the method in reference 12[12].

1) Translation-standard deviation transformation

Index of the triangular fuzzy number corresponding to the decision index p can be calculated by:

$$a_{i,t}^{p'} = \frac{|\alpha_{i,t}^{p} - \overline{\alpha}_{t}^{p}|}{S_{t}^{p}} \quad (i = 1,2,\ldots,N; t = 1,2,3) \tag{1}$$

where N is the number of customers.

$$\overline{\alpha_{t}^{p}} = \frac{\sum_{i=1}^{N} \alpha_{i,t}^{p}}{N} \tag{2}$$

$$S_{t}^{p} = \left[\frac{\sum_{i=1}^{N} (\alpha_{i,t}^{p} - \overline{\alpha_{t}^{p}})^{2}}{N-1}\right]^{\frac{1}{2}} \tag{3}$$

2) Translation-range transformation

$$\hat{a}_{i,t}^{p} = \frac{a_{i,t}^{p'} - \min\limits_{1\le n\le N}\left\{a_{n,t}^{p'}\right\}}{\max\limits_{1\le n\le N}\left\{a_{n,t}^{p'}\right\} - \min\limits_{1\le n\le N}\left\{a_{n,t}^{p'}\right\}} \tag{4}$$

The triangular fuzzy number of the standardized qualitative decision index can be expressed by Eq.(5).

$$\hat{\varphi}_{i}^{p} = [\hat{\alpha}_{i,1}^{\mathrm{p}}, \hat{\alpha}_{i,2}^{\mathrm{p}}, \hat{\alpha}_{i,3}^{\mathrm{p}}] \tag{5}$$

3) Calculating the similarity between customer i and customer j corresponding to the decision index p

$$X_{\mathrm{ij}}^{p} = 1 - \left(\sum_{t=1}^{3} |\hat{\alpha}_{i,t}^{\mathrm{p}} - \hat{\alpha}_{j,t}^{\mathrm{p}}|\right)/3 \quad (i, j = 1,2,\ldots,N) \tag{6}$$

2.1.3. Processing of quantitatine decision index. It is only necessary to standardize the data before calculating the similarity between the quantitative decision indicators X_{ij}^{p} .

$$X_{\mathrm{ij}}^{p} = 1 - \frac{|\hat{\varphi}_{i}^{p} - \hat{\varphi}_{j}^{p}|}{\max\limits_{1\le n\le N}\left\{\hat{\varphi}_{n}^{p}\right\} - \min\limits_{1\le n\le N}\left\{\hat{\varphi}_{n}^{p}\right\}} \tag{7}$$

2.1.4. Decision index comprehensive processing. The similarity between the customer i and the customer j can be calculated by:

$$S_{ij} = \sum_{p=1}^{m} \omega^p x_{i,j}^p \tag{8}$$

where ω^p represents the weight of the decision index p, which needs to satisfy:

$$\sum_{p=1}^{6} \omega_p = 1, \quad \omega_p \geq 0 \tag{9}$$

2.1.5. Generating fuzzy equivalence matrix. The fuzzy similar matrix R can be expressed by Eq.(10):

$$R = \left(S_{ij}\right)_{N\times N} = \begin{pmatrix} S_{11} & \cdots & S_{1N} \\ \vdots & \ddots & \vdots \\ S_{N1} & \cdots & S_{NN} \end{pmatrix} \tag{10}$$

According to theorem[13], the transitive closure of fuzzy similar matrix is the transitive fuzzy equivalent matrix. In this paper, we use the successive square method to find the fuzzy equivalent matrix. Firstly, we square the fuzzy similar matrix: $R \rightarrow R^2 \rightarrow R^4 \rightarrow \ldots$. When $R^K = R^{2k}$ appears, R^K is the transitive closure of the matrix R. Then it is necessary to select the appropriate confidence level $\lambda \in [0,1]$ and obtain the λ-cut matrix of the fuzzy equivalent matrix.

$$t(R) = \left(F_{ij}\right)_{N\times N} = \begin{pmatrix} F_{11} & \cdots & F_{1N} \\ \vdots & \ddots & \vdots \\ F_{N1} & \cdots & F_{NN} \end{pmatrix} \tag{11}$$

$$t(R)_\lambda = \left(F_{ij}(\lambda)\right)_{N\times N} \tag{12}$$

where, when F_{ij} not less than λ, $F_{ij}(\lambda)= 1$, when less than λ, $F_{ij}(\lambda)= 0$.

The grouping results are different when λ takes different values.

2.2. *Path optimization of electric distribution planning considering hybrid charging strategy*

2.2.1. Problem descrption. There is a distribution center which can provide electric distribution vehicles to meet customer demand for distribution services. During the distribution process, the vehicle can choose the charging station nearby to replenish the electricity.At the charging station, each vehicle can choose at most one of three charging strategies: Fast Full Charge, slow full charge, and battery replacement.

2.2.2. Basic assumption. (1) each customer point can only be delivered by a car, a distribution vehicle can be responsible for the distribution of multiple customers; (2) there is only one distribution center and the distribution vehicles depart from the distribution center, and return to the distribution center after completing all the distribution work; (3) each vehicle leaves the distribution center fully charged, regardless of the repeated use of the distribution vehicle; (4) the waiting time of the distribution vehicle at the charging station is ignored; (5) the distribution vehicle does not consume electricity when it stops at the customer point; (6) the power consumption of the distribution vehicle has a linear correlation with the driving distance; (7) it does not take into account the dynamic changes of traffic conditions and customer demands.

2.2.3. Notation definitions

The notation definitions are given in Table 3.

Table 1 Notation definitions

Notation	Definition
V	Customer point set, $V=\{1, \ldots, N\}$
F	Charging station set
U	Customer point and charging station set, $U=V \cup F$
d_{ij}	The distance from i to j
t_{ij}	Vehicle travel time from i to j
r	Battery power consumption rate
g_a	Average charging speed of fast charging
g_b	Average charging speed of slow charging
c_t	Unit distance transportation cost
c_f	Unit vehicle fixed cost
c_a	Unit fast charging cost
c_b	Unit slow charging cost
c_s	Each battery replacement cost
c_w	Unit time penalty cost
s_i	Service hours of customer point i
q_k	Battery power the vehicle will have when it reaches the charging station k
e_i	The earliest service start time for customer point i
l_i	The latest service start time for customer point i
TD	Range of an electric car
τ_i	Service start time for vehicle arrival at customer point i
Q	Rated capacity of vehicle battery
x_{ij}	0-1 variable, if the vehicle is traveling from point i to point j, Take 1; otherwise, take 0
A_{ik}	0-1 variable, if the vehicle at i choose to charge at station k quick charge, take 1; otherwise, take 0
y_{ik}	0-1 variable, if the vehicle at i choose to charge at station k slow charge, take 1; otherwise, take 0
z_{ik}	0-1 variable, if the vehicle at i choose to charge at station k to replace the battery, take 1; otherwise, take 0
c_{ik}	0-1 variable, if the vehicle at i choose to charge at station k, take 1; otherwise, take 0

2.2.4. Mathematical model. The mathematical model based on the above conditions is as follows.

$$\min C = c_f \sum_{j\in U} x_{0j} + c_t \sum_{i,j\in U} x_{ij} d_{ij} + c_a \sum_{i\in F} A_{ik}\left(Q-q_k\right) + c_b \sum_{i\in F} y_{ik}\left(Q-q_k\right) + c_s \sum_{i\in F} z_{ik} + c_w \sum_{i\in V} \{\max(e_i-\tau_i,0)+\max(\tau_i-l_i,0)\} \quad (13)$$

Constraints:

$$\sum x_{ij} = \sum x_{ji} = 1, \forall i, j \in U, \quad i \neq j \tag{14}$$

$$\sum_{j \in U, i \neq j} x_{ij} = \sum_{j \in U, i \neq j} x_{ji} \leq 1, \forall i \in F \tag{15}$$

$$\tau_{\mathrm{i}} - \tau_{\mathrm{j}} \geq \left(t_{\mathrm{ji}} + s_i\right)x_{\mathrm{ji}} - \left(t_{\mathrm{ij}} + s_j\right)x_{\mathrm{ij}} - (1 - x_{ij} - x_{ji}) * T_{\max}, \; i, j > 1, \mathrm{i} \neq j, \forall i, j \in U \tag{16}$$

$$\tau_j - \tau_i \geq \left(t_{ij} + s_i\right)x_{ij} - \left(t_{ji} + s_j\right)x_{ji} - (1 - x_{ij} - x_{ji}) * T_{\max}, i, j > 1, \mathrm{i} \neq j, \forall i, j \in U \tag{17}$$

$$\tau_i - \tau_0 \geq (t_{1i})x_{1i} - (1 - x_{1i}) * T_{\max}, i > 1, \quad \forall i \in U \tag{18}$$

$$\tau_n - \tau_i \geq (t_{i1})x_{i1} - (1 - x_{i1}) * T_{\max}, i > 1, \quad \forall i \in U \tag{19}$$

$$\sum_k \left(A_{ik} + y_{ik} + z_{ik}\right) \geq f_{ik}, f_{ik} \leq c_{ik}, i \in U, k \in F \tag{20}$$

$$\sum_{k \in F} \sum_{i \in V} f_{ik} \geq \frac{\sum_{i, j \subset U} dij}{TD} \tag{21}$$

$$A_{ik} + y_{ik} + z_{ik} \leq 1, k \in F, i \in U \tag{22}$$

$$f_{ik}, c_{ik}, x_{ij} \in \{0,1\}, \forall i, j \in U, i \neq j \tag{23}$$

$$A_{ik}, y_{ik}, z_{ik} \in \{0,1\}, \forall k \in F, \forall i \in U \tag{24}$$

The objective function (13) mainly considers the cost related factors: vehicle fixed cost, driving cost, vehicle power supply cost in charging station and penalty cost.

Constraint (14) ensure that each customer site has a delivery vehicle access once; constraints (15) ensure that each charging station only serves one delivery vehicle at a time; Constraints (16) to constraints (19) are time continuity constraints for the vehicle at each node; constraints (20) and constraints (21) determine the node and charging station to be charged; Constraints (22) limit the vehicle to choose at most one charging strategy at the charging station; constraints (23) and (24) represent 0-1 constraints of variables.

3. Test instance

3.1. Instance description

The logistics center needs to meet the distribution needs of 24 customers. The Logistics Center has a rated truck load of 2 tons. The locations of charging stations and customer points are shown in Table 4. The delivery order information for the day of the distribution center is shown in Table 5, where customer point 0 represents the distribution center.

Table 2 Location of charging station

Charging station	Location
F1	(20,55)
F2	(10,35)
F3	(35,30)
F4	(45,65)

Table 3 Attribute of customer requirement

Customer Point	Demand (0.01 t)	x	y	The earliest service time	The latest service time	Product's external similarity	Value	Service quality
0	0	40	50	0	1236	--	--	--
1	20	45	68	912	967	High	High	Medium
2	30	45	70	825	870	Very high	Very high	High
3	10	42	66	65	146	Medium	Medium	Low
4	10	42	68	727	782	Very low	Low	High
5	10	42	65	15	67	Low	Low	Very high
6	20	40	69	612	702	High	Very high	Medium
7	20	40	66	170	225	Very high	High	Low
8	20	38	68	255	324	Very high	High	High
9	10	38	70	534	605	Medium	High	Medium
10	10	35	66	357	410	Low	Low	Medium
11	20	35	69	448	505	High	High	Low
12	20	25	85	652	721	Very low	Very low	Low
13	30	22	75	30	92	Very high	High	Medium
14	10	22	85	567	620	Low	High	High
15	40	20	80	384	429	Very high	Low	Medium
16	40	20	85	475	528	Medium	Medium	Low
17	20	18	75	99	148	Medium	High	Medium
18	20	15	75	179	254	Very low	Low	Low
19	20	15	80	278	345	High	Very high	High
20	20	30	50	10	73	High	Very high	Low
21	20	30	52	914	965	Low	Medium	Low
22	20	28	52	812	883	Very high	High	Medium
23	30	28	55	732	777	Medium	Low	Medium
24	10	25	50	65	144	High	High	Very high

3.2. Customers grouping. The decision-making indicators 1,2,3,4,5 are the geographical location of the customer point, the latest service time, the external similarity of the product, value and the service quality of the customer point.

When $\lambda = 0.950$, 11 customer groups were obtained;

When $\lambda = 0.925$, 7 customer groups were obtained;

When $\lambda = 0.900$, 6 customer groups were obtained;

When $\lambda = 0.875$, 4 customer groups were obtained.

When the confidence level is 0.875, the utilization rate of vehicle load is the highest. Therefore, this paper selects the customer group when λ is 0.875. The comprehensive evaluation weight of the five decision indexes is (0.1,0.3,0.3,0.1,0.2) . The details of the grouping are shown in Table 6.

Table 4 Grouping result

Group number	Customers	Demand（0.01t）
1	(1,6,9,17,23)	100
2	(2,4,5,7,8,12,14,18,21)	160
3	(3,11,16,19,20,24)	120
4	(10,13,15,22)	100

3.3. Variable and parameter settings. In this paper, the maximum endurance of the electric distribution vehicle is set at 80 km, the average driving speed of the vehicle is 40 km/h, and the average service time of the distribution vehicle at the customer point is 15 minutes, the linear relationship between charging quantity and charging time is 0.22 kwh/km, the fast charging speed is 17.4 kwh, and the slow charging speed is 1.74 kwh/km. The battery capacity is 15 kwh. The relevant cost parameters of electric vehicles are shown in Table 7[14][15].

Table 5 Cost-related parameter setting

Cost parameter	Value
Fixed costs	¥80
Transport cost per unit distance	¥0.4/km
Penalty cost per unit time	¥24/h
Unit cost of quick charging	¥2/kWh
Unit cost of slow charging	¥1/kWh
Unit cost of battery replacement	¥180/time

3.4. Calculation result. Set the charge capacity of each transmission is 13kWh. Vehicle leaves from i to charging station Fk and go back to i after charging. 0 represents distribution center.The route optimization results as shown in Table 8.

Table 6 Route optimization result

Group number	Optimal path within the group	Travel distance（km）	Total cost(¥)
1	0→17→9→6→1→F4→23→0	101.65	484.49
2	0→5→F4→7→8→18→14→12→2→4→21→0	124.25	950.46
3	0→20→24→3→F4→11→16→19→0	120.34	452.08
4	0→10→13→15→22→0	79.25	387.35

According to the calculations, each of the first three customer groups had a single charge at the charging station F4; the total distance traveled by the Customer Group 4 was less than the maximum range of an electric vehicle, the vehicle therefore does not need to be recharged during the mission.

4. Conclusion

This paper presents a two-stage optimization method for electric vehicle routing problem based on hybrid charging strategy. In the first stage, all customer orders are grouped by fuzzy clustering, and in the second stage, electric vehicle routing is optimized based on the first stage. The feasibility of the method is proved by an example. However, the route optimization of electric vehicle is based on the static traffic network, but the real-time change of the traffic condition and the battery power consumption do not simply change linearly with the distance, this problem needs further study in the future.

References

[1] G. B. Dantzig,J. H. Ramser. (1959) The Truck Dispatching Problem. Management Science, 6(1).

[2] Golden Bruce,Assad Arjang,Levy Larry,Gheysens Filip. (1984) The fleet size and mix vehicle routing problem. Computers & Operations Research,11(1).
[3] Li Jun, Guo Yaohuang. (2001) Theory and method of vehicle optimal scheduling for logistics distribution. China Materials Publishing House, Beijing.
[4] M. W. P. Savelsbergh. (1985) Local search in routing problems with time windows. Annals of Operations Research, 4(1).
[5] Liu Zhishuo, Chai Yueyan, Shen Jinsheng. (2009) Ant colony algorithm and its application in vehicle routing problem with hard time Windows. Computer-integrated manufacturing, 12(4): 596-602.
[6] Zhang Kaiqing and Ji Qichun. (2021) Study on multi-center semi-open vehicle routing problem with time-varying velocity. Journal of Systems Simulation: 1-11.
[7] Erdoğan,S., Miller-Hooks, E. (2011).A Green Vehicle Routing Problem. Transportation Research Part E(1).
[8] Inês Frade,Anabela Ribeiro,Gonçalo Gonçalves,António Pais Antunes. (2011) Optimal Location of Charging Stations for Electric Vehicles in a Neighborhood in Lisbon, Portugal. Transportation Research Record, 2252(1).
[9] Guo, F., Huang, H., Yang, J. (2020) Rearch on route optimization and Service Strategy of electric vehicle considering customer pick-up radius. Journal of Management Engineering, 34(1): 154-163.
[10] Zhang, Q., Mao, Q., Liang, P.(2021) An improved moth algorithm for electric vehicle routing problem. Journal of Hubei University for Nationalities: Natural Science Edition, 39(1): 72-79.
[11] Wang, J., Lu, Z. (2000) Determination of membership function in fuzzy control. Henan science, 04: 348-351.
[12] Shyi-ming Chen. (1998) Aggregating fuzzy opinions in the group decision-making environment. Cybernetics and Systems(4).
[13] Qu, X, Zhang, Z. (2003) Fuzzy cluster analysis design based on fuzzy equivalence relation. Journal of Lanzhou Railway College, 03: 17-20.
[14] Xiao Qiaolin, (2018) The scale of electric vehicle charging service fee market and the charging standard of each province and city. http://www.istis.sh.cn/list/list.aspx?id=11621.
[15] Han, G., Cao, Y. (2018) Construction of Electric Vehicle Charging Infrastructure Planning System . City Construction, 08: 17-21.

An Accelerated Mixed-Height Cell Circuit Legalization Method

Chencan Zhou[1,2,a], Gengchen Yang[3,b], Quan Shi[2,c*]

[1]School of Information Science and Technology, Nantong University, Nantong, 226019, P.R. China

[2]School of Transportation and Civil Engineering, Nantong University, Nantong, 226019, P.R. China

[3]School of Sciences, Nantong University, Nantong, 226019, P.R. China

[a]email: zhouchencan@ntu.edu.cn, [b]email: 975720478@qq.com,

[*]Correspondence author: [c]email: sq@ntu.edu.cn

Abstract. In Very Large-Scale Integration Circuit (VLSI) advanced node design, the standard cells are usually designed with mixed height to meet various requirements. In this paper, benefit from the initial placement from global placement, we first align cells to the power rail type matching rows and divide multi-row-height cells into single-row-height ones. Next, quadratic programming model is constructed and converted into linear complementarity problem (LCP). Last, we choose two matrix splits and apply an accelerated modulus-based matrix splitting iteration method to solve LCP. Experimental results indicate that we achieve better performance than two published state-of the-art works.

1. Introduction

With the ever-increasing scale of integrated circuits, the physical design phase is complicated and extremely dependent on automated tools. In the circuit design of advanced nodes, standard cells are usually designed with different heights. Usually, simple cells (e.g. inverters) and complex cells (e.g. triggers) are devised as single-row-height cells and multi-row-height ones, respectively. Accordingly, mixed-height cell design has come into vogue to meet various demands. The placement stage of VLSI includes global placement (GP), legalization and detailed placement (DP). The goal of GP stage is to minimize the wire length. The legalization stage aims to eliminate the overlap and minimize the displacement of cells. The results of pre-layout are incrementally improved in DP stage.

The mixed-height cell legalization is an NP-hard problem. Many scholars have devoted to legalization research, Tetris [1] and Abacus [2] are the most classic methods, both of which are based on greedy algorithms. There are network flow algorithm [3-4], heuristic algorithm [5], greedy algorithm [6] and so on applied in the placement. Additionally, chen [7] convert the quadratic programming (QP) problem in legalization into Linear complementarity problem (LCP), and utilize the modulus-based matrix splitting (MMS) iteration method [8] to solve it. To solve the LCP, some iteration methods are proposed, for example, the projected method [9], series of MMS method[8,10-11], Shift-splitting preconditioners[12]. Among the approaches, experimental results show that MMS and its relaxation methods are more efficient. In this paper, we proposed a novel MMS method to speed up the solution of the LCP involved in the legalization problem.

Content from this work may be used under the terms of the Creative Commons Attribution 3.0 licence. Any further distribution of this work must maintain attribution to the author(s) and the title of the work, journal citation and DOI.
Published under licence by IOP Publishing Ltd

2. Preliminaries

In this section, we will formulate the legalization of mixed-height cell and introduce the LCP and the accelerated modulus-based matrix splitting (AMMS) iteration method.

2.1. Problem Formulation

Given a rectangle placement area with its coordinates of lower left corner $(0,\ 0)$ and upper right corner $(W,\ H)$. Legalization problem is to place a set of cells $C=(c_1,c_2,\cdots c_n)$ in this area. The initial lower left corner coordinate , width and height of cell c_i are denoted by (x_i^0, y_i^0), w_i and h_i, respectively. Cells should meet the requirements: a) Cells must be in the placement area. b) Cells cannot overlap. c) Cells should be placed on the site in row. d) Cells must match the power rail type of the row. If cells are aligned in the nearest row with power rail type matching, then the mixed-height cell legalization problem can be formulated as following QP problem:

$$\begin{aligned}&\min \ \frac{1}{2}\sum_{i=1}^{n}(x_i-x_i^0)^2\\&s.t.\ \ 1)\, x_j-x_i\geq w_i,\ \ if\ x_j-x_i\geq 0\\&\qquad\quad x_i,x_j\ \ are\ in\ the\ same\ row;\\&\qquad 2)\, x_i\geq 0.\end{aligned}\tag{1}$$

2.2. Linear Complementarity Problem

The linear complementarity problem, abbreviated as $LCP(q,A)$, to find a pair of vectors $(z,w)\in\mathbb{R}^n$ such that

$$w=Az+q\geq 0,\ z\geq 0 \text{ and } w^T z=0,\tag{2}$$

where $A=(a_{ij})^{n\times n}\in\mathbb{R}^{n\times n}$, $q\in\mathbb{R}^n$ are given, the notation '$\geq$' denotes the componentwise defined partial ordering between two vectors and the superscript $(\cdot)^T$ means the transpose of a vector.

2.3. Accelerated Modulus-Based Matrix Splitting Method

The accelerated modulus-based matrix splitting iteration method for solving $LCP(q,A)$ is described as follows [10]. Let $A=M_1-N_1=M_2-N_2$ be two splitting of the matrix $A\in\mathbb{R}^{n\times n}$. Given an initial vector $v^{(0)}\in\mathbb{R}^n$, for $k=0,1,2\cdots$ until the iteration sequence $\{z^k\}_{k=0}^{+\infty}$ is convergent, compute $v^{(k+1)}\in\mathbb{R}^n$ by solving the liner system

$$(M_1+\Omega)v^{(k+1)}=N_1v^{(k)}+(\Omega-M_2)|v^{(k)}|+N_2|v^{(k+1)}|-\gamma q\tag{3}$$

and set

$$z^{(k+1)}=\frac{1}{\gamma}(|v^{(k+1)}|+v^{(k+1)}).\tag{4}$$

Here, Ω is an $n\times n$ positive diagonal and γ is a positive constant.

3. The Proposed Method

3.1. Method Overview

The mix-height cell legalization flow is depicted in Figure 1. Taking the results of global placement as input, we pre-process the cells. In this stage, we first align standard cells to rows that match the power rail, then we move divide multi-height-row cells into single-height-row ones. In the second stage, we establish the QP model for legalization procedure, and transform the QP problem into an LCP problem. Furthermore, we use the accelerated modulus-based matrix splitting iteration method to solve the $LCP(q,A)$.

3.2. Mixed-Height Cell Legalization Modeling

There exists different height of cells in the standard cell library, such as single-row-height, double-row-height, triple-row-height and so on. Cells of different height should be treated consistently during modelling, to this end, we divide the multi-row-height cells into single-row-height ones. Hence, the x-coordinate of these cells change from x_i to $(x_{i1}, x_{i2}, \cdots x_{it})$, t is the ratio of cells height to row height. All cells in each row are sorted according to their x-coordinates.We can rewrite the formula (1) into QP problem standard formulation in the following.

$$\begin{aligned} \min \quad & \frac{1}{2}x^T Bx + c^T x \\ s.t. \quad & Wx \geq d, \\ & Rx = 0, \\ & x \geq 0, \end{aligned} \tag{5}$$

where B is an identity matrix, $c = (-x_1^0, -x_2^0, \cdots, -x_n^0)^T$. Since any adjacent cells can't overlap with each other, so they must satisfy the restriction that $x_j - x_i \geq w_i(x_j \geq x_i)$. Then we build the inequality equations $Wx \geq d$ for each pair of adjacent cells. W is a matrix which only contains -1 and 1 in each row, representing the left cell and right cell, respectively. The width of each left cell is the corresponding component of d. The equality constraints $Rx = 0$ are introduced to ensure that the x-coordinates of the subcells of each multi-row-height cell are equal. We will use a simple example to illustrate the matrix construction process. The initial positions of cells are shown in Figure 2. The matrices and vectors are given as follows.

$$W = \begin{pmatrix} -1 & 1 & 0 & 0 & 0 & 0 & 0 \\ 0 & 0 & -1 & 1 & 0 & 0 & 0 \\ 0 & 0 & 0 & -1 & 1 & 0 & 0 \\ 0 & 0 & 0 & 0 & 0 & -1 & 1 \end{pmatrix}, \quad R = \begin{pmatrix} -1 & 0 & 1 & 0 & 0 & 0 & 0 \\ 0 & 0 & 0 & 0 & -1 & 0 & 1 \end{pmatrix},$$

and $d = (d_1, d_1, d_3, d_5)^T$, $c = (-x_1^0, -x_2^0, -x_1^0, -x_3^0, -x_4^0, -x_5^0 - x_4^0)^T$, $x = (x_{11}, x_{21}, x_{12}, x_{31}, x_{41}, x_{51}, x_{42})^T$.

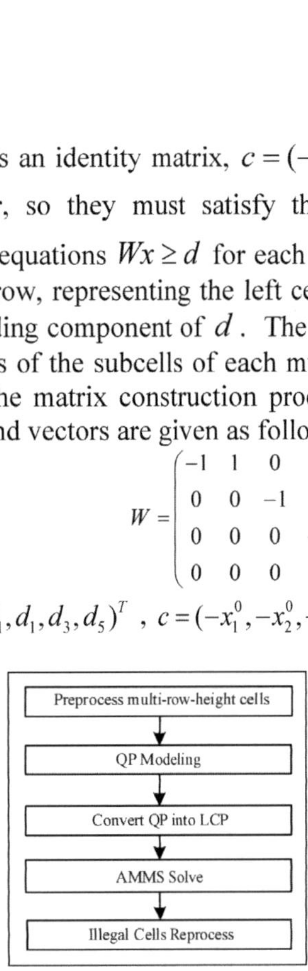

Figure 1. mix-height cell legalization flow

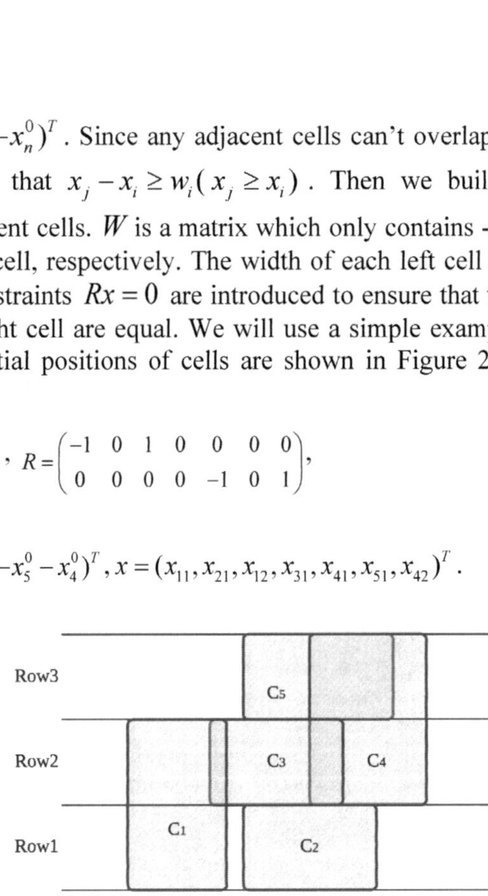

Figure 2. The initial position of cells

Based on the Lagrange Multiplier Method, adding the equality constraints into the objective function, the problem (5), can be rewritten as the following model (6).

$$\begin{aligned} \min \quad & \frac{1}{2}x^T Bx + c^T x + \lambda x^T R^T Rx \\ s.t. \quad & Wx \geq d, \\ & x \geq 0. \end{aligned} \tag{6}$$

where λ is the Lagrange multiplier. Obviously, B is a symmetric positive definite matrix, then (6) is convex quadratic programming. According to the Karush-Kuhn-Tucker (KKT) conditions, if x minimizes the QP problem, there exists vectors r, u such that the KKT conditions holds true:

$$\begin{cases} Bx + c + \lambda R^T Rx - W^T r - u = 0, \\ d - Wx \le 0, \\ r^T (d - Wx) = 0, \\ u^T x = 0, \\ x, r, u \ge 0. \end{cases} \tag{7}$$

Import a relaxation variable e into the inequality constraint $d - Wx \le 0$ such that $d - Wx + e = 0$, then $e = Wx - d$ and $e > 0$. The system of equations (7) is equivalent to the $LCP(q, A)$:

$$w = \begin{pmatrix} u \\ e \end{pmatrix} = \begin{pmatrix} B + \lambda R^T R & -W^T \\ W & 0 \end{pmatrix} \begin{pmatrix} x \\ r \end{pmatrix} + \begin{pmatrix} c \\ -d \end{pmatrix}, \tag{8}$$

where $A = \begin{pmatrix} B + \lambda R^T R & -W^T \\ W & 0 \end{pmatrix}, z = \begin{pmatrix} x \\ r \end{pmatrix}, q = \begin{pmatrix} c \\ -d \end{pmatrix}$ and satisfies

$$w = Az + q \ge 0, z \ge 0 \text{ and } w^T z \ge 0. \tag{9}$$

3.3. Solving the LCP (q, A)

To solve the $LCP(q, A)$, an implicit fixed-point equation

$$(M_1 + \Omega) v = N_1 v + (\Omega - M_2) |v| + N_2 |v| - \gamma q \tag{10}$$

is established by letting $z = \gamma^{-1}(|x| + x)$, $w = \gamma^{-1}\Omega(|x| - x)$ and utilizing matrix splitting technology.

We employ the accelerated modulus-based matrix splitting iteration method to solve the (10). By using the latest estimates $|v^{(k+1)}|$ on the right of the subsequent calculations, and choose the splitting matrices M_1, N_1, M_2, N_2 as follows:

$$M_1 = \begin{pmatrix} \frac{1}{\beta}(B + \lambda R^T R) & 0 \\ W & \frac{1}{\theta} D \end{pmatrix}, N_1 = \begin{pmatrix} (\frac{1}{\beta} - 1)(B + \lambda R^T R) & W^T \\ 0 & \frac{1}{\theta} D \end{pmatrix} \tag{11}$$

$$M_2 = \begin{pmatrix} B + \lambda R^T R & -W^T \\ 0 & 0 \end{pmatrix}, N_2 = \begin{pmatrix} 0 & 0 \\ -W & 0 \end{pmatrix} \tag{12}$$

where β, θ are two prescribed positive constants, $D = tridiag(W(B + \lambda R^T R)^{-1} W^T)$ is a tridiagonal approximation to the Schur complement $W(B + \lambda R^T R)^{-1} W^T$ of the matrix A. The split matrix N_2 should be a lower triangular matrix to ensure the feasibility of computations. Since the inversion of matrix is very time-consuming, the Sherman-Morrison formula [13] can be applied to simplify the calculation complexity. Given that B is an identity matrix, RR^T is a diagonal matrix with diagonal entries of 2, $(B + \lambda R^T R)^{-1}$ can be expressed as

$$(B + \lambda R^T R)^{-1} = I - \lambda R^T (I_1 + \lambda RR^T)^{-1} R = I - \frac{\lambda}{2\lambda + 1} R^T R, \tag{13}$$

then $D = tridiag(WW^T - \frac{\lambda}{2\lambda + 1} WR^T RW^T)$. Algorithm 1 shows the process of solving $LCP(q, A)$ with the AMMS method.

Algorithm 1 AMMS Method

Input: $M_1, N_1, M_2, N_2, \Omega,\ v^{(0)}, z^{(0)}, \gamma, \varepsilon$ **Output:** $z^{(k+1)}$

Begin: k=0;

do

Iteratively computing $v^{(k+1)}$ by solving the equation

$$(M_1+\Omega)v^{(k+1)} = N_1 v^{(k)} + (\Omega - M_2)|v^{(k)}| + N_2|v^{(k+1)}| - \gamma q;$$

$$z^{(k+1)} = \frac{1}{\gamma}(|v^{(k+1)}| + v^{(k+1)});$$

k++;

$z^{(k+1)} = z^{(k)}; v^{(k+1)} = v^{(k)};$

until $\| \min(Az^{(k)}+q, z^{(k)}) \|_2 < \varepsilon$

return $z^{(k)}$;

4. Experimental results

The AMMS-based legalization was implemented in C++ to validate our proposed method. We tested our algorithm on partial of the benchmark of ISPD 2015 Detailed Routing-Driven Placement Contest [14] provided by the authors of [15]. Since the standard cell library is single-row cell, the authors [15] randomly selected 10% of cells to double their height while halving their width. The experiments were performed on Intel(R) Core (TM) with i7-10710U CPU and 16GB RAM, under Windows 10 operating system. We compared our experimental results with two state-of the-art works [7] and [15]. The number of iterations was not involved in the legalization algorithm of [15], therefore we only compare displacement indicators with them. The binary file was obtained from the authors [15]. The termination criterion in [7] is not published, with the binary code not available, hence we programmed the MMS method to solve the $LCP(q, A)$. Under the condition of the same stopping criterion, our AMMS method can compare with it evaluated by performance indicators of iteration steps and runtime. Note that the coefficient matrix and parameters taken by the two methods were the same.

Table 1 lists the characteristics of benchmarks and results of experiments. In this table, "#S. Cell" , "#D. Cell" denote the number of single-row-height cells and double-row-height-cells, respectively. "Total Disp.(sites)" represents the total displacement of cells measured by the number of site widths. "IT. Step" is the iteration steps of MMS or AMMS method and "Runtime" is in seconds.

Table 1. Experimental results.

Benchmark	#S. Cell	#D. Cell	Total Disp. (sites)			It. Step		Runtime (s)	
			[15]	[7]	ours	[7]	ours	[7]	ours
des_perf_a	99975	8513	81452	72561	**72236**	160	**129**	2.122	**1.666**
des_perf_b	103842	8802	81540	71888	**71349**	105	106	1.443	1.456
fft_2	30297	1984	25697	20979	**20113**	53	**51**	0.143	**0.139**
fft_a	28718	1907	19613	18304	18555	104	108	0.27	0.293
fft_b	28718	1907	28461	21671	**21335**	89	**78**	0.231	**0.225**
matrix_mult_a	146837	2813	46001	40298	40611	119	**95**	1.906	**1.553**
matrix_mult_b	143695	2740	40059	37215	**37107**	138	**103**	2.223	**1.675**
pci_bridge32_a	26268	3249	27832	26289	**26252**	66	**58**	0.185	**0.171**
pci_bridge32_b	25734	3180	27864	26028	26358	188	**152**	0.528	**0.426**
N. Average	-	-	1.13	1.00	1.00	1.16	1.00	1.19	1.00

It can be seen from Table 1 that our method makes some progress than the other two. The total displacement results of our experiments on all benchmarks are less than [15], most of the results are less than [7]. Our results are reduced by 0. 009% compared with [7], and 11% compared with [15], in terms of "Total Disp." indicators. Our AMMS algorithm achieves 14% fewer iterations than [7] and is 1.19 times faster than theirs. It can be concluded that the proposed legalization algorithm is effective.

5. Conclusions

We have put forward a new mixed-height cell legalization method. First, cells are aligned to rows that match the power rail type. Then we divide multi-row-height cells into single-row-height ones and construct a quadratic programming model. The QR model is converted into linear complementarity problem by KKT conditions. To solve the LCP rapidly, we introduce the latest valuation to the right side of the implicit fixed-point equation and use accelerated modulus-based matrix splitting iteration method to compute it. Experimental results verify the validity of our method.

References

[1] Hill D. Method and system for high speed detailed placement of cells within an integrated circuit design: U.S. Patent 6,370,673[P]. 2002-4-9.
[2] Spindler P, Schlichtmann U, Johannes F M. Abacus: fast legalization of standard cell circuits with minimal movement[C]. In: Proceedings of the 2008 international symposium on Physical design. 2008: 47-53.
[3] Darav N K, Bustany I S, Kennings A, et al. Eh? Legalizer: A High Performance Standard-Cell Legalizer Observing Technology Constraints[J]. ACM Transactions on Design Automation of Electronic Systems (TODAES), 2018, 23(4): 1-25.
[4] Monteiro J, Johann M, Behjat L. An Optimized Cost Flow Algorithm to Spread Cells in Detailed Placement[J]. ACM Transactions on Design Automation of Electronic Systems (TODAES), 2019, 24(3): 1-16.
[5] Heo S, Kahng A B, Kim M, et al. Diffusion break-aware leakage power optimization and detailed placement in sub-10nm VLSI[C]. In: Proceedings of the 24th Asia and South Pacific Design Automation Conference. 2019: 550-556.
[6] Singhar S S, Ray B N B, Dash A K, et al. Optimizing Mixed Size & Large Scale Block Placement Using Greedy Approach[C]. In: 2019 International Conference on Information Technology (ICIT). IEEE, 2019: 442-447.
[7] Chen J, Zhu Z, Zhu W, et al. Toward optimal legalization for mixed-cell-height circuit designs[C]. In: Proceedings of the 54th Annual Design Automation Conference 2017. 2017: 1-6.
[8] Bai Z Z. Modulus-based matrix splitting iteration methods for linear complementarity problems[J]. Numerical Linear Algebra with Applications, 2010, 17(6): 917-933.
[9] Koulisianis M D, Papatheodorou T S. Improving projected successive overrelaxation method for linear complementarity problems[J]. Applied numerical mathematics, 2003, 45(1): 29-40.
[10] Zheng N, Yin J F. Accelerated modulus-based matrix splitting iteration methods for linear complementarity problem[J]. Numerical Algorithms, 2013, 64(2): 245-262.
[11] Zhang L L. Two-step modulus-based matrix splitting iteration method for linear complementarity problems[J]. Numerical Algorithms, 2011, 57(1): 83-99.
[12] Cao Y, Du J, Niu Q. Shift-splitting preconditioners for saddle point problems[J]. Journal of computational and applied mathematics, 2014, 272: 239-250.
[13] William P H, Teukolsky S A, William V T, and Flannery B P. Numerical Recipes: The Art of Scientific Computing (3rd ed.). New York: Cambridge University Press, 2007.
[14] Bustany I S, Chinnery D, Shinnerl J R, et al. ISPD 2015 benchmarks with fence regions and routing blockages for detailed-routing-driven placement[C]. In: Proceedings of the 2015 Symposium on International Symposium on Physical Design. 2015: 157-164.
[15] Chow W K, Pui C W, Young E F Y. Legalization algorithm for multiple-row height standard cell design[C]. In: 2016 53nd ACM/EDAC/IEEE Design Automation Conference (DAC). IEEE, 2016: 1-6.

Feature Analysis Method of Voltage Sag based on Sag Slope

Lixun Wang[1,a]*, Bin Zhang[1,b], Peng Gao[1,c], Jianchao Zhang[1,d], Caiya Chen[1,e]

[1]College of Automation and Electrical Engineering, Lanzhou Jiaotong University, Lanzhou, Ganshu, China

[a]email: 1476545997@qq.com, [b]email: zbwr2311@yahoo.com.cn,

[c]email: 578541548@qq.com, [d]email: 3055083460@qq.com,

[e]email: 2608155066@qq.com

*Newsletter author: [a]email: 1476545997@qq.com

Abstract. At present, most of the analysis of the amplitude and duration of the voltage sag is based on the probability distribution map of a large number of measured data, in order to solve this problem, a calculation method is proposed to transform the sag amplitude and duration into the sag slope and sag amplitude time, which can quantitatively express the tilt degree of the voltage sag waveform and the persistence at the lowest amplitude. After analyzing the simulation waveforms of short circuit fault and transformer excitation by slope calculation method, the values of sag slope and sag amplitude time obtained are more perfect than the original sag amplitude and duration description, which is beneficial to the follow-up study of voltage sag.

1. Introduction

In recent years, due to the influence and infiltration of new energy, electronic power system and rail transit electrification, power lines and power grid structures are subject to higher and higher power quality requirements, especially the impact caused by voltage sags; therefore, at this stage, a series of issues related to voltage sags have become the focus and hotspot of research. At present, the research on voltage sag has been involved in various aspects, such as cause identification of voltage sag[1], location detection method[2], equipment tolerance test[3-4], severity assessment[5], sensitivity assessment[6], risk assessment[7], feature extraction[8]and analysis[9-11], and some progress and achievements have been made in all aspects.

In some literature research on characteristic quantity analysis, most of them are based on the field measured data to analyze the characteristic indexes under different voltage levels in the form of probability distribution, which requires high accuracy of field monitoring equipment; but not all areas have the conditions to achieve real-time monitoring. In the study of characteristic quantity analysis, the sag amplitude and duration are important characteristics used to characterize the characteristics and severity of voltage sag[11], most of the related literatures use these two characteristics to analyze voltage sag events. Therefore, based on the above two aspects, this paper improves a combined analysis method for the two characteristics of sag amplitude and duration, which can characterize the characteristic distribution of a voltage sag event according to the waveform data of a voltage sag event.

Content from this work may be used under the terms of the Creative Commons Attribution 3.0 licence. Any further distribution of this work must maintain attribution to the author(s) and the title of the work, journal citation and DOI.
Published under licence by IOP Publishing Ltd

2. Slope calculation method

2.1. Thought

Due to the occurrence of voltage sag, there is a sudden influx of large current in the power line, and based on the original voltage sag waveform and characteristic analysis, with the help of the idea of slope, the sag amplitude and duration are transformed into sag slope and sag amplitude time to characterize the degree of voltage sag.

2.2. Definition

The waveform of the root mean square U_{RMS} of the voltage according to the voltage sag event (interphase short circuit fault) is shown in Fig.1 below. A voltage sag event is divided into five stages[12]:before the sag occurs, the transition stage of the sag (the initial stage of the sag),the duration of the sag, the transition stage of the sag (the recovery phase of the sag)and after the sag is restored, when there is a second sag, there will be two stages in the transition phase of the sag (the initial stage of the sag) and the duration of the sag, as shown in Fig.1.

The sag slope is defined as: the ratio of the voltage amplitude difference to the time used in the sag transition stage (the beginning stage or the recovery stage) is used to describe the tilt degree and rise and fall degree of the sag waveform; the sag amplitude time is defined as the time it takes to stabilize the voltage sag amplitude at the lowest value in the duration phase, which is used to describe the duration of the sag event. Therefore, the amplitude of the original voltage sag is part of the solution of the amplitude difference of the sag slope and the amplitude of the temporary duration phase, and the duration of the original voltage sag is equal to the sum of the transition period of the sag plus the duration of the sag.

3. Calculation

3.1. Short Circuit Fault

The waveform of the root mean square U_{RMS} of the voltage sag event caused by short-circuit fault is shown in Fig.1 The slope calculation method is used to analyze the second sag in the AB phase, so it includes two sag slopes (the initial stage of the sag) and two sag amplitude times. The waveform data are shown in Table.1, where Ai represents the coordinate values of the ith point of phase A in Fig.1, and Bi is the same. According to the data, through the definition of the sag slope, we can get the sag slope of each sag transition section of the AB phase as shown in Table.2, and according to the simple calculation, we can get the sag amplitude time.

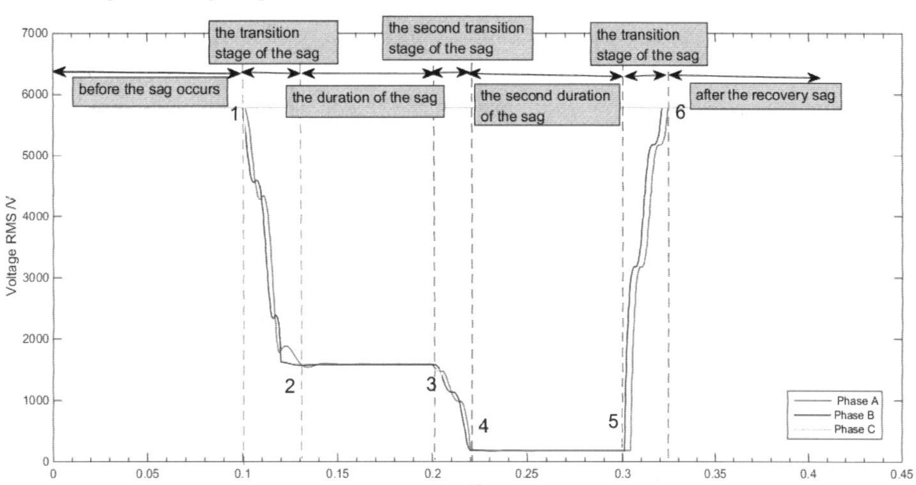

Fig.1 U_{RMS} waveform diagram of short-circuit faults

It is concluded that the sag slope of the first initial stage of phase A is -2.073×10^5, and the sag amplitude time is 0.081s; the sag slope of the second initial stage of phase A is -0.6643×10^5,and the sag amplitude time is 0.081s; the sag slope of the recovery phase of the sag of phase A is 2.7696×10^5. The sag slope of the first initial stage of phase B is -1.381×10^5, and the sag amplitude time is 0.0605s; the sag slope of the second initial stage of phase B is -0.707×10^5, and the sag amplitude time is 0.0845s; the sag slope of the recovery phase of the sag of phase B is 2.811×10^5.

Table.1 U_{RMS} waveform node data for short-circuit faults

Coordinate	U_{RMS}/V	Time/s
A1	5773	0.1
B1	5774	0.1002
A2	1626.6	0.12
B2	1590	0.1305
A3	1586.5	0.201
B3	1594	0.2
A4	191.5	0.22
B4	181	0.22
A5	179	0.301
B5	179	0.3045
A6	5773.5	0.3212
B6	5773	0.3244

Table.2 Sag slope of the short-circuit faults

Coordinate	Transition time/s	Sag Slope	Stage
A1~A2	0.02	-2.073×10^5	the initial transition stage of the sag of phase A
A3~A4	0.019	-0.6643×10^5	the second initial transition stage of the sag of phase A
A5~A6	0.0202	2.7696×10^5	the recovery transition stage of the sag of phase A
B1~B2	0.0303	-1.381×10^5	the initial transition stage of the sag of phase B
B3~B4	0.02	-0.707×10^5	the second initial transition stage of the sag of phase B
B5~B6	0.0199	2.811×10^5	the recovery transition stage of the sag of phase B

3.2. Transformer Excitation

The waveform of the root mean square U_{RMS} of the voltage sag event caused by transformer excitation is shown in Fig.2. The slope calculation method is used to analyze the voltage sag event caused by transformer excitation. Phase A in the figure has a voltage sag, the voltage drop of BC phase does not reach the national standard of voltage sag[13-14]. The waveform data is shown in Table.3, where Ai represents the coordinate values of the ith point of phase A in Fig.2, and Bi and Ci are the same. According to the data, through the definition of the sag slope, we can get the sag slope of the A phase transition section, and the tilt degree of the BC phase is also described according to the fluctuation slope as shown in Table.4.

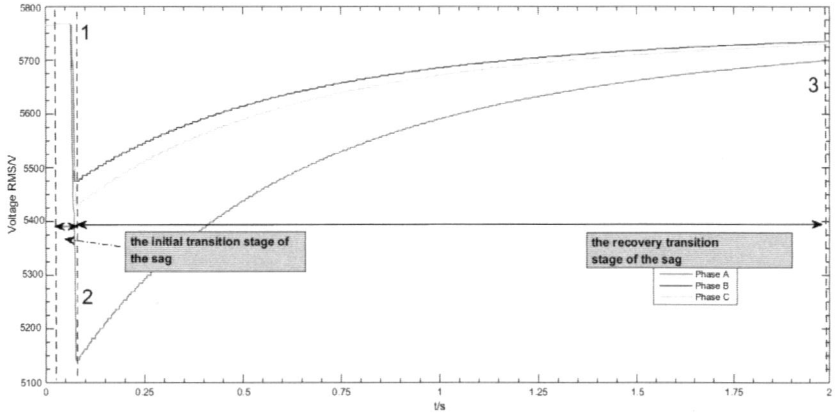

Fig.2 U_{RMS} waveform diagram of transformer excitation

It is concluded that the sag slope of the initial stage of phase A is -55.583×10^3, and the sag amplitude time is too small to be ignored; the sag slope of the recovery phase of the sag of phase A is 0.282×10^3; the voltage fluctuation slope of phase B is -38.74×10^3, and the voltage fluctuation amplitude time is too small to be ignored; the voltage fluctuation slope of the recovery phase of phase B is 2.894×10^3; the voltage fluctuation slope of phase C is -45.03×10^3, and the voltage fluctuation amplitude time is too small to be ignored; the voltage fluctuation slope of the recovery phase of phase C is 0.163×10^3.

Table.3 U_{RMS} waveform node data for transformer excitation

Coordinate	U_{RMS}/V	Time/s
A1	5773.2	0.0645
B1	5773.3	0.0645
C1	5773.2	0.0678
A2	5157	0.076
B2	5475	0.0722
C2	5426.5	0.0755
A3	5700	2
B3	5735	2
C3	5740	2

Table.4 Sag slope of the transformer excitation

Coordinate	Transition time/s	Sag Slope	Stage
A1~A2	0.0115	-55.583×10^3	the initial transition stage of the sag of phase A
A2~A3	1.924	0.282×10^3	the recovery transition stage of the sag of phase A
B1~B2	0.0077	-38.74×10^3	the initial transition stage of the fluctuation of phase B
B2~B3	1.9278	2.894×10^3	the recovery transition stage of the fluctuation of phase B
C1~C2	0.0082	-45.03×10^3	the initial transition stage of the fluctuation of phase C
C2~C3	1.924	0.163×10^3	the recovery transition stage of the fluctuation of phase C

4. Analysis and comparison

4.1. Short Circuit Fault Analysis

The second sag occurred in the voltage sag event caused by the short circuit fault. From the sag data of phase A, we get that the time used for the two the transition phase of the sag (the initial stage of the sag) is the same, but due to the different degree of amplitude drop, the sag slope is different, and the absolute value of the sag slope is small, so the degree of inclination is relatively gentle, alone, the impact on the equipment is not as large as the first sag. For phase B, because the time of the transition phase of the sag (the initial stage of sag) is about 1.5 times that of phase A, although the degree of voltage amplitude drop is approximately the same, the slope of phase B is much smaller than that of phase A in the same period, and the slope is slower than that of phase A; and the theoretical analysis of short circuit fault is consistent with [15].

4.2. Excitation Analysis of Transformer
In the voltage sag event caused by transformer excitation, from the ABC three-phase data, we can know that the three-phase voltage drops in varying degrees during the transformer switching process, but only phase A achieved the national standard of voltage sag, and the sag amplitude is between 10% and 20% of the rated voltage. The sag slope is smaller than the short-circuit fault, and the absolute value of the sag slope is also smaller, so the degree of inclination is gentle, and the impact on the equipment impact is not as great as the short-circuit fault.

Because the instantaneous values of the three-phase voltage are not equal at the same time, the saturation degree of the three-phase core is different, resulting in the difference in the amplitude of the three-phase voltage sag[16], and there is no the duration of the sag, the transition directly from the transition phase of the sag in the initial stage to the transition phase of the sag in the recovery stage, the absolute difference of the sag slope of the two transition stage is very large, and the tilt of the waveform in the recovery stage is more gentle, and the time used is much longer than that at the beginning of the sag. It is consistent with the mechanism analysis of transformer excitation.

4.3. Comparative Analysis
Based on the voltage sag events caused by the above two different reasons, the sag slope and sag amplitude time are analyzed according to the U_{RMS} waveform, and the conclusions are consistent with the theoretical analysis. For the single voltage sag event caused by short-circuit fault, using the national standard to describe the amplitude and duration of the sag is as follows: a short-term interruption event occurred in phase A, the amplitude of the sag is 0.03p.u.and the duration is 0.0081s; and phase B and phase C is the same. It can be seen from the above that there is no detailed description and analysis of this sag event, ignoring the problem and impact of the secondary sag, which will mislead the subsequent governance and evaluation, so the use of slope calculation method can be more detailed and intuitive analysis of the waveform.

For the single voltage sag event caused by transformer excitation, in the process of the sag, the speed of the sag is fast, and the amplitude drop is large, while the sag recovery stage is carried out gradually, the time is long, and the tilt degree of the waveform is gentle. However, the result described by the national standard is that only phase A has voltage sag time, the amplitude of voltage sag is 0.89p.u.and the duration is 1.94s. From the waveform diagram and the actual data, we can see that the description of the voltage sag in this method is not perfect, ignoring the influence of the other two phases, and the recovery phase takes a long time, because it is not described separately, it will produce a deviation in the understanding of the duration of the voltage sag and affect the implementation of subsequent improvement measures; therefore, the slope calculation method can be used to describe the sag event more completely and avoid the error caused by one-sided understanding.

5. Conclusion and prospect
In this paper, based on the idea of slope, the amplitude and duration of voltage sag are converted and combined, and the sag slope and sag amplitude time are defined. Different from the two characteristic values of the amplitude and duration of the voltage sag, starting from the characteristics of the U_{RMS} waveform of the voltage sag, after segmenting the voltage sag event, the rise and fall slope of the sag transition stage is quantified by the sag slope, and the duration of the sag event is subdivided into the time occupied by the transition stage and the duration stage, the sag event can be well separated, and the voltage sag can be analyzed more comprehensively according to each stage by making use of the characteristics ignored in the description of the amplitude and duration of the sag.

The waveform characteristics of voltage sags caused by different reasons are classified, and the causes of voltage sags are preliminarily determined by using the data such as the positive and negative magnitude and gap of sag slope, the duration of sag transition stage and the duration of sag. Because of the stage division of the voltage sag event, how to more accurately divide the transition stage and the duration phase of the sag is the next research point; how to make the sag slope more accurate? And how to divide the interval of the sag slope is also worth studying.

Acknowledgment
This paper is one of the phased achievements of the National Natural Science Foundation project: Research on the Modeling and Task Allocation Strategies of Distributed Model Predictive Control for Multiple Active Power Filters(61863023).

References

[1] K. X. Wang, Z. X. Song, D. G. Chen, J. H. Wang, and Y. S. Geng. (2003) The interference sourc e identification of the power distribution voltage of the distribution network based on wavelet transform.Proceedings of the CSEE, 21: 29–34.
[2] C. X. Zhao, S. Tao, X. N. Xiao. (2016) The method of fault location estimation based on power quality detection system voltage temporary drop information. Power System Technology, 40: 642–648.
[3] Y. H. Xu, W. S. Hong, Q. Q, Lan. (2016) Analysis on the impact of the starting point and phase jump on the ac contactor. Automation of Electric Power Systems, 40: 72-97.
[4] S. Z. Jokiae, J.V.S.Milanovic, K. Stockman. (2003) Voltage-tolerance curves of PWM drives comparison of simulation and measurement results. Scandinavian Journal of Gastroenterology, 4: 425-433.
[5] W. J. Fan, X. N. Xiao, S. Tao. (2019) A Muiti-index Evaluation Method of Voltage Sag Based on Comprehensive Weight. Power Capacitor and Reactive Power Compensation, 40: 137-144.
[6] W. Chen, J. B. Cui, X. P. Pei, T. Huang, Y. Lu. (2020) Assessment of Equipment Sensitivity to Voltage Sags Based on Equivalent Random Variables. Proceedings of the CSU-EPSA, 1:1-9.
[7] X. N. Liu, X. Y. Xiao, Y. Liu, Y. Wang. (2014) Multilevel fuzzy comprehensive evaluation of industrial process voltage drop risk management. Power System Technology, 38: 1984-1988.
[8] Y. Wang, B. Azam, M. Bollen. (2017) Singleevent characteristics for voltage dips in three-phases systems.IEEE Transactions on Power Delivery, 32: 832-840.
[9] K. M. He. (2020) Research on Characteristic Analysis and Detection Technology of Voltage Sag. Xi'an: Xi'an University of Science and Technology.
[10] X. Z. Si, Q. L. Li, J. L. Yang, Y. H. Xu, B. Zhang. (2017) Analysis of voltage sag characteristics based on measured data. Electric Power Automation Equipment, 37: 144-149.
[11] C. Q. Hu, A. M. Cao, Y. L. Huang, J. B. Wang, C. R. Liu. (2020) Analysis of Voltage Sag Characteristic Distribution in Shanxi Power Grid Based on Measured Data. Big Data Application, 48: 069-074.
[12] Wang Ying, Bollen M H J, X. Y. Xiao. (2015) Calculation of the phase-angle-jump for voltage dips in three-phase systems. IEEE Transactions on Power Delivery, 30: 480-487.
[13] GB/T 30137. (2013) Power quality voltage sag and short-term interruption. IEEE. (2014)IEEE Guide for Voltage Sag Indices IEEE Std 1564.
[14] Y. H. Xu, Q. Q. Lan, X. Y. Kong, W. S. Hong. (2016) Statistical analysis of voltage sag characteristics and research on sag propagation property. Transactions of China Electrotechnical Society, 31: 165-175.
[15] C. Wang. (2019) Typical Characteristic Quantity Analysis and Propagation Characteristics Study of Voltage Sag. Zhengzhou :Zhengzhou University.

Research on Power Quality Prediction Based on BiLSTM Optimized by Bayesian Algorithm

Wenhui Zhang[1,a], Baohua Wang[1,b*], Dafei Wang[1,c], Jiacheng Yu[1,d], Chi Zhang[1,e]

[1]School of Automation, Nanjing university of science and technology, Nanjing, China

[a]email: 2837013836@qq.com, [c]email: dafei.wang@njust.edu.cn,

[d]email: 251778438@qq.com, [e]email: zhangc2259@njust.edu.cn

*Corresponding author: [b]email: wangbaohuaa@163.com

Abstract - The steady-state indexes of power quality are obviously nonlinear and non-stationary under the influence of many factors, leading to the existing algorithms not to achieve high-precision prediction for power quality. Considering the excellent performance of Bi-directional Long Short-Term Memory model in time series analysis and the outstanding advantages of Bayesian algorithm in global optimization, a prediction model for power quality based on BiLSTM optimized by Bayesian is proposed. The actual data of power quality are used to test the performance of the model, and the results show that the proposed prediction model has higher accuracy and better applicability.

1. Introduction

With the rapid development of the power industry, the power quality problem is becoming more and more serious. Power quality has become an increasingly concerned issue for the power industry and electricity users [1]. Through the existing monitoring equipment of power quality, the historical steady-state indexes of power quality in different periods can be obtained, but the steady-state indexes in the future period cannot be known [2]. Therefore, scientific prediction of steady-state indexes is necessary.

At present, experts on power quality prediction have carried out active exploration, but systematic research and consensus results are still few [3]. Reference [5] uses the least square method to find the fitting function for prediction, and verifies the effectiveness of the proposed method with actual data. Reference [6] uses dynamic time warping and Monte Carlo algorithm for prediction, but the results have certain randomness. Reference [7] established a random forest prediction model by quantifying the correlation of the main influencing factors, but this algorithm is only applicable to some specific occasions. The above methods have their own characteristics. However, it is still a great challenge to better predict the steady-state indexes of power quality in the short- and medium-term.

A prediction model of power quality based on Bi-directional Long Short-Term Memory model optimized by Bayesian (BO-BiLSMT) is proposed. Bi-directional Long Short-Term Memory model (BiLSTM) can fully tap the internal relationship between historical data and future data, which can improve the prediction accuracy of the model. Machine algorithms have hyper-parameters, and how to automatically determine the value of the optimal hyper-parameters is an important issue. Bayesian algorithm is proposed to optimize the hyper-parameters of the BiLSTM model in this paper. The rationality and effectiveness of the proposed prediction model are verified by simulation analysis of data in actual power grid.

Content from this work may be used under the terms of the Creative Commons Attribution 3.0 licence. Any further distribution of this work must maintain attribution to the author(s) and the title of the work, journal citation and DOI.
Published under licence by IOP Publishing Ltd

2. Materials and Methods

2.1. LSTM Model

Long Short-Term Memory (LSTM) is a type of recurrent neural network proposed by Hochreiter and Schmidhuber in 1997 [8]. LSTM model can solve the problem of long-term dependence and make up for the defects of RNN to a certain extent. For the historical data, it can have a good function of information mining. The structure of LSTM is shown in Figure 1.

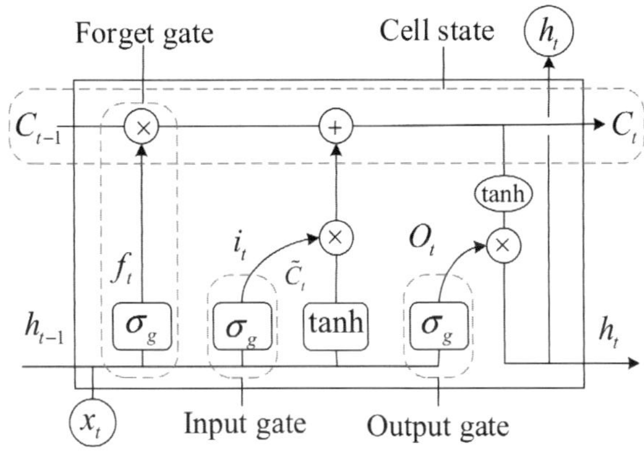

Figure 1. Classical LSTM Memory Cell Structure

2.2. BiLSTM Model

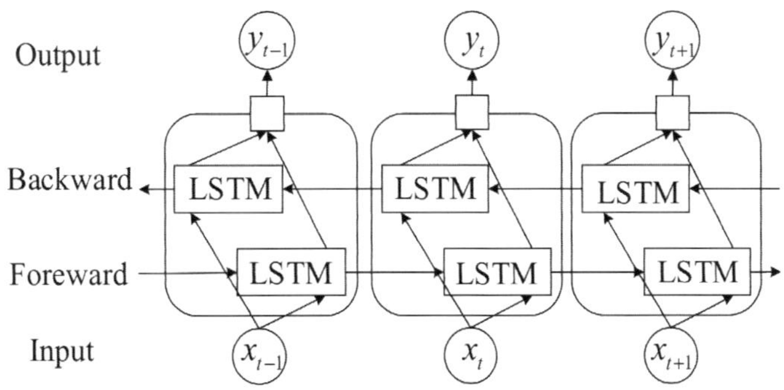

Figure 2. BiLSTM model Structure

LSTM model is trained in one direction, which can only refer to the characteristics of historical data and can not tap the potential relationship between current data and future data. Therefore, BiLSTM is proposed based on LSTM network. BiLSTM model is composed of forward pass and backward pass LSTM layers [9]. Two opposite LSTM hidden layers are connected to an output layer. Information can be transmitted simultaneously. So better results can be obtained for time-series prediction. prediction. The structure of BiLSTM is shown in Figure 2.

2.3. Bayesian Algorithm

In general, how to automatically determine the value of the optimal hyper-parameters is an important issue in deep learning. Therefore, this section proposes a Bayesian algorithm to optimize the hyper-parameters of BiLSTM model based on the above sections.

In recent years, Bayesian algorithm has been widely used in various fields [10]. Bayesian algorithm selects the most potential evaluation points according to the results of the surrogate model fitting the real objective function in each iteration. So it can obtain the global optimal solution of the complex objective function with less evaluation times.

2.4. Overall Prediction Model Framework

Aiming at the nonlinear and temporal characteristics of power quality data, a prediction model of power quality is proposed, as shown in Figure 3. Firstly, the dataset of power quality is standardized, and it is segmented according to the proportion to obtain the training set and the verification set. Then, a prediction model of power quality based on BiLSTM optimized by Bayesian algorithm is constructed to predict the final prediction data. Finally, the performance of the prediction model is evaluated by the evaluation indexes.

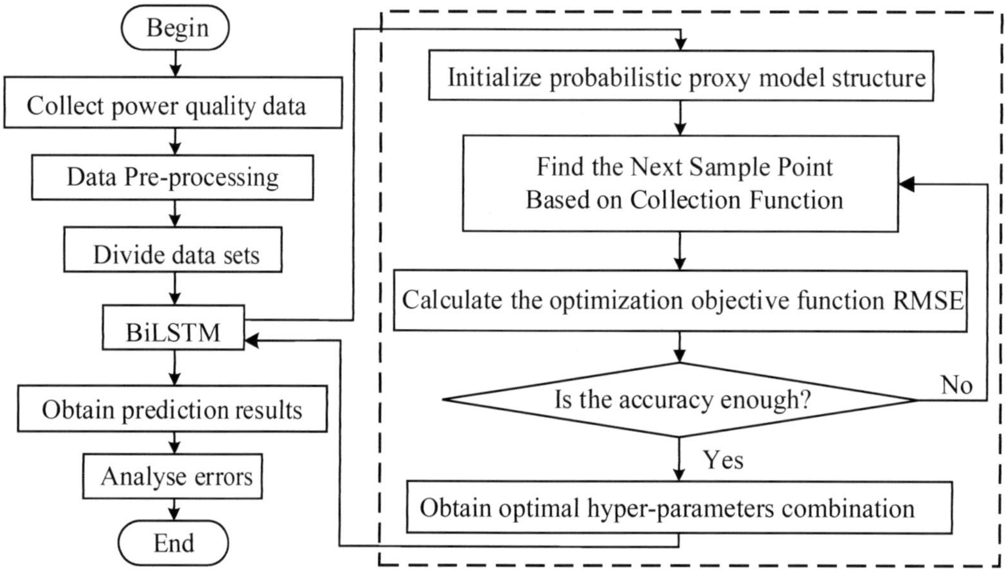

Figure 3. Overall Prediction Flow Chart

3. Results and Discussion

3.1. Data Sources

The experimental datasets are one-week power quality data provided by a monitoring point in a city, with a total of five power quality indicators and each of which has 1008 sets of data. In this chapter, the data of the first six days are used to train the network model, and the data of the seventh day are used as the test sample set to ensure the effectiveness of the obtained prediction model.

3.2. Evaluating Indicators

In this paper, the Mean Absolute Error (MAE), the Mean Absolute Percentage Error (MAPE) and Theil's Inequality Coefficient (TIC) are selected to evaluate the accuracy and fitting degree of the prediction model from multiple perspectives. The calculation formulas are as follows :

$$MAE=\frac{1}{n}\sum_{k=1}^{n}\left|\hat{y}_k-y_k\right| \tag{1}$$

$$MAPE=\frac{1}{n}\sum_{k=1}^{n}\left|\frac{\hat{y}_k-y_k}{\hat{y}_k}\right| \tag{2}$$

$$TIC=\frac{\sqrt{\frac{1}{n}\sum_{k=1}^{n}(\hat{y}_k-y_k)^2}}{\sqrt{\frac{1}{n}\sum_{k=1}^{n}(\hat{y}_k)^2}+\sqrt{\frac{1}{n}\sum_{k=1}^{n}(y_k)^2}} \tag{3}$$

In the above formula, y_k is the actual values, $\hat{y}_k$ is the predicted values, and *n* is the number of the test set.

3.3. Prediction results analysis

The accuracy and validity of the proposed model are verified by analyzing the deviation between the prediction value and the actual value of power quality under different models. In view of the limited space, this paper takes the total harmonic distortion of voltage at monitoring points as an example, and uses three prediction models to train and predict the dataset respectively. The prediction results and errors are shown in Figure 4 to Figure 7.

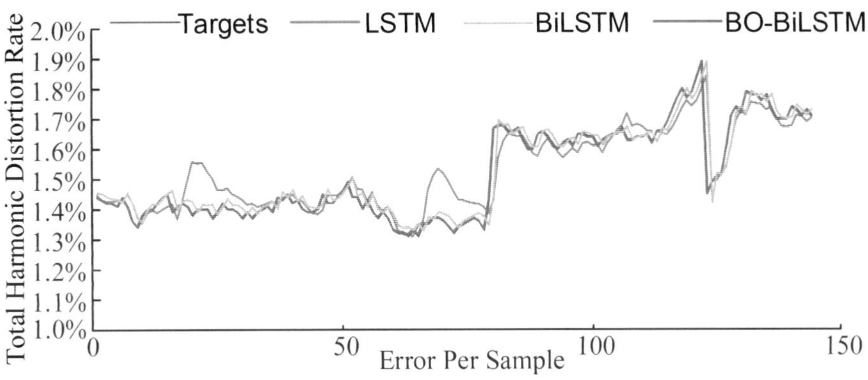

Figure 4. Comparison of prediction curves and actual curves of each model

Figure 4 shows the comparison between the predicted values and the actual values of the three models. The prediction accuracy of the three models can be seen intuitively from the figure. The predicted value of the BO-BiLSTM model is basically the same as the real value, and the predicted output is closer to the expected output. It can be seen from Figure 4 to Figure 7 that the BO-BiLSTM model has higher prediction accuracy.

It can be seen from table 1 that the evaluation indexes of LSTM model are ideal and the prediction accuracy is high, which verifies that LSTM model has advantages in time series prediction. The MAE, MAPE and TIC of BiLSTM model are reduced, indicating that the BiLSTM model has higher prediction accuracy than LSTM model. Compared with the LSTM model that can only refer to the characteristics of historical data, the BiLSTM model can simultaneously transmit information forward and backward, form a strong mapping relationship between historical data and future data, and improve the prediction accuracy of the model. Among the three prediction models, MAE, MAPE and TIC of the BO-BiLSTM model are the lowest, while R^2 is the highest, indicating that BO-BilSTM model has the best prediction effect, and proving the optimization on BiLSTM model by Bayesian algorithm is effective.

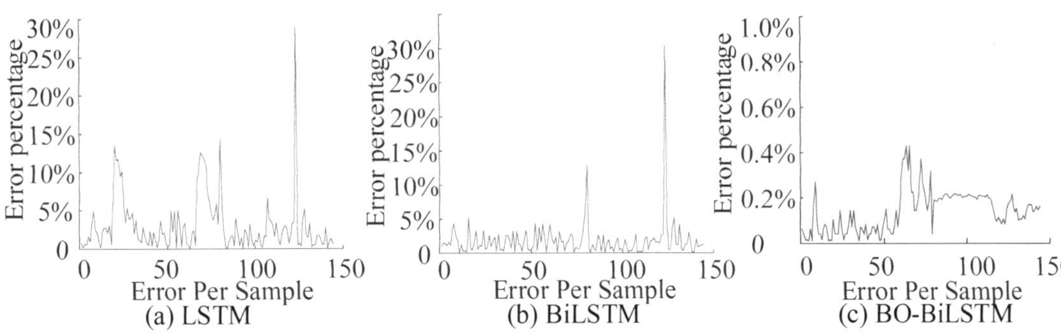

Figure 5. Percentage diagram of error of various prediction results

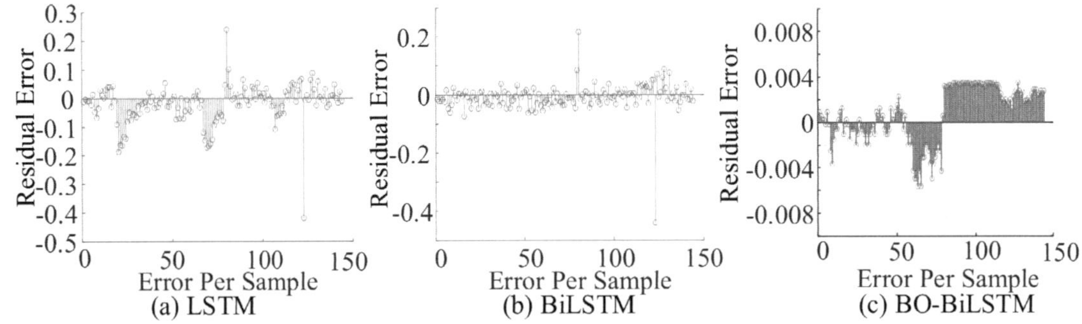

Figure 6. Residual distribution maps of various prediction results

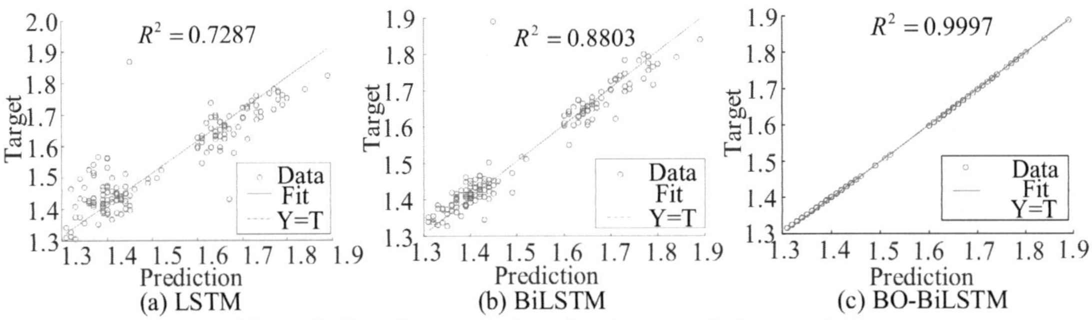

Figure 7. Sample scatter plot of various prediction results

Table 1. Comparative Analysis of Evaluation Indexes of Various Models

Model	LSTM	BiLSTM	BO-BiLSTM
MAE	0.0451	0.0280	0.0022
MAPE	3.07%	1.86%	0.14%
TIC	0.0229	0.0166	0.0008
R^2	0.7287	0.8803	0.9997

4. Conclusions

This paper proposes a prediction model of power quality based on BiLSTM optimized by Bayesian Algorithm from the perspective of data mining. Compared with the LSTM and BiLSTM models, the Bo-BiLSTM model can effectively improve the prediction accuracy and enhance the generalization ability of the neural network model. The predicted results are basically coincident with the actual values of power quality, which verifies that the BO-BiLSTM has stronger performance . The proposed model effectively improves the stability of the model and reduces the volatility, which provides a new idea for the study of power quality prediction.

References

[1] Shukl, P. , & Singh, B. . (2019). Delta-bar-delta Neural Network (NN) Based Control Approach for Power Quality Improvement of Solar PV Interfaced Distribution System. IEEE Transactions on Industrial Informatics, vol. 16, pp. 790-801.

[2] Yildirim, O. , Eristi, B. , Eristi, H. , Unal, S. , Erol, Y. , & Demir, Y. . (2018). Fpga-based Online Power Quality Monitoring System for Electrical Distribution Network. Measurement, vol. 121, pp. 109-121.

[3] Gao, P. , Fang, J. , Y Han, & Wang, J. . (2019). Analysis and Prediction of Power Quality Trend of

Electrified Railway. Computer & Digital Engineering. vol. 47, pp. 1521-1527.
[4] Ding, Z. , Liu, P. , Ouyang, S. , Zeng, J. , & Huang, R. . (2015). Mechanism of Power Quality Forecast and Early Warning and Their Application. Proceedings of the CSU-EPSA, vol.27, pp. 86-92.
[5] Lu, M. (2017). Power Quality Prediction Based on Least Squares Method. Vibroengineering PROCEDIA, vol. 12, pp. 153–158.
[6] Bai, J., Gu, W., Yuan X., & Wang X. . (2015). .Power Quality Prediction, Early Warning, and Control for Points of Common Coupling with Wind Farms. Energies, vol. 12, pp. 153-158.
[7] Tomá Vantuch, Stanislav Miák, Tomá Jeowicz, Tomá Buriánek, & Václav Snáel. (2017). The Power Quality Forecasting Model for Off-grid System Supported by Multiobjective Optimization. IEEE Transactions on Industrial Electronics, vol. 64, pp. 9507-9516.
[8] Greff, K. , Srivastava, R. K. , Jan Koutník, Steunebrink, B. R. , & Jürgen Schmidhuber. (2016). Lstm: a Search Space Odyssey. IEEE Transactions on Neural Networks & Learning Systems, vol. 28, pp. 2222-2232.
[9] Xu, G. , Y Meng, Qiu, X. , Yu, Z. , & Wu, X. . (2019). Sentiment analysis of comment texts based on bilstm. IEEE Access, vol. 7, pp. 51522-51532.
[10] Peralta, F. , D Gutiérrez, Toral, S. L. , Lopez, M. A. , & Gregor, D. O. . (2021). A Bayesian Optimization Approach for Water Resources Monitoring Through An Autonomous Surface Vehicle: the Ypacarai lake case study. IEEE Access, vol. 9, pp. 9163-9179.

The interference cancellation of non-dispersive UV based on active noise control

Bo-Qiang Fan[1,2,a], Yu-Jun Zhang[1,b,*], Kun You[1,c] and Ying He[1,d]

[1]Key Laboratory of Environment Optics and Technology of Chinese, Academy of Sciences, Anhui Institute of Optics and Fine Mechanics, Hefei, Anhui, 230031, China

[2]University of Science and Technology of China, Hefei, Anhui, 230026, China

[a]email: bqfan@aiofm.ac.cn, [b]email: yjzhang@aiofm.ac.cn, [c]email: kunyou@aiofm.ac.cn, [d]email: heying@aiofm.ac.cn

[*]email: yjzhang@aiofm.ac.cn

Abstract. In the process of gas measurement using non-dispersible UV detection technology, the quality of optical signal is directly related to the inversion of concentration. Traditional methods of improving the quality usually processed the detector signal directly, which destroy useful information, and is difficult to eliminate the overall interference of the environment to the detector. The analog circuit system for interference cancellation is designed based on active noise control (ANC) in the paper. The interference channel is added to obtain external noise signal, which through inverting, amplitude and phase adjustment superimposed on the measurement signal by the ANC system, and the digital potentiometer AD5263 is control to optimize by the MCU. The simulation results show that noise standard deviation decrease from 75.16mv to 10.27mv. And the correlation between the processed signal and the NO_2 concentration reached 0.99768, which shows the system can be applied suitably in the non-dispersible technology.

1. Introduction

Non-dispersive UV detection technology [1, 2] is based on gas has specific absorption characteristics in specific UV bands, thereby the light intensity and concentration show a certain relationship. However, the detection accuracy of the method is usually limited by the external interference, such as thermal noise, 1/f noise, electrical noise and temperature, humidity interference etc. [3] on the sensors. Therefore, how to reduce the interference is a problem that we urgently need to solve. Traditional methods such as RC filter, adjacent averaging, Savitzky–Golay (SG) filter, fast Fourier transform (FFT) filter, least-square fast transverse filtering, etc. [4-6] directly processes the measured signal, which lost some useful information and cannot eliminate the entire interferences of the environment on the detector.

ANC principle [7, 8] is to use the ambient noise obtained by the noise sensor to generate noise with equal amplitude and opposite phase through the ANC controller to superimpose and cancel the noise to achieve an improvement in the signal-to-noise ratio. Fig. 1 is a schematic diagram of residual noise generated by the matching and superposition of main noise and anti-phase noise. Therefore, the accuracy of the amplitude and phase of the inverse noise generated by the ANC system directly affects the elimination effect of the main noise. Currently, the method has been applied to the smart meters, headphones, microphones and loudspeakers, etc. in many fields.

Content from this work may be used under the terms of the Creative Commons Attribution 3.0 licence. Any further distribution of this work must maintain attribution to the author(s) and the title of the work, journal citation and DOI.
Published under licence by IOP Publishing Ltd

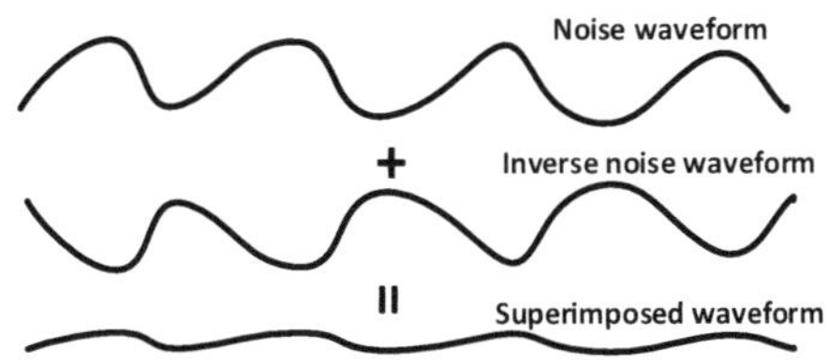

Fig. 1. The schematic diagram of ANC.

In this paper, combined with non-dispersive UV technology, an ANC system based on analog circuit is introduced. The function of the circuit is to reverse, amplitude and adjust the phase of the signal of the noise channel, and then accumulate it with the signal of the measurement channel to achieve noise cancellation. And the mode of phase and amplitude adjustment is fed back by digital potentiometer AD5263 and microcontroller STM32. Finally, the simulation and results show that the method is feasible to improve the resolution of the system.

2. Active Noise Control

The ANC system can be divided into two schemes: feedback and feedforward. [9, 10] The feedback system only analyzes the noise of the measurement signal itself and adjusts the control strategy of the system, which often results in system instability. The feedforward system is shown in Fig. 2. The noise sensor is used to detect the interference noise N2 related to the noise N1 in the measurement channel. The controller uses the residual noise of the output and the noise N2 to generate the inverse noise signal superimposed on the measurement channel to achieve noise cancellation. In normal practical applications, inverse noise can be generated by the adaptive filters, analog devices, and digital signal processing modules.

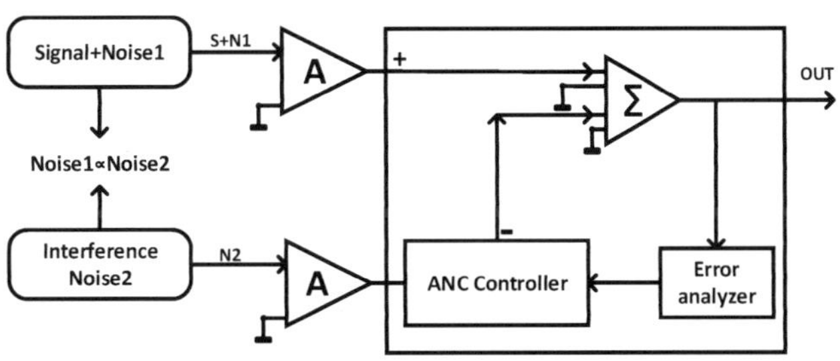

Fig. 2. The schematic diagram of the feedforward system.

Similarly, in terms of optical gas detection, we can also try to use the feedforward ANC system to eliminate the interference of the correlation of sensors. For example, Fig. 3 shows the relationship between the temperature and the photodiode amplitude of the two channels under the same light source and adjacent position conditions. Obviously, the ambient temperature directly affect the amplitude of the sensors with an inverse relationship, while the output of two channel detectors have the related trends. Therefore, the noise channel can be established to extract the related interference to cancel the noise in the measurement channel. And the influence of these interferences on signal and concentration inversion can be reduced.

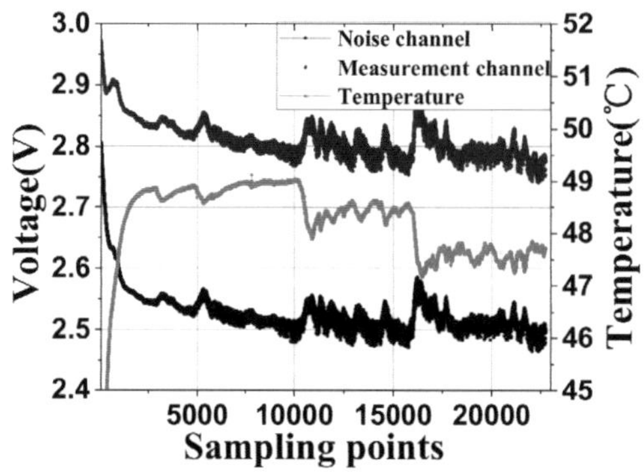

Fig. 3. The relationship between the temperature and the photodiode amplitude.

3. CIRCUIT DESIGN

The schematic diagram of the non-dispersive ultraviolet system based on the ANC is shown in Fig. 4. The three photodiodes of the same manufacturer and the same batch are placed in the same environment are welded on the one PCB. The functions of the three detectors are to receive modulated light signal with gas absorption, modulated light signal without gas absorption, and interference signal. The three-channel signal is amplified by the preamplifier and input to the ANC system. After the interference cancellation, the main control unit is used for signal acquisition, digital filtering and inversion of concentration. Also, the unit sets the standard deviation threshold for the feedback control ANC system.

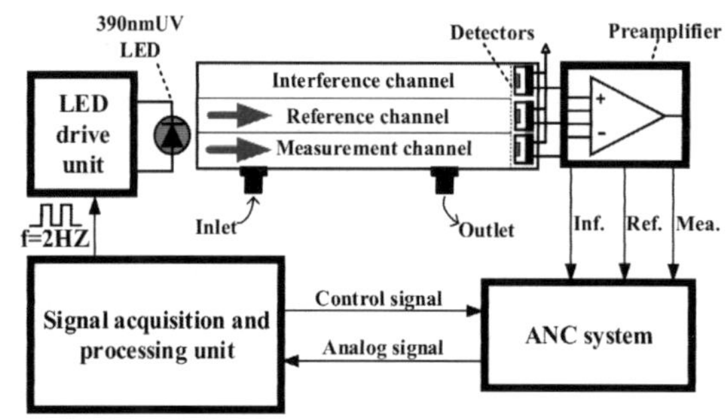

Fig. 4. The schematic diagram of the non-dispersive ultraviolet system based on the ANC.

In the experiment, the light source is 390nm ultraviolet LED with FWHM of 12nm, and its band covers the absorption area of NO2. The optical path of the cell is 20cm. The detector uses a silicon photodiode (S1226-HAMAMATSU), of which the spectral response range is 190-1000nm and the noise equivalent power is $5.0\times10^{-15}(W/Hz^{1/2})$. The pre-amplification adopts the transimpedance circuit, which converts current to voltage, designed by the ultra-low input current amplifier LMC6001. LMC6001 has the input current noise of $0.13(fA/Hz^{1/2})$ under typical 1kHz conditions, which can reduce electrical noise doping into the input weak signal.

The ANC system is built based on analog devices as shown in Fig. 5. The figure is only to eliminate the interference of the measurement channel, and the reference channel is the mirrored circuit. The circuit mainly consists of three parts: the reverse amplifier circuit, the phase adjustment circuit and the summing circuit. The low noise rail-to-rail op amp AD4841 with dual power supply is used, of which the voltage noise is 2.1 $(nV/Hz^{-1/2})$.

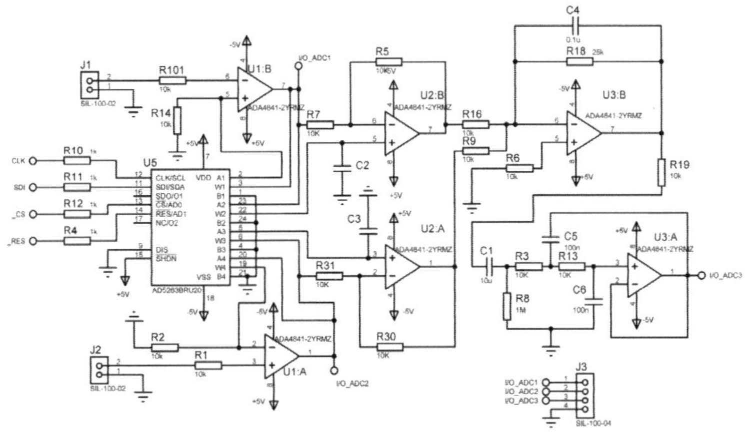

Fig. 5. The circuit schematic of the ANC system.

J1 and J2 input the signal of the interference and measurement channel respectively, which are amplified by transimpedance circuit. J1 is connected with the inverting amplifier circuit to obtain the inverse noise and J2 is connected with the same direction amplifier circuit to match the amplitude. The gains of the two circuits are controlled by the resistance R_{AW1} and R_{AW4} of the digital potentiometer AD5263. The digital potentiometer is controlled by the MCU (STM32F103) with the SPI. Here we choose the nominal resistance as 20K with 198Ω minimum adjustment resistance. Note that this stage does not add blocking capacitors, compensation capacitors, filter capacitors and inductors that have the delays and filters, which affects the subsequent cancellation process.

The next stage is the lag-type first-order all-pass filter with an amplitude gain of 1. Also, the function is to subtly adjust the phase deviation for two signals. Among them, the resistors (R7, R5, R31 and R32) connected with the negative terminal of the op amp must be equal, the value of which also should be as close to the potentiometer (R_{AW2} and R_{AW3}) as possible. The gain and phase formula are as follows:

$$U_o(\omega) = U_i(\omega) \times \frac{2}{1 + j\omega R_{AW2}C2} - U_i(\omega) = U_i(\omega) \times \frac{1 - j\omega R_{AW2}C2}{1 + j\omega R_{AW2}C2} \tag{1}$$

$$A(\omega) = \frac{U_o(\omega)}{U_i(\omega)} = \frac{1 - j\omega R_{AW2}C2}{1 + j\omega R_{AW2}C2} \tag{2}$$

$$|A(\omega)| = \sqrt{\frac{1 + (-\omega R_{AW2}C2)^2}{1 + (\omega R_{AW2}C2)^2}} = 1 \tag{3}$$

$$\varphi(\omega) = \tan^{-1}\left(\frac{-2\omega R_{AW2}C2}{1 - \omega^2 R_{AW2}^2 C2^2}\right) = -2\tan^{-1}(\omega R_{AW2}C2) \tag{4}$$

where $U_0(\omega)$, $U_i(\omega)$, $A(\omega)$ and $\varphi(\omega)$ are the output voltage, input voltage, gain and phase at the frequency ω, respectively. According to the Eq. (3) and Eq. (4), the gain in the full frequency range is 1, and the negative phase angle indicates that the output lags behind the input, which is regulated by R_{AW2} and C2. C2 is usually chosen to be of the μF level and the phase matching is controlled by R_{AW2}. The next stage uses the reverse summing circuit to sum the two signals that have undergone amplitude and phase matching to achieve noise cancellation, the amplification gain of which is determined by R16, R9 and R18.

In the last stage, the residual noise is processed by the DC blocking circuit and the second-order Sallen-Key low-pass filter. The frequency of blocking is set according to the luminous frequency (2Hz). The second-order SK low-pass filter consists of R3, R13, C5, C6 and op amp U3, the cut-off frequency is set by means of equal resistance and capacitance (R3=R13, C5=C6). And the maximum gain is 1. By setting R3=10K and C5=100nf, the cut-off frequency is 100Hz. The gain $A(\omega)$, characteristic frequency f_0 and cut-off frequency f_c are as follows:

$$A(\omega) = \frac{U_o(\omega)}{U_i(\omega)} = \frac{1}{1 + 2j\omega R3C5 - \omega^2 R3^2 C5^2} \tag{5}$$

$$|A(\omega)| = \frac{1}{1+\omega R3C5} \tag{6}$$

$$f_0 = \frac{1}{2\pi R3C5} \tag{7}$$

$$f_c = 0.6435 \times f_0 = \frac{0.6435}{2\pi R3C5} \tag{8}$$

Finally, the concentration-related signal are collected by the bipolar 16-bit AD chip AD7606. And the digital filtering and concentration calculation is processed by the main control unit, which adjusts the digital potentiometer in the ANC circuit by the interference signal threshold to find the optimal signal at the same time. Note that the above resistance and capacitance values need to be subtly adjusted for the actual experiment.

4. SIMULATION AND EXPERIMENTAL RESULTS

In order to verify that the designed analog ANC system can realize the cancellation of interference, the Proteus 8 software was used to simulate. The strong Gaussian white noise was simulated as the interference, of which noise mean voltage is 1.998mv and the standard deviation is 75.16mv. And the square signal with interference was simulated as the received signal of the detector, of which frequency is 2Hz, the duty cycle is 50% and the peak-to-peak value is 1V. The interference of the two input channels had a certain correlation.

Fig. 6(a) and 6(b) are respectively the input noise and the inverse noise after reverse processing. From the small picture, the reverse amplifier circuit composed of resistors and op amps can keep signal characteristics without excessive delay and filter. The simulation results showed that the front-stage reverse circuit can obtain the inverse noise, and the amplitude level can be guaranteed to be consistent.

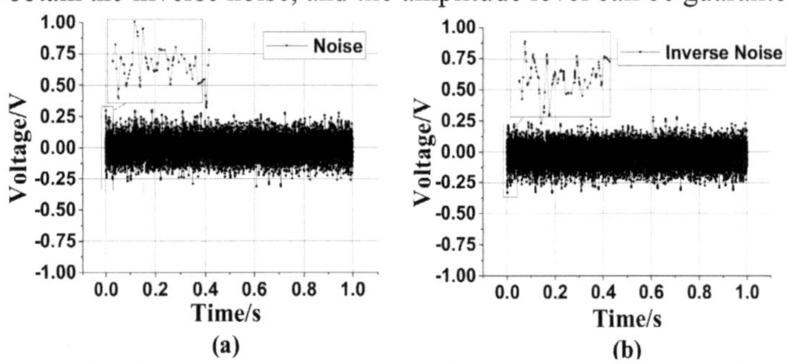

Fig. 6. shows the input noise (a) and the inverse noise (b).

Fig. 7(a) and 7(b) are respectively the input signal with interference and the output of the ANC system. Obviously, the interference on the input is largely eliminated. The standard deviation of the output is reduced to 10.27mv. Fig. 8(a) and 8(b) are respectively the input power spectrum and the output power spectrum, which shows the power is highlighted at each frequency points by the ANC system.

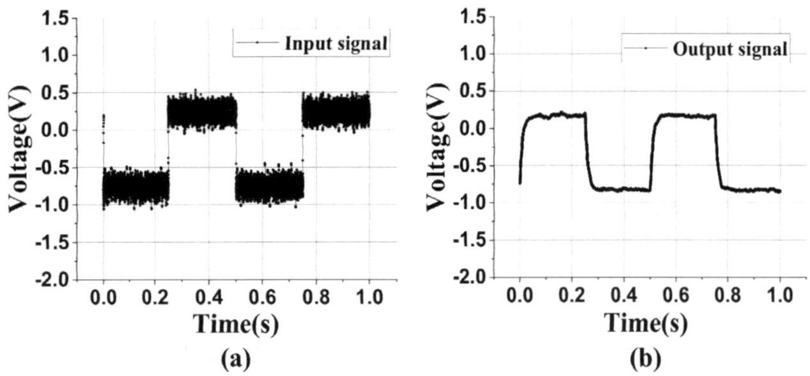

Fig. 7. shows the input (a) with interference and the output (b).

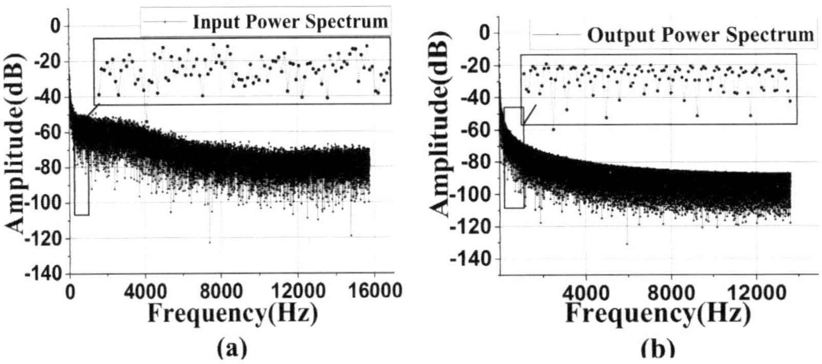

Fig. 8. shows the input power spectrum (a) and the output power spectrum (b).

The actual NO_2 concentration signals have been analyzed as shown in the Fig. 9(a), which shows the ratio of measure channel to reference channel based on ANC system. And a few ppm level concentrations can be clearly distinguished. Fig. 9(b) shows the relationship between concentration and ratio, and the correlation is 0.99768.

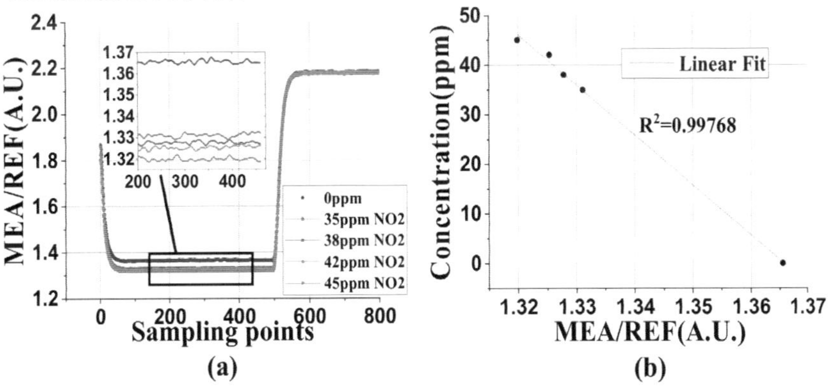

Fig. 9. shows the input noise (a) and the inverse noise (b).

5. Conclusion

The interference cancellation system of non-dispersive UV technology based on ANC has been proposed in this paper. By adding interference channel and building the analog circuits, the system can be simple and be appropriate for the complex environment. The experimental results show that the system can also be applied to strong ambient noise conditions. And it is a good prospect to the interference cancellation, which is applied not only for the non-dispersive technology.

Acknowledgment

Strategic Priority Research Program of the Chinese Academy of Sciences (NO.XDA23010200); National Natural Science Foundation of China (NO. 62033012); Major Subject of Science and Technology of Anhui Province (NO. 202003a07020005).

References

[1] Tirpitz, J. L., et al. (2019) Non-dispersive uv absorption spectroscopy: a promising new approach for in-situ detection of sulfur dioxide. Front. Earth, Sci., 7: 1-15.

[2] Davenport, J. J., et al. (2018) Non-dispersive ultra-violet spectroscopic detection of formaldehyde gas for indoor environments. IEEE Sens. J., 18: 2218-2228.

[3] Liu, H. M., (2015) Temperature effect on signal to noise ratio of NDIR gas analyzer. Instrum. Instrument Tech. Sens., 4: 49-50.

[4] Mahbub, P., et al. (2020) Continuous and real-time indoor and outdoor methane sensing with

portable optical sensor using rapidly pulsed IR LEDs. Talanta, 218: 1-9.
[5] Fellows, A. P., et al. (2019) Spectral analysis and deconvolution of the amide I band of proteins presenting with high-frequency noise and baseline shifts. Appl. Spectrosc., 74: 1-40.
[6] Ye, W. L., et al. (2011) Design and performances of a mid-infrared CH4 detection device with novel three-channel-based LS-FTF self-adaptive denoising structure. Sens. Actuators, B, 155: 37-45.
[7] Miljkovic D. (2016) Active noise control: from analog to digital-Last 80 years. In: 2016 39th International Convention on Information and Communication Technology, Electronics and Microelectronics (MIPRO). Opatija. pp. 1151–1156.
[8] Elliott, S. J., et al. (1993) Active noise control. IEEE Signal Proc. Mag., 10: 12-35.
[9] George, N. V., Panda, G. (2013) Advances in active noise control: A survey, with emphasis on recent nonlinear techniques. Signal Process., 93: 363-377.
[10] Pawelczyk, M. (2009) Analog active control of acoustic noise at a virtual location. IEEE T. Contr. Syst. T., 17: 465-472.

Investigation on the Critical Wind Speed of Instability Failure of UHV Transmission Tower-line Coupling System under Typhoon

Tao Li[1,2a*], **Lei Zheng**[1,2b*], **Zhihui Feng**[1,2], **Jian Li**[1,2]

[1] Wuhan NARI Limited Company, State Grid Electric Power Research Institute, Wuhan, Hubei, 430074, China

[2] NARI Group Corporation (State Grid Electric Power Research Institute), Nanjing, Jiangsu, 211106, China

[a*]litao3@sgepri.sgcc.com.cn, [b*]zhenglei@sgepri.sgcc.com.cn

Abstract—Wind is one of the most destructive external loads of transmission lines, and typhoon-caused severe damage or even destruction of electric power facilities occurred frequently. In order to research the critical wind speed of instability failure of UHV transmission tower-line coupling system under typhoon, the UHV transmission line in Zhejiang is chosen as the research subject and a finite element model of the UHV tower-line coupling system is established. By using the improved generalized force calculation method, the dynamic response of the transmission tower is converted to the static response. Then the LRC method is used to calculate the equivalent static wind load of the tower and obtains the wind response of the UHV transmission tower-line coupling system. The results show that under the typhoon wind field, the component at the junction of the tower head and the body first reaches the supercritical state. Specifically, when the wind speed at 10-meters height reaches 31m/s, the profile component of the tower head will be damaged. With the increase of wind speed, the damaged area of the tower head also expends. When the wind speed at 10-meters height reaches 37m/s, the main load-bearing component of the tower head-body junction will lose its stability, which will directly threaten the safety of the transmission tower.

1. Introduction

China was one of the countries most affected by typhoons in the world. Especially in coastal provinces such as Zhejiang and Fujian, typhoon-caused severe damage or even destruction of electric power facilities occurred frequently. For instance, Lekima, the 9th typhoon in 2019, made landfall in Wenling, Zhejiang province, caused damage in 72 substations and 4,823 lines, and frequent power outages to 7,722,300 households [1]. Yuna, the 14th typhoon in 2004, made landfall in Zhejiang, damaged 3,342 kilometers of transmission lines [2].

Due to the high towers and large spans, UHV transmission lines are sensitive to horizontal loads. Under heavy load conditions such as typhoons, the resistance of UHV transmission towers determines the structural safety of the entire line. Therefore, exploring the mechanical properties of transmission tower-line coupling system under various harsh conditions are the basis for improving the safety of UHV power transmission. This paper studies the static and dynamic characteristics of transmission towers under typhoons and summarizes the wind-induced response characteristics of the tower-line

Content from this work may be used under the terms of the Creative Commons Attribution 3.0 licence. Any further distribution of this work must maintain attribution to the author(s) and the title of the work, journal citation and DOI.
Published under licence by IOP Publishing Ltd

coupling system, which has important theoretical and engineering significance for improving the security of the power grid in coastal areas.

2. Establishing of Finite Element Model of UHV Transmission Tower-line Coupling System

The tower-line coupling system in this paper is based on the 1000kV Antang line. The split conductor of the target section uses 8×JLK/G1A-530(630)/45 steel core expanded aluminum stranded wire, and both ground wires use JLB20A-185 aluminum-clad steel stranded wire. The side and middle-phase insulator are respectively double-connected suspension I shape and V shape string. The target transmission tower is a single-circuit cup-type tower with a total height of 91.6m. The main interface form is hot-rolled equilateral angle steel. The elastic modulus of the angle steel is 206 GPa, the mass density is 7860 kg/m^3, and the Poisson's ratio is 0.3.

The transmission tower is a rigid body formed by multi-specification angle steel and eccentrically connected by bolts. The finite element method can be used to quickly realize the force analysis. In terms of the choice of the structural model, the reference [3] and [4] respectively adopted the space truss model (Link unit), the hybrid unit model (Beam unit + Link unit), and the space rigid frame model (Beam unit) to establish the finite element model of the transmission tower. The calculation results show that the space truss model is in good agreement with the field test results.

Considering that the transmission tower-line coupling system is a complex system composed of towers, lines, and insulators, to improve computational efficiency, the following simplifications and assumptions need to be made when establishing the coupling model in this paper:

(1) The connection of each rod is regarded as separated and rigidly connected, and ignores the specific structure of the connection; (2) Regarding the transmission line as a rational flexible component, which conforms to Hooke's law within the elastic range and does not consider the change in the cross-sectional area before and after deformation; (3) The split conductor is reduced to a single conductor, and its cross-sectional area is the sum of each split conductor.

Transmission towers are subjected to bending moment, torque, and axial force during operation. In this paper, the Beam188 spatial beam element is used to establish the rigid frame model of transmission tower, and Link10 and Link8 elements are used for wires and insulators. The angle steel tower is divided into 390 nodes and 1153 units. The finite element model of the two-span tower-line coupling system established in this paper is shown in Fig.1.

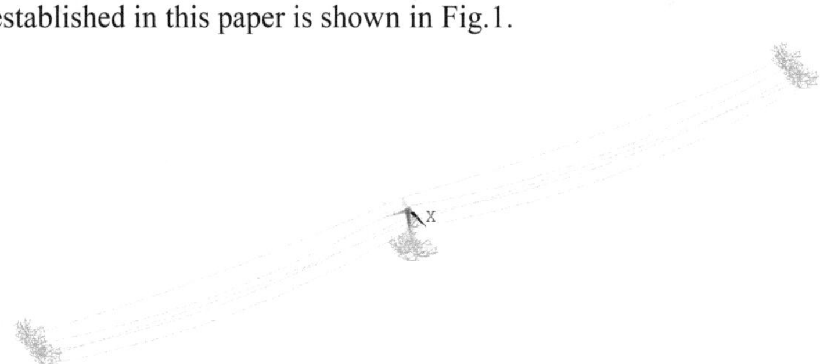

Fig.1 The finite element model of transmission tower-line coupling system established in this paper

3. Calculation of Wind-induced Response of Tower-line Coupling System

3.1. Calculation method of equivalent wind load on transmission tower and its response

Due to the obvious nonlinearity of the transmission tower mode shape, this paper converts the dynamic response of the transmission tower to the static response by using the improved generalized force calculation method. Then the LRC method is used to calculate the equivalent static wind load of the tower and obtains the wind response of the UHV transmission tower-line coupling system, where the equivalent static wind load is expressed as:

$$F_s(z) = W_{Rb}\sigma_{Rb,s}(z) + \sum_{j=1}^{N} W_{jRr}\sigma_{jRr,s}(z) \quad (1)$$

where $\sigma_{Rb,s}(z)$, $\sigma_{jRr,s}(z)$ are root mean square of background term and resonance term of wind load, $W_{Rb} = g_b\sigma_b/R_{\max}$ and $W_{jRr} = g_r\left(\sum \sigma_{\ddot{q}_k} m_k \ddot{r}_{jk}\right)/R_{\max}$ are the combination coefficients of background term and resonance term, where $R_{\max}$ can be expressed as:

$$R_{\max} = \sqrt{g_b^2\sigma_b^2 + g_r^2 \sum_{j=1}^{N}\sum_{k=1}^{N} \ddot{r}_{jk}\sigma_{\ddot{q}_j} m_j \sigma_{\ddot{q}_k} m_k} \quad (2)$$

where $\ddot{r}_{jk}$ is the acceleration correlation coefficient, $\sigma_{\ddot{q}_j}$ is the root mean square of the modal acceleration response, and m_j is the modal quality.

The simulation of the typhoon wind field in this paper is based on the measured data of the Donghaitang wind tower for the 20th typhoon "Kanu" in 2017 [5]. The wind profile index is fitted, and the turbulence intensity settings under level A landform are modified to better match the actual typhoon wind field. Finally, the fluctuating wind speed at various points on the transmission line is simulated based on the harmonic superposition method.

3.2. Calculation method of wind load at the hanging point of conductor and ground wire

The wind load of the tower in the tower-line coupling system is mainly divided into two parts: one exerts on the tower, the other exerts on the lines, and then is transmitted to the tower body through the hanging point of the conductor and ground wire. To fully consider the influence of the latter part of the wind load, this paper takes the insulator string and wire as a system to construct the finite element model, then applies the time-history of wind load and extracts the supporting reaction force of the hanging point. Taking into account the influence of resonance response, this paper chooses to directly apply the reaction force to the corresponding hanging point to calculate the time-history power of the transmission tower.

3.3. Equivalent method of wind-induced response of tower-line coupling system

Based on the coordinates of each node of the tower-line coupling system, the fluctuating wind load at each point is simulated considering the influence of spatial correlation. Combined with the stress response of the transmission tower under the converted static wind load, the total stress response of the transmission tower at a specific wind speed can be obtained. The expression for the step is as follows:

$$\sigma_Z = \overline{\sigma_Z} \pm 2.5 \cdot \Delta\sigma_Z \quad (3)$$

$$\overline{\sigma_Z} = \overline{\sigma_T} + \overline{\sigma_D} \quad (4)$$

$$\Delta\sigma_Z = \sqrt{\Delta\sigma_T^{\,2} + \Delta\sigma_D^{\,2}} \quad (5)$$

where σ_z, $\overline{\sigma_z}$ and $\Delta\sigma_z$ are the maximum, average, and root mean square of the response of the transmission tower component; $\overline{\sigma_T}$ and $\Delta\sigma_T$ are the average and root mean square of the response of the transmission tower component under the tower load; $\overline{\sigma_D}$ and $\Delta\sigma_D$ are the average and root mean square of the response of the transmission tower component under the lines load.

After the above equivalence, the calculation efficiency of the response can be improved. Based on the calculation criteria for tower components specified in the technical guideline for the design of overhead transmission lines, taking the instability damage as the basis for judging the critical wind speed of the transmission tower [6].

4. Calculation results and analysis

Fig.2 shows the axial stress cloud diagram of each component of the tower under the conductor and ground wire load when V_{10} = 37m/s. It can be seen that the most vulnerable force-bearing component is the transverse components of the tower head (*MX*, *MN* marked). However, because the transmission tower is a statically indeterminate structure, the individual rods of the tower head is not the main load-bearing components, and their damage is not enough to destroy the overall structure.

It can be seen from the cloud diagram that the tensile stress of the main load-bearing components is still within the effective stress range, but the compression situation is dangerous. The red line on the right identifies the main compression area of the tower components. The circle on the left marks the component with the largest average compressive stress (925#), and the component is located between the tower head and the tower body. If the junction is damaged, it will directly threaten the safety of the transmission tower. The number and position of the main component of the tower are shown in Fig.3.

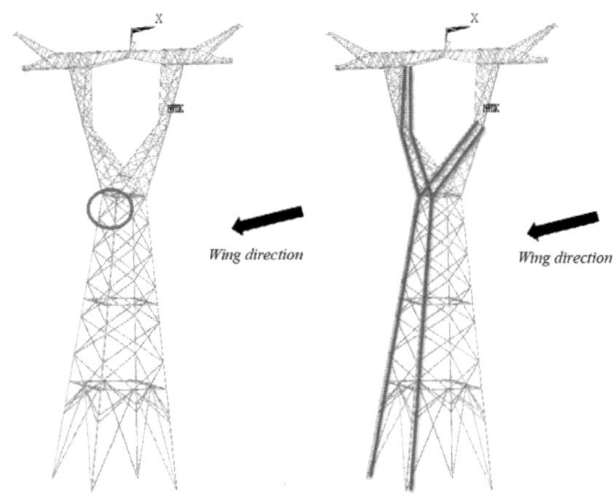

Fig.2 The stress cloud diagram of component under the lines load when V_{10} = 37m/s.

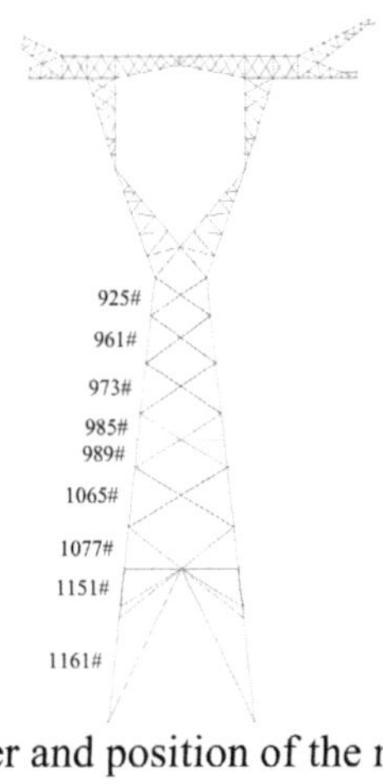

Fig.3 The number and position of the main component

It can be seen from Fig.4 that the force on component 925# is threatened. The component 925# lies in the junction of the tower head and the tower body. If damage occurs, the tower body cannot bear the force transmitted from the tower head, which directly threatens the safe operation of the transmission line. Fig.5 shows that under the wind direction angle of 0°, the stress of the component is mainly caused by the load from the conductor and ground wire. The influence of the load exerting on the tower is limited.

5. Discussion on critical wind speed of instability failure

For the typhoon early warning of the transmission tower, the main concern is the strength damage that may occur to the component under the influence of short-term strong winds. Due to the transmission tower being a spatially statically indeterminate structure, the damage of a certain component does not lead to the destruction of the overall structure. Only when the number of components damaged reaches a certain number and turns the structure into a mechanism, the tower structure will be truly destroyed.

Fig. 6 shows the locations where the tower head and the main component of the tower body are unstable (marked by the red line) under different wind speeds. It is worth mentioning that some components in the elliptical area have also suffered instability failures, but these components do not determine the stability of the overall structure. Under the basic wind speed of 31m/s, the outer components of the tower head will be destroyed first. As the wind speed increases, the scope of instability damage expands. Although component 925# suffers the maximum stress under each wind speed, it is still within the safe range due to its high strength level. When the wind speed reaches 37m/s, the junction component between the tower head and body will lose stability (925#). Fig.7 shows the stress cloud diagram of component 925# under different basic wind speeds. Compared with the total cross sections stress, the maximum compressive stress at the corner will reach the allowable stress earlier.

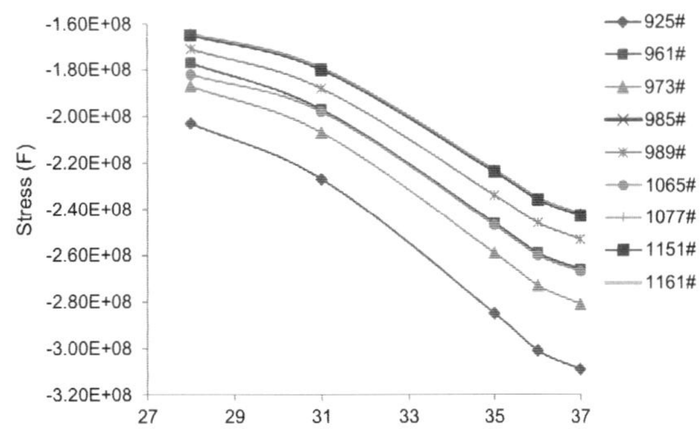

Fig.4 Stress changes of main components under different wind speeds

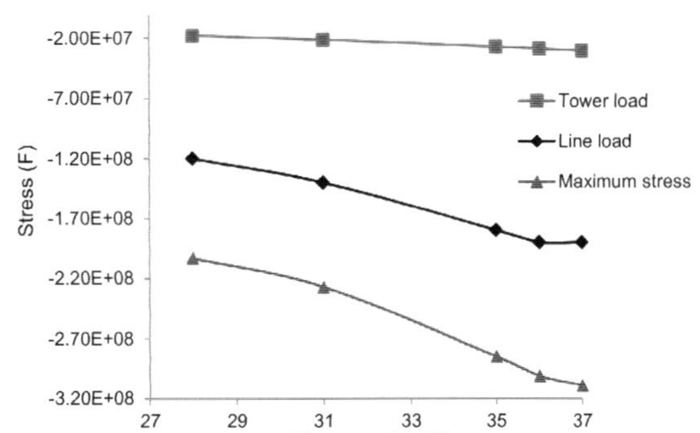

Fig.5 Stress changes of 925# components under different wind speeds

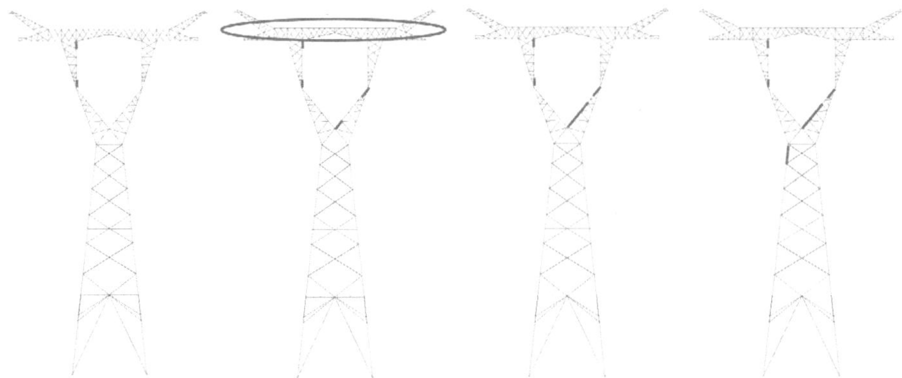

(a) V_{10}=31m/s (b) V_{10}=35m/s (c) V_{10}=36m/s (d) V_{10}=37m/s

Fig.6 Instability and failure position of components under different wind speeds

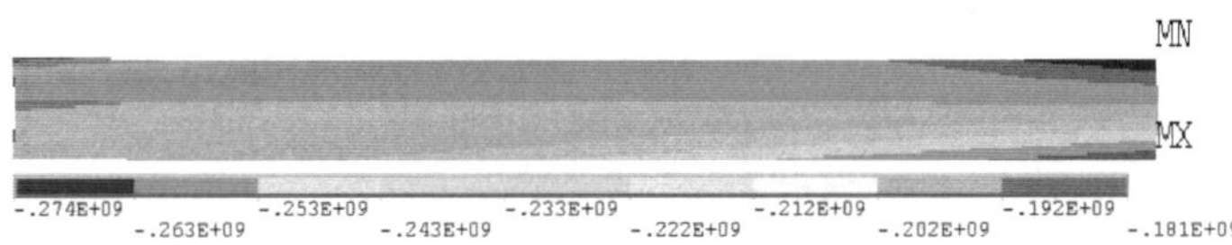

(a) V_{10}=35m/s, t=872.5s

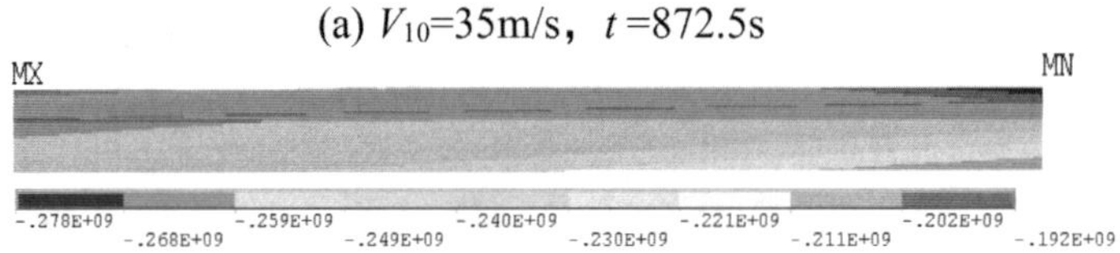

(b) V_{10}=36m/s, t=1011.5s

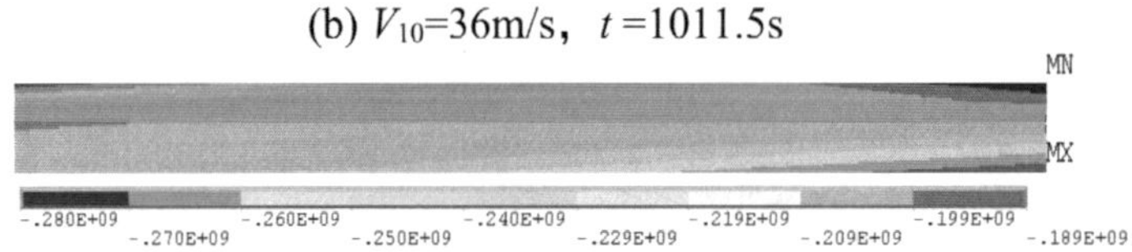

(c) V_{10}=37m/s, t=1018.5s

Fig.7 Stress cloud diagram of component 925# under different basic wind speeds

6. Conclusion

In this paper, the wind response of the UHV transmission tower-line coupling system is calculated and the critical wind speed of instability failure is discussed. The main conclusions can be summarized as follows:

(1) The stress of the components is mainly caused by the wind load from the conductors and ground wires. The damage caused by the wind load exerting on the tower is limited

(2) The junction components between the tower head and body are the most severely stressed area under the typhoon wind field. Furthermore, since the junction components are the most important bearing points of the tower, if damaged, it will directly threaten the safety of the transmission tower.

(3) Under the wind speed of 31m/s, the contour components of the tower head will be destroyed first. As the wind speed increases, the scope of instability damage expands. When the wind speed reaches 37m/s, the junction component between the tower head and body will lose stability.

Acknowledgments

This work was supported by a grant from Key Research and Development Program of Hubei (2020BAB108).

References

[1] Gu, W., Zhu, B., Shen, S. (2020) Experience and enlightenment of dispatching emergency

response of super typhoon Lekima. Zhejiang Electric Power, 39: 35-41.
[2] Zhang,F., Wu Q., Li, J. (2005) The damage of typhoon Yunna to Zhejiang power grid and its preventive measures. Electric Power, 38: 39-42.
[3] Yang, J., Han, J., Zhou, Q. (2010) Selection of Calculation Model for Steel Tubular Tower of UHV Power Transmission Line. Power System Technology, 34: 1-5.
[4] Chen, Z., Tang, K., Zhou, Q. (2018) Calculation Models of Ultimate Bearing Capacity and Failure Modes for Transmission Tower Under Wind Load. Journal of Architecture and Civil Engineering, 35: 47-55.
[5] Li, T., Wang, Q., Wu, D. (2021) Wind tunnel test and wind-induced response analysis of 1000 kV transmission line under typhoon wind fields. Journal of Electric Power Science and Technology, 36: 76-82.
[6] Lou, W., Wu D., Su, J. (2019) Discussion on Wind-induced Swing Flashover and Conductor Wind Load of EHV Transmission Lines. High Voltage Engineering, 45: 1249-1255.

Estimation methods for States of Charge and Health of Lithium-ion battery

Zhenxiao Yi[1], Zhenhua Zhao[2], Lingjun Li[3], Honghong Wang[1], Hongtao Wu[4], Kai Wang[1*]

[1] School of Electrical Engineering, Qingdao University, Qingdao 266071, China

[2]Weihai Rong Kang Real estate Co., Ltd (The Subsidiaries of Sunac China Holdings Limited), Weihai 264401, China

[3]State Grid Weihai Wendeng Power Supply Company, Weihai 264400, China

[4]Huadian Qingdao Power Generation Co., Ltd, Qingdao 266070, China

[*]wkwj888@163.com

Abstract— A battery energy storage system is one of the practical and effective ways to achieve carbon neutrality. Lithium-ion batteries are widely used because of their long cycle life, high energy density, high rated voltage, and low self-discharge rate, etc. Their large-scale application has put forward higher requirements for the battery management system (BMS) to ensure the reliable operation of batteries. However, due to the electrochemical reactions and complex operating conditions inside the battery, it is difficult to accurately assess the internal state of the battery. The state of charge (SOC) and state of health (SOH) are two important parameters in the battery management system. This paper systematically summarizes the current commonly used SOC and SOH estimation methods and analyzes the characteristics of each method. The problems and future research trends of battery SOC and SOH estimation techniques are also described, which are expected to provide a reference for further research in this field.

1. Introduction

In modern times, various countries around the world have paid particular attention to ecological sustainability, with China, the European Union, the United Kingdom, and more than 110 other countries pledging to achieve net-zero emissions in the next 30-40 years [1]. The large-scale application of battery energy storage technology can promote the electrification of transportation and the access and consumption of renewable energy, and play an important role in the process of achieving carbon neutrality [2]. Lithium-ion batteries have significant advantages such as long cycle life and high energy density and are therefore widely used in energy storage devices [3]. However, the safety of battery operation is an issue that cannot be ignored. A battery management system (BMS) is an important component to ensure safe, reliable, and efficient operation of the battery pack, and the battery state of charge (SOC) and battery state of health (SOH) are two important parameters in the battery system. Therefore, accurate and timely access to SOC and SOH information is essential to improve battery life and safety. However, as internal parameters of the battery, they cannot be measured directly and accurately, and can only be estimated accurately by a combination of various direct and indirect methods. In this paper, we review the characteristics, problems, and future research trends of the current major SOC and SOH estimation methods.

Content from this work may be used under the terms of the Creative Commons Attribution 3.0 licence. Any further distribution of this work must maintain attribution to the author(s) and the title of the work, journal citation and DOI.
Published under licence by IOP Publishing Ltd

2. Estimation methods for state of charge

2.1 Direct method

The direct method mainly includes the ampere-time integration (Ah) method and open-circuit voltage (OCV), where the ampere-time integration method estimates SOC by calculating the integration of the measured current against time when the battery is charged/discharged, but the process of soc estimation leads to the accumulation of errors thus making the results deviate from the true value. Also, the influence of battery degradation factors can lead to incorrect SOC estimation results [4]. The open-circuit voltage method can determine and establish the OCV-SOC relationship table by measuring OCV at different SOC step by step and inferring the battery SOC by finding this relationship table. The open-circuit voltage is very smooth when the SOC value of the Li-ion battery is between 20 % and 80 %. If we rely only on the open-circuit voltage for SOC estimation, the error is large [5]. Also, the open-circuit voltage method requires the battery to be left for a long time without load to measure OCV, which limits its application in practice.

Therefore, using direct methods for real-time online estimation of SOC is not an optimal option, and they usually need to be combined with model-based methods to achieve better estimation results.

2.2 Model-based estimation methods

The principle of the model-based SOC estimation method is that the model output value of the measured signal can be calculated based on the input signal and the measurement equation of the state-space model, and the deviation is obtained by comparing it with the actual measured value, and then the filter or observer will update the model parameters and the system state based on this information to give the SOC estimation result, which is shown in Figure 1.Based on the battery state-space model, different filter algorithms or observers can be used to realize the closed-loop estimation of battery SOC. The commonly used methods include the Kalman filter method, double Kalman filter method, extended Kalman filter method, unscented Kalman filter method, etc. all kinds of algorithms have their advantages and disadvantages. The lithium battery SOC estimation method combined with the Ensemble Kalman filter [6] effectively avoids the operation of a high-dimensional matrix and improves the efficiency and accuracy of the algorithm. By constructing the nonlinear state equation of lithium battery, a lithium battery SOC estimation method based on RC equivalent circuit model is proposed [7], which can effectively reduce the error of SOC estimation. Two commonly used RC equivalent circuit models are shown in Figure 2.

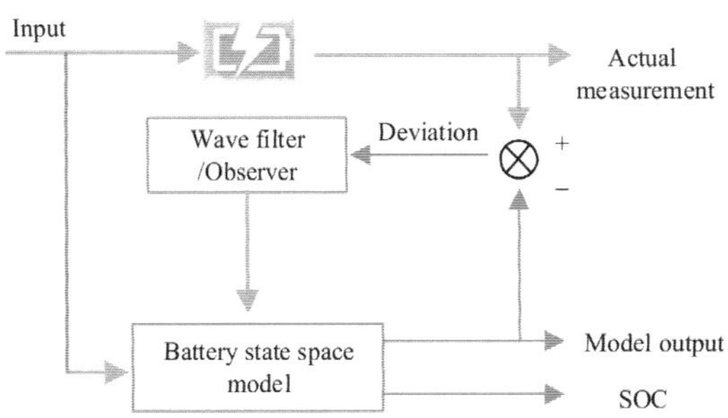

Figure.1 Schematic diagram of the model-based SOC estimation method.

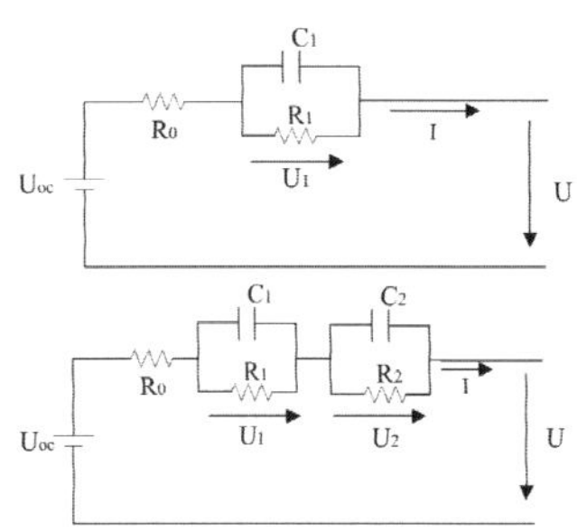

Figure.2 The equivalent circuit-1 RC and the equivalent circuit-2 RC.

Literature [8] proposed an extended Kalman filter (EKF) algorithm. Based on the Kalman filter (KF) algorithm, the higher-order term of KF is ignored and only the first-order term is used, so the nonlinearity is simplified to linear. Considering the influence of unknown noise on the estimation accuracy of battery SOC, reference [9] proposed a fuzzy Kalman filter algorithm (FUKF) combining the unscented Kalman filter method and fuzzy reasoning, which further improved the estimation accuracy. In reference [10], the least square method with a dynamic forgetting factor is introduced to identify the model parameters.

Based on the extended Kalman filter, the maximum likelihood estimation criterion is used to make the model noise covariance have self-learning ability, and a life-cycle SOC estimation algorithm based on parallel Kalman filter is proposed, it ensures the accuracy of estimation and has fast convergence and real-time performance.

The model-based SOC estimation method can give accurate online SOC estimation results on the premise that the model is accurate enough. However, the working characteristics of the battery are highly nonlinear, and it is not easy to establish an accurate battery model. Therefore, this method does not apply to all kinds of batteries.

2.3 Estimation methods based on machine learning

In recent years, several DNN-based SOC estimation methods have been proposed. In the literature, [11] constructed a SOC estimator using a multi-layered perceptron (MLP) network and trained the estimator with signals measured at different ambient temperatures, and the trained model could reduce the SOC estimation error. Long Short-Term Memory (LSTM) has the advantage of capturing time information, and an LSTM-based estimator was developed in the literature [12] to further improve the estimation accuracy and achieve good results. The literature [13] proposed an estimation model based on a simple architecture of Gate Recurrent Unit (GRU) and applied the estimation model to two Li-ion battery datasets, which was able to accurately estimate SOC at different ambient temperatures. Reference [14] proposed a lithium battery SOC estimation method based on PSO-RBF neural network. Using radial basis function neural network and particle swarm optimization algorithm, the efficiency and adaptability of the algorithm are effectively improved, and the difference between the predicted value and the actual value is less than 4%.

The SOC estimation method based on machine learning does not require any prior knowledge, and the performance of the model can be improved under various operating conditions by adding other inputs to the model during the training process, and it applies to all types of batteries, the advantages and disadvantages of each method are shown in Figure 3.

Methods		Advantages	Disadvantages
Direct methods	Ah	It is less affected by the battery itself and has high on-line measurement accuracy	The measurement error is accumulated
	OCV	The calculation is accurate when the battery is at rest	It can not be detected online
Model-based estimation methods		High precision, can measure accurate dynamic SOC	The accuracy of battery model requires accurate dynamic SOC and complex operation
Estimation methods based on machine learning		High precision, can measure accurate dynamic SOC	It needs a lot of reference data and accurate mathematical model, which is greatly affected by the training data

Figure.3 The advantages and disadvantages of each method.

3. Estimation Methods for State of Health

3.1 Model-based estimation methods

Equivalent circuit modeling is a hot topic in the study of SOH of lithium-ion batteries, in the literature [15], a nonlinear non-Gaussian system remaining life prediction method for lithium-ion batteries based on Gaussian Hermite particle filtering was proposed based on extended Kalman filtering. In the literature [16], a new support vector regression-based SOH state-space model for batteries was developed using the battery capacity as the state variable, the estimated impedance as the output variable, and the particle filter to estimate the impedance decay parameter. The literature [17] used the particle filter algorithm as the core, the double exponential model as the state equation, and the artificial neural network as the observation equation to fit the degradation curve of the battery well, and the prediction accuracy will gradually improve with the increase of cycles.

Although this method can achieve more accurate battery SOH estimation, the estimation results are highly dependent on the quality of the model. In addition, to keep the model performance from degrading as the battery ages, the parameters of the model need to be continuously updated during use, and a large number of parameters present in the model makes the computational load heavy.

3.2 Estimation methods based on machine learning

A neural network is used to establish the relationship between health index and SOH, estimate the health state, and a swarm intelligence optimization algorithm is used to optimize the network model parameters, so as to improve the accuracy of the model. Reference [18] proposes a back-propagation neural network optimized by firefly algorithm to estimate the SOH of lithium-ion batteries. The first-order RC circuit model is selected, and the recursive least square method is used for online parameter identification. Literature [19] designed SOH estimation method based on Long and short-term memory (LSTM) with constant voltage charging equal current drop time and constant voltage charging capacity as characteristic parameters to improve SOH estimation accuracy. Reference [20] proposed an attention mechanism decoder model for lithium-ion battery SOH estimation, encoded the data into a set of feature sequences, and established the model combined with a gated loop unit neural network. Literature [21] realizes the joint estimation of SOC and SOH of lithium-ion batteries under different temperature and working conditions based on a Gru cyclic neural network and convolutional neural network. In reference [22], aiming at the problem that the capacity regeneration phenomenon affects the modeling accuracy of lithium-ion battery health state, the energy weighted Gaussian process regression (GPR) method based on empirical mode decomposition (EMD) is proposed. The energy distribution of samples is obtained by EMD decomposition, and energy weighted GPR-based lithium-ion battery SOH prediction model is established. GPR has high prediction accuracy, The prediction results are in the form of probability density. The paper [23] used gated recurrent neural network (GRU-RNN) to establish the mapping law between voltage timing features and battery SOH and achieved a prediction accuracy of MAE less than 1.25%. A schematic diagram of the GRU-RNN structure is shown in Figure 4

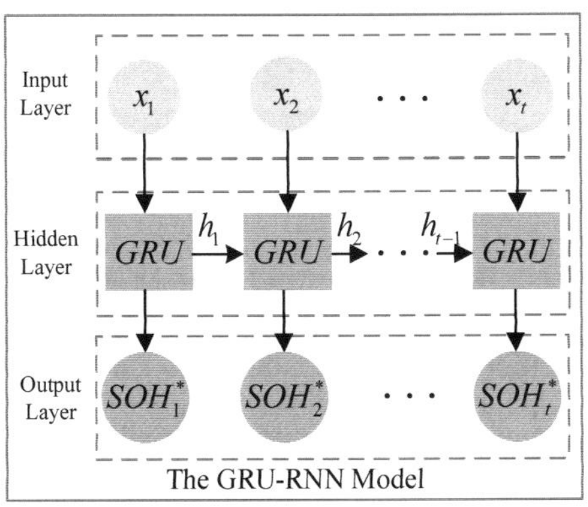

Figure.4 The GRU-RNN Model.

Methods	Advantages	Disadvantages
Model-based estimation methods	High precision Real-time estimation Good robustness	Large computational volume Accuracy is susceptible Requires extensive experimental validation
Estimation methods based on machine learning	High precision Strong adaptability Real time estimation	Large amount of calculation Dependent data Accuracy is easily affected

Figure.5 The advantages and disadvantages of each method.

The advantage of the machine learning-based estimation method is that it does not require a complex study or systematic analysis of the degradation process of Li-ion batteries, nor does it require an equivalent model of the battery or an electrochemical model and the identification of the parameters of the model, and it also has excellent algorithmic interpretability. The advantages and disadvantages of each method are shown in Figure 5.

4. CONCLUSION

With the wide application of lithium-ion batteries in various energy storage systems, SOC and SOH estimation of batteries is a hot topic to improve the safety of BMS. Among them, the limitation of the direct method makes it impossible to realize online estimation. Although the model-based estimation method has high accuracy, it depends too much on the accuracy of the model. The estimation method based on machine learning is simple to apply, does not need to understand the system prior knowledge, and can capture the nonlinear relationship between the measurement signal and SOC and SOH. It has high accuracy and convergence and has a good application prospect. However, there are high requirements for the sample size of training data and system hardware. In the future, we can consider combining transfer learning technology to reduce dependence. With the development of 5G technology, a large amount of data can be collected and processed through cloud platforms and cloud computing technology, which provides strong support for the application of machine learning algorithms in battery state estimation.

Acknowledgment

This work was supported by the Youth Fund of Shandong Natural Science Foundation (No. ZR2020QE212), Key Projects of Shandong Natural Science Foundation (No. ZR2020KF020), Shandong Natural Science Foundation (No. ZR2020MF068).

References

[1] RUAN M H, ZHENG J P, LIU R, et al. Optimizationconfiguration method for hybrid energy storage capacity of independent DC microgrid cluster [J]. Electric Power Engineering Technology, 2021,40(3): 99-105.

[2] YAN H B, KANGLX, ZHOU D. Optimal model of the day. -ahead dispatching and energy storage for micro-grid considering randomness[J]. Power System and Clean Energy,2019, 35(11): 61-65.

[3] GENG J, YANG D M, GAO Z P, et al. Optimal operation of distributed integrated energy microgrid with CCHP considering energy storage[J]. Electric Power Engineering Technology, 2021, 40(1): 25-32.

[4] KIM T, WANG Y, SAHINOGLU Z, et al. State of charge estimation based on a real-time battery model and iterative smooth variable structure filter[C]//2014 IEEE Innovative Smart Grid Technologies-Asia (ISGT ASIA), Kuala Lumpur, Malaysia,20-23 May, 2014:132-137.

[5] ATTANAYAKA A M, KARUNADASA J P , HEMAPALA K T. Estimation of state of charge for lithium-ion batteries: a review[J]. AIMS Energy, 2019,7(2):186-210.

[6] Chen Gang, Chu Jianxin, pan xuanlin, et al. Estimation of lithium battery SOC based on joint Ensemble Kalman filter [J]. Zhejiang electric power, 2021, 40 (1): 123 – 130.

[7] Yin Fujia. SOC estimation of lithium battery based on RC equivalent circuit model [J]. Software guide, 2021, 20 (1): 117 – 122

[8] PLETT G L. Extended Kalman filtering for battery management systems of LiPB-based HEV battery packs: Part 3:Stateand parameter estimation[J]. Journal of Power Sources, 2004, 134(2): 277-292.

[9] An ChangZu, Zhang Ruiping, Zhang Xiaozhou, et al. SOC estimation of lithium battery based on Fuzzy unscented Kalman filter algorithm [J]. Power technology, 2020, 44 (3): 333 – 336.

[10] Zhu Yinan, LV Taolin, Zhao Zhiyun, Yang Wen. State of charge estimation of lithium ion battery based on parallel Kalman filter [J]. Energy storage science and technology, 2021,10 (06): 2352-2362.

[11] Ragone M , Yurkiv V , Ramasubramanian A , et al. Data driven estimation of electric vehicle battery state-of-charge informed by automotive simulations and multi-physics modeling[J]. Journal of Power Sources, 2021, 483:229108.

[12] Hu X , Li S , Peng H , et al. Robustness analysis of State-of-Charge estimation methods for two types of Li-ion batteries[J]. Journal of Power Sources, 2012, 217(none):209–219.

[13]Yang F, Zhang S, Li W, et al. State-of-charge estimation of lithium-ion batteries using LSTM and UKF [J]. Energy, 2020, 201: 117664.

[14] Gao Junling, Zhang Yizhe. SOC estimation of lithium battery based on PSO-RBF neural network [J]. Journal of Chongqing University of technology and technology: Natural Science Edition, 2020 (2): 37-41.
[15] Ma Y, Chen Y, Zhou X W, et al. Remaining Useful Life Prediction of Lithium-Ion Battery Based on Gauss-Hermite Particle Filter[J]. Ieee Transactions on Control Systems Technology, 2019, 27(4): 1788- 1795.
[16] Wei J W, Dong G Z, Chen Z H. Remaining Useful Life Prediction and State of Health Diagnosis for Lithium-Ion Batteries Using Particle Filter and Support Vector Regression[J].leee Transactions on Industrial Electronics, 2018, 65(7): 5634-5643.
[17] Qin W,Lv H C, Liu C L, et al. Remaining useful life prediction for lithium-ion batteries using particle filter and artificial neural network[J]. Industrial M anagement & Data S ystems, 2020, 120(2): 312-328.2011, 36(5): 3531-3540.
[18] Wu Tiezhou, Liu Sizhe, Zhang Xiaoxing, Wu Linzhang. Estimation of SOH of lithium ion battery based on FA-BP neural network [J]. Battery, 2021, 51 (01): 21-25.
[19] Liu Zhiming. Research on SOH estimation algorithm of power battery based on lstm-rnn [D]. Guilin University of Electronic Science and technology, 2020.
[20] Liu Haotian, Wang Ping, Cheng Ze. - Estimation of lithium ion battery health state based on a codec model [J]. Chinese Journal of electrical engineering, 2021, 41 (05): 1851-1860.
[21] Li Chaoran, Xiao Fei, fan Yaxiang, Tang Xin, Yang Guorun. Joint estimation of SOC and SOH of lithium ion battery based on deep learning [J]. Chinese Journal of electrical engineering, 2021, 41 (02): 681-692.
[22] RICHARDSON RR, OSBORNE MA, HOWEY DA.Gaussian process regression for forecasting battery state of health[J].Journal of Power Sources, 2017,357:209-219.
[23] Zhang Shaoyu, Wu Chunhui, Xiong Wenyuan. Estimation of health status of lithium ion battery using gated cyclic neural network [J]. Infrared and laser engineering, 2021, 50 (2): 236-243.

A Compact 60GHz Power Amplifier in 65nm CMOS Technology

Ling Liu[1,a], Song Ye[1,b], Weimin Wang[2,c], Wenying MA [1,d*]

[1] College of Communication Engineering, Chengdu University of Information Technology, Chengdu, China

[2] College of Optoelectronic Engineering, Chongqing University,Chongqing, China

[a]email: 1507615624@qq.com, [b]email: 461964222@qq.com, [c]email: weiminwang@qq.com, [d*]email: 273029443 @qq.com

Abstract. A compact 60GHz power amplifier chip in 65nm CMOS technology of three-stage common source structure is presented. The first two amplifiers offer sufficient gain to pre-amplify the small input power. The third stage amplifier uses two sets of differential pairs to achieve power synthesis. On-chip transformer coupling is adopted to realize inter-stage impedance matching as well as input and output matching. The compact structure of on-chip transformer can reduce the chip size and improve the integration degree. The measured small signal gain at 60GHz of 22.3dB, the saturation power of 20.2dBm, the output P_{-1dB} of 17.3dBm, the PAE of 14%, and the core area is 0.19mm^2. The power amplifier can be used in high-speed short-range wireless communication system to enhance communication data transmission capacity.

1. Introduction

With the rapid development of millimeter wave technology, the application of the internationally universal license-free 60GHz band in high-speed short distance communication links is capable of thousands of megabits of data transmission per second[1]. In wireless communication system, power amplifier plays a key role in communication distance and channel quality. Nowadays, the cut-off frequency of CMOS process devices has exceeded 200GHz, and process nodes are gradually reduced to ultra-deep sub-micron level[2]. Compared with group III-V compounds semiconductor technology, silicon-based CMOS technology has the unique advantages of high integration, low cost and easy integration with digital circuits. Millimeter-wave power amplifier based on silicon CMOS technology is faced with two problems: Firstly, in millimeter-wave band, various parasitic effects generated by transistors have many adverse effects on the gain and stability of the amplifier. Secondly, the low breakdown voltage brought about by smaller transistor sizes greatly limits the output power.

In this work, a high performance 60GHz power amplifier is designed based on 65nm CMOS technology and transformer matching method on chip. It has great application potential in high-speed short-range wireless communication, point-to-point wireless communication, high-precision radar detection and other fields.

2. Circuit topology

The circuit structure of the compact power amplifier designed in this paper is shown in Fig.1. The circuit topology of the three-stage common source amplifier cascade and the matching method of on-

Content from this work may be used under the terms of the Creative Commons Attribution 3.0 licence. Any further distribution of this work must maintain attribution to the author(s) and the title of the work, journal citation and DOI.
Published under licence by IOP Publishing Ltd

chip transformer are adopted to obtain sufficiently high gain and output power. The first two amplifiers were used as the driving stage, and the differential common source amplifiers were used to enhance the gain. The third-stage amplifier uses two groups of symmetrical differential pair transistor amplifiers to improve the output voltage swing through the power synthesis network of the four-input transformer structure, so as to improve the output power [3]. The input and output signals of the power amplifier are single-ended signals matched to the load of 50 ohm. The input impedance matching design increases the series differential inductor on the transformer to avoid the impedance mismatch problem. The standard on-chip transformer coupling structure is used between the first and second stage amplifiers. The output of the second stage amplifier and the input of the third stage amplifier are matched by four inputs on-chip transformers.

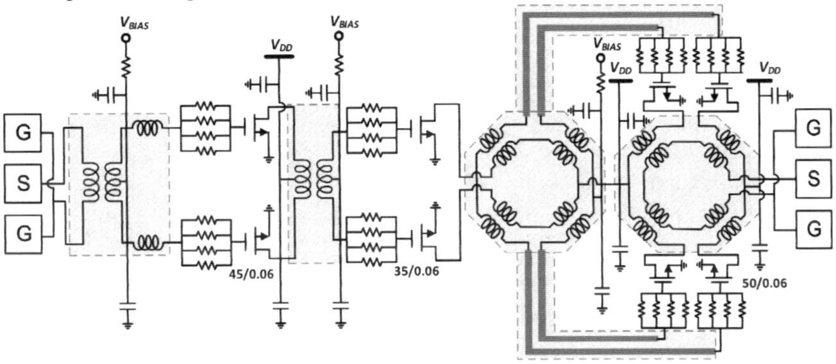

Fig.1 Schematic diagram of compact three stage power amplifier

2.1. *Amplifier section*

The active part of the power amplifier is constructed by the conventional amplifying circuit three stage topology cascade, which were all biased in class A. Based on the bias current density theory and load line theory combined with the formula, the transistor size of the power amplifier was deduced and calculated. Considering the transistor power gain and the loss and parasitic effect in the high frequency band, the transistor size and dc bias current are designed as follows: the transistor size of the first stage amplifier is set as 45/0.06μm, and the dc current is 20mA.The intermediate amplifier transistor size is set to 35/0.06μm, dc current 50mA. The third stage amplifier transistor size is set to 50/0.06μm, dc current 80mA.The transistor gate of each stage amplifier is in series, and the appropriate capacitance is added at each stage bias voltage and transformer tap to ensure the stability.

2.2. *Transformer on-chip section*

The 65nm CMOS process PDK used herein provides a total of 9-layer metal for a signal trace in which the top metal layer has the largest thickness and the lowest resistivity, and the insertion loss is relatively small [4]. The top two-layer metal is designed to have a tapped octagonal inductance, which is a state in which the space is vertically stacked. Using the primary coil, the vertical wide side coupling structure of the secondary coil can increase the coupling coefficient of inductance at a resonant frequency of 60 GHz. The central tap of the transformer primary coil and the secondary coil is a natural virtual ground, which is fed from this point to DC voltage, and the performance of the transformer is unaffected.

Power splitter uses a transformer matching network as shown in Fig.2(a). Two pairs of differential transmission lines are inserted between the power splitter and the output differential pair to compensate for the 1: 1 transformer impedance range. The output terminal impedance matching and power synthesis adopts the four-input transformer structure shown in Fig.2(b), the transformer secondary coil is grounded, and the power signal is completed by the differential-single-ended conversion, the synthesis of all the way signals. During the differential-single-ended conversion process, since the voltage swing of the matching network is effectively superimposed, the impedance seen in each input port is a quarter of the load, compared to the standard 1: 1 transformer compared to voltage gain.

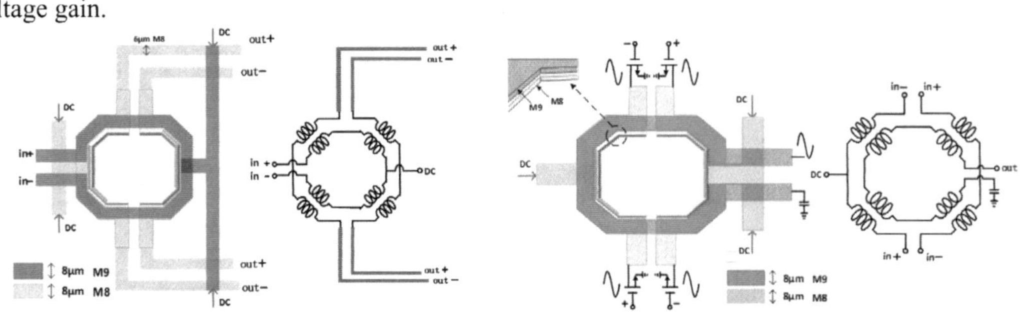

(a) (b)

Fig.2 On-chip transformer structure. (a) Power splitter, (b) Power combiner

3. Experimental results

The 60GHz power synthesizer chip is developed based on 65-nm CMOS technology. The on-chip transformer is used to make the circuit structure more compact and achieve sufficient power output while ensuring high gain and stability. The test conditions are power supply voltage 1.2V, each stage bias voltage 0.8V, input power is -25dbm.The measured S-parameters are shown in Fig.3. Under the condition of well-matched input/output, the power amplifier can achieve the working bandwidth of 8GHz in the range of 56GHz-64GHz, in which the small signal gain is better than 20dB and the reverse isolation is more than 40dB.

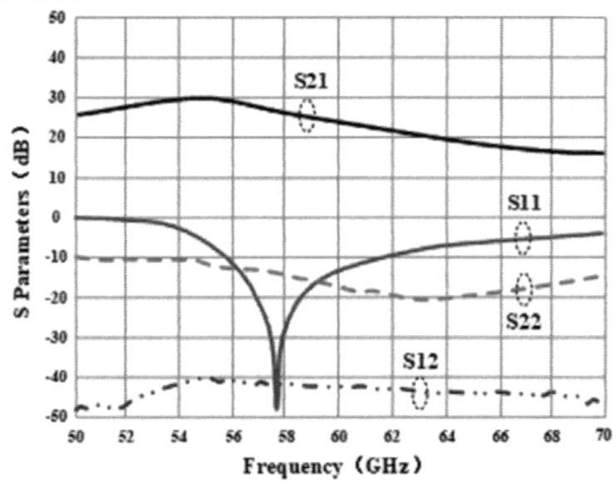

Fig.3 Measured S-parameters

As shown in Fig.4(a), the output power and gain test result at 60 GHz, wherein the PAE reaches 14% when the input power of 0 dBm. As shown in Fig.4(b), in the 60 GHz to 66 GHz frequency range, the P_{sat} of 20.2 dBm, and the P_{-1dB} of 17.3 dBm at 60 GHz.

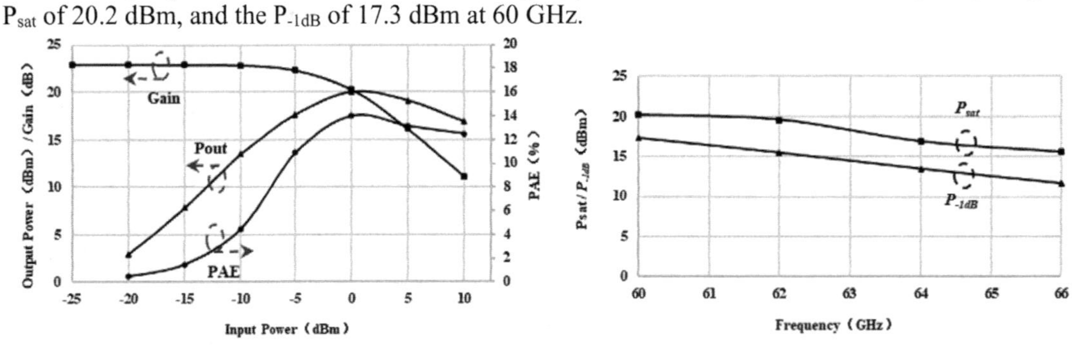

(a) (b)

Fig.4 (a)Measured gain and out power (b) Measured saturation power and output P_{-1dB}

The die photo of the PA is shown in Fig.5, the core area of 0.64 mm × 0.3 mm without the PAD. The Comparison of state-of-art 60GHz CMOS PAs list in Table. 1.

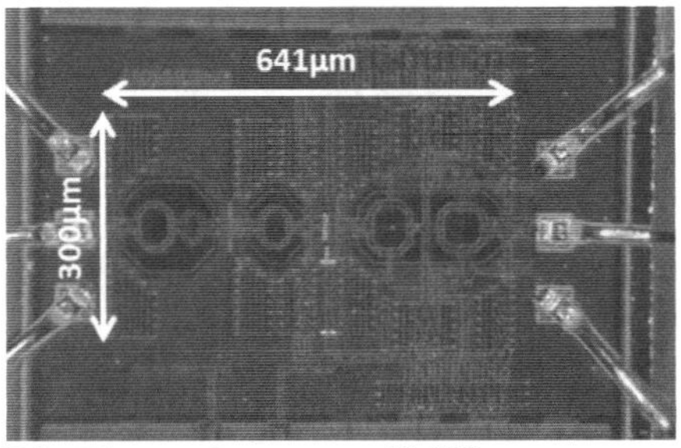

Fig.5 Die Photo of compact 60GHz power amplifier

Table.1 Comparison of state-of-art 60GHz CMOS PAs

Parameters	This work	[4]	[5]	[6]
CMOS/nm	65	65	65	65
Power Supply/V	1.2	1	1	1.2
Gain/dB	22.3	20.3	19.2	17.5
Psat/dBm	20.2	18.6	17.7	17.2
OP_{-1dB}/dBm	17.3	15	15.1	16.6
PAE/%	14	15.1	11.1	11.3
Core Area/mm^2	0.19	0.28	0.83	0.48

4. Conclusions

A high performance compact 60GHz power amplifier chip based on 65nm CMOS technology is designed. Three-stage common source amplifier cascade structure and on-chip transformer impedance matching networks are adopted to achieve sufficient power output while realising high gain and stability. The test results show that the power amplifier achieves a high gain of 19.5dB-27.5dB in the frequency range of 56GHz-64GHz, and the output saturation power is higher than 15dBm, which can meet the international universal application of 60GHz wireless communication system.

Acknowledgments

This work was sponsored by Central University Basic Research Business NO.2021CDJQY-041 and supported by the introduction plan of "Western Light" Talent Training Program.

References

[1] DI S W, WANG S, ZHANG J, et al. (2015) A CMOS 60 GHz Power Amplifier Utilizing Capacitance Neutralization. Chinese Journal of Electron Devices, 38(6): 1253-1256.

[2] HUANG Z Q, ZHANG X, ZHAO C X, et al. (2021) Design of a W-band two-way current-combining power amplifier in 65-nm CMOS. Journal of Nanjing University of Information Science & Technology, 13(4): 425-430.

[3] CHEN J, CHIU W, CHIANG Y.(2017) A transformer-coupled power amplifier in 90-nm CMOS process for V-band applications. In: IEEE Asia Pacific Microwave Conference (APMC). Kuala Lumpur, Malaysia: IEEE, pp: 318-321.

[4] CHAN J, NIKNEJAD A M.(2011) A compact 1V 18.6dBm 60GHz power amplifier in 65nm CMOS. In: IEEE International Solid-State Circuits Conference (ISSCC). San Francisco, CA, USA: IEEE, pp: 432-433.

[5] LAI J, Valdes-Garcia A..(2011) A 1V 17.9dBm 60GHz power amplifier in standard 65nm CMOS. In: IEEE International Solid-State Circuits Conference (ISSCC). San Francisco, CA, USA: IEEE, pp: 424-425.

[6] LIANG Y, LI N, WEI F, et al.(2015) A 60GHz digitally-assisted power amplifier with 17.2dBm Psat, 11.3% PAE in 65nm CMOS. In: IEEE MTT-S International Microwave Symposium. Phoenix, AZ: IEEE, pp: 1-4.

Non-intrusive Load Monitoring Based on Self-attention

Tuanjie Gan[1a], Yongchang Chen[1b], Yan Zhang[1c], Cong Cheng[2d*], Min Huang[2e*], Zhong Li[3f]

[1]Department of Jiangmen Power Supply Bureau, Guangdong Power Grid Co, Guangdong 529000, China

[2]South China University of Technology, Guangdong 510000, China

[3]Department of Electronic and Communication Engineering, North China Electric Power University, Hebei 071003, China

[a*]1125893346@qq.com, [b*] 7648625@qq.com, [c] 14137416@qq.com
[d] 957815154@qq.com, [e] minh@scut.edu.cn, [f] ncepulz@126.com

Abstract—Non-intrusive load monitoring is an important means to monitor the power consumption of end-user equipment in real time and timely. At present, there are still some problems in non-invasive load identification, such as the analysis of feature quantity is not detailed enough, the subjectivity of feature selection is strong, and the effect of identification algorithm is not good because it does not fully consider the information expression difference of feature quantity. In order to mine local features and the advantage of the latter in capturing the global relationship between sequence elements, a deep self-attention network method based on self-attention calculation is proposed. The research shows that this method has higher accuracy in solving the problem of load decomposition, and the training speed is faster than the traditional deep learning method. It performs well in the power decomposition of continuous variable state equipment and multi state equipment.

1. Introduction

Most of the world's electric energy comes from non-renewable fossil fuels, and its reserves are decreasing year by year. With the continuous growth of population around the world, the progress of civilization and the urbanization of rural areas, the demand for energy from all aspects of social production and life is increasing. The energy crisis, global warming and high-power costs are all historical omens for the power industry at present and in the future [1-3]. Non-intrusive load monitoring (NILM) is the main tool to analyse the power consumption behaviour of residential users. It completes the collection, storage and analysis of important electrical information through the data acquisition and communication device installed at the user's power supply inlet, so as to realize the accurate perception of the operation status and energy consumption of various electrical equipment within the user [4]. The classification and identification of power load is helpful to understand the load composition of power system, optimize the planning and dispatching of power system, and strengthen the ubiquitous relationship among power grid, users and equipment. It is of great significance for the development of ubiquitous power net of things.

Content from this work may be used under the terms of the Creative Commons Attribution 3.0 licence. Any further distribution of this work must maintain attribution to the author(s) and the title of the work, journal citation and DOI.
Published under licence by IOP Publishing Ltd

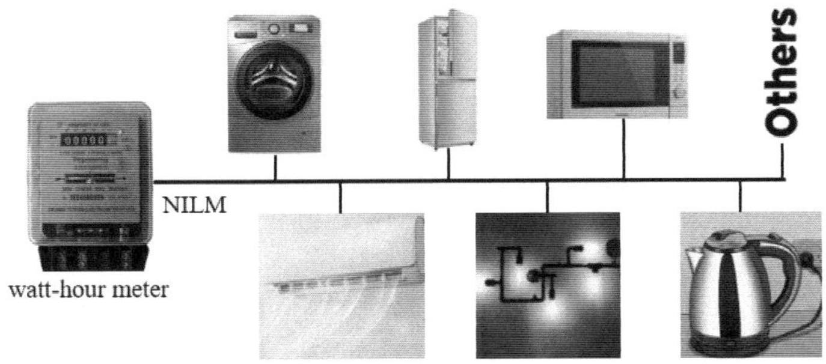

Fig.1. Schematic diagram of non-intrusive load decomposition

The existing load identification methods are mainly divided into invasive and non-invasive [1-8, 10,13]. By comparing the invasive and non-invasive load identification methods, the researchers found that the invasive load identification method has relatively great advantages in accuracy, but it is slightly insufficient compared with the non-invasive load identification method in terms of convenience, maintainability and user acceptance [9]. In recent years, more and more researchers have made great progress in the research of non-invasive load identification. The operation process of each electric equipment is modeled, and then the load model of each electric equipment is formed [12]. Finally, the time series P (or other electrical quantities) of the total electric load is decomposed through the decomposition algorithm, and the decomposed electric equipment is identified, as shown in Figure 1. The essence of non-invasive load monitoring is non-invasive load decomposition, that is, according to the total power consumption information of a family, the total power consumption is decomposed into the power consumption of each appliance in a family, and the user's power consumption behavior is analyzed according to the power consumption of each appliance at each time, in order to better guide users to make reasonable power consumption and reduce users' power consumption cost. Recently, the application of deep learning in non-invasive load decomposition has attracted more and more attention [13]. Different from traditional pattern recognition, deep learning can automatically extract features without manual extraction. Huang [14] proposed to realize high-frequency power decomposition through unsupervised deep learning method and identify the composition of power signal. Mauch et al [6]. Proposed a two-layer bidirectional recursive neural network (LSTM) architecture and a combination of HMM and deep neural network (DNN) The combined scheme performs load decomposition, which is improved compared with the traditional FHMM, but the generalization ability of the algorithm has not been fully verified due to insufficient data used in training the model. In 2019, some scholars proposed a structure based on convolutional neural network (CNN) [19] , it takes the input and output as data sequences, and takes into account the previous state of the equipment to better estimate its current state. Kelly et al. [16] used the deep neural network method to train a model for each electrical appliance separately, and the exploratory attempt in NILM verified the superiority of deep learning. Literature [18] Through the combination coding of the running state of electrical appliances, the mapping training is carried out on the gated cycle unit, and the time information is mined by it, so as to realize the decomposition of load energy.

Although the research system of non-invasive load has been relatively perfect, there are still many problems to be further studied. Although the methods based on deep learning (such as CNN, RNN) have made great progress compared with the traditional methods, the performance of the network model is still unsatisfactory because the current network model has insufficient learning ability for long sequences, that is, it is easy to forget the memory of the past time. Combined with self-attention, this paper constructs a method based on deep learning framework, and uses the powerful global learning ability of self-attention to further improve the performance of NILM. In conclusion, the contributions of this paper mainly include the following three points:

1) As far as our best efforts are concerned, this paper is the first network model to introduce self-attention into NILM task.

2) According to the characteristics of small inter class spacing and large intra class spacing in NILM tasks, combined with the concept of metric learning, a metric based loss function is proposed.

3) Experiments on public data sets show that the proposed method has better performance than the existing methods.

2. Related works

2.1. Seq2seq based on CNN and LSTM

A non-invasive load decomposition algorithm (seq2seq based on CNN and LSTM, seq2seq BCL) based on convolutional neural network (CNN) and short-term memory network (LSTM) is proposed. The deep learning model takes the power time series as the input of the network and extracts the features through CNN. The model is shown in Figure 2 [12]. Firstly, the total power of household electricity is input to one-dimensional convolutional neural network (conv1d) for feature self extraction. The extracted distributed power features are stored in a fixed length full connection layer (dense). The features integrated into the sample space are output to the next layer through the activation function to enhance its nonlinear expression ability. The lower layer uses LSTM to identify the load of electrical appliances, which can mine the value information in the time relationship hidden in the power data.

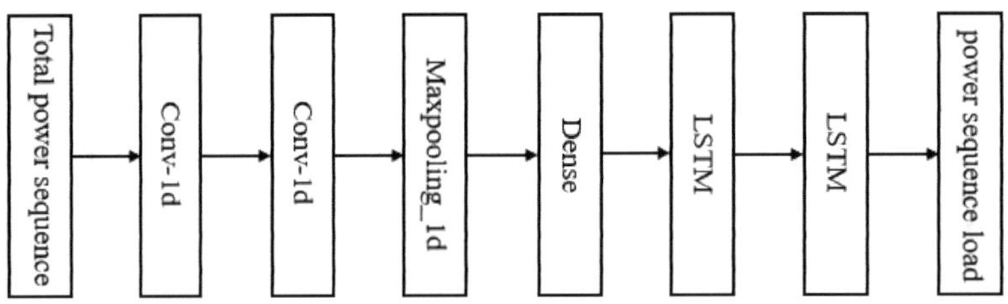

Fig.2. seq2seqBCL non-intrusive load decomposition model

The input vector is obtained by preprocessing the household power data, which is convoluted by convolution check, and then the distributed characteristics of power data are obtained by activation function. The convolution operation is shown in Formula 1.

$$x_i = f(x_{i-1} \otimes W_i + b_i) \quad \mathbf{(1)}$$

where x_i represents the input vector of layer i. f stands for activation function. The introduction of activation function can make the model have nonlinear processing and enhance the expression ability of the model. $\otimes$ represents convolution operation. W_i represents the weight matrix of the convolution kernel of layer i. b_i represents the offset value of the weight matrix in the convolution kernel of layer i.

The distributed features are further pooled and mapped to the full connection layer to obtain the final feature vector. The mathematical formula of pooling operation is as follows:

$$x_i = f(x_{i-1} \otimes W_i + b_i) \quad \mathbf{(2)}$$

where x_i represents the pooled vector. x_{i-1} represents the vector before pooling. Maxpooling represents the maximum pooling operation.

The non-invasive load decomposition algorithm of seq2seq based on the combination of CNN and LSTM has a simple network structure and takes into account the data feature extraction and the data characteristics based on time series. However, the algorithm performance needs to be further improved due to the forgetting of long series input.

2.2. The method based on colour encoding and deep Learning

Non intrusive load identification is a method that does not rely on users' internal devices and can realize the automatic perception of users' power consumption behavior only by relying on external analysis tools and means. Firstly, based on the traditional voltage current (V-I) grey trajectory method, bilinear interpolation technology is used to effectively solve the problem of pixel discontinuity [15]. Then considering the complementarity between features, the numerical features are embedded into the gray V-I trajectory by constructing three channels of current, voltage and phase, so as to obtain the color V-I image containing rich electrical features. Finally, Alex net deep learning algorithm is used to supervise the learning of color V-I images and corresponding equipment labels, so as to realize the effective identification of different types of electrical equipment.

The network model of this method takes the pre-training of AlexNet [17] on ImageNet as the backbone, modifies the last layer of AlexNet to the category description required by the task, and fine tune it by using the PLAID data set, so as to achieve better results. The test results show that, compared with the traditional load recognition methods, the load recognition based on image coding and deep learning can deeply integrate the shape features and numerical features in high-dimensional space with different channels, so as to more comprehensively and stereoscopically reveal the essential attributes of various electrical equipment, and finally achieve a significant improvement in the accuracy of load recognition.

However, in the process of converting power sequence into image, it is bound to destroy the original signal and lose some information, which may lead to the lack of model performance.

3. Methods and Materials

3.1. Self-attention

Data will be generated in people's daily life. With the increase of the amount of data, it is particularly important to clean, analyze and model these data. In the process of modeling, accelerating model training can save a lot of time and cost [7]. Therefore, some scholars put forward the self-attention mechanism according to the attention mechanism of human brain, which has been successfully applied to the field of natural language processing. The idea of the model comes from the attention mechanism, and the self-attention mechanism can realize parallel computing more easily than the attention mechanism. Its basic structure is shown in Figure 3.

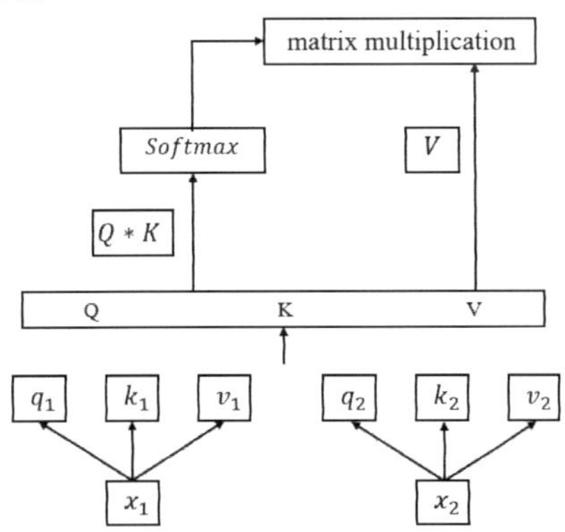

Fig.3. The basic structure of self-attention

Firstly, the input information is multiplied by the corresponding weights to obtain q_1, k_1 and v_1. The calculation process is as follows:

$$q_i = W^q x(i) \quad (3)$$
$$k_i = W^k x(i)$$
$$v_i = W^v x(i)$$

Where W^q, W^k and W^v correspond to the weight matrices of q, k and v respectively. $i \in [0, N]$, N is the size of the thesaurus. Carry out dot product operation on the obtained q_1 and k_1, then normalize the results, and finally multiply the corresponding weight $v(i)$ to obtain the output content, that is:

$$\alpha(1,i) = \frac{q(1)\cdot k(i)}{\sqrt{d}} \quad (4)$$
$$\tilde{\alpha}(1,i) = \frac{\exp(\alpha(1,i))}{\sum_j \exp(\alpha(1,j))}$$
$$b(1) = \sum_i \tilde{\alpha}(1,i)v(i)$$

From the calculation results of $b(i)$, it can be seen that the result value of each $b(i)$ is related to the whole input sequence, which is also a reason why the self attention mechanism can accelerate the calculation in parallel. The above calculation process is represented by matrix as follows:

$$self\_attention(Q,K,V) = \mathrm{softmax}(\frac{QK^T}{\sqrt{d_k}})V \quad (5)$$

where (Q,K,V) are matrices formed by splicing the above q_i, k_i and v_i respectively. Therefore, another reason for the acceleration of computing speed is that the essence of self attention mechanism is matrix computing.

3.2. *The method based on self-attention*

At present, the performance of non-invasive load detection based on deep learning is much better than that of traditional methods. This is because the traditional methods rely on the manual selection of features, which is more subjective. Based on the deep learning method, the network can automatically learn the distinguishing features, so as to achieve better results.

Although the method based on deep learning can achieve good results, it is better than CNN network and pays more attention to local areas, such as convolution operation with size of 3 * 3, and the receptive field is very small. Although the purpose of increasing the receptive field can be achieved by increasing the size of the convolution kernel or deepening the number of layers of the network, it is easy to bring a huge amount of calculation, and these methods are not desirable. Although the previous RNN or LSTM can retain memory, the ability of these models is insufficient for long sequences.

The recently proposed self attention not only focuses on local features, but also takes into account global features, so it has made remarkable achievements in the field of natural language processing. Subsequently, self attention was applied to various fields and achieved good performance. Therefore, this paper introduces self attention into nilm task to solve the problem of memory forgetting.

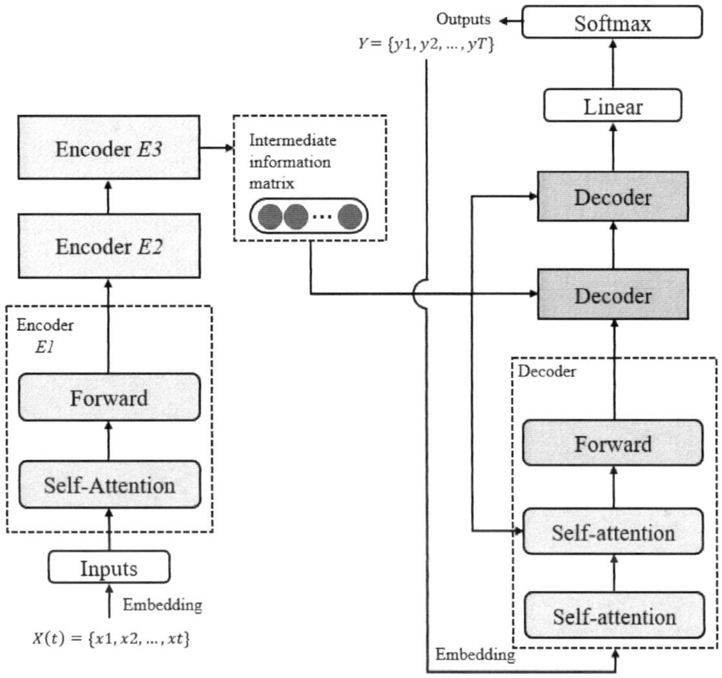

Fig.4. The network based on self-attention

The network model proposed in this paper is shown in Figure 4, which is mainly composed of encoder and decoder. In the decoder and encoder, the most important part is the self-attention module. We use three decoders and encoders, and of course we can use more. The internal structure of each encoder is exactly the same, including one self-attention calculation unit and one feedforward unit. The power sample sequence $X(t)$ of the input appliance is obtained through the feature embedding action, and then the intermediate information matrix is formed after the action of three encoders. It is a matrix composed of the feature sequence obtained after "coding" the input. The decoder is responsible for predicting the output of the next step according to the feature sequence encoded by the sequence and the model output of the previous step. Each decoder contains two self-attention calculation units and one feedforward unit. *H* (*intermediate information matrix*) will participate in the calculation of the second layer attention calculation unit in each decoder. We use residual connection in both feedforward unit and self-attention, as shown in Figure 5.

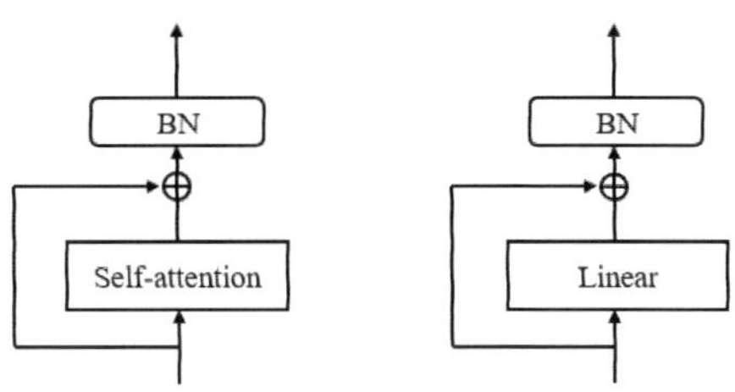

Fig.5. Details in our network

3.3. *Metric loss function*

In the NILM task, although the power of different devices is different, the power range is large, which may lead to different devices having the same power, such as microwave oven and air conditioner, there is likely to be the possibility of overlapping power ranges. Therefore, there is a phenomenon of small inter class spacing and large intra class spacing between samples, resulting in the use of the same cross entropy loss function, which can not effectively restrict the network parameters. In view of this phenomenon, this paper proposes a loss function based on measurement, which makes the samples of the same category more clustered, and the samples of different categories can have greater distance between classes.

By mapping the Euclidean distance to the exponential kernel, the triplet focal loss will automatically increase the weight of the hard example, which is much more penalized than the simple example. We add a regular term to the original focal loss to reduce the over fitting of the network, so as to make the network more robust.

$$L_{tf}(\theta) = \sum_{i=1}^{b} \max\left(0, e^{\frac{D_{a_i,p_i}}{\sigma}} - e^{\frac{D_{a_i,n_i}}{\sigma}} + m\right) \tag{6}$$

$$L_{ms} = -\sum_{i=1}^{b} log \frac{D_{a_i,p_i}}{D_{a_i,p_i} + D_{a_i,n_i}}$$

$$L = L_{tf}(\theta) + \lambda \, L_{ms}$$

Among them λ It is a super parameter that controls the trade-off between triple focus loss and SoftMax loss.

4. Experiment

4.1. Experiment details and dataset

In this paper, the Redd data set is used for the experiment. In order to ensure the reliability and authenticity of the experimental data, first find out the 7 kinds of electrical appliances with the most power consumption in the 6 rooms, and then select the electrical appliances used in the experiment according to the number of distributed rooms. The data of the selected electrical appliances are preprocessed, and the training set and test set are found for experiments. The experiment is divided into two parts. One part of the training set and the test set belong to the same room, and the other part of the training set and the test set belong to different rooms.

The data set used in this experiment is REDD data set, which is the first publicly available data set specially collected to help NILM research. It contains several weeks of power supply data of six different houses and high-frequency current / voltage data of main power supply of two houses. High frequency data is often used for traditional transient feature extraction, so the low-frequency data is used in this experiment. The sampling frequency of main power data is 1s / time, and the sampling frequency of sub table is 3S or 4S / time. The load activation is obtained through change point identification, the training data is expanded through data enhancement, and then the load model is trained. Finally, the process of load identification.

Table 1. Electrical parameters

Electric	furnace	Refrigerator	Washing machine	dishwasher	Microwave Oven
Time/min	50	20	60	60	5
max Power/W	1400	200	2800	1200	1800

After the extracted load is activated, the total power load of the corresponding time stamp is taken as its input data x, so as to form the corresponding input-output training set. Then, the remaining time interval in which the load does not run is counted as the training data opposite to the load activation, which is also uniformly expanded into the window length.

4.2. Results

After training the network, apply the test data set to the load model. Table 2 is the obtained results. Experiments show that the network structure proposed in this paper can give full play to the training speed of CNN network and process timing data like RNN network, so that the network proposed in this paper has excellent recognition performance index scores for each electrical appliance in NILM application, and has good robustness. Moreover, the algorithm proposed in this paper can be applied not only to Redd data, but also to the actual data recognition in the laboratory.

Table 2. Experimental load identification results

Equipment	PTE[20]	PTE+[20]	Seq2seq[12]	Ours
Microwave oven	0.981	0.990	0.924	0.992
Hot water kettle	0.985	0.991	0.918	0.994
Electric lamp	0.947	0.974	0.884	0.981
Dehumidifier	0.920	0.921	0.841	0.934
Air conditioner	0.871	0.905	0.826	0.911
Fan	0.861	0.913	0.834	0.905
Exhaust fan	0.882	0.926	0.819	0.924
Notebook	0.734	0.791	0.741	0.833
Desktop	0.773	0.787	0.792	0.862

In the experimental part of this paper, the performance of many different methods, such as PTE, PTE + and seq2seq. Through data analysis, it can be seen that the performance of the method proposed in this paper is greatly improved compared with other methods.

5. Conclusion

Based on the results and discussions presented above, the conclusions are obtained as below:

(1) It is shown that compared with traditional methods such as combinatorial optimization, the performance of NILM has been greatly improved after the introduction of deep neural network.

(2) Only low-frequency steady-state active power sampling data is required.

(3) Experiments show that the introduction of self-attention can better preserve the global information, so as to further improve the performance.

Acknowledgments

This work was financially supported by the Natural Science Foundation Project of research on multi-source perceptual information fusion method and application based on D-S evidence theory of Guangdong Province (2021A1515011496) and Power grid science and technology project of Jiangmen Power Supply Bureau of Guangdong Power Grid Co of Guangdong Province (Fund Project No: GDKJXM20198068(030700KK52190190)).

References

[1] Zoha A, Gluhak A, Imran M A, et al. Non-intrusive load monitoring approaches for disaggregated energy sensing: A survey[J]. Sensors, 2012, 12(12): 16838-16866.

[2] Batra N, Kelly J, Parson O, et al. NILMTK: An open source toolkit for non-intrusive load monitoring[C]//Proceedings of the 5th international conference on Future energy systems. 2014: 265-276.

[3] Beckel C, Kleiminger W, Cicchetti R, et al. The ECO data set and the performance of non-intrusive load monitoring algorithms[C]//Proceedings of the 1st ACM conference on embedded systems for energy-efficient buildings. 2014: 80-89.

[4] Tabatabaei S M, Dick S, Xu W. Toward non-intrusive load monitoring via multi-label classification[J]. IEEE Transactions on Smart Grid, 2016, 8(1): 26-40.

[5] Figueiredo M B, De Almeida A, Ribeiro B. An experimental study on electrical signature identification of non-intrusive load monitoring (nilm) systems[C]//International Conference on Adaptive and Natural Computing Algorithms. Springer, Berlin, Heidelberg, 2011: 31-40.

[6] Egarter D, Sobe A, Elmenreich W. Evolving non-intrusive load monitoring[C]//European Conference on the Applications of Evolutionary Computation. Springer, Berlin, Heidelberg, 2013: 182-191.

[7] Wu Q, Wang F. Concatenate convolutional neural networks for non-intrusive load monitoring across complex background[J]. Energies, 2019, 12(8): 1572.
[8] Klemenjak C, Makonin S, Elmenreich W. Towards comparability in non-intrusive load monitoring: On data and performance evaluation[C]//2020 IEEE power & energy society innovative smart grid technologies conference (ISGT). IEEE, 2020: 1-5.
[9] Kaselimi M, Doulamis N, Doulamis A, et al. Bayesian-optimized bidirectional LSTM regression model for non-intrusive load monitoring[C]//ICASSP 2019-2019 IEEE International Conference on Acoustics, Speech and Signal Processing (ICASSP). IEEE, 2019: 2747-2751.
[10] Liu Q, Kamoto K M, Liu X, et al. Low-complexity non-intrusive load monitoring using unsupervised learning and generalized appliance models[J]. IEEE Transactions on Consumer Electronics, 2019, 65(1): 28-37.
[11] Himeur Y, Alsalemi A, Bensaali F, et al. Effective non-intrusive load monitoring of buildings based on a novel multi-descriptor fusion with dimensionality reduction[J]. Applied Energy, 2020, 279: 115872.
[12] Wang A L, Chen B X, Wang C G, et al. Non-intrusive load monitoring algorithm based on features of V–I trajectory[J]. Electric Power Systems Research, 2018, 157: 134-144.
[13] Lu M, Li Z. A hybrid event detection approach for non-intrusive load monitoring[J]. IEEE Transactions on Smart Grid, 2019, 11(1): 528-540.
[14] Pereira L, Nunes N. An empirical exploration of performance metrics for event detection algorithms in Non-Intrusive Load Monitoring[J]. Sustainable Cities and Society, 2020, 62: 102399.
[15] Pereira L, Nunes N. An empirical exploration of performance metrics for event detection algorithms in Non-Intrusive Load Monitoring[J]. Sustainable Cities and Society, 2020, 62: 102399.
[16] Renaux D P B, Pottker F, Ancelmo H C, et al. A dataset for non-intrusive load monitoring: Design and implementation[J]. Energies, 2020, 13(20): 5371.
[17] Wu X, Gao Y, Jiao D. Multi-label classification based on random forest algorithm for non-intrusive load monitoring system[J]. Processes, 2019, 7(6): 337.
[18] Çimen H, Çetinkaya N, Vasquez J C, et al. A microgrid energy management system based on non-intrusive load monitoring via multitask learning[J]. IEEE Transactions on Smart Grid, 2020, 12(2): 977-987.
[19] Houidi S, Fourer D, Auger F. On the use of concentrated time–frequency representations as input to a deep convolutional neural network: Application to non intrusive load monitoring[J]. Entropy, 2020, 22(9): 911.
[20] YiLingZheng, Yongqiang Liu, et al.Application of Improved Least Squares Changepoint Recognition Method in Load Disaggregation [J]. Computer measurement and control. 2019, 27(6): 226-230.

Experimental Study on the Efficiency of RF Energy Transfer System

Jing Guo [1,a*], Dongkun Lu [2,b], Weige Zheng[3,c]

[1]Department of Automation Foshan University, Foshan, China

[2]Department of Automation Foshan University, Foshan, China

[3]Department of Automation Foshan University, Foshan, China

[a]e-mail:guojing_cc@163.com,[b]e-mail:wh.fish@qq.com, [c]e-mail:1019950988@qq.com

*the corresponding author: [a]e-mail: guojing_cc@163.com

Abstract: This paper presents a 915 MHz radio frequency (RF) wireless energy transfer system which contains RF energy transmitters and RF energy harvesting nodes. Firstly, Advanced Design System (ADS) is used to design and optimize the monopole voltage doubler rectifier circuit. Secondly, an energy harvesting node is designed by a commercial RF/DC rectifier and a 915 MHz antenna. Finally, the RF energy transfer experiment between RF energy transmitter and RF energy harvesting node is demonstrated. Experimental data fits well with theoretical analysis and the harvested energy show a non-linear relationship with the distance and angle between the receiving node antenna and the transmitter antenna. The experimental data is valuable for designing RF-powered nodes and optimizing the movement trajectory of the mobile RF energy transmitter, such as an unmanned aerial vehicle (UAV).

1. Introduction

Energy harvesting technology is an opportune solution for wireless sensor network energy supply [1]. In the circumambient environment, energy sources that can be used primarily contain kinetic energy [2,3], radio frequency (RF) energy [4,5], differential pressure [6,7] and solar energy [8,9]. Electromagnetic radio waves of different frequency bands are omnipresent, which can be recovered infinitely. These are available even in inaccessible places. Available environmental RF sources or RF sources dedicated to specific harvesting purposes can be used as RF energy sources for energy harvesting devices. In the light of application scenarios and limiting factors, the specific or the environmental RF sources may be used for harvesting, which is expectable and controllable [10,11].

Traditional wireless energy transmission systems that use fixed-location RF sources face many challenges. Owing to the high long-distance propagation loss of RF signals, the end-to-end wireless energy transmission efficiency is universally low while the distance from the RF source to the collector increases [12]. Through distributed or mobile RF energy sources, the conversion efficiency of the receiving end can be effectively improved. A lot of current work assumes that the conversion efficiency of the receiving end has a linear relationship with the distance, and then the UAV path planning [13,14]. In actual situations, the relationship between RF-DC conversion efficiency and received signal strength is non-linear. Therefore, this paper conducts research on RF energy transmission experiments at different distances and angles between the RF source and the harvester, which is valuable for path optimization of mobile RF source.

Content from this work may be used under the terms of the Creative Commons Attribution 3.0 licence. Any further distribution of this work must maintain attribution to the author(s) and the title of the work, journal citation and DOI.
Published under licence by IOP Publishing Ltd

The rest of this paper is organized as follows: Section 2 presents the RF energy harvesting system for this study. Section 3 presents the simulation and analysis of the rectifier circuit. Section 4 presents the RF energy transmission experiment in detail and experimental results. Section 5 draws conclusions about the RF wireless energy transfer system.

2. RF wireless energy transfer system

Figure 1 shows the block diagram of the RF energy transfer system. The RF energy transfer system is mainly composed of RF source, receiving antenna, matching network, rectifier and boost converter. The receiving antenna is used for receiving RF energy in the surrounding environment. The rectifier is serve as a converter which transforms RF energy into a DC voltage output. The function of the matching network matches the input impedance of the rectifier circuit to the output impedance of the receiving antenna, improving energy transmission efficiency. The boost converter can increase the output voltage of the rectifier to provide a suitable voltage for the load.

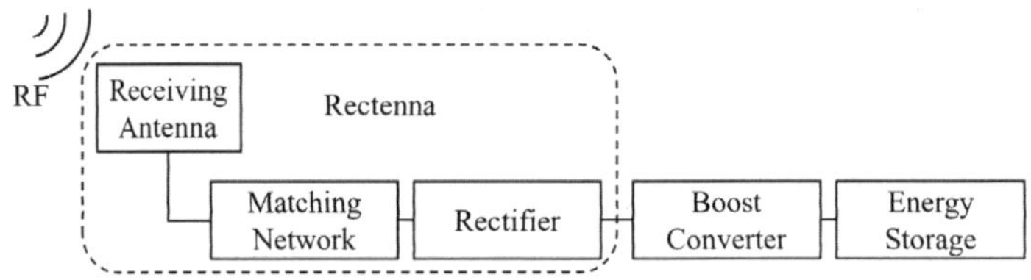

Figure 1 Block diagram of the RF energy transfer system

3. Simulation results and analysis of rectifier circuit

Figure 2 shows a simulation schematic diagram of monopolar voltage doubler rectifier circuit. The rectifier is connected to the source by an inductor L1. The rectifier consists of two diodes HSMS-2860, two capacitors C and load impedance $\boldsymbol{Z_{Load}}$. The internal resistance of the power source is 50**Ω**. ADS is used to design and optimize the rectifier circuit through S-parameter simulation and harmonic balance simulation.

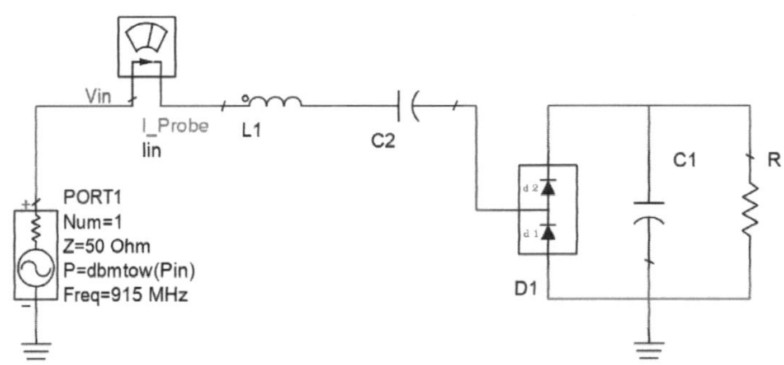

Figure 2 Schematic diagram of simulation of monopolar voltage doubler rectifier circuit

The initial value of the load impedance is set to 1000Ω. Six groups of capacitors (1pF,10pF, 50pF, 100pF, 200pF and 300pF) were selected for the values of the two capacitors. Figure 3 shows that when the value of the capacitor increases from 1pF to 10pF, the rectification efficiency of the circuit increases significantly, while when the value of the capacitor is greater than 10pF, the corresponding rectification efficiency curve remains stable. Therefore, in order to obtain higher rectification efficiency, a capacitor with a larger capacitance value should be selected. Considering comprehensively, the simulation chooses a capacitance of 100pF.

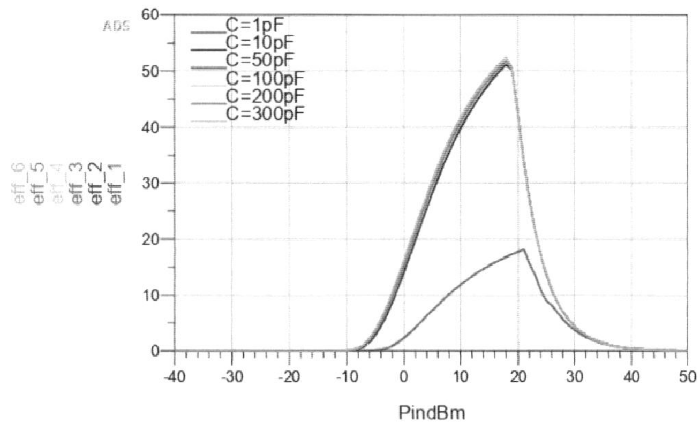

Figure 3 The curve of diode rectification efficiency with input power

The values of the two capacitors are both set to 100pF, and the load impedance is selected from six groups of resistances of 100Ω, 200Ω 300Ω, 500Ω, 700Ω and 1000Ω. Figure 4 shows that the efficiency of the rectifier circuit does not always increase with the increase of load resistance. The relationship between diode rectification efficiency and input power is non-linear. Considering comprehensively, the simulated load impedance is 800Ω.

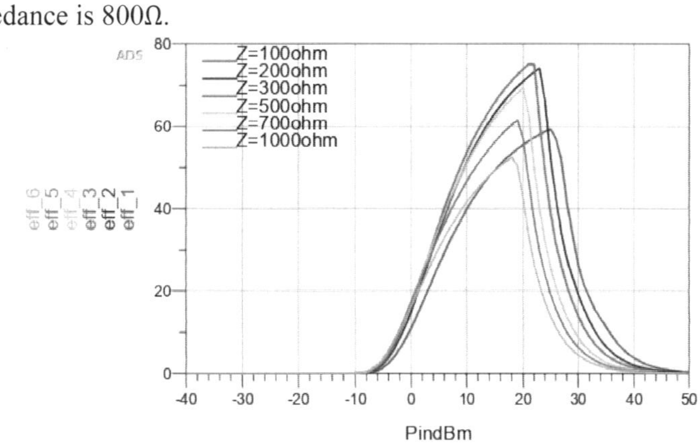

Figure 4 The curve of diode rectification efficiency with input power

4. Experimental results and analysis

Figure 5 shows the block diagram of the 915 MHz RF wireless energy transfer system which contains RF energy transmitter and RF energy harvester. The RF energy harvester is composed of RF receiving node antenna, RF-DC rectifier and power management module. The RF receiving node antenna adopts a circularly polarized patch antenna. The RF-DC rectifier is connected to the SMA connector through a 50**Ω** impedance line. The DC input end of the BQ25570 [15] power management module is connected to the DC output end of the rectifier, and the energy storage device used is a super capacitor.

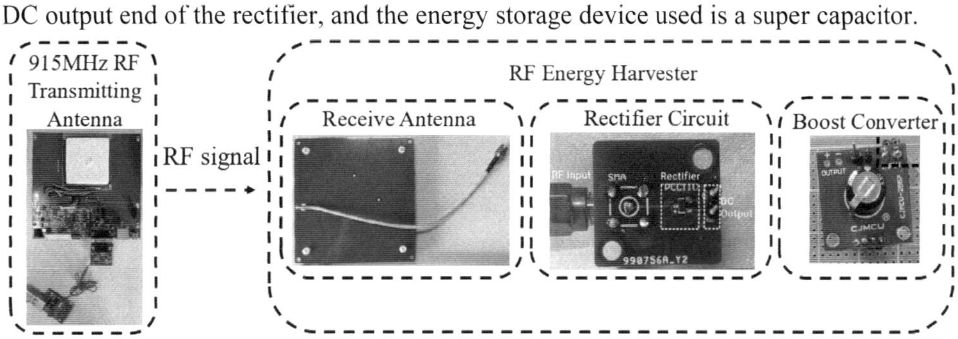

Figure 5 Block diagram of the 915 MHz RF wireless energy transfer system

Figure 6 is the plan and elevation view of experimental set-up, using a RF energy transmission and a RF energy harvester. d is the distance between the receiving antenna and the transmitting antenna. θ is the relative angle between the normal of the transmitting antenna and the receiving antenna.

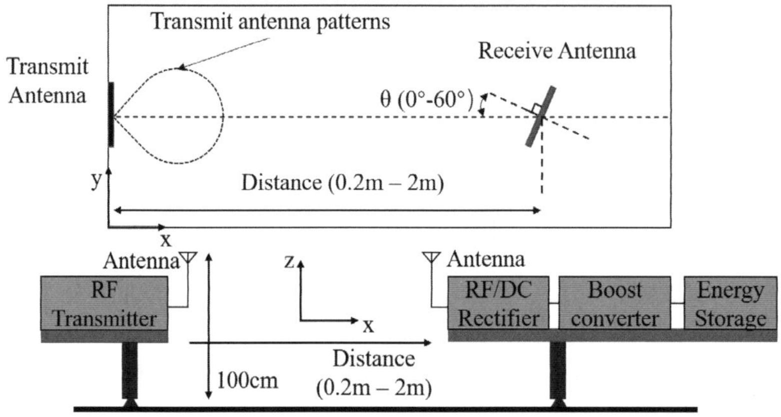

Figure 6 Plan and elevation view of experimental set-up, using a RF energy transmission and a RF energy harvester

Figure 7 shows the curves of the harvesting power of the harvester with the distance d and the angle θ. When the distance from the transmitter antenna to the receiving node antenna is equal to d = 0.2m, and θ changes from 0° to 60°, the harvesting power gradually decreases from 9.65mW to half of it. When the distance d is less than 1m, the harvesting power varies greatly with distance and angle. When the distance d is greater than 1m, the harvesting power varies little with distance and angle. Experimental data fits well with theoretical analysis and the harvested energy show a non-linear relationship with the distance and angle between the receiving node antenna and the transmitter antenna.

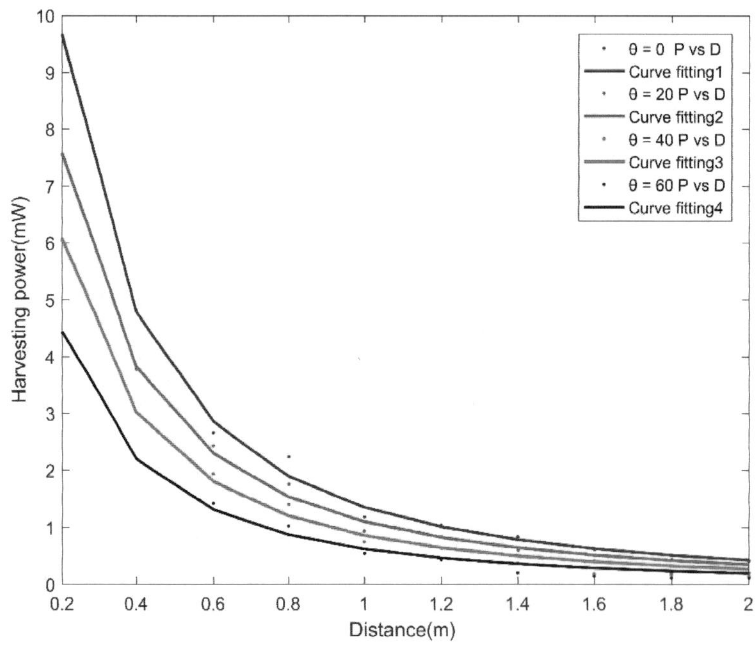

Figure 7 The harvesting power of the harvester varies with d and θ.

5. Conclusions

A 915 MHz RF wireless energy transfer system which contains RF energy transmitters and RF energy harvesting nodes. The RF energy transfer experiment between RF energy transmitter and RF energy harvesting node is demonstrated. Experimental data fits well with theoretical analysis and the harvested energy show a non-linear relationship with the distance and angle between the receiving node antenna and the transmitter antenna. The experimental data is valuable for designing RF-powered nodes and optimizing the movement trajectory of the mobile RF energy transmitter, such as an UAV.

Acknowledgement: This work was supported by National Natural Science Foundation of China Grant number 61803087.

References

[1] Singh J., Kaur R., Singh D. (2021) Energy harvesting in wireless sensor networks: A taxonomic survey. Int J Energy Res , 45:118-140.
[2] P. D. Mitcheson, E. M. Yeatman, G. K. Rao, A. S. Holmes, T. C. Green. (2008) Energy Harvesting From Human and Machine Motion for Wireless Electronic Devices. In Proceedings of the IEEE ,96:1457-1486.
[3] P. K. Illenberger, S. Rosset, U. K. Madawala, I. A. Anderson. (2021) The Integrated Self Priming Circuit: An Autonomous Electrostatic Energy Harvester With Voltage Boosting. IEEE Transactions on Industrial Electronics, 68:6982-6991.
[4] P. Li Z. Long, Z. Yang. (2021) RF Energy Harvesting for Batteryless and Maintenance-Free Condition Monitoring of Railway Tracks. IEEE Internet of Things Journal, 8:3512-3523.
[5] G. Verma; V. Sharma. (2021) A Novel RF Energy Harvester for Event-Based Environmental Monitoring in Wireless Sensor Networks. IEEE Internet of Things Journal.
[6] J. H. Kim, J. Y. Cho, J. P. Jhun, G. J. Song, J. H. Eom, S. Jeong, W. Hwang, M. S. Woo, T. H. Sung. (2021) Development of a hybrid type smart pen piezoelectric energy harvester for an IoT platform. Energy, 222,119845.
[7] G. Shi et al. (2021) A Sensorless Self-Tuning Resonance System for Piezoelectric Broadband Vibration Energy Harvesting. IEEE Transactions on Industrial Electronics, 68:2225-2235.
[8] Rokonuzzaman, M., Mishu, M.K., Amin, N., Nadarajah, M., Roy, R.B., Rahman, K.S., Buhari, A.M., Binzaid S., Shakeri M., Pasupuleti J. (2021) Self-Sustained Autonomous Wireless Sensor Network with Integrated Solar Photovoltaic System for Internet of Smart Home-Building (IoSHB) Applications. Micromachines, 12:653.
[9] Saji M. Antony,S. Indu,Rajeshwari Pandey.(2020)An efficient solar energy harvesting system for wireless sensor network nodes.Journal of Information and Optimization Sciences,41:39-50.
[10] S. Shieh, M. Kamarei. (2018) Transient Input Impedance Modeling of Rectifiers for RF Energy Harvesting Applications. IEEE Transactions on Circuits and Systems II: Express Briefs, 65: 311-315.
[11] H. Zhang, N. Lu, R. Song, J. Li. (2020) Stochastic Analysis on Ambient RF Energy Harvesting. In: 2020 IEEE 6th International Conference on Computer and Communications (ICCC). Chengdu, China. pp. 771-775.
[12] L. Xie, J. Xu, R. (2018) Zhang. Throughput Maximization for UAV-Enabled Wireless Powered Communication Networks. IEEE Internet of Things Journal, 6: 1690-1703.
[13] P. Wu, F. Xiao, H. Huang, R. Wang. (2020) Load Balance and Trajectory Design in Multi-UAV Aided Large-Scale Wireless Rechargeable Networks. IEEE Transactions on Vehicular Technology, 69: 13756-13767.
[14] W. Feng et al. (2020) Joint 3D Trajectory Design and Time Allocation for UAV-Enabled Wireless Power Transfer Networks. IEEE Transactions on Vehicular Technology, 69: 9265-9278.
[15] Texas Instruments. (2019) bq25570 nano power boost charger and buck converter. https://www.ti.com/lit/ds/symlink/bq25570.pdf?ts=1634988791716&ref.

Methods for increasing the conversion efficiency of solar cells

Yuxuan Wang[1,*]

[1]School of Electronic and Information Engineering, Soochow University, Suzhou,Jiangsu Province 215000, China

[*]Corresponding author e-mail: yxwang1129@stu.suda.edu.cn

Abstract. Solar energy is the most promising new energy. It is inexhaustible, and the use of solar energy will not damage the earth's ecology and environment; Safe, no pollution, low utilization cost, and not restricted by geographical conditions. The solar cell converts light energy into electrical energy and transmits it to the load. When the non-uniform semiconductor (PN junction) is irradiated with light of appropriate wavelength, the electromotive force is generated inside the semiconductor due to the action of the internal field. Solar cell is a semiconductor photovoltaic element that converts light energy into electrical energy, using the photovoltaic effect of semiconductor. The main focus of solar cell design is to increase their conversion efficiency. This paper focuses on methods for increasing the conversion efficiency of solar cells, which can lead to more efficient energy for space technology. The conversion efficiency of a solar cell is defined as the ratio of the output electrical energy to the incident light energy. This paper focuses on the following methods to increase the conversion efficiency: enabling the solar cells to correspond to a wider spectrum and gradient doping. We have found that some of these methods can be applied to different solar cell materials. They are also quite realizable and can be put into practical production.

1. Introduction

The main focus of the design of solar cells dedicated to various applications is to increase their conversion efficiency. A number of relevant studies have already been conducted in this field. In Ref. [1], authors showed that the materials for each layer of a multifunctional layer p-n junction cell should be selected so that their respective different forbidden bandwidths match the different frequency bands in the solar spectrum. It tried to make the cell respond to a wider spectrum and thus improve the conversion efficiency. Improving conversion efficiency through the application of quantum wells is also being investigated, where the depth of the quantum well affects the solar cell's absorption of different wavelengths of sunlight. In Ref. [2], the authors found that a less common conversion efficiency enhancement technique is band gap grading by applying material composition grading within a specific layer of the device. More recently, it was mentioned in Ref. [3] that optical lenses with larger dimensions could be used to concentrate sunlight onto solar cells, increasing the light intensity to improve conversion efficiency.

Although a number of methods to increase the conversion efficiency of solar cells have been identified in past research, there are still some methods that are worth discussing. In fact, there are many factors that affect the conversion efficiency of solar cells. Based on the relevant studies, we have researched the effects of multifunctional materials p-n junction cells and gradient doping on increasing the conversion efficiency of solar cells. We found that multifunctional layer p-n junction solar cells can respond to different spectrum, which means that multifunctional layer p-n junction

Content from this work may be used under the terms of the Creative Commons Attribution 3.0 licence. Any further distribution of this work must maintain attribution to the author(s) and the title of the work, journal citation and DOI.
Published under licence by IOP Publishing Ltd

solar cells have a higher conversion efficiency; the increasing gradient doping of the donor impurity and the acceptor impurity in the n region and p region respectively with increasing distance from the depletion layer is beneficial to increase the conversion efficiency.

2. Mechanism

For the maximum power output of solar cell, we have the following equation

$$\eta = \frac{P_m}{P_{in}} \times 100\% = \frac{I_m V_m}{P_{in}} \times 100\% \qquad (1)$$

The maximum possible current in a solar cell is I_{oc}, the maximum possible voltage is V_{oc}, and the duty cycle of a solar cell is defined as the ratio $\frac{I_m V_m}{I_{oc} V_{oc}}$. The duty cycle, which is a measure of the power that can be achieved by a solar cell, usually ranges from 0.7 to 0.8. The ratio of the product of the current and voltage of a solar cell at maximum output power to the product of the short-circuit current and open-circuit voltage is called the fill factor (FF) [4]. It is an important parameter for evaluating the output characteristics of solar cells. The larger the value of the fill factor, the more the output characteristics of the solar cell tend to be rectangular, and the higher the photoelectric conversion efficiency of the solar cell. Figure 1 shows the maximum power rectangle of the current-voltage characteristic curve of a solar cell.

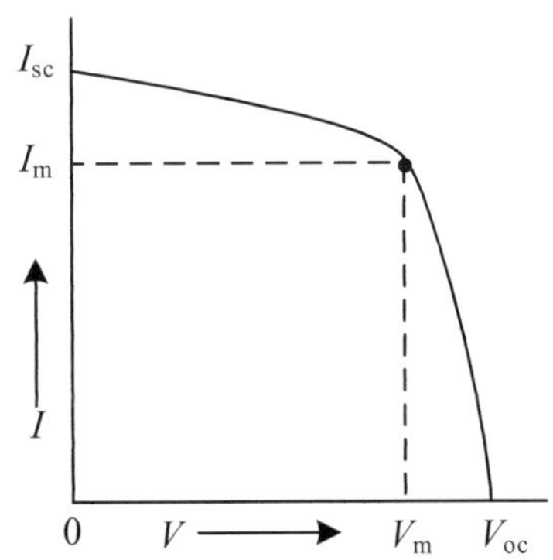

Figure 1. Maximum power rectangle of the current-voltage characteristic curve of a solar cell

2.1 Because the output power of a solar cell is $V_{oc}I_{sc}$, to increase the conversion efficiency, it is necessary to increase the three physical quantities of open circuit voltage V_{oc}, closed circuit current I_{sc}, and fill factor [6]. Increasing the open circuit voltage V_{oc} plays an important role in improving the performance of solar cells. The relationship with V_{oc} and the reverse saturation current I_o is: V_{oc} increases as I_o decreases. As the forbidden bandwidth E_g of the semiconductor increases, I_o decreases rapidly and therefore V increases as E_g increases. I_{sc} decreases because the number of photons with energy greater than E_g absorbed by the semiconductor material decreases as E_g increases. This means that there is an optimum E_g for the highest energy conversion efficiency, and therefore the forbidden bandwidth of solar cells is one of the most important parameters in the study of conversion efficiency. Multifunctional layer p-n junction cells are those in which the different forbidden bandwidths of the layers match different bands in the solar spectrum, thus allowing the solar cell to respond to a wider spectrum and thus increase its conversion efficiency. The same principle is applied to heterojunction solar cells. A heterojunction is a p-n junction formed by two semiconductors with different forbidden bandwidths, and its energy band diagram at thermal equilibrium is shown in Figure 2.

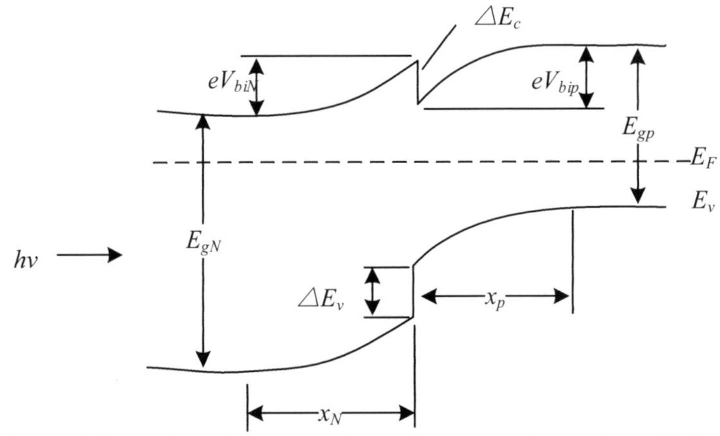

Figure 2. Heterojunction energy band diagram at thermal equilibrium

2.2 In the case of conventional p-n junction solar cells, since the p-n junction is uniformly doped, carrier diffusion is only present in the depletion layer (about 1 millimeter) at the interface between the p and n regions where a small linearity is formed, with zero electric field strength in the rest of the region [7]. When sunlight is incident on the p-n junction, only the photogenerated electron-hole pairs within and near the depletion layer are collected by the electrodes in the presence of an electric field and form a photogenerated current. Due to the absence of electric fields in other regions, the photoelectron-hole pairs cannot be separated efficiently and the compounding rate of photoelectrons and holes is larger, so this part of the photogenerated carriers cannot produce photogenerated currents most efficiently. For this conventional uniformly doped solar cell, we consider gradient doping in the p and n regions, when the p-n junction can form a certain charge distribution under light-free equilibrium conditions due to the diffusion of carriers in the homotypic region, thus generating an electric field. The efficient collection of photon-generated carriers is key to increasing the efficiency of solar cells [8,9]. Therefore we need to rationally design the electric field distribution within the semiconductor so that the electrodes collect as many photogenerated carriers as possible, thereby increasing the closed circuit current I_{sc} and increasing the conversion efficiency of the solar cell.

We consider the effect of gradient doping on conversion efficiency for a single p-n junction, as shown in Figure 3, assuming that the length of the n-region of the p-n junction is L.

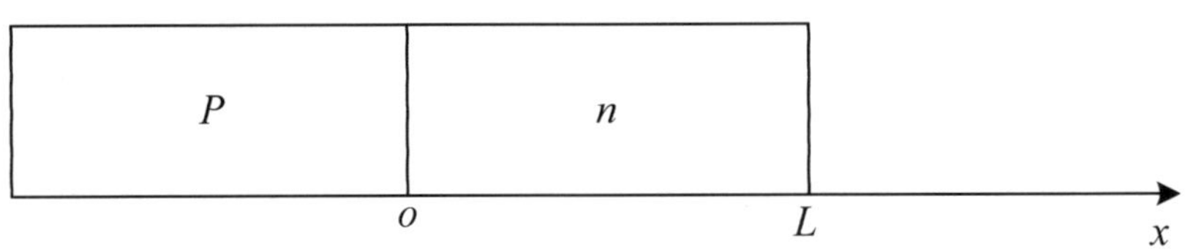

Figure 3. Simplified structure diagram of a p-n junction

We know that the ionization energy of doped impurities is much smaller compared to that of intrinsic semiconductors, and therefore doped impurities have more electrons excited from the bound state to the conduction band compared to intrinsic semiconductors. The theoretical studies show that exponential doping is more efficient than linear doping to obtain high intensity additional electric field [10]. We assume that the doped impurities in a solar cell are doped as an exponential function(As shown in Equation 2, the concentration of donor impurity increases exponentially with x, a and b are constants).

$$N_d(x) = a \cdot \exp(bx) \qquad (2)$$

We only need to discuss the electric field distribution within the p-n junction under gradient doping conditions to obtain a method for increasing the closed-circuit current I_{sc}. Then we can obtain the method for increasing the conversion efficiency of the solar cell.

Because the donor impurity is gradient doping in n region, there is an impurity concentration gradient in the n region, so the carriers will diffuse in the opposite direction along the x-axis, creating a diffusion current in the p-n junction. Since the ionized impurity cannot move, the diffusion movement of carriers will make the carriers gradually and uniformly distributed, which makes the p-n junction semiconductor interior no longer remain electrically neutral everywhere, so when the carriers are uniformly distributed, there must be an electrostatic field inside the semiconductor. This electrostatic field in turn generates carrier drift currents inside the p-n junction. For non-simplex semiconductors, when the internal electric field is balanced, the electron concentration satisfies equation (3), where $n_e(x)$ is the electron concentration and $E_x(x)$ is the component of the electric field strength on the x-axis.

$$\frac{dN_e(x)}{dx} = -n_e(x)\frac{q}{k_B T} \times E_x(x) \qquad (3)$$

The electric field strength also satisfies Poisson's equation (Equation (4)), where ρ is the net charge density and $n_+(x)$ is the concentration of ions at that location on the x-axis after ionization of the impurity.

$$\frac{dE_x(x)}{dx} = \frac{\rho}{\varepsilon} = \frac{q}{\varepsilon}[n_+(x) - n_e(x)] \qquad (4)$$

For general semiconductors, we consider that at room temperature the impurities in the nondegenerate semiconductor will be completely ionized, i.e.

$$n_+(x) = N_d(x) \qquad (5)$$

Substituting equation (3) into equation (4) yields,

$$\frac{1}{n_e(x)}\frac{dn_e(x)}{dx} - \frac{1}{n_e(x)}\frac{d^2 n_e(x)}{dx^2} = \frac{q}{\varepsilon k_B T}[N_d(x) - n_e(x)] \qquad (6)$$

As we discussed earlier, it is only necessary to solve for the electric field distribution $E_x(x)$ inside the p-n junction under gradient doping conditions to obtain a method for improving the conversion efficiency of solar cells. From the previous analysis, we know that the key to solving for an expression for $E_x(x)$ lies in solving for an expression over $n_e(x)$. However, by looking at equation (6), we find it difficult to solve for an analytic expression for $n_e(x)$ directly.Our idea is to simplify the analysis by reducing it to the case of a single-crystal silicon semiconductor, for which the carrier diffusion length is known to be several hundred microns, and the carrier transport is mainly by diffusive motion, while the electric field has little effect on the diffusive motion. Therefore, we approximate the distribution of majority carrier in the n region to be uniformly distributed, which simplifies the analysis of the problem and thus the calculation. For the simplified problem, we have the following analysis.

For a typical solar cell with a large area and a small thickness [11], we consider dividing the entire n region of a p-n junction into an infinite number of uniformly charged thin plates of thickness dx. We consider the electric field generated by each thin plate in the space around it to be a uniform electric field with the magnitude of the electric field strength as shown in equation (7), where here the ε is the dielectric constant of the semiconductor material.

$$dE_x = \frac{\rho(x)}{2\varepsilon}dx \qquad (7)$$

Since the total charge in the depletion and p region is zero, the electric field generated by these charges at any point in the n region has zero strength. Therefore, we consider that the total electric field at x in the n region is only generated by the gradient-doped n region charges. We assume that the width of the n region is L. Since the width of the depletion layer is much smaller than the entire n region, we assume that the homotypic diffusion carriers diffusion carriers in the depletion layer are negligible and the total charge of the n region other than the depletion layer is zero. The number of charges in the range 0 to x in n region must be the same as the number of charges in the range x to L with opposite electrical properties, and the charges in both ranges produce an electric field of equal magnitude and direction at x. As shown in equation (8), where $\overline{n_e}$ is the average concentration of the major carrier.

$$E_x(x) = 2\int_0^x \frac{\rho(x)}{2\varepsilon}dx = \frac{q}{\varepsilon}\left[\int_0^x N_d(x)dx - \int_0^x \overline{n_e}dx\right] \qquad (8)$$

3. Result

3.1 For multifunctional layer p-n junction solar cells, suppose a photon is incident on a p-n junction with a large forbidden bandwidth. Then photons with energy less than E_{gN} will pass through the semiconductor material with the larger forbidden bandwidth, photons with energy greater than E_{gP} will be absorbed by the semiconductor material with the smaller forbidden bandwidth, and photons with energy greater than E_{gN} will be absorbed by the material with the larger forbidden bandwidth (rather than passing through it as photons with energy less than E_{gN} do), with the resulting excess carriers in the diffusion length range are collected. If the value of E_{gN} is sufficiently large, photons with high energy will be absorbed in the spatial electric region by semiconductor materials with a small forbidden bandwidth.

3.2 For the gradient doping method to increase the conversion efficiency of solar cells, from equation (2) we can obtain that

$$\overline{n_e} = \frac{1}{L}\int_0^L a\exp(bx)dx = \frac{a}{bL}[\exp(bL)-1] \qquad (9)$$

From equation (8) and equation (9) we can solve for $E_x(x)$, as shown in equation (10)

$$E_x(x) = \frac{aq}{b\varepsilon}\left\{[\exp(bx)-1] - \frac{x}{L}[\exp(bL)-1]\right\} \qquad (10)$$

From the above analysis, we find that for gradient doped single crystal silicon semiconductors, the impurities are completely ionized at room temperature conditions and in the nondegenerate condition. Due to the diffusive motion of the carriers, the net charge distribution in the n region outside the depletion layer should change from negative to positive with increasing x if we use exponentially increasing doping. From equation (8) and equation (10) we can intuitively see that $E_x(x)$ is constantly non-positive, which means that the direction of the electric field intensity on the x-axis points in the negative direction of the x-axis. This is in the same direction as the electric field intensity within the depletion layer, which facilitates the collection of photoelectrons by the electrode in the n region and the drift of holes to the p region, where they are collected by the electrode in the p region under the acceleration of the gradient-doped p region electric field, thus increasing the closed-circuit current I_{sc} of the p-n junction and improving the photoelectric conversion efficiency of the solar cell. If the n region is not doped in increasing gradients, the electric field is not guaranteed to be the same everywhere, and the electric field pointing in the positive and negative direction of the x-axis is not conducive to the collection of photon-generated carriers by the electrodes on both sides, which will reduce the photoelectric conversion efficiency. Based on the above analysis the gradient doping and electric field distribution in the p region can be calculated and discussed in the same way.

We use AFORS-HET to simulate gradient doping of solar cells and compare it with uniform doping. The main parameters used in the simulations are shown in Table 1 and the photovoltaic performance of uniformly and gradient doped solar cells is shown in Table 2 [12].

Table 1. Main parameters used in the AFORS-HET simulations

Parameters	a-Si:H(n)	a-Si:H(i)	c-Si:H(p)
Layer thickness (nm)	5	5	3×10^5
Dielectric constant	11.9	11.9	11.9
Electron affinity (eV)	3.8	3.8	4.05
Band gap (eV)	1.72	1.72	1.12
Optical band gap (eV)	1.72	1.72	1.12
Effective conduction band density (cm^{-3})	10^{20}	10^{20}	2.8×10^{19}
Effective valence band density (cm^{-3})	10^{20}	10^{20}	1.04×10^{19}
Electron mobility ($cm^2V^{-1}s^{-1}$)	5	5	1040
Hole mobility ($cm^2V^{-1}s^{-1}$)	1	1	412
Doping concentration of acceptors (cm^{-3})	0	0	1.5×10^{16}
Doping concentration of donators (cm^{-3})	$1\times10^{18}\sim9\times10^{18}$	0	0
Thermal velocity of electrons (cm/s)	1×10^7	1×10^7	1×10^7
Thermal velocity of holes (cm/s)	1×10^7	1×10^7	1×10^7
Layer density (g/cm^3)	2.328	2.328	2.328

Table 2. Enhanced photovoltaic performance of solar cells with gradient doping compared to uniform doping

	V_{oc} (mV)	J_{sc} (mA/cm^{-2})	FF (%)	E_{FF} (%)
Uniform doping	568.00	28.60	80.74	13.12
Gradient doping	710.20	41.40	82.57	24.27
Increments	25.0%	44.7%	2.3%	84.9%

4. Conclusion

After analysis, we have shown that especially for the short wavelength sunlight, multifunctional layer p-n junction solar cells can respond to a different spectrum than single material p-n junction solar cells, which means that multifunctional layer p-n junction solar cells have a higher conversion efficiency.

For the gradient doping method to increase the conversion efficiency of solar cells, after analysis we find that only the increasing gradient doping of the donor impurity and the acceptor impurity in the n region and p region respectively with increasing distance from the depletion layer is beneficial to increase the conversion efficiency. This conclusion applies not only to monocrystalline silicon, but also to other solar semiconductor materials.

It is worth noting that the total doping content of solar cells is higher when they are uniformly doped than when they are gradient doped. Therefore, when studying the effect of gradient doping, we should compare two gradient doped and uniformly doped solar cells with the same doping content. In addition, our analysis does not consider the effect of doping content on the conversion efficiency of the solar cell. It has been shown that at moderate doping concentrations ($<10^{19}\ cm^{-3}$), the higher the doping content, the better the performance of the solar cell. An excessively high doping concentration can lead to a reduction of the open-circuit voltage [13]. If the total doping content has an effect on the conversion efficiency of solar cells, the positive effect of gradient doping on the conversion efficiency will be different for solar cells with different total gradient doping content, and this will be the direction we will focus on in our subsequent research.

Reference

[1] Li G , Li G . Theoretical investigation on new materials for raising efficiency of energy conversion in PEC solar cells[J]. Solar Energy Materials and Solar Cells, 1993, 30(1):61-64.
[2] Joanna P , Regina P , Ryszard K , et al. Solar cells conversion efficiency enhancement techniques[J]. Optica Applicata, 2007, 37(1):93-99.
[3] Donald A. Neamen. Semiconductor Physics and Devices:Basic Principles,Fourth Edition[M]. Beijing Publishing House of Electronics Industry, 2018.
[4] Ryu S , Noh J H , Jeon N J , et al. Voltage output of efficient perovskite solar cells with high open-circuit voltage and fill factor[J]. Energy & Environmental Science, 2014, 7(8):2614-2618.
[5] Green M A . Solar cell fill factors: General graph and empirical expressions[J]. Solid-State Electronics, 1981, 24(8):788-789.
[6] He Z , Zhong C , Xun H , et al. Simultaneous enhancement of open-circuit voltage, short-circuit current density, and fill factor in polymer solar cells.[J]. Advanced Materials, 2011, 23(40):4636-4643.
[7] Quarto F D , Mantia F L , Santamaria M . Physicochemical Characterization of Passive Films and Corrosion Layers by Differential Admittance and Photocurrent Spectroscopy[M]. Springer New York, 2009.
[8] Walsh K M . A new approach to the effective use of intrinsic aSi : H in thin film solar cells[J]. Journal of Physics D Applied Physics, 2007, 40(4):1007-1009.
[9] Inthisang S , Krajangsang T , Sichanugrist P , et al. Effect of Hydrogen Dilution on the Metastability of Hydrogenated Amorphous Silicon Oxide Solar Cells[J]. Japanese Journal of Applied Physics, 2011, 50(11Issue1):1401.
[10] Wang A K , Ren Q H , Xue J H , et al. Electric Field Strength Numerical Calculation of the Gradient Doping Semicondouction[J]. Journal of Hebei Normal University(Natural Science Edition), 2009.
[11] Gao Z , Yongdong L I , Sun M . Comparison and Analysis of Some Typical Solar Cell Simulators[J]. Power Supply Technologies and Applications, 2010.
[12] Zhang C L , Du H J , Zhu J Z , et al. Enhanced Photovoltaic Properties of Gradient Doping Solar Cells[J]. Chinese Physics Letters, 2012, 29(12):127305.
[13] Korte L , Schmidt M . Investigation of gap states in phosphorous-doped ultra-thin a-Si:H by near-UV photoelectron spectroscopy[J]. Journal of Non-Crystalline Solids, 2008, 354(19-25):2138-2143.

Resolution technology of lithography machine

Yuanxuan Fang[1, a, †] **, Yunfei He**[2, b, †]

[1]School of Electrical and Automation Engineering,Nan Jing Normal University, Nan Jing, Jiang Su 210023, China;

[2]School of electrical engineering, North China University of Technology, Beijing, Beijing 100144, China;

[*]Corresponding author e-mail: [a]21190304@njnu.edu.cn, [b]yhe131@ucr.edu

†These authors contributed equally.

Abstract. Photolithography is one of the core methods in the semiconductor industry for the mass production of integrated circuits (IC). It is also the driving force behind Moore's Law, which predicts the number of transistors in an integrated circuit to double every two years. This paper aims to overview the photolithography process and its current situations, starting with the rationale behind it and its advantages. We review the photolithography process in individual steps and gave typical process parameters when applicable. Then we introduce the major photolithography system manufacturers of interest, followed by an overview of techniques used to improve the resolution of photolithographic systems, namely immersion lithography, Extreme-Ultraviolet (EUV) lithography, and Resolution Enhancement Techniques (RETs). Finally we discuss the challenges encountered in lithography technology.

1. Introduction

Lithography refers to a graphic micro machining technology that uses the chemical sensitivity of photoresist under the action of visible light, ultraviolet light and electron beam to transfer the graphics designed on the mask to the substrate through the process of exposure, development and etching [1]. Lithography was first applied to the processing of semiconductor devices. It is the foundation of semiconductor industry, electronic industry and information industry. Now it has been widely used in Pan semiconductor industry and microelectronics industry.

There are sufficient historical reasons for the emergence and development of lithography technology. Traditional circuits are composed of transistors, resistors, capacitors and other electronic components. Their inherent size makes simple circuits occupy a large space. Moreover, due to the need for welding, the reliability of the circuit is poor and the quality is difficult to be guaranteed. The integrated circuit manufactured by lithography can connect all electronic components in a small silicon wafer. This reduces the volume of the circuit, improves the efficiency of the circuit and reduces the power consumption of the circuit [2]. In addition, as the fastest medium, light can bring the highest efficiency to the production of integrated circuits. For chip manufacturers operating 24 hours, time is money, and lithography can undoubtedly bring the greatest economic benefits. Therefore, lithography has been highly sought after and fully utilized in countless fields since its inception.

Since Moore's law was put forward, lithography has experienced rapid iteration. According to the law that the number of integrated transistors is doubled or the cost is reduced by half every 18 months, lithography methods and materials are constantly renovated [3]. At present, the resolution of common

Content from this work may be used under the terms of the Creative Commons Attribution 3.0 licence. Any further distribution of this work must maintain attribution to the author(s) and the title of the work, journal citation and DOI.
Published under licence by IOP Publishing Ltd

lithography is up to 7Nm. Advanced Semiconductor Material Lithography(AMSL) and Taiwan Semiconductor Manufacturing Company(TSMC) have put forward the idea of reducing the resolution to less than 3nm. In addition, many new lithography methods, such as electron beam lithography, laser direct writing lithography, micro led lithography, nanosphere lithography and immersion lithography, have been proposed one after another, which enriches the applied range of lithography [4].

Two main advantages of the photolithography process are the ability to have parallel processes and its high precision. Parallel process means multiple copies of one design could be patterned onto the photomask; this allows one print run to produce many equivalent patterns, which is essential for large-scale production. The minimum feature size of photolithography is limited to half of the wavelength of the light used, making photolithography a suitable technique for fabricating nano-scale devices [5]. Thus the most prevalent use of photolithography is in the patterning process of integrated circuits such as solid-state drives and microprocessors. Another common application of photolithography is in the multi-layer manufacturing process of printed circuit board, where the technique is used to pattern copper layers [6]. Researches are being conducted to develop photolithography systems using, e.g., digital projectors and optical microscope, aiming to produce more cost-effective and viable photolithography at less specialized facilities. These researches could help explore possible new applications of this technology [7].

2. Brief description of lithography process

The complete lithography process can be divided into ten parts： 1. Wafer Clean,2. Pre-bake and Primer Vapor,3. Photoresist Coating,4. Soft Bake,5. Alignment,6. Exposure,7. Development,8. Hard Bake,9. Etching,10. Photoresist removal. These ten parts will be briefly introduced below.

2.1. Wafer Clean

Silicon wafer may be partially oxidized or stained due to improper storage. Cleaning silicon wafer can remove pollutants, remove particles, reduce pinholes and other defects, and improve the adhesion of photoresist. Generally, residues include grease, organic residues and silica. The methods of wafer cleaning are physical cleaning and chemical cleaning, for example: Solvent clean, HF dip, RCA-1 clean and ultrasonic cleaning [8].

2.2. Pre-bake and Primer Vapor

In order to remove the residual water in the previous step and provide a more suitable environment for photoresist attachment, the temperature of the silicon wafer needs to be heated to 80 degrees during coating. In this situation, add hexamethyldisilazane (HDMs). This step can replace the hydrophilic hydroxyl groups on the surface of the silicon wafer into hydrophobic groups, so that the silicon wafer can better adsorb photoresist [9]. The reaction process has been shown in Figure 1.

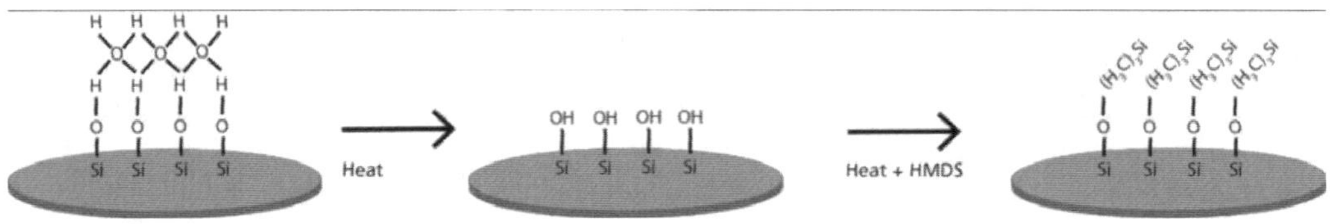

Figure 1. reaction process of HDMs Photoresist Coating

2.3. Photoresist Coating

This is the initial and one of the most important steps in the graphics conversion process. The quality of coating directly affects the defect density of the processed devices. The standard photoresist thickness is 20mil ~ 30mil. The methods of photoresist coating include spin coating, spray coating, dip coating and roller coating, among which dip coating is the most commonly used [10].

2.4. Soft Bake
This dehydrates the photoresist by drying at high temperature, which can increase the adhesion between photoresist and silicon wafer and increase the stability of adhesion [11]. According to the thickness of the lithography machine after photoresist coating, the temperature of the process is generally set at 80 to 100 degrees. Because the photoresist is not resistant to high temperature, the process generally lasts only one minute.

2.5. Alignment
This is one of the three core technologies of lithography. It is a method to align the silicon wafer with the mask. When the mask tilts or rotates, the accuracy of the linear scale will be greatly affected [12]. Therefore, the accuracy of alignment directly determines the accuracy of the pattern projection of the mask.

2.6. Exposure
This is the most critical step in the photolithography process. After registering the photoresist-coated wafer with the alignment system of the lithography machine and the lithography plate, it is irradiated with UV light, and the irradiated photoresist undergoes a photochemical reaction. The purpose of exposure is to fully sensitize the photoresist in the shortest possible time, and obtain the highest possible film retention rate, approximately vertical photoresist sidewalls and controllable strip width after development. The photoresist is mainly sensitive to ultraviolet light sources, and the commonly used ultraviolet light source is a high-pressure mercury lamp. Among them, the main effect of photoresist sensitivity is the spectral lines with wavelengths of 435.8nm (g line), 365nm (i line), 248nm (DUV far ultraviolet) [9]. The purpose of exposure is to fully sensitize the photoresist in the shortest possible time and obtain the highest possible film retention rate after development. Approximately vertical photoresist sidewalls and controllable strip width.

2.7. Development
Development is the process of using a developer to remove the exposed part of the photoresist to form the desired pattern on the silicon wafer. The photoresist in the exposed part interacts with the developer and dissolves in water, and the unexposed part does not interact with the developer and remains in its original state.

2.8. Hard Bake
In order to make the properties of the photoresist more stable, the hard film can remove the solvent remaining in the photoresist after development, and proper post-baking can improve the adhesion and etching resistance of the photoresist [11].

2.9. Etching
Etching is a process that uses chemical methods to selectively remove specific parts of the deposited layer in the manufacture of semiconductor devices. The pattern on the mask is transferred to the photoresist on the surface of the silicon wafer by photocopying.

2.10. Photoresist removal
After etching or ion implantation, the photoresist can be removed from the surface of the substrate, and the step is changed to a de-glue. The degumming methods in the microelectronics process include wet and dry degumming. Among them, wet degumming is divided into organic solution degumming and inorganic solution degumming [13].

3. Current situation and problems of lithography machine
Lithography machine is the aggregate of the above lithography technology applications. It is the equipment most widely used and involved in most fields. Now, the author will focus on the current

situation of lithography machine, analyze the bottlenecks encountered in the development of lithography machine, and put forward some ideas for future development.

3.1. Current situation

3.1.1Development aspects

There are now three giant lithography machine companies worldwide, namely ASML in the Netherlands, Nikon in Japan and Canon in Japan. When the early lithography machine was just emerging, ASML almost fell into bankruptcy, because its technology was backward, the market was saturated that year, the competition was extremely fierce, and there was not enough financial support [14]. However, in 2021, ASML has grown to be the absolute leader in the field of lithography machines, with a global market share of nearly 70%, and it monopolizes 100% of the market for high-end EUV lithography machines. As early as 2000, the Twins can dual-stage lithography machine established its dominant position in one fell swoop. It can be said that ASML's leading road is not only closely related to the industrial environment, but also the inevitable result of its own emphasis on research and development. With the continuous shrinking of modern chip sizes, EUV equipment that can etch smaller resolutions will soon become a rigid demand in the industry. Behind the success of ASML is its research team and its suppliers [15]. It wisely chose to cooperate with the world's top scientific research institutions and universities, adopting an open innovation model, and allowing experts to innovate in their areas of expertise. Share the patent achievements and R&D risks with extremely high efficiency and flexibility. Table 1 below summarizes the resolution changes of lithography machines produced by amsl, Nikon, Canon and Smee from 2000 to 2020.

Table 1. Resolution comparison of mainstream manufacturers

Comparison（Resolution）							
	00-02	03-05	06-08	09-11	12-14	15-17	18-20
ASML	284	110	60	38	20	10	3
Nikon	100	65	38	38	38	38	38
Canon	500	500	350	90	90	90	90
SMEE			1000	800	600	200	90

3.1.2 Technical aspects

At present, the efficiency of lithography is no longer the focus of attention. As the representative of the most advanced technology, high resolution can still quickly occupy the high-end market even if the efficiency is low. The development of ASML in recent years supports this conclusion. In recent years, a large amount of technical investment is also inclined to improve the resolution of lithography. The calculation formula of resolution in lithography is $CD = K1^* \lambda / Na$. Among them, this paper mainly focuses on λ and Na, for example: Electron beam exposure, Extreme Ultraviolet(EUV) , Immersion Lithography and resolution enhancement techniques.

Lithography and electron beam exposure are currently the two most commonly used micro-structure processing methods in industrial production and scientific and technological research fields. The electron beam exposure can be regarded as an extension of lithography. Although it was proposed by G. Mollenstedt and R. Speidel as early as 1960, it was not until the 1990s that breakthroughs in the resolution limit of lithography became more and more difficult. It is regarded as one of the potential alternative technologies of lithography and has been vigorously developed [15].

At present, photolithography can achieve high exposure resolution, and can be used in the processing of 5nm and below-node integrated circuits, while ensuring high processing efficiency. Taking the most advanced EUV lithography machine (NXE: 3400C, ASML) as an example, its throughput can reach 170wafers/h [16]. However, the cost of high-resolution lithography is also very high, not only from the high price of lithography machines (over 100 million euros), but also from the extremely high requirements for material quality and process level. In contrast, e-book exposure has higher flexibility. On the one hand, there is no need to prepare a mask in advance according to the exposure layout, and rapid iteration can be carried out during small-batch pattern testing. On the other hand, the relatively low equipment cost (No more than millions of RMB) can get very high processing resolution [17]. However, the main disadvantage of electron beam exposure lies in its exposure method, which makes its processing efficiency lower than that of photolithography. Therefore, in the current research fields of nanotechnology, the mainstream processing scheme is to process high-resolution micro-structure with e-book exposure technology, and to process large-area low-resolution micro-structure with photolithography [18].

Extreme Ultraviolet Lithography (EUV) is thought to have increased efficiency, reduced manufacturing cost, and better supports for the development of processing power [19]. EUV normally uses light of 13.5 nm wavelength generated by a plasma source. The generated light is collected and guided onto the photomask, and then the reflected image from the photomask is focused on the wafer. The investigations on this technology were relatively few until the end of the 1980s [20]. This is due to the high threshold of this technology, which is caused by extremely short absorption length of EUV light in both solid (~nanometers) and gases (~microns). As a result the investigations on EUV process involved speciallized and expensive apparatus such as high vacuum and hot plasma, and consequently its application were used to be limited to astrophysics. Higher threshold of this technique, consequently, causes issues in EUV machines such as high tool costs, complexity and reliablity issues caused by, e.g., stochastic phenomena [21]. Currently ASML is the only supplier of EUV volume production machines. Their state-of-art EUV system, NXE:3600D, has resolution of 13nm, a numerical aperture of 0.33, and can process >160 wafers per hour.

Immersion lithography is a process in which the liquid with high refractive index is filled between the lower surface of the last lens of the projection objective of the lithography machine and the photoresist on the silicon wafer to improve the resolution of the finished product. Because the numerical aperture formula of the projection objective is: $Na = n\sin\theta$, Therefore, $CD = K1 * \lambda/n\sin\theta$ can be obtained by bringing it into the calculation formula of resolution, it can be seen that there is a linear relationship between the resolution and the refraction of light [22]. At present, the immersion liquids involved in commercial applications are pure water. Although there are studies on other media, they cannot be applied in practice for many reasons [23]. John H. Burnett has done a lot of research on the optical parameters of water, and the specific values are shown in the table2 below [24]:

Table 2. Main optical parameters of water

Main Optical Parameters	
	measurement result
Refractive index	1.4362(2)
Temperature coefficient	-1.00*E-4/k
Dispersion coefficient	2.109(17)*10-3/nm

Therefore, compared with the case without immersion lithography, immersion lithography can further reduce the resolution by 30.3%. Immersion lithography can further provide effective help from the physical level in the case of reducing the resolution by focusing on reducing the wavelength of the light source.

ASML and Nikon are the two big manufacturers of immersion lithography systems. Today, ASML's 193nm (ArF) immersion lithographic system can have a numerical aperture up to 1.35 and a resolution of 38nm. Numerous efforts are being made to develop new immersion fluids, coatings, and matching photoresists to reach even higher resolutions. However, the fact that liquid is used in immersion lithography also brings many problems with it. The absorption coefficient of immersion liquid is heavily influenced by impurities, which must be carefully controlled. Temperature variation within the liquid must be minimal in order to maintain a constant refractive index. Watermarks, air bubbles tend to generate defects on the wafer. Handling these problems often results in a reduced wafer throughput [25].

An immersion lithography system alone, however, cannot satisfy the need for modern IC manufacturing. Modern IC has dimensions much smaller than half of the wavelength of the light source, and are still being pushed down following Moore's law. At the sub-wavelength level effects such as diffraction and interference become significant in quality control [26]. These effects can give irregularities or different dimensions to the printed shape on the wafer, which in turn will affect the properties and performance of the final product. To counter this resolution enhancement techniques (RET) are developed, aiming to improve wafer quality through manipulating the wave [27].

One of the most important RET method is optical proximity correction (OPC) method, which compensate for the loss by diffraction by adding extra patterns onto the photomask. The precise shape and dimensions of the extra patterns are determined by simulations based on computer model, called model based OPC. Sometimes they can be determined by searching pre-calculated look-up table (rule based OPC), but this is relatively rare. OPC method can also improve lithography process window by including additional assist patterns such as serif, hammerhead and sub-resolution assist features [28]. The algorithm used for model OPC is also crucial, as a better algorithm can save computation time and improve pattern fidelity.

3.2. Problems

Although the resolution of lithography has been reduced to 3nm by a series of methods and is continuing to 1nm, there still remain many problems in the development direction of lithography. They cannot be solved once and for all. They can only be iterated with the continuous refinement of technology, so as to deal with the new problems accompanied by new technology. This paper mainly analyzes the existing and long-term problems of lithography technology from the perspectives of light source and mask.

The light source problem is mainly reflected in the difficulty of obtaining a new light source. The following mainly compares deep ultraviolet and extreme ultraviolet to reflect the improvement of the difficulty of obtaining the light source.

Deep ultraviolet lithography, using the non-imaging method of reflecting mirror, using the non-imaging method of reflecting mirror, using reflection to pass deep ultraviolet light with a wavelength of 180 to 260 through the flat tube nanoplate point lithography, so that the image of the partial plate touches Lithography technology onto the substrate. Can be processed with sub-micron fine patterns [29].

The 193nm light source of the deep ultraviolet lithography machine, take the ASML 1400 F model TWINSCAN as an example. It has <= 57nm resolution, 0.93NA projection optics, and >=300 wafers per hour. Moreover, its cost is less than 20 million U.S. dollars, and it only needs one aircraft for transportation, and the installation and commissioning time of the machine is one to two months. The light source of the DUV is an excimer laser (the manufacturers are cymer and gigaphoton, and its output power is also 90W). It improves the resolution by shortening the wavelength, which is probably doubled. The DUV mask is to selectively project and transmit light, the material is quartz and chromium, and the projection lens is used. Its numerical aperture is 1.2 to 1.35, and a machine can produce 6,000 wafers a day. DUV wafer stage uses air-bearing air-bearing [30]. If contamination enters the stage, it can be cleaned with IPA, ACETONE or water. Machine maintenance is done by the fab's own engineers.

The extreme ultraviolet lithography machine currently uses 13.5nm light source. Take ASML's NXE 3400C as an example; it has 13nm resolution, 0.33NA projection optics, and >=125 wafer per hour. The EUV costs nearly 120 million U.S. dollars, weighs 180 tons, has more than 100,000 parts, and requires three aircraft for transportation, and installation and commissioning takes one year. It needs a light source with an output power of 500W. It uses laser produced plasma LPP. This plasma can stably emit 13.5nm soft X-rays. The chemical formula is soft X-rays = CO2 Laser Light * Sn(tin) Droplet. An EUV consumes 3 watts of electricity a day, which is very high, and it was once in the news. Its power conversion power is 5%, and the wavelength is directly reduced to less than one-tenth of the original, but the resolution is increased by almost ten times [31]. The photomask used in EUV is opaque and uses the principle of reflection (the photomask with composite multi-layer mirrors is used, and its layers are up to 50 layers). At the same time, the photomask material is molybdenum and silicon. It is a Bragg reflector (a multi-layer mirror that can concentrate small reflections into a single strong reflection) to produce an EUV for the mirror processing roughness requirement is less than 50pm. The numerical aperture is about 0.55, so in terms of DUV, EUV can increase the resolution by 5 times. In order to prevent the light source from being absorbed, the EUV internal optical path is vacuum, including the reticle stage and the wafer stage of the reticle stage, and the vacuum degree needs to reach the magnitude of the minus 7th power of 10. Its lithography machine can only produce 1,000 wafers a day. EUV uses a magnetic levitation plane motor because it has a vacuum inside. Once pollution enters the workbench, a load lock in the excessive transmission cavity is used to control the pollution particles or particles from entering the main vacuum cavity. Uptime Vacuum breaking and evacuation are very time-consuming. The extreme ultraviolet lithography machine is maintained by ASML engineers. The warranty fee is very expensive, but its chip performance is better, the cost has not increased, the overall process cost can be reduced by 10%, and the simplification of the process has also increased the yield by about 10 %. So for TSMC and Samsung's major manufacturers to buy EUV profitable, small manufacturers should forget it [32].

Industrial machine manufacturing is the timely and cradle of the entire industrial system. It is at the core of the industrial chain and determines the industrial level of a country. ASML's lithography machine has been a state-of-the-art lithography machine from TWINSCAN. The state of art is not exaggerated [33].

The problem of mask is mainly reflected in optical proximity effects. The graphics projected on the wafer are inconsistent with those on the mask plate due to diffraction and interference between adjacent graphics on the mask plate. Because the light wave will diffract when passing through the mask, and interference will occur at different positions of the mask, the light intensity distribution actually projected on the silicon wafer is the result of the superposition of these diffracted interference light waves, which is not exactly the same as the mask pattern [34]. Therefore, this problem cannot be completely eliminated on the premise of using today's widely used lithography technology.

For transistors, proximity effect will significantly affect their threshold voltage, saturation current and other device electrical parameters, and then affect the device performance. For 45nm process devices, the saturation current difference of 30% and the threshold voltage difference of 100mV can be achieved for two transistors with the same channel length and channel width due to the proximity effect [35].

4. Conclusion

At this stage, many companies are also promoting nanoimprinting, mask less lithography, or an emerging technology called self-assembly. But EUV lithography is still considered to be the best process for next-generation CPUs. Because the 193nm immersion process is very extensible, and EUV technology is costly and progress is slow. EUV (Extreme Ultraviolet Lithography) is the next generation lithography technology (lithography technology at the <32nm node). It is a technology that uses soft x-rays with a wavelength of 13.4nm for photolithography. The basic equipment of EUV lithography still needs a lot of development work to reach a mature level suitable for mass production. The following challenges currently exist: 1) Develop a light source with high enough power and make the system have enough transmittance to achieve and maintain high throughput. 2) The maturity of mask technology includes manufacturing reflective mask substrates with sufficient flatness and yield, photochemical inspection of reflective masks, and difficulty in meeting the requirements for defect-free operation due to the lack of a protective film on the mask surface. 3) Develop high sensitivity and low line edge roughness photoresist. Make the possibilities of the lithography machine broader.

References

[1] Yulian Pang Yingquan Zou. Development and application of photoresist [J]. Information recording materials,2015,16(01):36-51.

[2] QIhong LouZhijun Yuan,Haibo Zhang.History and current situation of lithography technology [J].science,2017,69(03):32-36.

[3] Jianjun Ma. Historical evolution of optical lithography [J].Special equipment for electronic

industry,2008(04):28-32.
[4] Liwen He,Le Luo,Gang Meng,Jingzhen Shao,Xiaodong Fang.Research progress of new lithography technology [J].Laser technology,2019,43(01):30-37.
[5] Ishihara, T., & Luo, X. (2006). Nanophotolithography based on surface plasmon interference. Nanoplasmonics - from Fundamentals to Applications, Proceedings of the 2nd International Nanophotonics Symposium Handai, 305–312.
[6] Shamkhalichenar, H., Bueche, C. J., & Choi, J. W. (2020). Printed Circuit Board (PCB) Technology for Electrochemical Sensors and Sensing Platforms. Biosensors, 10(11), 159.
[7] Haldar, S., Vashisht, G., Ghosh, U. K., Jaiswal, A. K., Porwal, S., Khakha, A., Sharma, T. K., & Dixit, V. K. (2019). Development of a simple cost-effective maskless-photolithography system. DAE SOLID STATE PHYSICS SYMPOSIUM 2018.
[8] UCLRVINE sop-wet-silicon-solvent-clean
[9] Xiao Ming Hu Photolithography technology in electronic fabrication International Power, Electronics and Materials Engineering Conference (IPEMEC 2015)
[10] Nga Phuong Pham, E. Boellaard, Joachim N. Burghartz, Fellow, IEEE, and Pasqualina M. Sarro, Senior Member Photoresist Coating Methods for the Integration of Novel 3-D RF Microstructures JOURNAL OF MICROELECTROMECHANICAL SYSTEMS, VOL. 13, NO. 3, JUNE 2004
[11] Duta Jaizul-Aziz Principles of Photolithography in Semiconductor Manufacturing 260690285 May 1, 2020
[12] Dongxu Ren 1, Jianpu Xi 1, Zhengfeng Li 1, Bin Li 1, Zexiang Zhao 1, Huiying Zhao 2, Lujun Cui 1 and Hang Xu 1 The Alignment Method for Linear Scale Projection Lithography Based on CCD Image Analysis †15 November 2017
[13] Centro Algoritmi.Centro de Física.Optimized SU-8 Processing for Low-Cost Microstructures Fabrication without Cleanroom Facilities.Micromachines 2014, 5, 738-755; doi:10.3390/mi5030738
[14] Guo Qiantong, Li Bo. Analysis of the breakthrough path of my country's lithography machine based on the development of the global industry of lithography machine[J]. Integrated Circuit Applications, 2021, 38(09): 1-3. DOI: 10.19339/j.issn. 1674-2583.2021.09.001.
[15] Zhang Xia, Liu Hongbo, Gu Wen, Zhou Xiying, Yu Zhishui. Global lithography machine development and localization of lithography equipment[J]. Wireless Internet Technology, 2018, 15(19): 110-111+ 118.
[16] Hu Chao, Wang Xingping, You Chun, Sun Feng. Application of high-precision electron beam lithography in micro-nano processing[J]. Electronics and Packaging, 2017, 17(05): 28-32+36. DOI :10.16257/j.cnki.1681-1070.2017.0060.
[17] Xie Ying. Preparation and characterization of Au wire grid array by electron beam lithography[J]. Information Recording Materials, 2019, 20(05): 158-159. DOI: 10.16009/j.cnki.cn13-1295/tq. 2019.05.099.
[18] Pan Junchen, Lang Fengchao, Wang Shiyu, Zhang Weiguang, Jiang Aifeng, Li Jijun, Xing Yongming. Research on the micromechanical properties of electron beam photoresist based on nano-scratches[J]. Surface Technology, 2021, 50(03): 219-224 +260.DOI:10.16490/j.cnki.issn.1001-3660.2021.03.021.
[19] Van de Kerkhof, M., Benschop, J., & Banine, V. (2019). Lithography for now and the future. Solid-State Electronics, 155, 20–26. https://doi.org/10.1016/j.sse.2019.03.006
[20] Hasan, R. M. M., & Luo, X. (2018). Promising Lithography Techniques for Next-Generation Logic Devices. Nanomanufacturing and Metrology, 1(2), 67–81. https://doi.org/10.1007/s41871-018-0016-9
[21] De Bisschop, P., van de Kerkhove, J., Mailfert, J., Vaglio Pret, A., & Biafore, J. (2014). Impact of stochastic effects on EUV printability limits. Extreme Ultraviolet (EUV) Lithography V. Published. https://doi.org/10.1117/12.2047827
[22] Yuan Qiongyan, Wang Xiangchao, Shi Weijie, Li Xiaoping. Research progress of immersion

lithography [J]. Progress in laser and optoelectronics, 2006 (08): 13-20
[23] Soichi Owa, Katsushi N Akano, et al. Current status of water immersion lithography and prospect of higher index method [J]. Proceedings of SPIE, 2007, 6533
[24] John H. Burnett and Simon Kaplan. Measurement of the refractive index and thermo-opticcoefficient of water near 193 nm [J]. Proceedings of SPIE.2003, 5040:1742-1749
[25] Lin, B. J. (2006). Optical lithography—present and future challenges. Comptes Rendus Physique, 7(8), 858–874.
[26] Ahmed Awad, Atsushi Takahashi and Chikaaki Kodaman (December 24th 2017). Optical Proximity Correction (OPC) Under Immersion Lithography, Micro/Nanolithography - A Heuristic Aspect on the Enduring Technology, Jagannathan Thirumalai, IntechOpen, DOI: 10.5772/intechopen.72699. Available from: https://www.intechopen.com/chapters/58480
[27] Shibuya, M. Resolution enhancement techniques for optical lithography and optical imaging theory. Optical Review 4, 151 (1997)
[28] Y. Hou and Q. Wu, "Optical Proximity Correction, Methodology and Limitations," 2021 China Semiconductor Technology International Conference (CSTIC), 2021, pp. 1-5, doi: 10.1109/CSTIC52283.2021.9461507
[29] Feng Boru, Yao Hanmin, Institute of Optoelectronics Technology, Academy of Sciences. The development of deep ultraviolet lithography [J]. The Ninth National Electron Beam. Ion Beam. Photon Beam Academic Annual Conference, 1997.
[30] Tao Li, Li Qiang. Asim's "Foresight": the battle of lithography machines and the rise of the Chinese market [N]. 21st Century Business Herald, 2021-11-10 (006). DOI: 10.28723 /n.cnki.nsjbd.2021.004651
[31] Chen Bingxin. DUV lithography machine is very promising [N]. China Electronics News, 2021-10-12 (001). DOI: 10.28065/n.cnki.ncdzb.2021.001254
[32] Tan Enmin, Shi Tingting, Zhang Zhigang, Zhang Lin. Adaptive feedforward control applied to the mask stage of the lithography machine[J]. Control Engineering, 2021, 28(06):1069-1074. DOI :10.14107/j.cnki.kzgc.20190021
[33] Zhao Xiaoyang. Research on the motion control technology of the micro-moving stage of the lithography machine work table [D]. University of Electronic Science and Technology of China, 2021. DOI: 10.27005/d.cnki.gdzku.2021.002291
[34] Ke shunkui. Review and Reflection on the development of optical proximity effect correction technology [J]. Shandong industrial technology, 2018 (10): 207. Doi: 10.16640/j.cnki.37-1222/t.2018.10.187
[35] LINXW,MOROZ V.Layout proximity effects and modeling alternatives for IC designs [J]. IEEE Design&Test Comput， 2010， 27(2):18-25.

Welding defect signal extraction technology based on GAMP-SBL algorithm

AiLing Qi[1], JiaSen Wang*[2]

[1]School of Computer Science and Technology, Xi'an University of science and technology, Xi'an, Shanxi, 710600, China

[2]School of Computer Science and Technology, Xi'an University of science and technology, Xi'an, Shanxi, 710600, China

* Corresponding author: 15877558337@163.com

Abstract: Flip chip technology has been rapidly developed and widely used in the field of microelectronic packaging, and defect detection has also received more and more attention. Aiming at the problem that noise affects the location and extraction of signal defects in ultrasonic testing, the sparse Bayesian learning based on generalized approximate message passing (GAMP-SBL) algorithm is used to extract signal defects, and the over-complete Gabor dictionary is used to reconstruct signal defects to effectively improve Sparse decomposition algorithm. The precision experiment tested the defect simulation signal and the actual ultrasonic signal respectively, and compared with the greedy algorithm. The experimental results show that the GAMP-SBL algorithm can more effectively extract the defect signal under the noise background.

1. Introduction

As a rapidly developing and widely used microelectronic packaging technology, flip-chip bonding technology has the advantages of high alignment accuracy, short interconnection lines, and high input and output density. It is an important means to reduce package size and increase package density [1-2]. The ultrasonic echo estimation carries interface defect information through each ultrasonic reflection echo. In ultrasonic nondestructive testing, noise is often accompanied by noise interference, which will change the true information returned by ultrasonic echo. In severe cases, it will bring difficulties to subsequent signal processing and defect identification. [3] In 2001, Taping proposed a sparse Bayesian learning method, which is a machine learning algorithm used to solve classification and regression problems. [4] In 2004, due to the excellent performance of the algorithm, Wipf and Rao of the University of California, San Diego, et al. The SBL algorithm based on Bayes' theorem is introduced into the compressed sensing sparse reconstruction algorithm [5]. In order to solve the huge problem of matrix inversion in the calculation process of SBL algorithm, Al-Shoukairi et al. proposed a sparse Bayes algorithm based on generalized message transmission[6].

In this paper, a GAMP-SBL algorithm is used to extract welding defect signals. Under the condition of ensuring the accuracy of the reconstructed signal, it has higher robustness, thereby effectively extracting the defective signal.

Content from this work may be used under the terms of the Creative Commons Attribution 3.0 licence. Any further distribution of this work must maintain attribution to the author(s) and the title of the work, journal citation and DOI.
Published under licence by IOP Publishing Ltd

2. Ultrasonic signal extraction based on GAMP-SBL algorithm

2.1 Sparse Bayesian signal reconstruction method

In sparse Bayesian learning, the ultrasonic reflection coefficient c satisfies the Gaussian distribution with zero mean variance σ^2, and the corresponding likelihood function is:

$$p(y|c,\sigma^2) = (2\pi\sigma^2)^{-N/2}\exp\{-\frac{1}{2\sigma^2}||y-\Phi c||^2\} \quad (1)$$

Where $y = (y_1 \dots\dots y_N)^T, c = (c_0 \dots\dots c_{M-1})^T$, and Φ is an over-complete dictionary with a scale of N×M.

In order to avoid over-fitting of the maximum likelihood estimates of c and σ^2, we adopt the prior probability distribution of Bayesian view to "constrain" the parameters. The universal choice code for the zero-mean Gaussian prior distribution of c:

$$p(c|\alpha) = \prod_{i=0}^{M-1} N(c_i|0,\alpha_i^{-1}) \quad (2)$$

α is an M-dimensional hyperparameter vector. Each weight has a separate hyperparameter, which reduces the strength of the prior. Need to define the super prior of α and the noise variance σ^2. The appropriate prior for these parameters is the Gamma distribution:

$$p(\alpha) = \prod_{i=0}^{M} Gamma(\alpha_i|a,b) \quad (3)$$

$$p(\beta) = Gamma(\beta|c,d) \quad (4)$$

Where $\beta \equiv \sigma^2$ and

$$Gamma(\alpha_i|a,b) = \Gamma(\alpha)^{-1}b^a\alpha^{a-1}e^{-b\alpha} \quad (5)$$

Where $\Gamma(\alpha) = \int_0^\infty t^{a-1}e^{-t}dt$ is the "gamma function", a=b=c=d=0. If the evidence of the data supports such a hypothesis, then the extensive prior alpha of the hyperparameters will become very large, so that the posterior probability weights will be concentrated to zero, so they are regarded as "irrelevant".After defining the prior, Bayesian inference is performed by calculating the posterior of all unknowns of the given data according to the Bayes rule:

$$p(c,\alpha,\sigma^2|y) = \frac{p(y|c,\alpha,\sigma^2)p(c,\alpha,\sigma^2)}{p(y)} \quad (6)$$

We cannot directly calculate the posterior $p(c,\alpha,\sigma^2|y)$ in (9), because we cannot calculate $p(y) = \int p(y|c,\alpha,\sigma^2)p(c,\alpha,\sigma^2)\, dc\, d\alpha\, d\sigma^2$ Perform normalized integration. Therefore, the posterior is decomposed into:

$$p(c,\alpha,\sigma^2|y) = p(c|y,\alpha,\sigma^2)p(\alpha,\sigma^2|y) \quad (7)$$

And note that we can analyze and calculate the posterior distribution of weights, because its normalized integral $p(y|\alpha,\sigma^2) = \int p(y|c,\sigma^2)\,p(c|\alpha)\,dc$ is Gaussian convolution. Therefore, the posterior distribution of weights is given by:

$$p(y|c,\alpha,\sigma^2) = \frac{p(y|c,\sigma^2)p(c|\alpha)}{p(y|\alpha,\sigma^2)} = (2\pi)^{-M/2}|\Sigma|^{-1/2}\exp\{-\frac{1}{2}(c-\mu)^T\Sigma^{-1}(c-\mu)\} \quad (8)$$

The covariance and mean are respectively:

$$\Sigma = (\sigma^2\Phi^T\Phi + A)^{-1} \quad (9)$$

$$\mu = \sigma^2\Sigma\Phi^T y \quad (10)$$

Where $A = diag(\alpha_0\alpha_1 \dots\dots \alpha_{M-1})$.The second term $p(\alpha,\sigma^2|y)$ of the posterior can be understood as $p(\alpha,\sigma^2|y) \propto P(y|\alpha,\sigma^2)p(\alpha)p(\sigma^2)$ To obtain the maximum value of α and σ^2, for the uniform super-priority, we only need to maximize the term $P(y|\alpha,\sigma^2)$, which is computable and is given by the following formula:

$$P(y|\alpha,\sigma^2) = \int p(y|c,\sigma^2)p(c|\alpha)\,dc = (2\pi)^{-N/2}|\sigma^2 I + \Phi A^{-1}\Phi^T|^{-1/2}\exp\{-\tfrac{1}{2}y^T(\sigma^2 I + \Phi A^{-1}\Phi^T)^{-1}y\} \quad (11)$$

The final parameter estimates are:

$$\alpha_i^{new} = \frac{\gamma_i}{\mu_i^2} \quad (12)$$

$$(\sigma^2)^{new} = \frac{||y-\Phi\mu||^2}{N-\sum_i \gamma_i} \quad (13)$$

Where μ_i is the i-th element in the mean vector μ, $\gamma_i \equiv 1-\alpha_i\sum_{ii}$, $\sum_{ii}$ represents the i-th diagonal element of the covariance matrix $\sum$. By using the above method to continuously update α and μ, the corresponding sparse signal reconstruction problem is completed. The reconstructed signal $Y_{rec} = \Phi * u$.

2.2 Signal reconstruction method based on GAMP-SBL

In GAMP algorithm, the input vector is $q = [q_1, q_2 \dots\dots, q_M]^T \in Q^N$, and the random vector $x = [x_1, x_2 \dots\dots, x_M]^T \in Q^M$ is generated through the input channel with conditional density function $p_{X|Q}(x_j|q_j)$ element by element. X is linearly transformed:

$$z = Dx \quad (14)$$

Where D is the over-complete dictionary, and the scale is N×M. Each element of z passes through an output channel with a conditional density probability of $p_{Y|Z}(y_j|z_i), i = 1,2 \dots\dots N$ to obtain a set of output vectors $y = [y_1, y_2 \dots\dots, y_M]^T$. The GAMP algorithm converts vector estimation into a series of scalar estimation problems, and finally estimates the posterior distribution of parameters effectively through the prior distribution and likelihood function.

Let each element in X be distributed independently of each other, and its conditional probability density is $p(x_i|q_i)$. Where q_i is a known hyperparameter, assuming that it is also independent of each other, and $p(y_a|x)$ is its likelihood function. The posterior probability density function of x can be obtained through Bayesian Theory:

$$p(x|y) \propto p(y|x)p(x) = \prod_i p(y_i|x)\prod_j p(x_j) \quad (15)$$

Where $\propto$ represents the positive proportional relationship.

The steps of the GAMP algorithm are described as follows. Each element in the complete dictionary D is set as d_{ij}.

1) Input: over-complete dictionary D, input variable q, output variable y, scalar estimation functions g_{in} and g_{out}, number of iterations t and maximum number of iterations t_{max}.
2) Initialization conditions: assign initial values to $\hat{x}_j(t)$ and $\tau_j^x(t)$.
3) Output variable: For each i there are:

$$\tau_i^p(t) = \sum_j |d_{ij}|^2 \tau_j^x(t) \quad (16)$$

$$\hat{p}_i(t) = \sum_j d_{ij}\hat{x}_j(t) - \tau_i^p(t)\hat{s}_i(t-1) \quad (17)$$

$$\hat{z}_i(t) = \sum_j d_{ij}\hat{x}_j(t) \quad (18)$$

When t=0, $\hat{s}_i$=0.

4) Input variable: For each j there are:

$$\tau_j^r(t) = [\sum_i |d_{ij}|^2\, \tau_i^s(t)]^{-1} \quad (19)$$

$$\hat{r}_j(t) = \hat{x}_j(t) + \tau_j^r(t)\sum_i d_{ij}\,\hat{s}_j(t) \quad (20)$$

5) Let t=t+1, return to step 3), until the maximum number of iterations terminate the algorithm.

In order to describe the maximum likelihood estimation of the GAMP algorithm, assume that the conditional density function is

$$p(x|q,y) = \frac{1}{Z(q,y)} \exp\left(F(x,Dx,q,y)\right) \tag{21}$$

Where

$$F(x,Dx,q,y) = \sum_{j=1}^{M} f_{in}(x_j,q_i) + \sum_{j=1}^{N} f_{out}(z_j,y_i) \tag{22}$$

$$f_{in}(x,y) = \log_{P_{X|Q}}(x|q) \tag{23}$$

$$f_{out}(z,y) = \log_{p_{Y|Z}}(y|z) \tag{24}$$

Where $Z(q,y)$ is a regularization constant, then the maximum posterior probability of x is:

$$\hat{x}_{MAP} = \arg\max_{x} F(x,Dx,q,y), \qquad \hat{z} = D\hat{x} \tag{25}$$

For each element j, the maximum marginal value of the posterior probability is

$$\Delta_j(x_j) = \max_{a/aj} F(x,z,q,y),\ \hat{z} = D\hat{x} \tag{26}$$

The input function of equation 26 maximum confidence propagation is approximately:

$$g_{in}(\hat{r},q,\tau^r) = argmaxF_{in}(\alpha,\hat{r},q,\tau^r) \tag{27}$$

Where $F_{in}(\alpha,\hat{r},q,\tau^r) = f_{in}(\alpha,q) - \frac{1}{2\tau^r}(\hat{r}-\alpha)^2$.

The input function of maximum confidence propagation is approximately:

$$g_{out}(\hat{p},y,\tau^p) = \frac{1}{\tau^p}(\hat{z}-\hat{p}) \tag{28}$$

3. Experiment analysis

3.1 Analog signal test

In order to observe the influence of the proposed method on signal defects, we compared it with several mainstream greedy sparse algorithms, including MP algorithm, OMP algorithm, SMP algorithm and SBL algorithm. In the simulated ultrasound signal, Gaussian white noise of different decibels is added, and then a variety of sparse decomposition techniques are used for reconstruction.

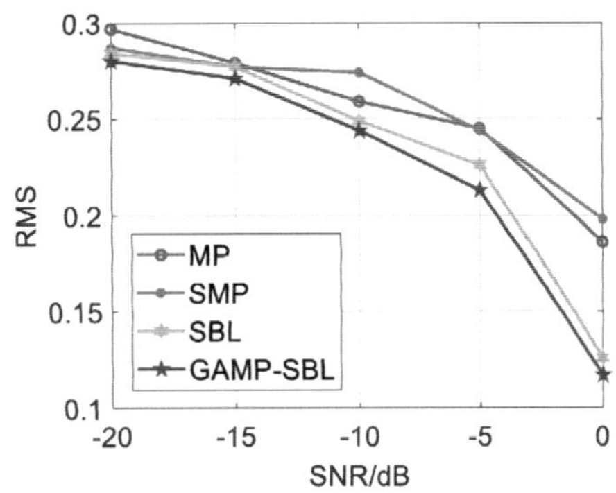

Fig 1. Denoising effect diagram of several sparse decomposition algorithms

It can be seen from Fig 1 that the GAMP-SBL algorithm is better than other sparse algorithms, so it can extract the defect signal more accurately and is closer to the source signal.

3.2 Measured signal experiment

The experimental data are the commercial AMI system developed by Sonoscan, Inc. Scanning the flip chip to obtain the acoustic volume data as the experimental test data. The sensor frequency used is 230MHz.

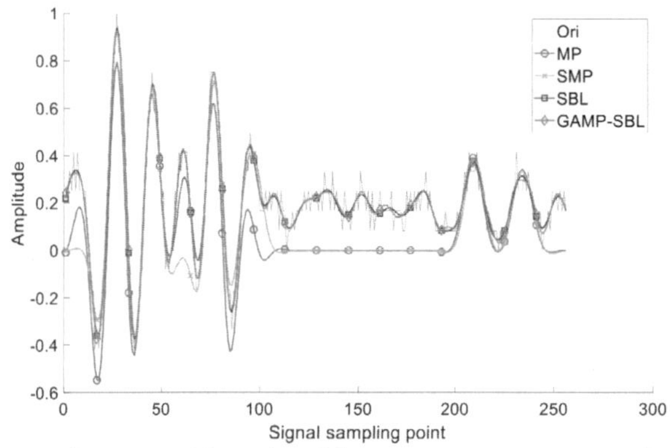

Fig 2. Several sparse decomposition algorithms to reconstruct actual ultrasonic signals

It can be seen from Fig 2 that the GAMP-SBL algorithm performs best in reconstructing the actual sampled signal. The reconstruction technology based on the greedy algorithm loses the information of the defect signal to a certain extent. This is very detrimental to our actual research.

4. Conclusion

This paper proposes a sparse reconstruction method of ultrasound signal based on generalized approximate message passing Bayesian learning algorithm. This method has higher robustness under the condition of ensuring reconstruction accuracy. Therefore, the method proposed in this paper can be used to study the defects of actual ultrasonic signals.

Acknowledgments

This work was supported in part by the National Natural Science Foundation of China (61674121).

References

[1] Lei S A , Xy A , Ke L A , et al. Defect inspection of flip chip solder joints based on non-destructive methods: A review[J]. Microelectronics Reliability, 110.

[2] Zhang G M , Harvey D M , Braden D R . Advanced acoustic microimaging using sparse signal representation for the evaluation of microelectronic packages[J]. IEEE Transactions on Advanced Packaging, 2006, 29(2):271-283.

[3] Semmens J E , Kessler L W . Application of Acoustic Frequency Domain Imaging for the Evaluation of Advanced Micro Electronic Packages[J]. Microelectronics Reliability, 2002, 42(9/11):1735-1740.

[4] Tipping M E . Tipping, M.E.: Sparse bayesian learning and the relevance vector machine. JMLR 1[J]. 2001.

[5] Wipf D P , Rao B D . Sparse Bayesian learning for basis selection[J]. IEEE Transactions on Signal Processing, 2004, 52(8):2153-2164.

[6] Al-Shoukairi M , Schniter P , Rao B D . A GAMP Based Low Complexity Sparse Bayesian Learning Algorithm[J]. IEEE Transactions on Signal Processing, 2017, PP(99):1-1.

Linked Data Fusion Based on Similarity Calculation and K-Nearest Neighbor

Yiming Chen[1a*]

College of Computer Science & Technology, Nanjing University of Aeronautics and Astronautics, Nanjing, Jiangsu, China

*e-mail: 244139157@qq.com

Abstract. The development of semantic web technology supports the continuous development of linked data and its applications. In order to make effective use of the growing linked data on the web, multi-source data should be fused, which is a key step in multi-source large-scale data analysis and management. Currently, merging these data can be difficult, because various sources usually provide multiple conflict descriptions for entities in the same real world. To complete the fusion, we need to solve the problem of object conflict. This paper proposes a linked data fusion method based on similarity calculation and k-nearest neighbor. This method has two contributions. Firstly, a similarity calculation method of linked data is proposed, which can effectively integrate URI nodes and blank nodes in linked data; Secondly, a literal type node fusion strategy based on k-nearest neighbor classification method is proposed, which realizes the automation of fusion and has the independence of data source. The results show that compared with other methods, this method can improve the conciseness and consistency and precision by up to 12.9%, 30.6% and 12%.

1. Introduction

The development and standardization of semantic web has led to an unprecedented large amount of data published on the network as linked data. However, it should be noted that due to the diversity of data sources, in linked data, different resources may point to the same entity in the real world [1]. To use these data, data fusion is required. The purpose of data fusion is to provide unified access to data residing in multiple, autonomous and heterogeneous data sources. Fusing these data sources can be tricky because data from various sources often provides multiple conflicts for a predicate of the same real-world entity.

For linked data, because its structure is a graph based data model, the fusion of two nodes can not simply merge the data, but also consider the fusion between their adjacent graphs. We need to find the conflicting data in the adjacency graph and resolve the conflict. The solution of this problem is also called object conflict resolution [2].

Linked data is mainly described in the form of resource description framework (RDF). The node types in RDF are mainly divided into URI node, blank node and literal node. There are still many problems in the current methods for RDF data fusion. First, most methods ignore the conflicts caused by some URI nodes and blank nodes in the semantics of the nodes to be fused. Most of these conflicts exist in the form of entity coreference[3]. If these conflicts are not resolved, the fused data will contain a large amount of redundant data. Second, the fusion strategy for literal type nodes is too simple, or too dependent on the association between expert experience and data sources, and lacks a method with strong independence and adaptability to multiple scenarios.

Content from this work may be used under the terms of the Creative Commons Attribution 3.0 licence. Any further distribution of this work must maintain attribution to the author(s) and the title of the work, journal citation and DOI.
Published under licence by IOP Publishing Ltd

The main process of linked data fusion is that data conflict will occur when the object values of RDF triples sharing the same subject and predicate are inconsistent. The task of fusion is to resolve conflicts and remove low-quality values to obtain a clean representation of resources [4].In order to solve the object conflict problem in RDF fusion, we propose the corresponding solution strategy. First, for the conflict data generated by URI nodes and blank nodes, we propose RDF semantic similarity algorithm *RDFsim*. The second is the conflict data generated for literal type nodes, In the final experiment, we verify the performance of our system for different types of data sets. The final results show that our system has significant advantages in conciseness and consistency on the basis of ensuring completeness.

2. Semantic fusion of linked data

In order to solve the problem of semantic fusion of linked data, our method is to first propose the semantic representation of RDF and object conflict cluster. Semantic representation can effectively represent the semantic information of RDF. Object conflict clustering is to extract conflicting objects from semantic information. The conflict is resolved, the true value is selected, and then the false data is removed to complete the data fusion.

We give conflict resolution strategies for different types of data in the object conflict cluster. First, for the conflict data generated by URI nodes and blank nodes, we propose a semantic similarity algorithm based on RDF to resolve the conflict. Secondly, for the conflict data generated by literal type nodes, a conflict resolution strategy based on KNN classification is proposed to resolve the conflict.

2.1. *Semantic representation of RDF nodes*

This section introduces the semantic information representation of RDF. RDF is a logical model which describes data in the form of triples. Each triple contains a subject, a predicate and an object. The subject corresponds to the resource being described, the predicate corresponds to the attribute of the resource, and the object corresponds to the value of the attribute. A set of RDF triples can be regarded as a directed graph, in which subjects and objects are the nodes, and predicates label the edges from the subject nodes to the object nodes. RDF context refers to a graph composed of nodes and edges connected to the target node. It is an important basis for describing the semantic information of nodes in the dataset. In order to represent the context of RDF, we give the following definitions:

Definition 1 (Path). In an RDF graph, predicates are represented by edges, and subjects and objects are represented by vertices. Then a path in the RDF graph is represented as a sequence as follows.

$$path(i) = (i, predicate[1], node[1], predicate[2], node[2], \ldots predicate[k], node[k]) \quad (1)$$

Where i is an instance, node[i] ($i > 0$) is any other expanded node from the RDF graph, and predicate[i] is a predicate that connects two nodes in a path.

The depth of a path is the number of predicates in this path. The path direction represents the positive and negative of a predicate in the path. For a predicate, if its connected left and right are respectively a subject and an object, we say the direction of this predicate is positive and have (subject, predicate+, object); if its connected left and right are respectively an object and a subject, we say the direction of this predicate is negative and have (object, predicate−, subject).

The idea of obtaining the context information of the target RDF node is to find a subgraph, which takes the resource as the center and the specified distance as the radius. This subgraph represents the context of the target node. Our method is to construct a set of paths to represent subgraphs. We treat the subgraph as an undirected graph, and then construct a set of paths from the target node to other nodes with a specified distance. After the construction is completed, the direction of the edge in the subgraph is added to the predicate in the path. In algorithm 1, we describe the construction method of the context representation of RDF nodes.

Algorithm 1 GetContext(*i*, *G*, *limit_depth*)
Input: RDF graph *G*, the upper limit *limit_depth* of path depth, and target node *i* in *G*
Output: *path*

```
path ← <i>
expansion_set ← {path}
paths ← {}
for path in expansion_set do
  last ← path[-1]
  pre_p ← path[-2]
  if type(last) == Literal or depth(path) == limit_depth do
    paths ← union(paths, path)
  else if depth(path) < limit_depth do
    for all t = <last, p, o> in G do
      if not (p.v == pre_p.v and p.dirct != pre_p.dirct) do
        new_path ← path.append(p,o)
        expansion_set.add(new_path)
    for all t = <s, p, last> in G do
      if not (p.v == pre_p.v and p.dirct != pre_p.dirct) do
        new_path ← path.append(p-,s)
        expansion_set.add(new_path)
  expansion_set.remove(path)
return path
```

2.2. RDF object conflict cluster

In Section 3.1, we introduce the semantic representation of RDF, which is composed of paths constructed based on its context. Therefore, in this section, the semantic fusion of two consistent RDF nodes can be realized by fusing their contexts, that is, by fusing the paths in the context. For the two nodes participating in the fusion, only the path with path depth of 1 in their context needs to be fused, because as long as the path with depth of 1 is fused, the nodes on the path with depth higher than 1 will also be fused together as the context of their parent node.

Our goal is to match the triples that can be fused with each other in the context of two nodes to be fused, so as to determine the fused object. The main method is to replace the name of one of the two nodes to be fused with the other. As shown in Figure 1, replace node N2 with N1 to obtain the data shown in Figure 1 (c). After the replacement, we will select the objects linked by predicates in the figure that can be fused with each other into a set.

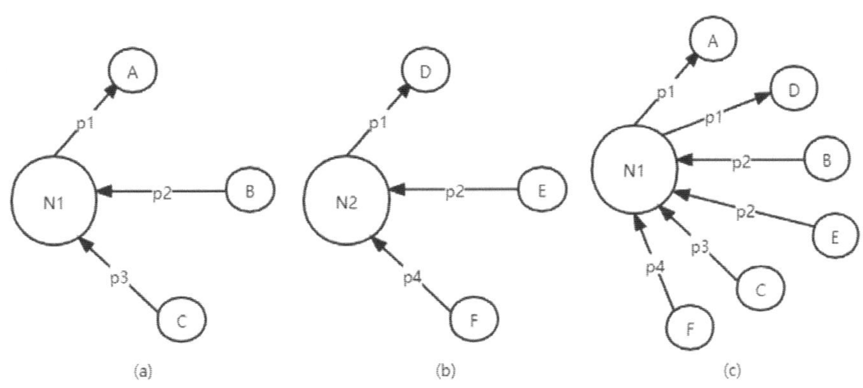

Figure 1 Context information for nodes N1 and N2

The method we choose is that all objects linked by comparable predicates are regarded as a set of mutually fusible object conflict sets. The cluster composed of these sets is also called object conflict cluster [5]. After constructing the cluster, our ultimate goal is to traverse all object conflict sets in the cluster, implement conflict resolution strategies for each set, and find the true value in the set. The comparability of predicates is defined as the two predicates have the same value and the same direction. If the values are different, but one of the predicates is the ancestor of the other predicate in the ontology model, the two predicates are also considered comparable.

2.3. Object conflict resolution

Errors inevitably occur during the creation of linked data. There are many linked data sources that contain noisy, outdated, missing, or incorrect data. Worse, multiple linked data sources often provide conflicting data [6]. Therefore, each conflict set in the RDF object conflict cluster contains many conflict descriptions. For conflicting data, most methods only focus on the conflict of literal type nodes, and fail to solve the conflict for URI nodes and blank nodes. In this section, we will introduce the conflict resolution methods we designed for these two types of conflicts.

2.3.1. Conflict resolution for URI node and blank node

At present, most methods only consider the conflict resolution of literal type nodes in the conflict set for the fusion of RDF conflict sets, and pay little attention to the conflict of URI nodes or blank nodes. Such conflicts usually result in data redundancy.

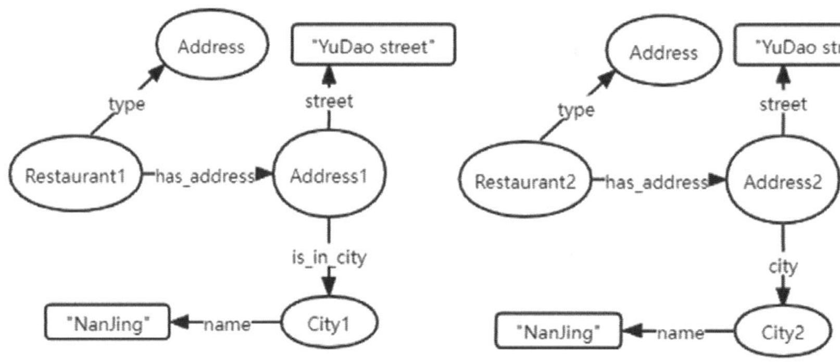

Figure 2 Triples in the context of Restaurant1 and Restaurant 2

As shown in Figure 2, the context is composed of triples associated with Restaurant1 and Restaurant2, respectively. We are now going to fuse the two. If we only consider the conflict of literal type nodes, we will introduce "Address1" and "Address2" URI nodes in the fusion process. However, it is not difficult to find from the figure that these two nodes obviously point to the same entity "Yudao Street", which brings redundancy to our data. So we must avoid this kind of situation.

Therefore, in order to solve the conflict for URI nodes and blank nodes, we design a context based RDF node similarity algorithm *RDFsim* to calculate the semantic similarity between RDF nodes, so as to identify the redundant items in the conflict. When the similarity of two conflict items is greater than the threshold, we believe that there is redundant data, and that the element with the most complete context description in the conflict set is the true value. If the similarity is less than the threshold, we think it is two independent data, which belong to the multivalued description of the entity. They are true values and need to be retained.

RDFSim algorithm calculates the similarity between the two nodes through the context information of RDF nodes to detect RDF data that may be semantically similar. We propose algorithm 2 to complete this process. For example, we want to calculate the similarity of *uri1* and *uri2* nodes. We first need to obtain their context information representation, which are represented by *paths1* and *paths2* respectively. then, for each *path* in *paths1* (such as *path1*), starting from *path1*, search *paths2* for all possible *paths* that can be compared with *path1* (such as *path2*) and calculate the similarity between *path1* and *path2*. There are three cases. First, if the last element in the two paths is of literal type, the text similarity score between the last elements of the two paths is directly calculated. Second, if the last element in the two paths is of URI type, directly compare whether the two URI nodes are equal. If they are equal, the score is 1, Not equal to 0. Third, if the last element in the two paths, one of which is of literal type and the other of which is of URI type, search whether there are literal type nodes in the context where the node depth of the URI type is 1. If so, calculate their text similarity score. We add the similarity score of the match to the *path_ score*, and calculate the weight of the two matching paths, which are added to the *path_ weight*.

Algorithm 2 GetSim(*uri1*, *uri2*, *G*, *limit_depth*).
Input: RDF graph *G*, the upper limit *limit_depth* of path depth, *uri1* and *uri2* in *G*
Output: The similarity between *uri1* and *uri2*

total_score ← 0
total_weight ← 0
paths1 ← Get_context(*G*, *limit_depth*, *uri1*)
paths2 ← Get_context(*G*, *limit_depth*, *uri2*)
for *path1* in *paths1* **do**
    path_score ← 0
    path_weight ← 0
    if type(*path1*[-1]) == Literal **do**
        for *path2* in *paths2* **do**
            if PathComparable(*path1*, *path2*) **do**
                if Sim(*path1*[-1], *path2*[-1]) > *path_score* **do**
                    path_score ← Sim(*path1*[-1], *path2*[-1])
                    path_weight ← (W(*path1*) + W(*path2*))/2
    total_score ← *total_score* + *path_score* * *path_weight*
    total_weight ← *total_weight* + *path_weight*
sim = *total_score* / *total_weight*
return *sim*

In algorithm 2, we only list the pseudo code when the last node of *path1* is of literal type. When the last node is of other types, the calculation method is similar. The *PathComparable* function is used to indicate whether two paths can be compared and calculate the similarity. The two paths are considered comparable only when they have the same depth and the predicates at the corresponding positions in the path are comparable. The *W* function is used to calculate the weight of the path, and the weight calculation method based on the classification ability of predicates in the path is adopted [1]. The *Sim* function is used to calculate the text similarity. If the calculated objects are two non-literal nodes, the value of *Sim* is 1 if the two nodes are identical, and 0 if they are different. Here, we use the character-level Levenshtein[7] similarity measure to calculate text similarity. The designed RDF similarity algorithm is called *RDFSim*.

2.3.2. Conflict resolution for literal nodes

Fusion of literal type nodes. We use the KNN based classification method to learn the common allocation strategies for conflict resolution between literal nodes, and then use the learned model to predict the conflict resolution strategies for the experimental data. This method can not only avoid a lot of labor costs, but also give a variety of solutions for complex scenes, and better solve the conflict problem caused by text type nodes.

KNN is a nonparametric classification method, which is simple and effective in many cases [8]. We use KNN classification method, which can classify the object conflict sets with different data characteristics and mark them as appropriate conflict resolution strategies, so as to complete the training of the model. In this paper, we will use Euclidean distance [9] to measure the distance between nodes.

We define a set of methods for parsing text to form vector representation, and will be used to test the basic objects of document classification. Parsing removes all control characters, spaces between words, dots, commas, and similar characters. In this paper, we consider a variety of data categories, such as name, phone number, date, address, title, summary and so on. The eigenvalue description for extracting given two conflicting texts is formulated. As shown below.

Type: we have defined three types of data features: time type data, number type data and text type data. For the type of current conflict text, the value of this type is 1 and the rest is 0.

Digital difference: for time type data, the characteristic value is the difference of timestamp, for digital type data, the characteristic value is the difference between numbers, and for text type data, the characteristic value is 0. After the eigenvalue is obtained, it may be quite different from other eigenvalues, so it needs to be normalized.

Similarity of character type data: Levenshtein Distance is used for calculation, and the calculated similarity value is the characteristic value.

Proportion of illegal words: calculate the proportion of illegal words in the two objects, find the absolute value of the difference between the two proportions, and take the absolute value as the final eigenvalue

After we get all the categories of object conflict clusters, we also need to mark a conflict resolution strategy for each category to explain which kind of conflict resolution strategy this kind of object conflict should apply to. After marking, when a new conflict set subsequently enters our system, the basic idea of KNN algorithm is: after a new conflict set is given, Consider k conflict sets closest (most similar) to the conflict set in the training conflict set, and determine the conflict resolution strategy of the new conflict set according to the conflict resolution strategy of the k conflict sets. The specific algorithm steps are as follows:

1) The trained conflict object vector is re described according to the feature set.

2) After the new conflict object arrives, the features of the new conflict object are extracted according to the eigenvalue extraction method, and the vector representation of the new sample is determined.

3) Select the k objects most similar to the new conflict object in the training conflict object set, and the calculation formula is:

$$Sim(d_i,d_j)=\frac{\sum_{k=1}^{M}W_{ik}\times W_{jk}}{\sqrt{(\sum_{k=1}^{M}W_{ik}^2)(\sum_{k=1}^{M}W_{jk}^2)}} \tag{2}$$

Where M represents the dimension of eigenvalue, W_{ik} represents the value of row k of vector d_i, and W_{jk} represents the value of row k of vector d_j.

4) Among the k neighbors of the new conflict object, the weight of each class is calculated in turn. The calculation formula is:

$$p(\vec{x},C_j)=\sum_{\overrightarrow{d_\iota}\in KNN} Sim(\vec{x},\overrightarrow{d_\iota})\,y(\overrightarrow{d_\iota},C_j) \tag{3}$$

Where, $\vec{x}$ is the feature vector of the new conflict object, $Sim(\vec{x},\overrightarrow{d_\iota})$ is the calculation formula of similarity, and $y(\overrightarrow{d_\iota},C_j)$ is the category function, that is, if$\overrightarrow{d_\iota}$ belongs to conflict resolution strategy C_j. Then the function value is 1, otherwise it is 0.

5) Compare the weights of several conflict resolution strategies, mark the conflict object as the conflict resolution strategy with the largest weight, and then implement the conflict resolution strategy to find the true value and complete the fusion.

Next, we will give the specific methods of conflict resolution strategies under different characteristics. The fusion strategy can be used to solve the possible conflicts and inequalities at the text level. Based on some existing common conflict resolution strategies [5], we have added some more complete and applicable strategies. They are:

Earliest time strategy: treat the value with the earliest timestamp in the conflict set as true. It is mainly used for conflict resolution that requires the latest time.

Similarity scoring strategy: when the similarity of two conflicting texts is lower than the threshold, it is considered that the predicate corresponding to the conflict set should be multi valued predicate. This strategy regards all conflicting objects as true values: if it is higher than the threshold, any one of them will be selected as true values.

Legitimacy strategy: when the proportion of illegal words in one of the two conflicting texts is much higher than that in the other. It is considered that the one with the smallest proportion of illegal words is the true value.

The scope of possible fusion strategies will not be limited to the above. The fusion strategy allows flexible management and targeted control of the creation of fusion data.

3. Results & Discussion

We used several real RDF data sets of different sizes from different fields to evaluate our method. In order to train our KNN algorithm model, we selected 1/4 data from each data set for training. At the same time, we introduce a set of evaluation indicators of data fusion to compare our method with several existing popular methods to prove that our method has better performance.

3.1. Real RDF dataset

We implement instance matching under the Ontology Alignment Evaluation Initiative (OAEI) platform Collect RDF data in instance match(IM). OAEI is a coordinated international initiative. Its goal is to evaluate the advantages and disadvantages of alignment / matching system and help improve ontology alignment / matching. The task goal of IM is to generate a set of mappings between matching instance pairs, and these matching instances are found to refer to the same real entity. Our task is to map the already formed images in IM Data fusion is carried out to form a complete and brief centralized data set. We from IM@OAEI (2010)[1] and IM@OAEI(2019)[2]has built two data sets with two types: restaurant and creative work. Each type of data has two heterogeneous data sets, and our task is to fuse the two data sets in each type.

Table 1 Details of two RDF datasets

Dataset	Number of triples	Number of instances	Number of equivalent entities
Restaurant	10380	865	112
Creative Work	22427	699	299

3.2. Evaluations

Improving completeness, conciseness and consistency are the three goals of data fusion [10]. In order to evaluate the fusion effect of our proposed method, we use these three standard measures, which can be calculated by equations 4, 5 and 6 respectively.

$$completeness = \frac{|\,unique\ objects\ in\ dataset|}{|unique\ objects\ in\ universe|} \tag{4}$$

$$conciseness = \frac{|unique\ objects\ in\ dataset|}{|all\ objects\ in\ dataset|} \tag{5}$$

$$consistency = \frac{|object\ conflict\ clusters\ without\ conflicts|}{|all\ object\ conflict\ set\ in\ clusters|} \tag{6}$$

[1] http://oaei.ontologymatching.org/2010/im/index.html
[2] https://project-hobbit.eu/challenges/om2019/

After data fusion, completeness is used to evaluate semantic loss. Conciseness is used to evaluate the introduction of redundant data. Consistency is used to evaluate the resolution of conflicting data. These three indicators can well evaluate the performance of system integration data.

We downloaded the ground truth of each data set from the IM@OAEI. This data can provide us with all common entities in the dataset through owl: sameas to facilitate us to prepare the objects to be fused. Our experiments are implemented on Python 3.6 and use rdflib package to process RDF data. All experiments are carried out on window10 system running 2.90 GHz Intel I-7 processor and 8GB memory.

3.3. Experimental results

We selected three popular RDF data fusion systems for comparative experiments. They are ObResolution [2], Truth Discovery [11], MINTE [12]. Our method is called RDFSim+KNN. ObResolution is a new algorithm for resolving object conflicts in linked data. In a unified framework, this method models all clues from source objects through a heterogeneous information network called source object network. The object conflict is solved by evaluating the quality of the source object. Truth Discovery is similar to the method proposed by ObResolution. The difference is that it uses the potential energy function between objects and proposes two more complete methods. One is the similarity algorithm for character data and the other is the similarity algorithm for digital data, which makes the system more complete. The fusion strategy proposed by MINTE is mainly based on reasoning. The advantage of this method is that the fusion index will be better, but it has a great impact on data search, because the digested data may cause search rewriting. In addition, the system does not give the object selection strategy after reasoning, which may lead to a significant reduction in the accuracy of the fused data.

We first conducted an experiment in the data set of IM@OAEI (2010), which included two data sets to be fused: dataset1 and dataset2. Then we name the data set that we add directly together the original name of the data. Such as Restaurant 1, Restaurant 2 and Restaurant. The experimental results are shown in table 2. This dataset is characterized by an object collision cluster with a similar pair of URI nodes in common URI nodes. In the current approach, few systems can identify and resolve such object conflicts. So our method is superior to other systems in resolving such conflicts. The results show that our system is 12.9% and 30.6% higher than the second place in conciseness and consistency on the basis of ensuring the completeness as much as possible. The improvement effect is very obvious, mainly because the object conflict cluster of restaurant data contains a large number of similar URI nodes, and the conflicts caused by URI nodes are difficult to be solved by other systems.

Table 2 Experimental results of Restaurant data

	completeness	consiseness	consistency
Restaurant1	66.0%	100%	100%
Restaurant2	52.4%	100%	100%
Restaurant	100%	78.1%	52.0%
ObResolution	93.8%	83.7%	62.8%
Truth Discovery	93.8%	83.7%	62.8%
MINTE	97.2%	78.8%	53.4%
RDFSim+KNN	93.8%	**96.6%**	**93.4%**

Finally, we conducted experiments in the dataset of IM@OAEI (2019). The experimental results are shown in Figure 3. We modified some objects of the predicate related to "modification time" in the data set. After modification, modify the "modification time" of the same subject in the data set In the conflict set of, the maximum timestamp is true. Accordingly, we formed a correct data set to verify the precision.

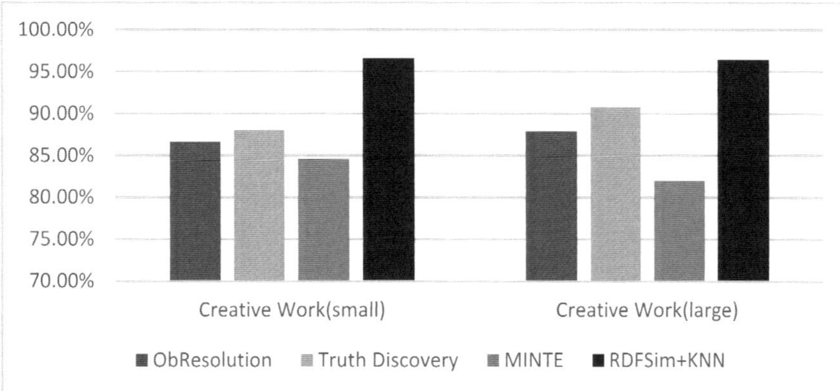

Figure 3 Comparison of precision of each system

We found that although MINTE system has good fusion performance in the previous group of experiments, this advantage is at the expense of precision. Therefore, in this group of experimental results, we can see that our system has 12% higher precision than MINTE system and obtains the highest precision of all systems. This is mainly due to our conflict resolution strategy based on KNN classification method. This method can effectively solve the problems of time conflict and illegal word conflict in data. And once the model is successfully trained, it can automatically judge the strategy, which greatly improves the accuracy of the system.

4. Conclusion

Aiming at the problem of linked data fusion, this paper first proposes the semantic representation of RDF data and object conflict cluster to extract the objects that need conflict resolution. Then, aiming at the conflict between URI node and blank node, we propose a context based similarity algorithm to identify the conflict items. For the conflict caused by literal type nodes, we propose a conflict resolution strategy based on KNN classification. Through experiments, we prove that compared with the existing methods, our system can effectively solve the conflict between URI nodes and blank nodes, and is suitable for the fusion of data sets from independent data sources, can effectively solve the conflict of literal nodes in multiple fields, and greatly improve the performance of the system to fuse RDF Data.

Data with good fusion effect can be provided to users for unified access to data residing in multiple, autonomous and heterogeneous data sources, which is conducive to the management and use of large-scale linked data.

References

[1] SONG, D., HEFLIN J. (2013) Domain-independent entity coreference for linking ontology instances. Journal of Data and Information Quality, 4 (2): 1-29.
[2] Liu W., Liu J., Duan H., Hu W., Wei B. (2017) Exploiting source-object networks to resolve object conflicts in linked data. In: European Semantic Web Conference. Cham. pp. 53-67.
[3] Song D., Luo Y., Heflin J. (2016) Linking heterogeneous data in the semantic web using scalable and domain-independent candidate selection. IEEE Transactions on Knowledge and Data Engineering, 29 (1): 143-156.
[4] Michelfeit J., Mynarz J. (2014) New directions in linked data fusion. In: Proceedings of the 2014 International Conference on Posters & Demonstrations Track. Aachen. pp. 397–400
[5] Michelfeit J., Knap T., Nečaskŷ M. (2014) Linked Data Integration with Conflicts. Computer Science, arXiv:1410.7990.
[6] Liu W. (2017) Truth discovery to resolve object conflicts in linked data. Computer Science, arXiv:1509.00104.
[7] Li Y., Bo L. (2007) Normalized levenshtein distance metric. IEEE Transactions on Pattern Analysis and Machine Intelligence, 29(6): 1091-1095.

[8] Guo G., Wang H., Devid B., Bi Y., Kieran G. (2003) KNN model-based approach in classification. In: OTM Confederated International Conferences "On the Move to Meaningful Internet Systems". Berlin. pp. 986-996.
[9] Su M.Y. (2011) Using clustering to improve the KNN-based classifiers for online anomaly network traffic identification. Journal of Network and Computer Applications, 34(2): 722–730.
[10] Dong X.L., Naumann F. (2009) Data Fusion: Resolving Data Conflicts for Integration. VLDB Endowment, 2(2): 1654–1655.
[11] Liu W., Liu J., Duan H., Hu W., Wei B. (2019) A new truth discovery method for resolving object conflicts over Linked Data with scale-free property. Knowledge and Information Systems, 59(2): 465-495.
[12] Diego C., Mikhail G., Ignacio T., Vidal M. (2017) In: MINTE: Semantically integrating RDF graphs[C]//International Conference on Web Intelligence, Mining and Semantics. New York. pp. 1-11.

Research on Multi-target Selection and Sequencing Based on Analytic Hierarchy Process

Jun Xie, Hongwei Zhang, Zhiqiang Wang, Xin Guo

Air Force Aviation University, Changchun 130022, China

xiejun1981@gmail.com

Abstract. Focusing on the characteristics of naval base targets, according to the conditions of the geographical location, the facilities size, the capacity of the station and the undertake task, the target level analysis were taken according to the comprehensive value, value and alternative target based on the Analytic Hierarchy Process(AHP). The hierarchical model which reflects the correlation of all factors was build. The quantitative analysis and evaluation of the multi-naval bases were carried out from combat capability, guarantee capability, command and control capability, reconnaissance and early warning capability and defensive capability. Multi-target selection analysis and comprehensive capability ranking of naval base are obtained.

1. Analytic hierarchy process of target selection

Objective selection method is a basic means of target selection activities. It is an effective way to achieve optimal ordering and reasonable grouping of objectives [1]. Analytic Hierarchy Process (AHP) [1-2] is a value-based target selection method. It is a systematic and hierarchical decision analysis method with qualitative and quantitative combination proposed by The American operational research scientist Satie in the early 1970s. The method takes a complex multi-objective decision making problem as a system, decomposes the target into multiple objectives or criteria, and then decomposes several levels of multiple indicators, and sorts them by calculating the weight of the levels.

When the target selection object is a target system composed of many interrelated and mutually restricted factors, there are many evaluation criteria, but there is no common scale to measure, and there is a lack of quantitative data, the AHP can be used to evaluate the target value and ranking. The method quantifies the decision maker's experience judgment, which is especially applicable to the problem of target selection which is difficult to complete quantitative analysis due to the lack of necessary data and complicated target structure.

The basic principle of using AHP to evaluate target value is as follows [2]: Firstly, hierarchical analysis is conducted on target value according to comprehensive value, sub-value and alternative target, and a hierarchical model reflecting the correlation of various factors is constructed, as shown in Figure 1. Then through the pairwise comparison of the influence of each factor on the upper factor, the relative importance of each factor is obtained. Finally, according to the hierarchical structure, the comprehensive ranking of alternative target value is obtained. The modeling process of AHP can be carried out according to the following four steps [3-5]:

(1) Establish the hierarchical structure model. According to the actual situation of the problem, related factors are decomposed into several layers from top to bottom according to different attributes, which are decision-making target layer(layer O), evaluation criteria layer(layer G), and target to be selected layer (layer M).

Content from this work may be used under the terms of the Creative Commons Attribution 3.0 licence. Any further distribution of this work must maintain attribution to the author(s) and the title of the work, journal citation and DOI.
Published under licence by IOP Publishing Ltd

(2)All pairwise comparison matrices in each level are constructed, and judgment matrices are constructed respectively by pairwise comparison of factors belonging to the same index. (Comparison of the importance of M layer elements to G layer, and Comparison of the importance of G layer elements to O layer).

(3)Hierarchical single sequencing. In order to ensure the rationality of the evaluation results, it is necessary to calculate the weight vectors of each matrix and conduct consistency test to form the hierarchical single ranking results.

(4)Hierarchical comprehensive sequencing. According to the evaluation index and weight of each factor, the total value of the target individual is calculated and the comprehensive value is sorted.

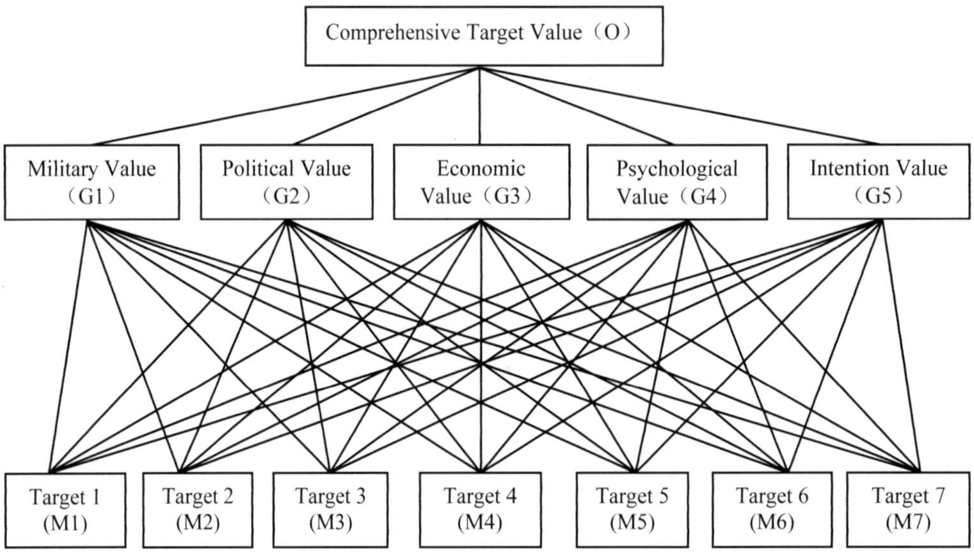

Figure 1 Hierarchical structure model of target value evaluation

2. Typical characteristics of naval base targets

A naval base is an important guarantee and support for safeguarding national territory, territorial sea and island sovereignty, carrying out various combat tasks, plundering Marine resources, safeguarding maritime economic interests and developing national economy. In recent years, many countries have formulated national policies to expand their maritime rights and interests, taking their navies as the focus of their military development and beginning to pay attention to the development and construction of coastal ports. After years of development and construction, naval bases have become the main body of the maritime defense operation system militarily and the main support for strategic countermeasures and offshore surprise operations.

Naval bases include large bases and ports, where various types of ships are stationed. They are equipped with complete port facilities and have strong maritime combat support capability. In addition, some ports and islands have built simple ports, ship berthing docks and ammunition depots and other logistics and replenishment facilities [6].

2.1 Highlight the main combat direction, important location configuration

The overall deployment of naval bases and ports is usually characterized by highlighting the main combat direction, point with area, and azimuth balance, and it can effectively control the open sea area. In terms of specific configuration, major military ports are generally located in strategic locations and military frontier areas as important strategic support points.

2.2 Relying on the construction of key cities, ensuring comprehensive protection

Most naval bases and ports rely on important coastal cities. The developed urban economy, industrial foundation and convenient transportation provide rich material and technical support for military ports, which is conducive to the rapid concentration and evacuation of various combat materials in wartime. After years of development, the naval ports have perfect general facilities and strong combat support capability. Major military ports are generally equipped with shipyards to repair all kinds of ships. There are various ammunition depots, oil depots and supply depots scattered throughout. They also have missile, mine, torpedo detection, assembly and maintenance facilities, which can provide more comprehensive logistics and technical service support for stationed vessels.

2.3 The terrain is hidden and the natural protection conditions are good

The ports of naval bases are mostly located on the peninsula and the harbors surrounded by mountains or built in estuaries. The terrain of the port area and its surroundings is hidden. Some of the harbors are surrounded by islands, which can form natural barriers. It is not only convenient for the organization of maritime and air defense, but also for mutual support between ports, which is conducive to covert attack by warships. In important ports, most of the important facilities such as battle command posts, communication stations, oil and ammunition depots are built underground or in tunnels. In and outside the port area, most of the radar stations against the sea are equipped with shore guns, anti-aircraft guns, and shore to ship and surface to air missile positions, with strong survivability.

3. Naval base multi-target selection and sequencing

Taking a certain naval base group as an example, the AHP is used to comprehensively analyze and sort the targets of the naval base group. The targets are analyzed from five aspects of combat ability, support ability, command and control ability, reconnaissance and early warning ability and defense ability, and seven naval bases in the naval base group are quantitatively analyzed and evaluated [5-7].

3.1 Operational assessment

The combat capability of a naval base can be evaluated by the type and number of its resident combat ships. Among them, the type of combat ships can reflect the ability of carrying out diversified tasks, and the number of combat ships can reflect the ability of striking. According to the supporting function of each element to the operational capability of the base, the weight distribution is that the weight of the ability to carry out diversified tasks accounts is 40%, and the weight of the striking ability accounts is 60%. By comparing the types of combat ships in each naval base, the ranking of the capability of carrying out diversified tasks is obtained, as shown in Table 1. By comparing the number of combat ships in each naval base, the ranking of striking capability is obtained, as shown in Table 2.

Table 1 Combat ship type comparison matrix

	TargetA	TargetB	TargetC	TargetD	TargetE	TargetF	TargetG	Mark
TargetA	×	0	0	0	0	1	0	1
TargetB	1	×	0	0	1	1	1	4
TargetC	1	1	×	0	1	1	1	5
TargetD	1	1	1	×	1	1	1	6
TargetE	1	0	0	0	×	1	1	3
TargetF	0	0	0	0	0	×	0	0
TargetG	1	0	0	0	0	1	×	2

Table 2 Comparison matrix of number of combat ships

	TargetA	TargetB	TargetC	TargetD	TargetE	TargetF	TargetG	Mark
TargetA	×	0	0	0	0	0	0	0
TargetB	1	×	1	1	1	1	1	6
TargetC	1	0	×	0	1	1	1	4
TargetD	1	0	1	×	1	1	1	5
TargetE	1	0	0	0	×	1	1	3
TargetF	1	0	0	0	0	×	0	1
TargetG	1	0	0	0	0	1	×	2

By weighted calculation, the ranking of naval bases' combat capability is obtained in Table 3.

Table 3 Ranking of combat capability

Ranking	Base	Mark
1	TargetD	5.4
2	TargetB	5.2
3	TargetC	4.4
4	TargetE	3
5	TargetG	2
6	TargetF	0.6
7	TargetA	0.4

3.2 Assurance capacity assessment

The support capacity of naval base can be evaluated by the size of wharfs, oil depots, ammunition depots and shipyards. Among them, the scale of wharf can reflect the docking capacity of ships, the scale of oil depot can reflect the support capacity of oil, the scale of ammunition depot can reflect the support capacity of ammunition, and the building (repairing) shipyard can reflect the maintenance capacity of bases for ships. According to the supporting function of each factor to the base support capability, the weight of ship berthing capability is 30%, the weight of oil support capability is 20%, the weight of ammunition support capability is 30% and the weight of ship maintenance capability is 20%. By comparing the size and number of docks in each naval base, the ranking of ship's berthing capacity is obtained, as shown in Table 4. By comparing the number of oil storage facilities in each naval base, the ranking of oil support capability is obtained, as shown in Table 5.

Table 4 Comparison matrix of wharf scale

	TargetA	TargetB	TargetC	TargetD	TargetE	TargetF	TargetG	Mark
TargetA	×	0	0	0	0	0	0	0
TargetB	1	×	1	0	0	1	1	4
TargetC	1	0	×	0	0	1	1	3
TargetD	1	1	1	×	1	1	1	6
TargetE	1	0	1	0	×	1	1	4
TargetF	1	0	0	0	0	×	0	1
TargetG	1	0	0	0	0	0	×	1

Table 5 Comparison matrix of oil storage facilities

	TargetA	TargetB	TargetC	TargetD	TargetE	TargetF	TargetG	Mark
TargetA	×	0	0	0	0	0	0	0
TargetB	1	×	1	1	1	0	1	5
TargetC	1	1	×	0	0	1	1	4
TargetD	1	1	0	×	0	1	1	4
TargetE	1	1	0	0	×	1	1	4
TargetF	0	0	0	0	0	×	0	0
TargetG	1	0	1	1	1	1	×	5

The ranking of ammunition support capability is obtained by comparing the number of ammunition storage facilities in each naval base, as shown in Table 6. By comparing the number of shipyards built (repaired) in each naval base, the ship maintenance capability ranking is obtained, as shown in Table 7.

Table 6 Comparison matrix of bomb storage facilities

	TargetA	TargetB	TargetC	TargetD	TargetE	TargetF	TargetG	Mark
TargetA	×	1	1	0	1	0	1	4
TargetB	0	×	0	0	0	0	0	0
TargetC	0	1	×	0	1	0	1	3
TargetD	1	1	1	×	1	1	1	6
TargetE	0	0	0	0	×	0	0	0
TargetF	1	1	1	0	1	×	1	5
TargetG	0	0	0	0	0	0	×	0

Table 7 Comparison matrix of construction (repair) shipyards

	TargetA	TargetB	TargetC	TargetD	TargetE	TargetF	TargetG	Mark
TargetA	×	0	0	0	0	0	0	0
TargetB	0	×	0	0	0	0	0	0
TargetC	0	0	×	0	0	0	0	0
TargetD	0	0	0	×	0	0	0	0
TargetE	0	0	0	0	×	0	0	0
TargetF	1	1	1	1	1	×	1	6
TargetG	0	0	0	0	0	0	×	0

In Table 8, the ranking of naval base support capability is obtained through weighted calculation.

Table 8 Ranking of support capability

Ranking	Base	Mark
1	TargetD	4.4
2	TargetF	3
3	TargetC	2.6
4	TargetB	2.4
5	TargetE	2
6	TargetG	1.5
7	TargetA	1.2

3.3 Capability assessment of command and control capability

The command and control capability of a naval base can be evaluated by the command structure and communication facilities in the base. Among them, command organization can reflect the ability of combat planning, command and battlefield control. Communication facilities can reflect the ability of communication, command and coordination. The ranking of combat planning ability, command ability and battlefield control ability is obtained by comparing the command organization levels of each naval base. By comparing the status, role and number of communication facilities in each naval base, the ranking of communication support capability and command coordination capability is obtained. The ranking of command and control capability of naval bases is obtained, as shown in Table 9.

Table 9 Comparison matrix of command and communication facilities

	TargetA	TargetB	TargetC	TargetD	TargetE	TargetF	TargetG	Mark
TargetA	×	0	0	0	0	0	0	0
TargetB	1	×	1	1	1	1	1	6
TargetC	1	0	×	0	0	1	1	3
TargetD	1	0	1	×	1	1	1	5
TargetE	1	0	1	0	×	1	1	4
TargetF	1	0	0	0	0	×	0	1
TargetG	1	0	0	0	0	1	×	2

3.4 Evaluation of reconnaissance and early warning capabilities

The reconnaissance and early warning capability of a naval base can be evaluated by the main radar facilities around the base. By comparing the geographical location, maximum detection range and detection altitude of radar facilities within 30km of each naval base, the score of reconnaissance and early warning ability is obtained, as shown in Table 10. The ranking of naval base reconnaissance and early warning ability is obtained.

Table 10 Radar facility comparison matrix

	TargetA	TargetB	TargetC	TargetD	TargetE	TargetF	TargetG	Mark
TargetA	×	0	0	0	1	0	1	2
TargetB	1	×	1	1	1	1	1	6
TargetC	1	0	×	1	1	1	1	5
TargetD	1	0	0	×	1	0	1	3
TargetE	0	0	0	0	×	1	0	1
TargetF	1	0	0	1	1	×	1	4
TargetG	0	0	0	0	0	0	×	0

3.5 Defence capability assessment

The defense capability of a naval base can be evaluated by the main air defense missile sites around the base. The defense capability score is obtained by comparing the missile performance and number of major air defense missile sites within 30km of each naval base, as shown in Table 11. The ranking of naval base defense capability is obtained.

Table 11 Comparison matrix of air defense missile positions

	TargetA	TargetB	TargetC	TargetD	TargetE	TargetF	TargetG	Mark
TargetA	×	0	0	0	1	0	1	2
TargetB	1	×	1	1	1	0	1	5
TargetC	0	1	×	1	1	0	1	4
TargetD	0	1	0	×	1	0	1	3
TargetE	0	0	0	0	×	0	1	1
TargetF	1	1	1	1	1	×	1	6
TargetG	0	0	0	0	0	0	×	0

Finally, according to the supporting function of each combat support capability to the comprehensive capability of the base, the weight distribution is as follows: combat capability accounts for 35%, support capability accounts for 20%, command and control capability accounts for 25%, reconnaissance, warning and defense capability accounts for 10% respectively. Finally, the score and ranking of the comprehensive capability of the naval base for multiple targets are obtained, as shown in Table 12. Under this condition, the comprehensive capability of naval bases is ranked as Target B, D, C, E, F, G, A.

Table 12 Comprehensive ranking list of naval bases

	Combat Ability	Support Ability	Command and Control Ability	Reconnaissance and Early Warning Ability	Defense Ability	Mark	Ranking
Target B	5.2	2.4	6	6	5	4.9	1
Target D	5.4	4.4	5	3	3	4.62	2
Target C	4.4	2.6	3	5	4	4.06	3
Target E	3	2	4	1	1	2.65	4
Target F	0.6	3	1	4	6	2.06	5
Target G	2	1.5	2	0	0	1.5	6
Target A	0.4	1.2	0	2	2	0.78	7

4. Conclusion

It is a very complicated and difficult work to select target from numerous targets, which requires a set of scientific and rigorous target selection methods. Analytic hierarchy process is a kind of assessment method based on the target value which is the combination of qualitative and quantitative, the method take the target value as starting point, evaluate the important degree of the target in target system. At the same time, the target value attribute mathematical calculation can be conducted into data, according to the target value and the political, economic, military, psychological and the relationship of the operational intention, the mathematical method were used to establish mathematical model for operation, by comparing the quantitative value to determine the choice of the target. The method is established on the basis of mathematical quantitative calculation of related factors, supported by relevant theories and technical means such as operations research and decision science. It is more complicated and difficult to use, but the conclusion obtained by using this method is more scientific and reliable, which is an important multi-objective analysis, selection and ranking method.

References

[1] Luofeng. Tracking Target Selection Based on AHP. Journal of Jining University, 2012, 33(6): 68~71.

[2] Zhao Hui,Jiang Teng, Feng Gang. Research on Target Selection of Smoke Screen Support Based on Analytic Hierarchy Process.Digital Ocean&Underwater Warfare,2018,1(1): 90~94.
[3] Wu xiaopeng, Li yaxiong, Huyao. Target Individual Value Assessment Based on DEA-AHP Model. Ship Electronic Engineering, 2020(8): 28~33.
[4] Tong haipeng, Liuxiaojing,Mayanming. Classified Sequencing Model of Target Based on Grey Relation Analysis. Fire Control&Command Control. 2010, 35(12): 93~96.
[5] Niu yuguang, Yan gaowei, Xie gang, et al. Research on granular computing based multi-objective ranking method. Computer Engineering and Applications, 2010, 46(9): 42-45.
[6] Hu Ping , Jiang Lihua. Multi-dimensional Analysis of Military Logistics Capability of Naval Bases. Logistics Technology, 2019, 38(10): 133~136.
[7] Wang Shoupeng, Liu Liang, Liu Wei. Research on Joint Operations Target Support. Command Control & Simulation, 2020, 42(4): 73~76.

Interference-Avoid Based Spectrum Sharing Approach for Loosely-Coupled Cognitive Radio Networks

Zheng Qiu, Luyong Zhang*

School of Information and Communication Engineering, Beijing University of Posts and Telecommunication, Beijing,100876, China

*Corresponding author's e-mail: lyzhang@bupt.edu.cn

Abstract. Cognitive radio (CR) network is a promising technology to solve the problem of spectrum scarcity arising from the growth of usage of wireless networks and mobile services. CR networks can find idle channels and access them opportunistically with spectrum sensing technology. In this paper, resource allocation with sensing-based interference recording is considered for several loosely-coupled cognitive radio networks, in which the base station controls the end-users' (EUs) transmission by making channel allocation strategies based on recorded interference information. We proposed a method to record this information and an allocation strategy with low computational complexity, to increase the capacity of tasks with fixed data size in the system. Numerical results show the effectiveness of our model for recording the statistical behaviour of the cognitive network interference and allocating channels for higher system capacity.

1. Introduction

With the rapid development of technologies such as the Internet of Things and 5G a large number of terminals are deployed in high-density places, but there are also a great number of spectrum holes in many frequency bands and regions. Therefore, opportunistic spectrum access together with a cognitive radio [1] technology has become a promising solution to resolve this problem.

When end-users access network resources in a hierarchical cognitive network [2], they need to monitor the surrounding channel information within a predetermined time cycle to avoid interference with the primary user. There are three main spectrum access modes for cognitive networks: opportunistic spectrum access [3-4], spectrum sharing [5-6], and perception-based spectrum sharing [7-8]. The system we proposed uses opportunistic spectrum access to allocate channels to EUs. But to take control of the dynamic interference in the area, the system also takes the advantage of perception [7].

As an emerging field, mobile edge computing provides new possibilities for the design of cognitive network systems. For example, the author of [9] has proposed that cognitive network and edge mobile computing can increase the working capacity of the cognitive network, reduce its work requirements, and increase the computing and storage capabilities in the cognitive network. Also, based on the proposed mobile edge computing concept [10], many tasks can be done with computing power at the edge, and in [11] the author jointly allocates computing resources and spectrum resources to increase the utilization rate in the mobile edge computing system and reduce its energy consumption.

In this paper, we proposed a new framework for allocating spectrum resources among multiple loosely coupled cognitive networks, accounting for the perception process, communication methods between networks, and the reuse of spectrum in space. In addition, our framework allows us to model

Content from this work may be used under the terms of the Creative Commons Attribution 3.0 licence. Any further distribution of this work must maintain attribution to the author(s) and the title of the work, journal citation and DOI.
Published under licence by IOP Publishing Ltd

and count the model of the cognitive network interference generated in a limited finite region while considering the influence of inter-network interference on the quality of the channel. Based on this framework, we use two strategies to evaluate and instruct spectrum allocation plans. Then, we proposed a spectrum resource allocation protocol for the cooperation of two methods of active interference-avoiding (AIA) and passive interference-avoiding (PIA) for specific application scenarios. By using heuristic algorithms, we optimize the complexity of the algorithm to a level acceptable to the system in the real world. The numerical results verify the effectiveness of our model in network spectrum allocation in different scenarios.

2. System model and problem descriptions

We consider a CRN consists of several loosely coupled subnetworks located in an adjacent area. Define K as the total number of frequency bands $\boldsymbol{FB} = \{\mathrm{fb}_1, \mathrm{fb}_2, \dots \mathrm{fb}_\mathrm{k}, \dots \mathrm{fb}_\mathrm{K}\}$ that are available for all base stations (BS) and EUs in this area. We assume that in this area, there are N BSs (BS)$\boldsymbol{BS} = \{\mathrm{bs}_1, \mathrm{bs}_2, \dots \mathrm{bs}_\mathrm{i}, \dots \mathrm{bs}_\mathrm{N}\}$, and M EU $\boldsymbol{EU_i} = \{\mathrm{eu}_1^\mathrm{i}, \mathrm{eu}_2^\mathrm{i}, \dots \mathrm{eu}_\mathrm{j}^\mathrm{i}, \dots \mathrm{eu}_\mathrm{M}^\mathrm{i}\}$ separately belongs to each BS bs_i. In this area, each BS bs_i and EU set $\boldsymbol{EU_i}$ managed by bs_i form a CRN network $\boldsymbol{RN_i} = \{bs_i \cup \boldsymbol{EU_i}\}$. Additionally, EU eu_j^i can only communicate with its BS ($\boldsymbol{BS_i} | eu_j^i \in \boldsymbol{EU_i}$), and at each time, eu_j^i can only use one frequency band.

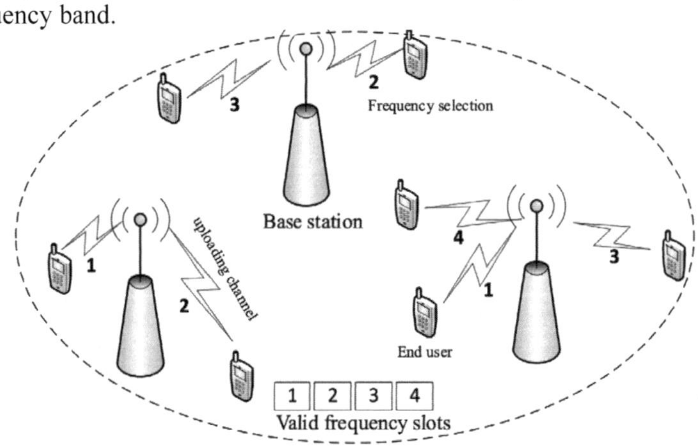

Fig. 1. Model of loosely coupled CR networks

By definition, bs_i and eu_j^i can communicate on a channel $cs_{ij}^k \in \boldsymbol{CS}$. $\boldsymbol{CS}$ means the available channel set determined by timeslot, frequency band, and location information. Leaving out timeslot that is irrelated to the channel allocation, we use bs_i, eu_j^i, fs_k to represent channel cs_{ij}^k, in which location information was indicated by bs_i and eu_j^i.

The system we proposed focuses on one-way transmission tasks from EU to BS. These tasks have strict latency requirements, so only when a pack of data is transferred to a BS in the same timeslot when the data was generated, will we consider it a successful task. The channel capacity matrix we mentioned above indicates the bitrate of the upload channel between an EU and a BS.

Reusing frequency bands in different networks will increase the usage rate of wireless spectrum. Despite that the number of channels used in the system can be multiplied by space-division multiplexing, collisions among different networks will disable many channels and significantly impact the communication process.

At the beginning of each communication time-slot, each BS will allocate a channel to each EU. We use $cc_{ij}^k \in \boldsymbol{CC_i}$ to indicate the allocation choice. When frequency band fb_k will be allocated to EU eu_i^j in network $\boldsymbol{RN_i}$ in next time-slot, $cc_{ij}^k = 1$. And $cc_{ij}^k = 0$ means channel cs_{ij}^k will not be used.

To focus on the time-sensitive uploading problem, we defined an indicator $\mathcal{R}_{valid}$ for these tasks. EUs who choose channels with a higher bitrate than $\mathcal{R}_{valid}$ will finish the task. We define $f(\mathcal{R})$ as the score of a channel in one time-slot.

$$f_{assess}(\mathcal{R}) = \begin{cases} 1, & if\ \mathcal{R} \geq \mathcal{R}_{valid} \\ 0, & if\ R < \mathcal{R}_{valid} \end{cases} \tag{1}$$

The signal sent by these EUs to their BS will interfere with other EUs in different networks. We use $\mathcal{R}^{k}_{real\ ij}$ to represent the bitrate of an uploading channel from EU eu_j to BS bs_i.

The interference power and bitrate of each channel can be defined as

$$I_i^k = P_N + P_I^k \tag{2}$$

$$\mathcal{R}^k_{real\ ij} = B_k * \log_2(1 + \mathrm{Pr}^k_{\mathrm{ij}}/\mathrm{I}^{\mathrm{k}}_i\ ) \tag{3}$$

where $\mathrm{Pr}^k_{\mathrm{ij}}$ is the received signal power and P_I^k represents the interference sent by EUs from other networks. So the utilization problem is thus:

Maximize:

$$Score = \sum_{i=1}^{N}\sum_{j}^{M}\sum_{k}^{K} cc^k_{ij} * f\left(\mathcal{R}^k_{real\ ij}\right) \tag{4}$$

Subject to:

$$\sum_{j}^{M} cc^k_{ij} \leq 1, \forall i, k, \sum_{k}^{K} cc^k_{ij} = 1, \forall i, j \tag{5}$$

3. Proposed method

We proposed a method that uses historical channel information and active interference-avoiding (AIA) schemes to instruct the BS to make channel allocation decisions. We will illustrate this method in 3 parts below.

3.1. Record and update historical information

As mentioned above, in the system, there are N*M*K potential available channels. For cognitive network $\boldsymbol{RN_i}$, we use a channel capacity matrix $\boldsymbol{\mathcal{R}_i}$ to indicate the capacity of each channel and more precisely, $\mathcal{R}^k_{ij} \in \boldsymbol{\mathcal{R}_i}$ indicates the potential transmit rate of channel cs^k_{ij}. Each matrix $\boldsymbol{\mathcal{R}_i}$ is individually maintained and used by network $\boldsymbol{RN_i}$ to make channel allocation decisions.

After each communication time-slot, BSs calculate the bitrate of each working channel based on observed signal and interference information. Using these newly generated channel statuses, we update the channel capacity matrix $\boldsymbol{\mathcal{R}_i}$ with rate α_{update}. Pseudocode is given below.

Algorithm 1: update $\boldsymbol{\mathcal{R}_i}$

Input: last channel decision matrix $\boldsymbol{CC_i}$
Input: historical and current channel capacity matrix $\boldsymbol{\mathcal{R}_i}$, $\widetilde{\boldsymbol{\mathcal{R}_i}}$
Output: historical channel capacity matrix $\boldsymbol{\mathcal{R}_i}$

for $j = 1$ to length ($\boldsymbol{\mathcal{R}_i}$) **do**
  for $k = 1$ to length ($\boldsymbol{\mathcal{R}^j_i}$) **do**
    if $cc^k_{ij} = 1$ **then**
      $\boldsymbol{\mathcal{R}^k_{ij}} = \mathcal{R}^k_{ij} * \alpha_{update} + \widetilde{\mathcal{R}^k_{ij}} * (1 - \alpha_{update})$
    end if
  end for
end for
return $\boldsymbol{\mathcal{R}_i}$

3.2. Allocate channel based on historical information

By monitoring and collecting channel connecting status, we have obtained the interference information between these networks. Next, we use this information to make the best channel allocation decision. We can use $Score$ as we proposed above to evaluate our decision, but matrix $\boldsymbol{\mathcal{R}_i}$ is just a prediction of channel status. To find best channels, we choose

$$f_p(\mathcal{R}) = \begin{cases} (\mathcal{R}/\mathcal{R}_{valid})^2, & if\ \mathcal{R}/\mathcal{R}_{valid} \leq 1 \\ 2 - \mathcal{R}/\mathcal{R}_{valid}, & if\ \mathcal{R}/\mathcal{R}_{valid} > 1 \end{cases} \tag{6}$$

as the criterion of single-channel allocation quality and

$$f_{sum}(\boldsymbol{CC_i}, \boldsymbol{\mathcal{R}_i}) = \sum_{i=1}^{N} \sum_{j}^{M} \sum_{k}^{K} cc_{ij}^{k} * f_p\left(\mathcal{R}_{ij}^{k}\right) \tag{7}$$

as the criterion of the total system. We choose this criterion to focus on choosing channels that can finish our proposed task, instead of increasing the overall signal rate of the system. Then we can get the most valid channels out of N*M*K potential available channels. This is a Non-deterministic Polynomial (NP)-hard problem, so we use simulated annealing (SA) algorithm to solve this problem:

Algorithm 2: channel allocation using SA algorithm

Input: historical channel capacity matrix $\boldsymbol{\mathcal{R}_i}$
Input: initial temperature $T_{initial}$ and cooling rate α_{cool}
Output: channel decision matrix $\boldsymbol{CC_i}$

initialize a valid channel decision matrix $\boldsymbol{CC_i}$
while $T \geq 2$ **do**
  $\boldsymbol{CC_{copy}} \leftarrow shuffle(\boldsymbol{CC_{copy}}, T)$
  $Score_{old} \leftarrow f_{sum}(\boldsymbol{CC_i}, \boldsymbol{\mathcal{R}_i})$
  $Score_{new} \leftarrow f_{sum}(\boldsymbol{CC_{copy}}, \boldsymbol{\mathcal{R}_i})$
  if $Score_{new} \geq Score_{old}$ **then**
    $\boldsymbol{CC_i} \leftarrow \boldsymbol{CC_{copy}}$
  else if $1 - Score_{new}/Score_{old} > random(0,1) * \alpha_{accept\ rate} * T$ **then**
    $\boldsymbol{CC_i} \leftarrow \boldsymbol{CC_{copy}}$
  end if
  $T_{initial} \leftarrow T_{initial} * \alpha_{cool}$
end while
return $\boldsymbol{CC_i}$

We use a function *shuffle* to indicate a random channel-reallocating process. It makes a copy of matrix $\boldsymbol{\mathcal{R}_i}$ and randomly swap between the allocation decision of T channels. If the new allocation decision is a better choice according to criterion $f_{sum}(.)$, we simply accept the new selection. And to avoid the system from being trapped by local optimizations, when we get a worse score out of the new allocation decision, we randomly accept the choice by chance. Thus, as temperature $T_{initial}$ goes down, the stride of *shuffle* and the chance of accepting worse allocation decisions will decrease and lead the system to a convergent solution.

3.3. AIA and PIA schemes

By using the A and B part of the proposed method, these networks can avoid certain interference from other networks by recording interference information and adjusting the channel allocation decisions based on this information. We call it a passive interference-avoiding (PIA) scheme.

Based on this, we proposed an active interference-avoiding (AIA) scheme. In a communication timeslot, when a BS senses interference and the corresponding channels have a lower signal rate than needed. A BS will actively emit signals in these frequency slots of specified power.

Based on the above model, the system flow and algorithm proposed in this paper are composed of the following steps:

1. According to the information of the historical channel matrix, the BS allocates channels to EUs in the network. Then the network enters the actual task data transmission stage 2. Each EU transmits tasks to the BS on the allocated frequency band. At this time, the BS will monitor the signal and interference of each channel and update the capacity matrix. 3. The BS starts active interference-avoiding measures for the channels with serious interference. 4. Each BS updates its channel interference status again and updates the capacity matrix.

4. Result and discussion

Our scenario deployed 3 networks in an area, where there are 100 available frequency slots for all BSs and EUs. This simulation does not consider power allocation, and each EU has the same transmission power. The result is represented by the number of valid channels in the CR network system. When we randomly allocated channels, there were about 160 channels available in each round of communication. By using the PIA algorithm, after 400 communication rounds, this number went up to 200 by choosing channels with less interference, increasing the utilization rate by 24%. When we applied the PIA algorithm on the system based on the AIA, we got another 13% improvement.

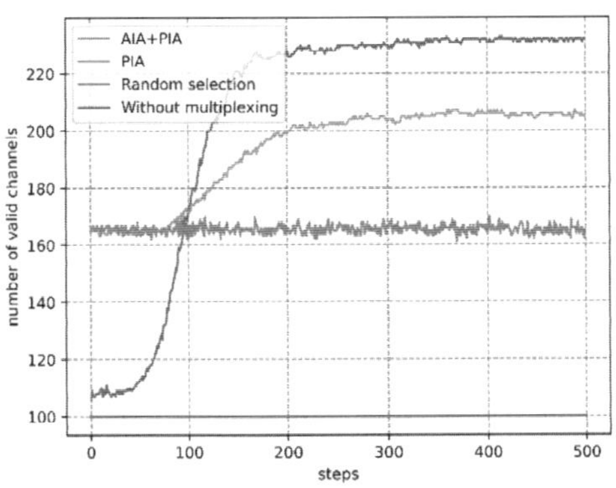

Fig. 2. Channel Allocation by AIA and PIA Algorithm

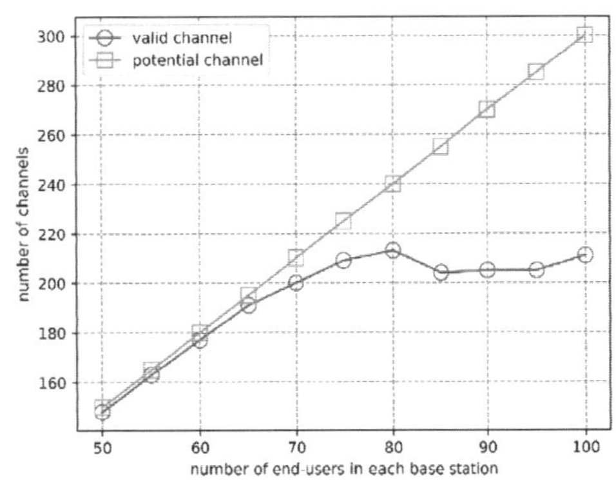

Fig. 3. Comparison of achieved under different number of EUs

As shown in figure 2 above, at the beginning of the simulation, because of the intended interference emitted by networks on interfered channels, the system started with low capacity. But after that, it converged to a higher capacity.

As shown in figure 3 above, in this simulation scenario, before we allocate more than 65 EUs for each network. After the system converged to a plateau, there was basically no interference happening in the system (the number of valid channels in the system are close to potential channels). As we increase EUs, after the system had found all available channels, the capacity of these networks was no longer related to the number of EUs.

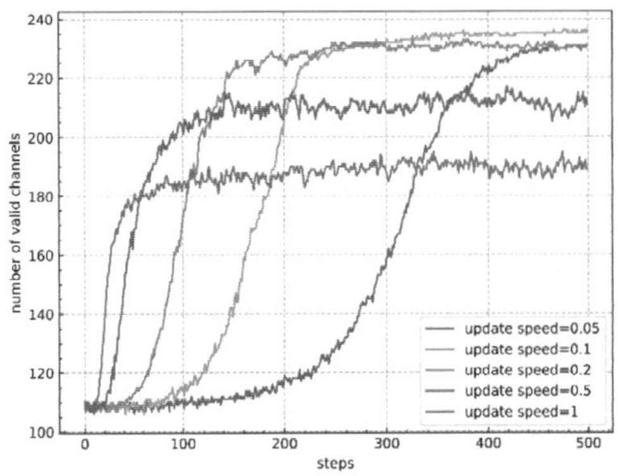

Fig. 4. Comparison of achieved capacity under different update speed

As we mentioned in algorithm 1, we use α_{update} to control the speed of updating the information in historical capacity matrix.

As shown in figure 4 above, when we set update speed as 1, we didn't record any historical information, and the system fluctuated greatly and ended with an unsatisfactory result. When we decreased the update speed, the system exchanged their information more slowly, so the processes of their convergence took more iterations. But they converged more smoothly and had better results. Better overall capacity between a period can be obtained by selecting update rate according to system transformation situations like the movement and status changes of EUs.

5. Conclusion

In this paper, we proposed two methods to reduce interference between several loosely-coupled CR networks. Without communicating about allocation strategies directly, BSs with proposed methods can share interference information using spectrum-sensing and actively emitting signals on disrupted channels. We used a capacity matrix to record historical interference information about channels and represent the future status of these channels. With historical information and simulated annealing algorithm, we managed to reduce the complexity of the channel allocation algorithm from NP to an acceptable standard and effectively increased the system capacity. Simulation results show that the proposed channel allocation algorithm improves about 30% in the achieved capacity over tasks with a fixed amount of data than the baseline algorithm with random allocation. Ideally, a system with proposed methods can find most of the valid channels in the system and maintain the performance when more EUs join the network after the system reaches its performance limit.

References

[1] Wang B, Liu K J R. (2010) Advances in cognitive radio networks: A survey. IEEE Journal of selected topics in signal processing, 5(1): 5-23.

[2] Ghosh G, Das P, Chatterjee S. (2014) Cognitive radio and dynamic spectrum access-A study. International Journal of Next-Generation Networks, 6(1): 43.

[3] Sharma M, Sahoo A. (2013) Stochastic model based opportunistic channel access in dynamic spectrum access networks. IEEE Transactions on Mobile Computing, 13(7): 1625-1639.

[4] Rawat D B, Bajracharya C, Grant S. (2017) nroar: Near real-time opportunistic spectrum access and management in cloud-based database-driven cognitive radio networks[J]. IEEE Transactions on Network and Service Management, 14(3): 745-755.

[5] Yan Z, Zhang X, Liu H L, et al. (2018) An efficient transmit power control strategy for underlay spectrum sharing networks with spatially random primary users. IEEE Transactions on Wireless Communications, 17(7): 4341-4351.

[6] Ahmad W S H M W, Radzi N A M, Samidi F S, et al. (2020) 5G technology: Towards dynamic spectrum sharing using cognitive radio networks. IEEE Access, 8: 14460-14488.

[7] Chen Z, Gao F. (2015) Cooperative-generalized-sensing-based spectrum sharing approach for centralized cognitive radio networks. IEEE Transactions on Vehicular Technology, 65(5): 3760-3764.
[8] Gao X, Feng S, Niyato D, et al. (2019) Dynamic access point and service selection in backscatter-assisted RF-powered cognitive networks. IEEE Internet of Things Journal, 6(5): 8270-8283.
[9] Kakalou I, Psannis K E, Krawiec P, et al. (2017) Cognitive radio network and network service chaining toward 5G: Challenges and requirements. IEEE communications Magazine, 55(11): 145-151.
[10] Mach P, Becvar Z. (2017) Mobile edge computing: A survey on architecture and computation offloading. IEEE Communications Surveys & Tutorials, 19(3): 1628-1656.
[11] Zhang, X., Mao, Y., Zhang, J., & Letaief, K. B. (2017). Multi-objective resource allocation for mobile edge computing systems. In: 2017 IEEE 28th Annual International Symposium on Personal, Indoor, and Mobile Radio Communications pp. 1-5.

Fault identification of fan blade based on improved ResNet-18

Hengfu Zhang[1,a], Fang Wang[1,b*]

[1]School of Electrical Engineering, Shanghai Dianji University, Shanghai, 201306, China

[a]email: 3230156194@qq.com

*Fang Wang: [b]email: wangf@sdju.edu.cn

Abstract-With the continuous development of the wind industry, wind turbine blade failure causes wind turbine shutdown, which damages the revenue of wind power. In order to more effectively detect the damage on the surface of wind turbine blade. By comparing ResNet-18 /50, VGG-19 and GoogLeNet models, ResNet-18 has the advantages of short training time and high accuracy. Therefore, ResNet-18 convolutional neural network is selected as the basic network, and a deep learning training and detection model is built based on MATLAB. In order to further optimize the model, multiple convolution kernel is selected for the main branch of residual structure of ResNet-18 to extract features, and add 3×3 pool layer, it is used to improve the learning ability of the model. In addition, Aadm, Sgdm and Rmsprop optimization algorithms are added to further improve the robustness of the model. Experiments show that the classification and recognition accuracy of the improved ResNet-18 model for fan blades is 94.17%, which is 2.5% higher than that of the original model. The improved ResNet-18 model has good recognition effect,it improves a faster and more accurate solution for fan blade surface damage identification in complex weather environment.

1. Introduction

With the rapid growth of the wind power industry, the failure of fan blades is also increasing, so that maintenance costs are rising, the failure detection of fan blades also needs higher accuracy, of which convolutional neural network in the fan blade image processing is the most widely used[1].Aiming at the problem of fan blade fault classification, based on the existing deep network ResNet-18 [2], this paper adds multiple convolution and maximum pool structure in the main branch, and adds optimization algorithm. The model has achieved good results on the collected four types of fan blade fault data sets

In recent years, deep learning has been more and more applied in fan blade detection, and has achieved good results.In document [3], LIU, etc. proposed deep learning methods based on convolutional neural networks,short-term and long-term memory networks and hybrid networks.Experiments have effectively predicted stiffness residues under leaf fatigue.In document [4],Sarkar Dipu, for all, uses ultra-resolution convolutional neural networks to reconstruct blurry images into high-resolution images. The results show that YOLOv3 is superior to the traditional model in precision and processing time.In document [5],Wang Yinan et al. propose a method to distinguish between normal and abnormal parts of the blade by combining a single type of support vector machine with the depth characteristics learned from the common image dataset

Content from this work may be used under the terms of the Creative Commons Attribution 3.0 licence. Any further distribution of this work must maintain attribution to the author(s) and the title of the work, journal citation and DOI.
Published under licence by IOP Publishing Ltd

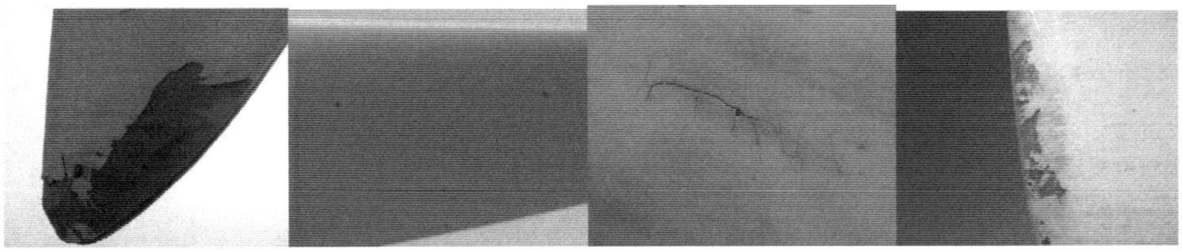

(a) trachoma (b) crack (c) corrode (d) coating shedding

Figure 1 Fan blade fault image

2. Data set

This paper screened and collated the photos taken by the UAV in the wind farm, creating a data set, including four faults: trachoma, crack, corrosion and coating shedding, each fault has 100 images. Some examples of wind turbine blades are shown in Figure 1.

In practical application, fan blades are usually in a complex environment, which is affected by factors such as weather and wind farm equipment noise. For the influence of complex background such as weather, the background factors affecting the algorithm are segmented by image segmentation method. For the influence of noise, Gaussian filter is used to remove image noise. At the same time, due to the geographical location of the wind farm, there are differences in the number of images of various faults of wind turbine blades, which is easy to cause the problem of over fitting in the model. In order to make the data set in this paper conform to the application in complex environment as much as possible, and improve the generalization ability of the network. In this paper, the images of the data set are flipped horizontally, vertically and randomly, so as to increase the diversity of the image data set. Finally, the image of 5184 pixels×3456 pixels is adjusted to 224 pixels×224 pixels of RGB images by MATLAB's Imresize function, and according to 7:3 division of the training set, test set.

3. Improve convolutional neural networks

3.1 The structure of convolutional neural networks

The main framework of convolutional neural networks consists of three layers: convolutional, pooled, and fully connected. Convolution layer is the characteristic extraction of fan blade image, pooling layer is the characteristic image after convolution to reduce dimension processing, avoid overfitting, the whole connection layer uses the quantitative parameters, so as to obtain a higher accuracy of the output[6].

3.2 Improve the ResNet-18 model structure

The error generated in the network operation becomes the gradient error in continuous training, and the gradient error will form a gradient explosion exponentially under the weight of continuous imbalance. The residual structure is beneficial to deepen the number of network layers and solve the problem of gradient explosion.This paper uses ResNet-18 residual network as the basic network, and the specific structure is shown in Figure 2.Firstly, the convolution layer feature extraction operation is used to extract image features, and then the maximum pool layer is used to reduce the feature size to prevent over fitting and speed up the calculation speed. Finally, the high-level feature information is extracted through the residual structure to prevent gradient explosion, and the extracted feature information is input to the fully connection layer for classification output[7].

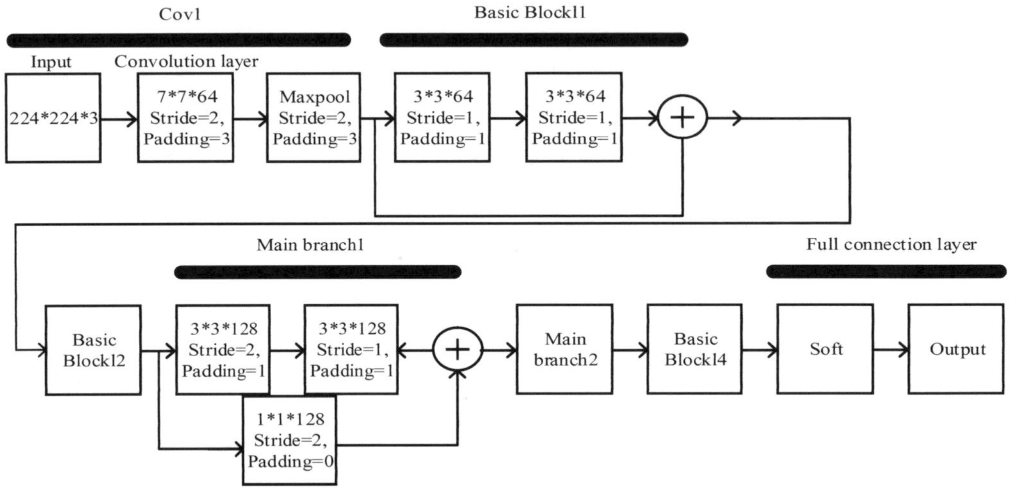

Figure 2 ResNet-18 model

In the ResNet-18 model, shortcut branch is used to change the feature dimension, this operationavoids the degenerative effect of the model, the main branch structure mainly uses the characteristics of different levels, further extracts the deep features, helps to solve the problem of gradient disappearance and gradient explosion, lets us train deeper network at the same time, but also can guarantee good performance.

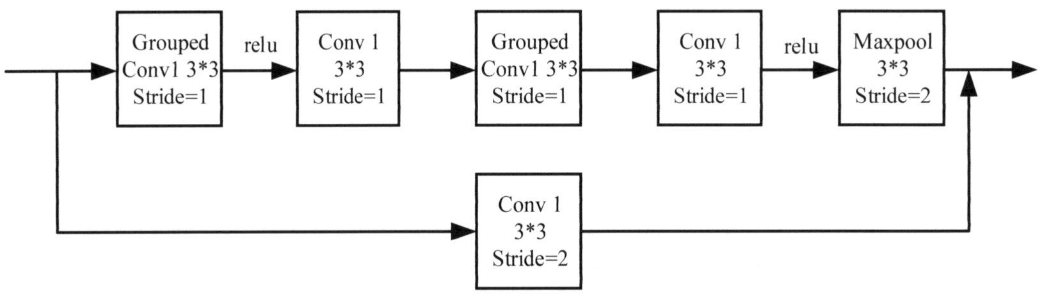

Figure 3 Main branch structure is improved

The main branch of the residual structure is optimized, as shown in Figure 3, by replacing the convolution core with a step of 1 of 3×3 and a convolution core of 1×1, replacing the 3×3 convolution core with a step of 1, and after the relu function to add a step of 2 3×3 pooling layer, through multiple convolution can effectively retain the characteristics of the image, and can effectively improve the calculation speed, only modify the structure of the main branch, can effectively improve, the generalization ability of the model is better, indicating a single 3×3 Convolution core feature extraction performance is insufficient.

3.3 Optimize the algorithm

The full name of Rmsprop[8] algorithm is Root Mean Square Prop. Based on the preliminary solution of the problem of large swing amplitude in optimization by momentum optimization algorithm, Rmsprop algorithm uses differential square weighted average for the gradient of weight and bias in order to further optimize the problem of excessive swing amplitude in the update of loss function and further accelerate the convergence speed of function.

The full name of Adam[9] is Adaptive Motion Estimation. Adam combines the optimal performance of Momentum and Rmsprop algorithms. Adam algorithm can solve the problem of sparse gradient and noise by calculating the exponential moving average of gradient and square gradient and controlling the attenuation of moving average.

4. Experimental results and analysis

4.1 Experiment with the test platform

The experimental platform is amD Ryzen 5 3600X CPU, 3.8GHz processor, and the system runs with 16GB of memory. The operating environment is Windows 10, the software platform is Matlab, and the classification model is implemented by Matalb Deep Learning Framework Training.

4.2 Analysis of experimental results

The experimental image input size is set to 224 pixels×224 pixels of RGB image, image data set randomly selected 70% of fan blades as training data set, the remaining 30% as the validation set. In order to verify the influence of optimization algorithm on the model, three optimization algorithms, Sgdm, Adam and Rmsprop, were used to test the optimization algorithm of ResNet-18/50, VGG-19, GoogLeNet and import ResNet-18. In order to reduce the impact of hardware on the experiment and ensure the effectiveness of the experimental results, the experiment is repeated three times at a time.

Table 1 Results of the experiment

The experiment number	The type of model	Optimal optimization algorithm	Verify the correct rate/%
1	ResNet-18	Sgdm	91.67%
2	ResNet-50	Adam	89.17%
3	VGG-19	Adam	67.50%
4	GoogLeNet	Sgdm	83.33%
5	Import ResNet-18	Sgdm	94.17%

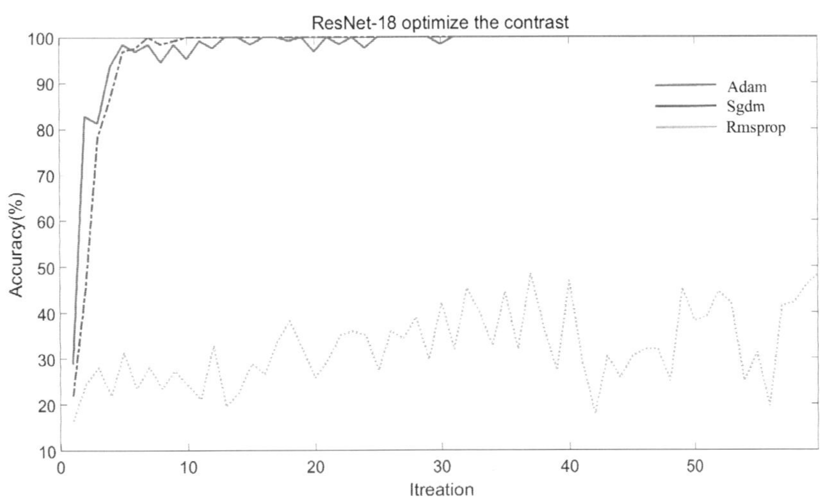

Figure 4 ResNet-18 optimizes algorithm comparison

As shown in Figure 4, in the ResNet-18 algorithm, Sgdm trains to converge faster than Adam and is more accurate than Rmsprop. Therefore, Sgdm was chosen as the optimization algorithm of ResNet. On this basis, the optimization algorithm screening of ResNet-50, VGG-19, GoogLeNet and import ResNet-18 models is carried out, and the optimization algorithm with the highest training convergence and accuracy is selected, as shown in Table 1.

In this paper, the improved ResNet-18 model compared with other models, as shown in Figure 5, it can be seen that the ResNet model with multiple convolution structure did not reach the first in the convergence of training, because multiple convolution increased the amount of calculation, through Sgdm optimization algorithm in Table 1 accuracy is the highest, compared with the existing algorithm accuracy has a certain improvement.

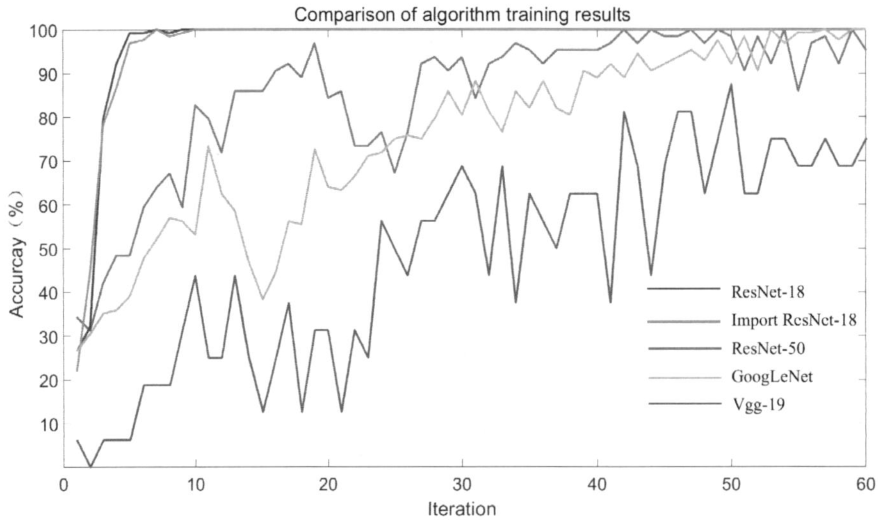

Figure 5 Comparison of various algorithms

5. Conclusion

In view of the problem of fault detection of fan blades, an improved ResNet-18 model is constructed, and the image features of each classification can be extracted efficiently by improving the main branch feature extraction method of the model, and the accuracy of 94.17 percent is obtained on the data set by constantly adjusting the optimization algorithm and learning efficiency. With the continuous development of wind power industry, in order to improve the accuracy of fan fault classification, it is necessary to further enrich the training data set and optimize the network structure and training methods.

References

[1] FLORIAN, MIHAI, SORENSEN, JOHN, DALSGAARD.(2017)Case Study for Impact of D-String (R) on Levelized Cost of Energy for Offshore Wind Turbine Blades. International journal of offshore and polar engineering., 27: 63-69.
[2] Wang, C. F., Ni, P. C., Mao, Y.(2021)Research on crop disease recognition based on Multi-Branch ResNet-18. Journal of Physics: Conference Series., 1961:
[3] Liu, H. W., Zhang, Z. C., Jia, H. B.(2020)A novel method to predict the stiffness evolution of in-service wind turbine blades based on deep learning models. Composite structures., 252:
[4] SARKAR, D., GUNTURI, S. K.(2020)Wind turbine blade structural state evaluation by hybrid object detector relying on deep learning models. Journal of Ambient Intelligence and Humanized Computing., 12: 8535-8548.
[5] WANG, Y. A., YOSHIHASHI, R. T., KAWAKAMI, R.(2019)Unsupervised anomaly detection with compact deep features for wind turbine blade images taken by a drone. IPSJ Transactions on Computer Vision and Applications., 11: 1-7.
[6] JANKE, J., CASTELLI, M., POPOVIC, A.(2019)Analysis of the proficiency of fully connected neural networks in the process of classifying digital images. Benchmark of different classification algorithms on high-level image features from convolutional layers. Expert Systems with Application., 135: 12-38.

[7] Moussa, G. S.,Owais, M.(2021)Modeling Hot-Mix asphalt dynamic modulus using deep residual neural Networks: Parametric and sensitivity analysis study. Construction and Building Materials., 294:
[8] Sang, K., Lee, H., Lee, J., Back, K., An, Y. Y., Kiho, Y., Sungdae, k., Sung, H.(2021)Prediction of tire pattern noise in early design stage based on convolutional neural network. Applied Acoustics., 172:
[9] Wang, Y. j., Zhou, P. Y., Zhong, W. Y.(2018)An Optimization Strategy Based on Hybrid Algorithm of Adam and SGD. Proceedings of 2018 2nd International Conference on Electronic Information Technology and Computer Engineering., 18: 630-633.

Authenticated encryption chip implementation against side-channel attack

Xiaoyu Ma[1,a], Fan Zhang[2,b*] and Yan Han[1,c]

[1]College of Information Science & Electronic Engineering, Zhejiang University, Hangzhou, Zhejiang, 310027, PRC

[2]College of Computer Science and Technology, Zhejiang University, Hangzhou, Zhejiang, 310027, PRC

[a]email: xymahust@163.com, [c]email: hany@zju.edu.cn

[b*]Corresponding author's e-mail: fanzhang@zju.edu.cn

Abstract. An authenticated encryption chip with novel nonce generation circuit was developed. This circuit associates the nonce generation with the contents and receiving time of plaintext. It will also generate overlapping power compensation for the whole chip. The integrated chip was fabricated under SMIC 180nm technology. Under 1.8V V_{DD}, 100MHz global clock, the power consumption is about 14mW within 50k gates. The test result exhibits more than 10 times the strength in resistance to side-channel attack than the unprotected version without increasing hardware cost.

1. Introduction

Side-Channel Attacks (SCA) is one of the most important security problems faced by authenticated encryption (AE) hardware implementation. When the cryptographic implementation works it is unavoidable that information leaks. An attacker can then exploit the leakage and destroy the confidentiality of data. Side-channels that can leak information for instance time, electromagnetic, and power. Among them, power analysis and electromagnetic analysis are most commonly used. Their principles are the same, so it is enough to analyse one of them. To reveal the secret key of hardware implementation, they both analyse a large number of traces that have been recorded while the device is processing[1]. They are universally applicable methods since they have the advantages that no in-depth knowledge of the device is required. When the implementation is functioning, the internal information processing will have a certain connection with the power consumption. The attacker can deduce the hardware internal data information by analysing the power or electromagnetic changes of the implementation, then crack the secret key[2]. The two most common way of power analysis are Differential Power Analysis (DPA) and Correlation Power Analysis (CPA). The former analyses the rationality of guessing the key by calculating the differential value. The latter cracks the key by calculating the correlation coefficient.

In this article, we chose the unprotected version of ASCON as authenticated encryption prototype. An authenticated encryption chip with power compensation and novel nonce generation circuit was generated. Two methods, "Reversed Grouping Padding (RGP)" and "Appended Random Delay (ARD)" were proposed to improve hardware resistance to SCA. These methods were realized by modifying the nonce generation mechanism without adding complex peripheral circuits. Our integrated chip was

Content from this work may be used under the terms of the Creative Commons Attribution 3.0 licence. Any further distribution of this work must maintain attribution to the author(s) and the title of the work, journal citation and DOI.
Published under licence by IOP Publishing Ltd

fabricated under SMIC 180nm technology. The test result exhibits 10 times the strength in resistance to side-channel attack than the unprotected version with few increase in hardware cost.

2. Side-channel attack on ASCON

2.1. Basic theory of ASCON

The mode of operation of ASCON is based on duplex sponge modes like MonkeyDuplex. It is a well-known authentication encryption algorithm with superior performance for lightweight cases. ASCON contains four phases: *Initialization*, *Associated Data*, *Encryption/Decryption*, and *Finalization*, as figure 1 shows. The round transformation is a sponge-like SPN-based permutation p. Each round p contains constant-addition operation p_C , substitution layer p_S and linear diffusion layer p_L . Since the implementation of ASCON is not the focus of this article, here is only a brief introduction. Those who are interested in related content can refer to [3].

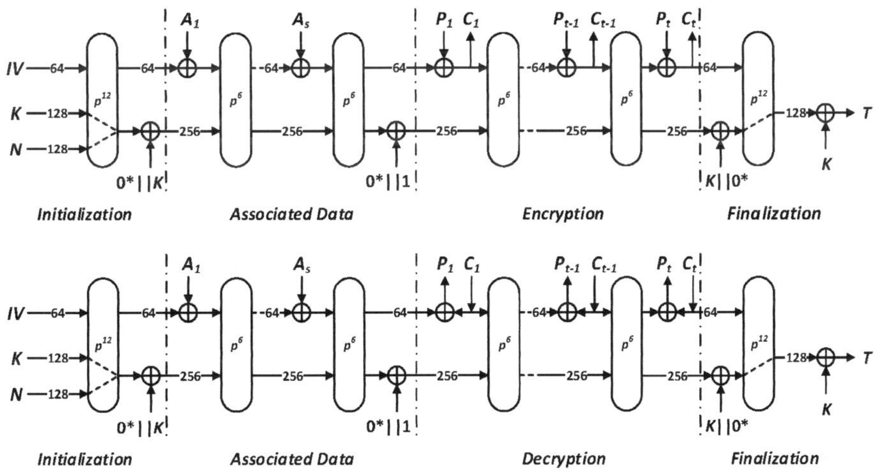

Figure 1. mode of operation of ASCON

2.2. Related work

An unprotected hardware implementation of ASCON and a Threshold Implementation (TI) variants are given in [4]. They were completed to the back-end routing stage under 90nm UMC technology with global clock of 1MHz and 1V power supply. For the unprotected version, about 1000 traces are needed for a successful attack. With about 3100 traces the success rate is higher than 99% on average. For the TI variants, SCA cannot seem to analyse the correct key. This masking method performs pretty well. However, the TI variants needs four times of power and area than the unprotected version.

The work of [1] is very systematic. It not only introduces the implementation of ASCON unprotected version, but also gives several SCA-protected variants, including Domain-Oriented Masking (DOM), Unified Masking (UMA) and Low-Latency Masking (LOLA). Designers can make a trade-off among area, throughput, randomness and latency according to the requirements of SCA protection and application scenarios. Since this work has a lot of crosswise tasks, each hardware implementation stays in front-end synthesis stage under 90nm UMC technology with global clock of 1MHz and 1V power supply. Although only netlist-level implementation results are available, this article still provides constructive guidance to designers who want to study the SCA protection of ASCON in detail.

[5] describes the SCA principle of ASCON. For the single S-box implementation, the non-linear transformation output is selected as the analysis point for the attack. For the multiple parallel S-box structure, the output of round transformation is taken as attack point. The masking protected version of ASCON needs about 150k traces to attack. Same as [4], about four times of power and hardware cost are needed for implementation.

Summarizing other state-of-the-art works, there have been many studies on the hardware implementation of ASCON, and there are also sufficient theories and schemes for its SCA research. These studies provide constructive support for our research. Our purpose is to proposed a chip-level implementation with certain safety improvements, but don't increase hardware cost.

2.3. Side-channel attack on ASCON

In order to improve hardware security, we need to analyse the SCA security risks of the algorithm. The most vulnerable phase is the *Initialization* for ASCON since three out of five input words (x_0, x_3, x_4) of state S are publicly known and the other two words (x_1, x_2) belong to the key. Therefore, the acknowledged power analysis attack on ASCON occurs in the *Initialization* phase. For 64 parallel S-boxes, it is impractical and inaccurate to analyse the output of a single nonlinear S-box separately. So, we chose the first whole round transformation of permutation p to analysis in a single clock cycle. Since the contents of the state at *Initialization* are known except for the key K (x_1, x_2), the attacker can vary the nonce N as different input and observe the output of the first-round transformation. Here we pick register x_0 as the analysis object to explain the attack on ASCON in detail.

The linear diffusion layer step for x_0 in ASCON is:

$$\sum_0 (x_0') = x_0' \oplus (x_0' >>> 19) \oplus (x_0' >>> 28) \tag{1}$$

Where x_0' is the input of linear step p_L. The output of the non-linear S-box for x_0 is:

$$x_0' = x_1x_4 + x_3 + x_1x_2 + x_2 + x_1x_0 + x_1 + x_0 \tag{2}$$

(2) can be rewritten to one quadratic term:

$$x_0' = x_1(x_4 + x_2 + x_0 + 1) + x_3 + x_2 + x_0 \tag{3}$$

We know that x_0 is constant of the IV, x_1 and x_2 are constant of the key K, x_3 and x_4 are variable of the nonce N. We remove the words that add a constant amount to the activity to the register and the equation turns to:

$$x_0' = x_1(x_4 + 1) + x_3 \tag{4}$$

Since each word x_i is 64 bits in ASCON, we combine equations (1) and (4) and change it into a register-like expression:

$$S_i(N, K^*) = k_0^*(n_{i+64} + 1) + n_i + k_1^*(n_{i+109} + 1) + n_{i+45} + k_2^*(n_{i+100} + 1) + n_{i+36} \tag{5}$$

From (5) we can see that one function contains 3 key bits thus there are 8 key guesses. In [6], a result shows that using fifty thousand traces results in a success rate of almost 1. That means we can obtain the key information of x_1 by attack on register x_0, which is half of the secret key.

Similar to this derivation, we further analyse register x_1. The linear diffusion layer step for x_1 in ASCON is:

$$\sum_1 (x_1') = x_1' \oplus (x_1' >>> 61) \oplus (x_1' >>> 39) \tag{6}$$

The output of the non-linear S-box for x_1 is:

$$x_1' = x_2x_3 + x_4 + x_1x_3 + x_3 + x_1x_2 + x_2 + x_1 + x_0 \tag{7}$$

We change (7) like x_0:

$$x_1' = x_3(x_2 + x_1 + 1) + x_4 + x_2x_1 + x_2 + x_1 + x_0 \tag{8}$$

$$x_1' = x_3(x_2 + x_1 + 1) + x_4 = x_3(x_{12} + 1) + x_4 \tag{9}$$

Combine (6) and (9), we get:

$$S_i(N, K^*) = n_i(k_0^* + 1) + n_{i+64} + n_{i+3}(k_1^* + 1) + n_{i+67} + n_{i+25}(k_2^* + 1) + n_{i+89} \tag{10}$$

Same as x_0, we can attack on register x_1 to get the key candidates k^*. The difference is this time k^* does not represent the key itself, but the XOR of the corresponding key bits in the register x_1 and x_2. The attacker cannot directly obtain the other half of the key x_2, but can calculate x_2 through XOR of the known x_1 and x_{12} mediately. [6] shows that also fifty thousand traces are needed to get key guesses at a success rate of almost 1. That means by using power analysis, an attacker can possibly get all the key information from ASCON.

3. Detail of realization

The block diagram and data path of our integrated chip is shown in figure 2.

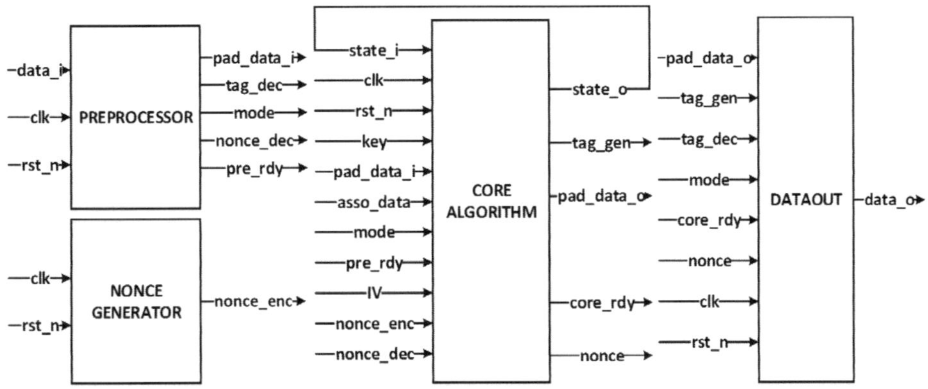

Figure 2. Block diagram of integrated chip

As analysed in section 2. The power analysis relies on the fact that the power dissipation of hardware is correlated to its activity, and associates with the key-dependent data being processed. Two mainstream countermeasures have been proposed in prior literature: hiding and masking[7]. After carefully studying the principles of these two methods, we noticed that the former is to hide the processed data into indistinguishable power patterns, which decreases the dependency between the power model and the physical dissipation[8]. The latter involves masking the key-dependent data with random variables, which attempts to reduce the dependency between the processed data and the power model mainly at algorithmic or gate level. The masking method usually needs several times hardware resources than the unprotected implementation[9]. In our design, hiding method was used as chip-level countermeasure since it saves hardware cost. It can also be used together with masking for further improvements.

We know that in practice, the *NONCE GENERATOR* module is needed to generate nonce N. This module contains a great deal of calculation processing, thus can provide the power patterns naturally. Therefore, we need *NONCE GENERATOR* to generate nonce and at the same time as a power compensation module. The real problem is that if the 128-bit nonce is directly generated, this part of the power compensation will have a simple repetitive rule, which can be easily eliminated by the attacker in a statistical way. Moreover, according to section 2, nonce needs to participate in ASCON's SCA as a known quantity. By analysing the law between power consumption and time, it is even possible for an attacker to infer the law of nonce generation, thereby inferring the value of nonce in each encryption process.

3.1. "Reversed Group Padding" method

In order to solve the existing problems, we came up with a "Reversed Group Padding (RGP)" method. The nonce generation is based on the principle of Linear Feedback Shift Register (LFSR). We didn't directly use 128-bit LFSR to form 128-bit Deterministic Random Bit Generators (DRBG) since the hardware cost is unfavourable. The method we take is shown in figure 3. Two 16-bit *DRBG* modules employs different seed and XOR logic. Once the chip is powered on, they start to work synchronously. Under the control of the *Counter*, the variational output of them are sent to 128-bit nonce simultaneously and reversely. That is why this method was named like this. In this case, hardware cost is reduced by about 70%. Two 16-bit generator's power is superimposed on each other, and the correlation with the nonce decreases. It is no longer possible for the attacker to infer the nonce generation rules by analysing the relationship between power consumption and time. Besides, this power compensation is of the same magnitude as the valid data, which is much larger than the noise magnitude. It is extremely difficult to eliminate the power compensation by statistical methods.

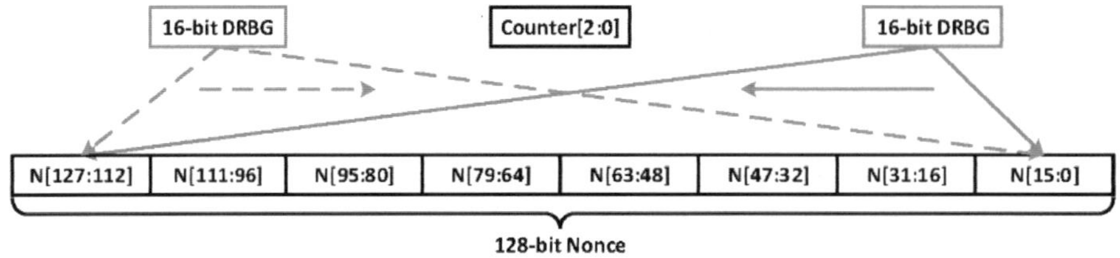

Figure 3. "Reversed Grouping Padding" method

3.2. *"Appended Random Delay" sampling solution*

On this basis, we further proposed an "Appended Random Delay (ARD)" sampling solution. We note that only one nonce value will be used in each encryption process. If the pseudo-random nonce is sampled at a fixed point each time and used for encryption, the attacker can easily find the rule. To deal with this problem, we proposed a nonce generation principle used in encryption as depicted in figure 4. Where, i, j, k are any three bits in the plaintext, which can be arbitrarily specified by the designer without affecting the throughput. During encryption, when the kth bit of plaintext is detected, the start point of nonce sampling, T_S, is set, but sampling will not start at this time. Meanwhile, i^{th}, j^{th}, k^{th} bits of plaintext form an initial value for the counter and decrease to 0, resulting in a nonce sampling delay time T_D. Therefore, the real nonce is extracted at $T = T_S + T_D$. This series of calculation will not affect the throughput of the chip because it is performed in parallel during the data reception process. It can be seen that the nonce used in the encryption process not only relies on the arrival time of the plaintext, but the value of plaintext as well. For the attacker, the nonce becomes uncontrollable and this greatly increases the attack complexity.

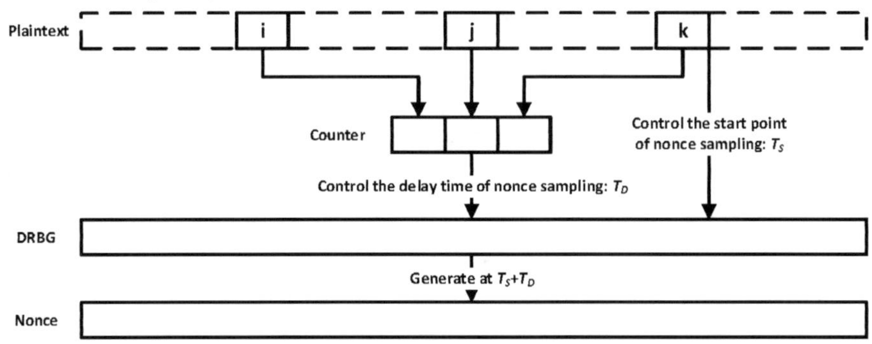

Figure 4. "Appended Random Delay" sampling solution

4. Results and discussion

We apply these designs to the unprotected ASCON AE hardware implementation. The whole AE chip with certain protection was fabricated under SMIC 180nm technology and packaged in SOP-28. The optical photo of chip is shown in figure 5. Under 1.8V V_{DD} and 100MHz global clock, the chip exhibits a power consumption of about 14mW within 50k gates. In this case, it reaches a high throughput of approximately 1Gbps, which is the same as the unprotected version. This proved that these methods will not affect throughput of the implementation. Besides, side-channel attack was presented on the implemented chip. As mentioned in section 3, we have made RGP and ARD treatment for power compensation and nonce generation. The attacker cannot directly obtain the nonce corresponding to each traces of power analysis. In order to proceed the attack, we lowered the difficulty by letting the attacker know the nonce corresponding to each round of trace generation. The test method was stated in subsection 2.3. We applied 64 attacks on x_0 and 64 attacks on x_1 to analyse all 128 bits of possible key. Each attack corresponds to 8 possible keys, which means 8 different correlation traces could be obtained. As an example, figure 6 shows the correlation traces corresponding to the first 9 bits analysis of x_0. The trace with highest correlation is marked in red, which correspond to the most likely correct key, and the others are marked in grey. It's easy to recognize that the trace with highest correlation have no significant differences with other traces. In addition, the maximum value in the red curve is not prominent, which indicates that the sampling point has little correlation with the statistical calculated value, thus cannot reliably crack the correct key.

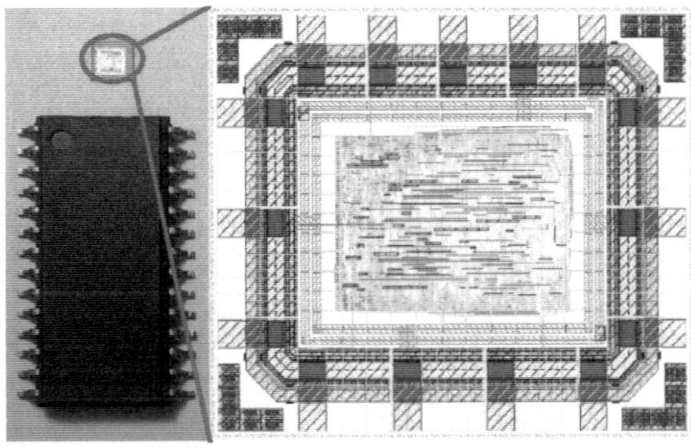

Figure 5. Optical photo of chip

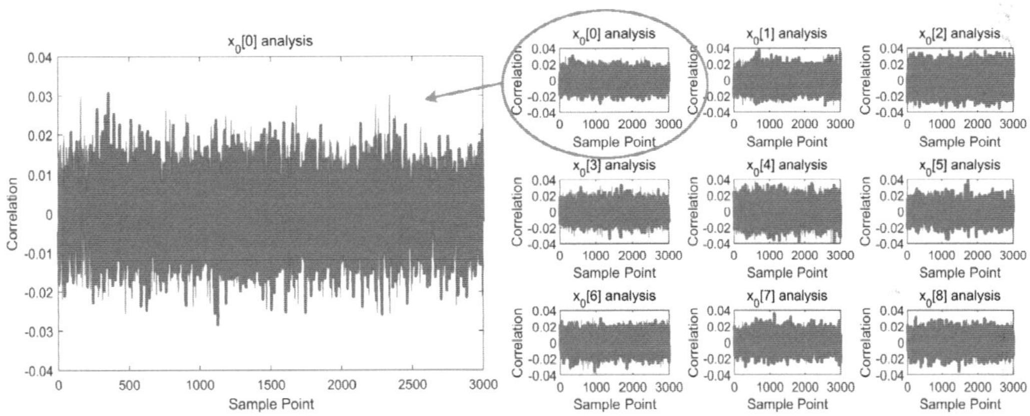

Figure 6. Correlation-based electromagnetic analysis

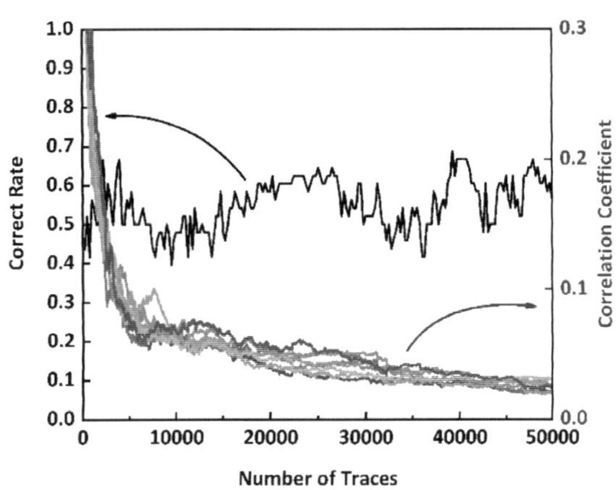

Figure 7. Correct rate and correlation coefficient vs. number of traces

We further calculated the correlation coefficient and guessing correct rate for all guessed key under different number of traces. The results are depicted in figure 7. The curve above the image shows the relationship between statistical speculation accuracy of all guess keys and trace number, which corresponds to the left Y axis. The curves group below the image reveals the relationship between correlation coefficient and trace number. Each curve corresponds to a key guess and the ordinate matches the right Y axis. It can be seen that the statistical accuracy of key guessing is maintained at about 50%, and there is no obvious difference between the correlation coefficients under different guessing keys. These results show that as high as 50k traces is not enough to cause the leakage of key information for our proposed chip. While in [4], however, the unprotected implementation only needs about 3100 traces to achieve crack success rate greater than 99%. By the way, these methods will not increase additional hardware consumption since the number of registers for nonce generation has been reduced from 128 to 38.

5. Conclusions

An authenticated encryption chip with novel nonce generation circuit was developed. This integrated chip was fabricated under SMIC 180nm technology. The power consumption is about 14mW within 50k gates under 1.8V V_{DD}, 100MHz global clock. Two methods, RGP and ARD was proposed and applied for power compensation and nonce generation. The test result exhibits 10 times the strength in resistance to side-channel attack than the unprotected version. This method is easy to implement. Although it cannot completely solve the SCA security problem in principle, it can create lots of difficulties for the attacker's analysis process. Most importantly, it does not cause additional hardware cost and can be shared with other SCA countermeasures. It has a wide range of application value in encryption and security hardware implementations.

References

[1] Groß, H. (2018). Domain-Oriented Masking: Generically Masked Hardware Implementations.
[2] Avital, M., Dagan, H., Levi, I., Keren, O., & Fish, A. (2014). DPA-secured quasi-adiabatic logic (SQAL) for low-power passive RFID tags employing S-boxes. IEEE Transactions on Circuits and Systems I: Regular Papers, 62(1), 149-156.
[3] Dobraunig, C., Eichlseder, M., Mendel, F., & Schläffer, M. (2016). Ascon v1. 2. Submission to the CAESAR Competition.
[4] Gross, H., Wenger, E., Dobraunig, C., & Ehrenhöfer, C. (2017). Ascon hardware implementations and side-channel evaluation. Microprocessors and Microsystems, 52, 470-479.
[5] Samwel, N. (2016). Side-Channel Analysis of Keccak and Ascon.
[6] Tezcan, C. (2020). Analysis of Ascon, DryGASCON, and Shamash Permutations. International Journal of Information Security Science, 9(3), 172-187.
[7] Zhang, F., Yang, B., Yang, B., Zhang, Y., Ren, X., Bhasin, S., & Ren, K. (2020). Design and Evaluation of Fluctuating Power Logic to Mitigate Power Analysis at the Cell Level. IEEE Transactions on Computer-Aided Design of Integrated Circuits and Systems, 40(6), 1063-1076.
[8] Levi, I., Bellizia, D., Bol, D., & Standaert, F. X. (2020). Ask Less, Get More: Side-Channel Signal Hiding, Revisited. IEEE Transactions on Circuits and Systems I: Regular Papers, 67(12), 4904-4917.
[9] Diehl, W., Abdulgadir, A., Farahmand, F., Kaps, J. P., & Gaj, K. (2018). Comparison of cost of protection against differential power analysis of selected authenticated ciphers. Cryptography, 2(3), 26.

Research on ultra-tight integrated navigation method based on CKF filtering algorithm

Jiafan Zhao [1,a], **Xin Zhao** [1,b], **Quanhu Li** [1,c], **Jingyu Zhang** [1,d]

[1]College of Electronic Information Engineering, Inner Mongolia University, Hohhot, Inner Mongolia, China

[a]email:13132257987@163.com,[b]email:zx_buaa@sina.com,[c]email:eelqgh@imu.edu.cn, [d]email:269253194@qq.com

Abstract. This paper takes the SINS/BDS ultra-tight integrated navigation system as the background of research. Based on the existing integrated navigation nonlinear filtering algorithm, an ultra-tight integrated navigation method based on CKF filtering algorithm is proposed. Compared with the EKF and UKF algorithms, the CKF algorithm reduces the errors caused by intercepting higher order terms in the EKF, overcomes the problem of difficult to determine adjustable parameters in the UKF in practical applications, and greatly reduces the computational effort. Because of the above advantages nowadays the research focus of filtering algorithms is more on CKF algorithm related methods. The EKF, UKF and CKF are applied to the SINS/BDS ultra-tight integrated navigation system for simulation and comparison analysis in order to verify the performance of the algorithm. The simulation results show that the CKF algorithm is an effective and feasible algorithm for integrated navigation method because its stability, filtering accuracy and adaptability are synthetically better than those of the EKF and UKF algorithms.

1. Introduction

As navigation technology continues to develop, navigation is widely adopted in various fields. However, in modern warfare, technological confrontation has become the key, and highly dynamic carriers such as missiles and fighter jets have put forward higher requirements on the accuracy, reliability and anti-jamming capability of modern navigation technology. Satellite navigation is susceptible to environmental influences, and the signal is easily interrupted in the city; inertial navigation system is not effected by the environment, and can be used to offer highly performance in short-term periods, but long periods of time working alone can cause error accumulation[1]. Based on the above defects and limitations of individual systems, the combined navigation has emerged. With the continuous improvement of China's self-developed Beidou navigation system, the research of SINS/BDS integrated navigation system has become a current trend. At present, integrated navigation is divided into loose, tight and ultra-tight integrated navigation based on the different degree of combination and the tightness of the connection. Compared with loose and tight integrated navigation, ultra-tight integrated navigation system selects better observation and has higher positioning accuracy as well as reliability, and its advantages are more obvious in the harsh environment of weak signal and high dynamics.

The SINS/BDS integrated navigation system is a nonlinear system, and in order to solve the nonlinear problem, many experts and scholars have proposed many filtering algorithms in recent years. The EKF can adjust the linear approximation at each filter update and is commonly used in some systems with less severe nonlinearity, but the EKF is an estimation theory that finds suboptimal solutions and suffers from truncation errors arising when Taylor series expansions are used, resulting in inaccurate estimation

Content from this work may be used under the terms of the Creative Commons Attribution 3.0 licence. Any further distribution of this work must maintain attribution to the author(s) and the title of the work, journal citation and DOI.
Published under licence by IOP Publishing Ltd

results. The UKF overcomes the shortcomings of EKF by UT transform, but the computational complexity of the UKF algorithm increases with the increasing number of sampling points as the dimension of the system increases, which greatly reduces the filtering effect. In contrast, the CKF has neither the problem of computational effort nor the issue of low filtering accuracy of UKF at high dimensions of the system, and is applicable to any form of nonlinear model, so the research of SINS/BDS ultra-tight integrated navigation method based on CKF has become a new hot spot[2].

2. SINS/BDS ultra-tight integrated navigation model

2.1. SINS/BDS ultra-tight integrated navigation system structure.

Unlike the loosely and tightly coupled integration modes, in the ultra-tight integrated navigation system, a more primitive data I/Q signal than the pseudo-range and pseudo-range rate information is used for signal fusion, and the output of the combined filter is then fed back to the satellite receiver's tracking module and the SINS system as error compensation. This involves the internal structure of the Beidou receiver, which leads to a tighter connection between the two navigation systems and a deeper degree of data fusion. The structure of the SINS/BDS ultra-tight integrated navigation system is shown in Figure 1.

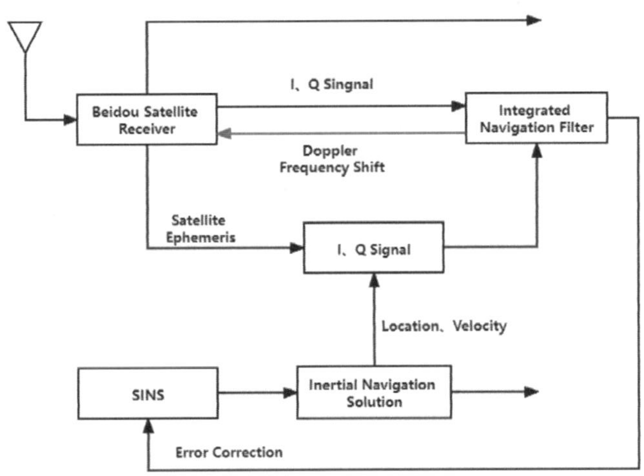

Figure 1 SINS/BDS ultra-tight integrated navigation system structure

2.2. SINS/BDS ultra-tight integrated navigation system model

2.2.1. Status Model. The SINS/BDS ultra-tight integrated navigation system model is composed of two sections: the measurement equation and the state equation. The equation of state can be taken as

$$\dot{X}(t) = F(t)X(t) + G(t)W(t) \tag{1}$$

where $X(t)$ is the state variable, The specific variables:

$$X = [\delta L, \delta\lambda, \delta h, \delta v_E, \delta v_N, \delta v_U, \varphi_E, \varphi_N, \varphi_U, \varepsilon_{b_E}, \varepsilon_{b_N}, \varepsilon_{b_U}, \nabla_{b_E}, \nabla_{bN}, \nabla_{b_U}, b_{clk}, d_{clk}]^T$$

in which, δL、$\delta\lambda$、δh denote longitude, latitude and altitude errors respectively; δv_E、δv_N、δv_U denote the velocity error components of the east, north, and sky axes respectively; φ_E、φ_N、φ_U denote the attitude angle error components on the three axes of ENU coordinate system respectively; ε_{b_E}、ε_{b_N}、ε_{b_U} denote the gyroscopic drift of the gyroscope on the ENU coordinate system respectively; ∇_{b_E}、∇_{b_N}、∇_{b_U} denote the components of accelerometer drift on the three axes of the ENU coordinate system respectively; b_{clk} and d_{clk} denote time-dependent errors respectively, that is clock error, clock frequency error.

2.2.2. Measurement model. For the SINS/BDS ultra-tight integrated navigation system, the difference between the calculated value of INS system and the measured value of BDS system is measured as the deviation of pseudo-range and pseudo-range rate. The pseudo-range measurement equation is

$$Z_\rho = H_\rho X + V_\rho \tag{2}$$

where $H_\rho = [0_{n\times3} \quad H_\rho \quad 0_{n\times9} \quad -1_{n\times2}]$ is the n×17 measurement matrix, and $H_{\rho1}$ and $H_{\rho2}$ take the values of

$$H_v = \begin{bmatrix} e_{11} & e_{12} & e_{13} \\ e_{21} & e_{22} & e_{23} \\ \vdots & \vdots & \vdots \\ e_{n1} & e_{n2} & e_{n3} \end{bmatrix} C_n^e$$

similarly, the pseudo-range rate measurement equation is obtained as

$$Z_{\dot\rho} = \delta\dot\rho_i = H_{\dot\rho} X + V_{\dot\rho} \tag{3}$$

where $\delta\dot\rho_i$ is the difference between the pseudo-range rate of SINS and the pseudo-range rate of BDS, $H_{\dot\rho}$ is the pseudo-range rate measurement equation, and $V_{\dot\rho}$ is the pseudo-range rate noise.

$$H_{\dot\rho} = [0_{n\times3} \quad H_{\dot\rho} \quad 0_{n\times9} \quad -1_{n\times2}]$$

$$H_{\dot\rho} = \begin{bmatrix} e_{11} & e_{12} & e_{13} \\ e_{21} & e_{22} & e_{23} \\ \vdots & \vdots & \vdots \\ e_{n1} & e_{n2} & e_{n3} \end{bmatrix} C_t^e$$

where $C_t^e = [C_e^t]^{-1} = [C_e^t]^T$

In summary, combining equations (2) and (3), the measurement equation for the SINS/BDS ultra-tight integrated navigation can be derived as

$$Z_t = \begin{bmatrix} H_\rho \\ H_{\dot\rho} \end{bmatrix} X + \begin{bmatrix} V_\rho \\ V_{\dot\rho} \end{bmatrix} = H_t X_t + V_t \tag{4}$$

3. CKF algorithm

3.1. Basic Principles of 12CKF

The cubature Kalman filter was originally developed by researchers in Canada, Arasaratnam and Haykin, in 2009. The CKF is theoretically the nearest current approximation algorithm to Bayesian filtering and it uses a group of cubature points to approximate the state mean and covariance of a nonlinear system with additional Gaussian noise on the basis of a spherical radial volume criterion to solve the state estimation of nonlinear systems.

3.2. The CKF implementation process

Assuming the state $N\left(x_k, \ \hat{x}_k, \ P_k\right)$ at moment k, the CKF algorithm is implemented as follows.

3.2.1. Time update process

Cholesky decomposition of $P_{k|k}$, (k=1,2,3,…∞)

$$P_{k|k} = S_{k|k} S_{k|k}^T \tag{5}$$

Forming cubature points, (i=1,2,3,…m)

$$X_{i,\ k|k} = S_{k|k}\xi_i + \hat{x}_{k|k} \tag{6}$$

Which, $\xi_i = \begin{cases} \sqrt{ne_i} & i = 1,2,\dots n \\ -\sqrt{ne_i} & i = n+1, n+2, \dots 2n \end{cases}$

where e_i is the i-th element of the unit column vector, which is the only one equal to 1 and the rest of the elements are equal to 0.

Calculate the propagation cubature points

$$X^*_{i,\ k+1|k} = f\left(X^*_{i,\ k|k}\right) \tag{7}$$

Calculate the predicted state

$$\hat{x}_{k+1|k} = \frac{1}{m}\sum_{i=1}^{2n} X^*_{i,\ k+1|k} \tag{8}$$

Calculate the predicted state error matrix

$$P_{k+1|k} = \frac{1}{m}\sum_{i=1}^{2n}\left(X^*_{i,\ k+1|k}\right)\left(X^*_{i,\ k+1|k}\right)^T - (\hat{x}_{k+1|k})(\hat{x}_{k+1|k})^T + Q_k \tag{9}$$

3.2.2. Measurement update process

Cholesky decomposition of $P_{k+1|k}$, (k=1,2,3,…∞)

$$P_{k+1|k} = S_{k+1|k}S^T_{k+1|k} \tag{10}$$

Forming cubature points, (i=1,2,3,…m)

$$X_{i,\ k+1|k} = S_{k+1|k}\xi_i + \hat{x}_{k+1|k} \tag{11}$$

Calculate the propagation cubature points

$$Z_{i,\ k+1|k} = h\left(X_{i,\ k+1|k}\right) \tag{12}$$

Calculate the predicted value of the measurement

$$\hat{z}_{k+1|k} = \frac{1}{m}\sum_{i=1}^{2n} Z_{i,\ k+1|k} \tag{13}$$

Estimated mutual covariance matrix

$$P_{zz,\ k+1|k} = \frac{1}{m}\sum_{i=1}^{m}\left(Z_{i,\ k+1|k}\right)\left(Z_{i,\ k+1|k}\right)^T - (\hat{z}_{k+1|k})(\hat{z}_{k+1|k})^T + R_{k+1} \tag{14}$$

Calculate the mutual covariance matrix

$$P_{xz,\ k+1|k} = \frac{1}{m}\sum_{i=1}^{m}\left(X_{i,\ k+1|k}\right)\left(Z_{i,\ k+1|k}\right)^T - (\hat{x}_{k+1|k})(\hat{z}_{k+1|k})^T \tag{15}$$

Estimated filter gain

$$W_{k+1} = P_{xz,\ k+1|k}P^{-1}_{zz,\ k+1|k} \tag{16}$$

Calculate the state estimate

$$\hat{x}_{k+1|k+1} = \hat{x}_{k+1|k} + W_{k+1}(Z_{k+1} - \hat{z}_{k+1|k}) \tag{17}$$

Calculate the filtered covariance matrix

$$P_{k+1|k+1} = P_{k+1|k} - W_{k+1}\left(P_{xz,\ k+1|k}\right)(W_{k+1})^T \tag{18}$$

4. Results & Discussion

In a bid to test the effectiveness of the ultra-tight integrated navigation method based on the CKF algorithm suggested in this paper, the standard EKF and UKF algorithms are taken for comparison. The simulation analysis will be performed in the environment of the SINS/BDS ultra-tight integrated navigation system with EKF, UKF and CKF filtering algorithms respectively.

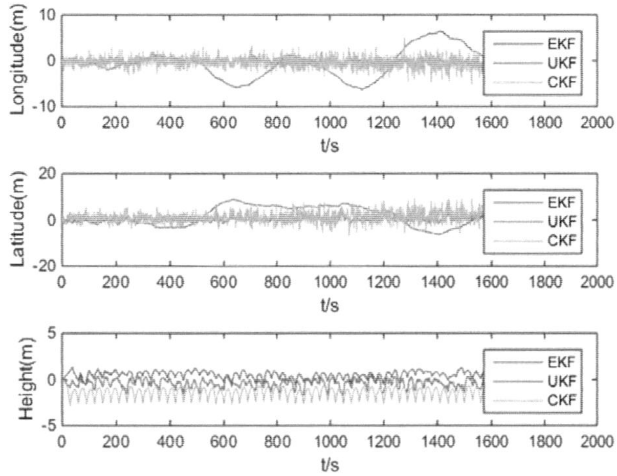

Figure 2 Longitude、Latitude and Height estimation error

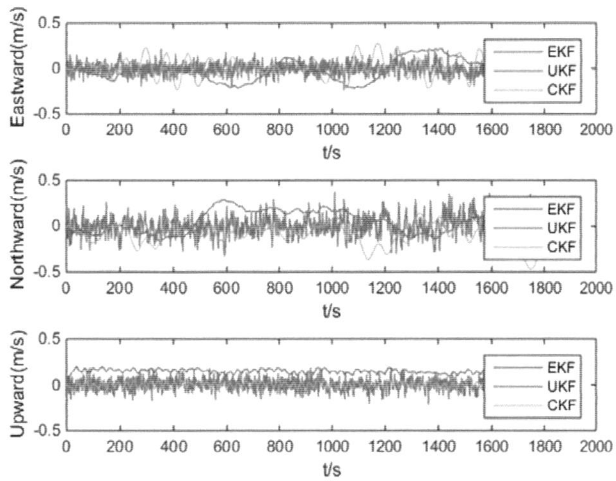

Figure 3 Eastward、Northward and Upward velocity estimation error

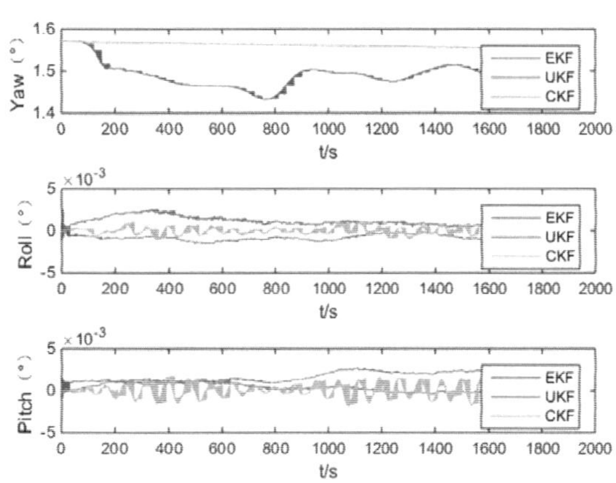

Figure 4 Yaw、Roll and Pitch angle error

Figure 2 shows the estimation error results of longitude, latitude and height for integrated navigation, Figure 3 shows the estimation error results of eastward, northward and upward velocity for integrated navigation, and Figure 4 shows the estimation error results of attitude for integrated navigation. Simulation performance indicates that compared to the EKF and UKF algorithms, the CKF algorithm is able to consistently converge and keep its longitude and latitude estimation errors floating around zero error in the ultra-tight integrated navigation system, while the EKF algorithm will have some fluctuations; The accuracy in the error results of the east, north and sky velocity estimation is significantly better than the EKF algorithm, and less different from the UKF algorithm; The EKF algorithm showed significant divergence in the attitude angle estimation error results, while the CKF algorithm was able to keep the error results of the cross-roll and pitch angle estimation basically stable and floating around zero error.

5. Conclusion

In this paper, the filtering using CKF filtering algorithm is proposed on the ultra-tight integrated navigation system. The specific implementation steps of the algorithm are similar to UKF, but the theoretical derivation of CKF is more rigorous. It is designed on the basis of the spherical-radial criterion and Bayesian theory, selecting a group of points with the same weight and then passing the variance of the nonlinear system, and using the transformed points to obtain the estimated system state at the next moment, which greatly reduces the computational effort. Simulation results indicate that the localization accuracy of the approach is synthetically better than that of the EKF algorithm and the UKF algorithm.

References

[1]Lu Wang,Xin Zhao,Quanhu Li,Weide Kang.(2021)Research on Fault Tolerant Algorithm of SINS/BDS/GPS Integrated Navigation System.In: 2021 33rd Chinese Control and Decision Conference (CCDC).Kunming.pp.1-6.

[2]Gao Zhongyu. (2012)Inertial Navigation System Technology. Beijing: Tsinghua University Press.

[3]Wei Xie, Zhao Huang, Weiguo He, Kangyuan Wang. (2020)A Square Root Cubature Kalman Filter Based Dynamic State Estimation of Distribution Network.In:2020 IEEE/IAS Industrial and Commercial Power System Asia (I&CPS Asia).Weihai.pp.1-6

[4]Ravindra Babu & Jinling Wang.(2009).Ultra-tight GPS/INS/PL integration: a system concept and performance analysis. GPS Solutions, 13(1):75-82.

[5]Yang Changlin, Ou Didong, Liu Yanfei. (2012).Integrated navigation fault diagnosis and fault tolerance research based on federated filtering.Science Technology and Engineering,12(32):8601-8605.

[6]E. Gallizia,G. Apicella,T. Cena,M. etc. (2017).The spine instability neoplastic score (SINS) in the assessment of response to radiotherapy for bone metastases.Clinical and Translational Oncology,19(11).

[7]Bing Xia, Yunlong Shang, Truong etc. (2017).A correlation based fault detection method for short circuits in battery packs.Journal of Power Sources,337.

[8]Kong Xuebo, Xu Jianxin, Xiong Zhi et al. (2016).Research on inertial-assisted Beidou navigation fault adaptive method.Aviation Computing Technology, 46(03):106-110.

Analysis and Research of Pin Void for High Voltage Intelligent Power Module

Qing Hua[1], Huanyu Wang[1], Zhijian Pan[2], Tuming Zhang[2] and Yuxiang Feng[2,a]

[1]Shandong Normal University, School of Physics and Electronics, 250358 Jinan, China

[2]Guangdong HIIC Semiconductor Co., LTD., 528216 Foshan, China

[a] Corresponding author: fengyx@hisemi.net

Abstract. Pin void is one of the key factors affecting the reliability of intelligent power module. It is difficult to accurately identify and measure whether it is a pin void and its size through X-ray equipment. This paper investigates the pin void analytical method of high voltage intelligent power module. An effective combinatorial analysis method is provided. Besides, pin void is compared and studied with different situations of reflow soldering, pin pressure bar and pin type in this research. The effects of different situations of these parameters on pin void are evaluated.

1. Introduction

With the development of electronic technology and the further improvement of users' requirements for electronic products, electronic products are developing towards miniaturization and multifunction. Since intelligent power modules can provide a compact, high-performance solution for appliance motor drive applications such as air conditioners, refrigerators and washing machines, it is widely used in rent years [1].

When the pins of intelligent power module are soldered, pin void will inevitably occur. The pin void will reduce the mechanical strength, affect the reliability and service life of pin solder joint. Therefore, the generation of pin void phenomenon must be studied and controlled [2-4].

In the quality evaluation standard of pin solder connection, pin void plays a decisive role, especially in large-scale solder joints. Poor pin soldering technologies can even lead to up to 20% void ratio. In terms of product reliability, void may lead to intelligent power modules dysfunction and even damage during normal operation. Therefore, quality control is necessary in the pin soldering process [5-6].

The pins soldering quality of intelligent power module is closely related to testing technology. Without effective testing methods, it is impossible to accurately evaluate the current production quality level and carry out the process improvement.

In this paper, combination of different analytical technologies is used to assessment the pin void of the intelligent power modules [7-8]. The pin void testing method presented in this research can facilitate the further product quality analysis studies on the mechanical strength, reliability, and service life of solder joints. As an instance, the pin void of the 600V/15A high voltage intelligent power module are characterized and analyzed. The X-ray, fracture and slice analysis for pin void are

Content from this work may be used under the terms of the Creative Commons Attribution 3.0 licence. Any further distribution of this work must maintain attribution to the author(s) and the title of the work, journal citation and DOI.
Published under licence by IOP Publishing Ltd

illustrated and discussed in section 2. Analysis of influence and improvement factors of pin void are illustrated and discussed in section 3. And the conclusions are stated in section 4.

2. X-ray, Fracture and Slice Analysis for Pin Void

The research object of this paper is the 600V/15A three-phase hybrid integrated high voltage intelligent power module which have been designed for motor control applications.

The detection technology of soldering void is continuously improved with the continuous development of power module manufacturing technology. But the most popular and economical method is the X-ray detection technology. Thus, in order to detect the general situation of void in the pins after soldering, an X-ray imaging test was used firstly and this is also the most used pin void defect detection method.

For this type of process quality problem, there is nothing to do with the detection methods such as visual inspection, optical microscope, or laser infrared. Therefore, in order to find out the real situation of pin void after pin soldering, it is necessary to select the X-ray inspection method which with the ability to penetrate non transparent objects for detection.

X-ray has strong penetrability which can clearly show the position, shape, and quantity distribution of the pin void, fully reflect the soldering quality of the pins of high voltage intelligent power modules.

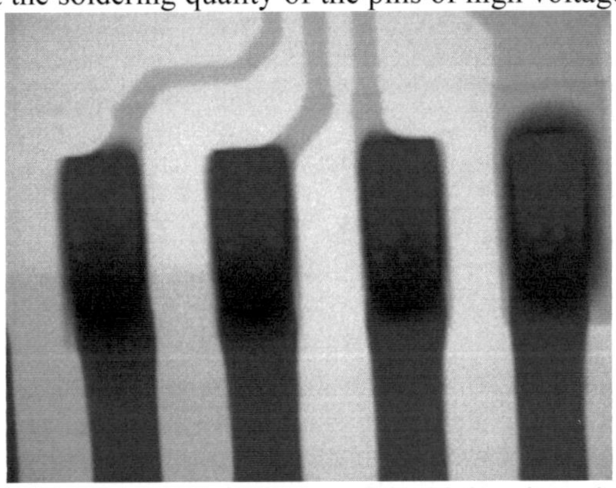

Figure 1. X-ray photograph of the void in the pins of the high voltage intelligent power module.

The X-ray image for the pins of the intelligent power module is shown in Figure 1. With the help of X-ray non-destructive testing technology, common problems such as internal physical structure characterization, defect detection and failure analysis of pin soldering can be well presented, which can meet the testing requirements of high voltage intelligent power modules mass manufacturing technology, help to improve the production process technology, and greatly improve the qualification rate. However, accurate positioning and analysis of pin void still require a combination of different technologies.

Figure 2. Fracture photograph of the void in the pins of the high voltage intelligent power module.

In this paper, scanning electron microscope is used to observe and analyze the fracture surface of the intelligent power module pins, find out the causes, put forward improvement and preventive measures, which is of great significance.

The fracture image for the pins of the intelligent power module is shown in Figure 2. Fracture means the disconnection surface between pins and solders. Because the fracture truly records all kinds of fracture related information in the whole process of void initiation, propagation, and finalization. Therefore, the information on the fracture surface is the result of the comprehensive action of many internal and external factors. The qualitative and quantitative analysis of fracture surface can provide a reliable basis for the determination of formation mechanism of void and provide clues for reduce the pin void ratio.

The metallographic slicing technology plays a key role in pin void process control. The slice image for the pins of the intelligent power module is shown in Figure 3. The solder layer thickness, soldering quality and void profile can be obtained by metallographic slice.

Figure 3. Slice photograph of the void in the pins of the high voltage intelligent power module.

3 Analysis of Influence and Improvement Factors of Pin Void

To evaluate the influence factors and propose improvement methods for pin void, relevant experiments have been carried out through the study of the impacts of reflow soldering, pressure bar and pin type on pin void ratio. The overall implementation scheme has been designed and summarized in Table 1 below.

Table 1. Overall implementation scheme of pin void analysis.

Sample NO.	Reflow Soldering	Pressure Bar	Pin Type
S1#	C1#/ C2#/ C3#	YES	Planar
S2#	C1#/ C2#/ C3#	NO	Planar
S3#	C1#/ C2#/ C3#	NO	Etch
S4#	C1#/ C2#/ C3#	YES/Meta 1	Etch
S5#	C4#	YES	Planar
S6#	C1#/ C2#/ C3#	YES	Arc

3.1 Influence of Reflow Soldering on Pin Void

The specific scheme to study the effects of reflow soldering on pin void has been designed and summarized in Table 2 below.

By comparing experimental results based on the reflow curves of C1#, C2#, C3#, it can be seen the pin void ratio was not improved by increasing the maximum temperature and reflow time.

Table 2. Reflow soldering experimental scheme.

Curve	Max. Temp. (℃)	Reflow Time (s)	Preheating Time (s)	
			50-150℃	150-200℃
C1#	254	90.2	63.7	107
C2#	249	88.6	63.5	110
C3#	254	108	53.1	121
C4#	249.8	82.7	65.8	125.8

Add a reflow curve C4# for verification, that is, reduce the maximum temperature and reflow time and increase the preheating time to confirm the impact of reflow curve on pin void ratio. By analyse the C4# reflow curve sample, it is confirmed that the reflow curve has no improvement on the pin void rate at this time.

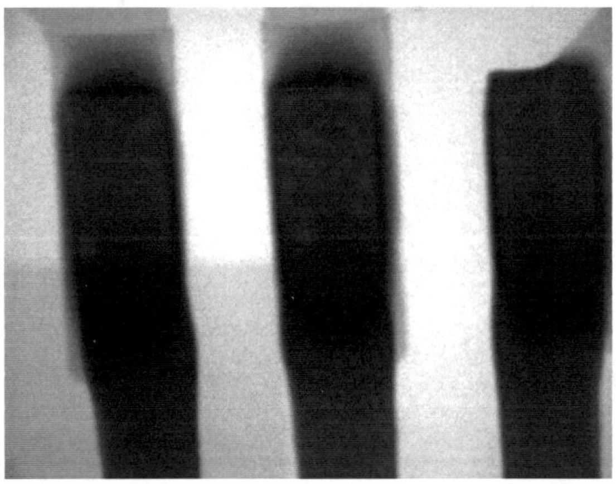

Figure 4. X-ray photograph of pin void based on reflow soldering curve C1#.

Figure 5. X-ray photograph of pin void based on reflow soldering curve C2#.

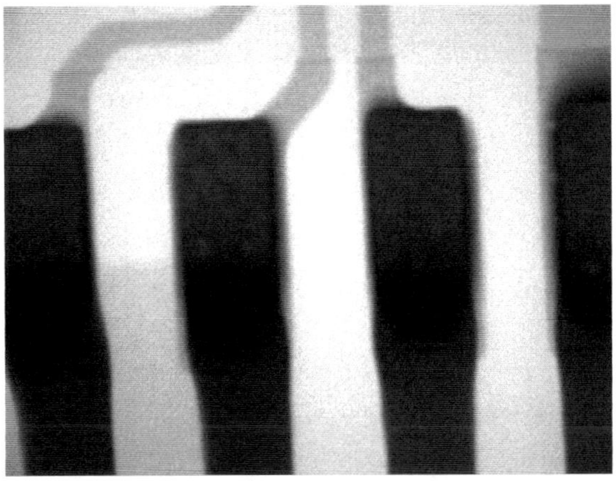

Figure 6. X-ray photograph of pin void based on reflow soldering curve C3#.

Figure 7. X-ray photograph of pin void based on reflow soldering curve C4#.

3.2 Influence of Pin Pressure Bar on Pin Void

The specific scheme to study the effects of with and without the pin pressure bar on pin void has been implemented and the experimental results are shown in the figures 8-11. By Comparing the pin void ratio of S1# and S2# samples, it can be seen that the void section is mostly circular or elliptical without pin pressure bar, and irregular with pin pressure bar. Both S3# and S4# pin void ratio morphologies are circular or elliptical, which may be that the pin pressure bar fails to fully press the pins. By comparing the pin void ratios of S2# and S3# samples, there is no significant difference between the etched pin and the stamped plane pin.

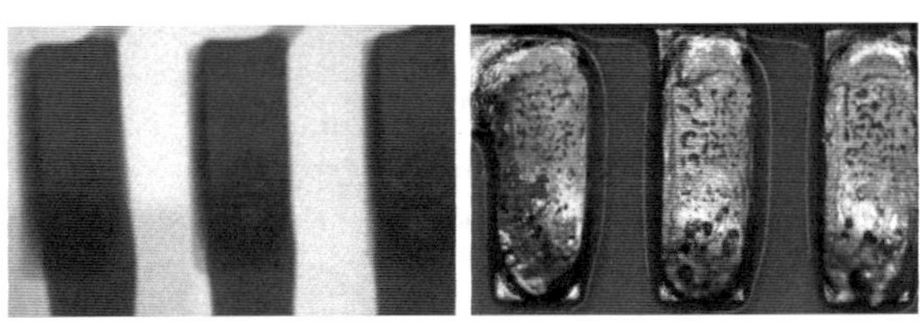

Figure 8. X-ray and fracture photographs of pin void of sample S1#.

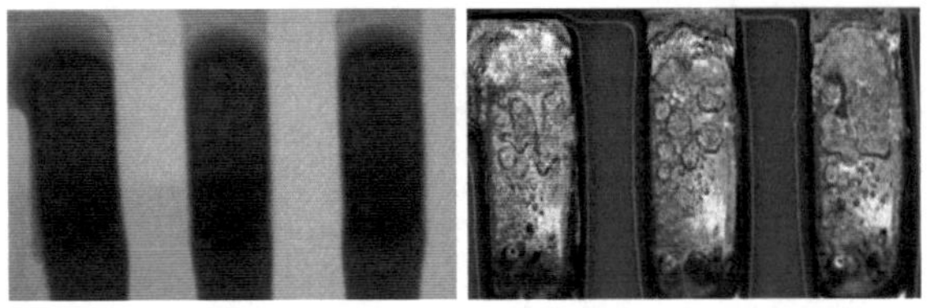

Figure 9. X-ray and fracture photographs of pin void of sample S2#.

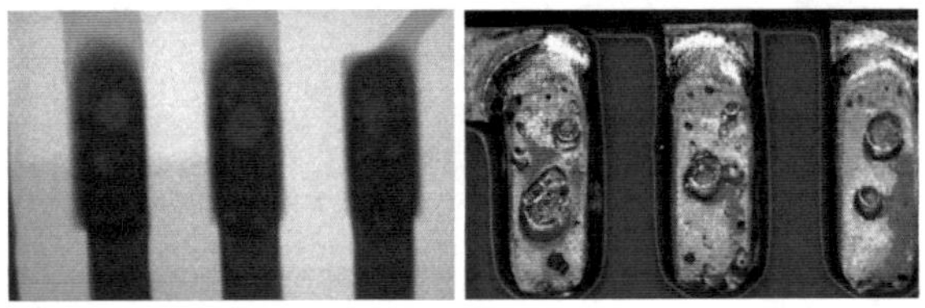

Figure 10. X-ray and fracture photographs of pin void of sample S3#.

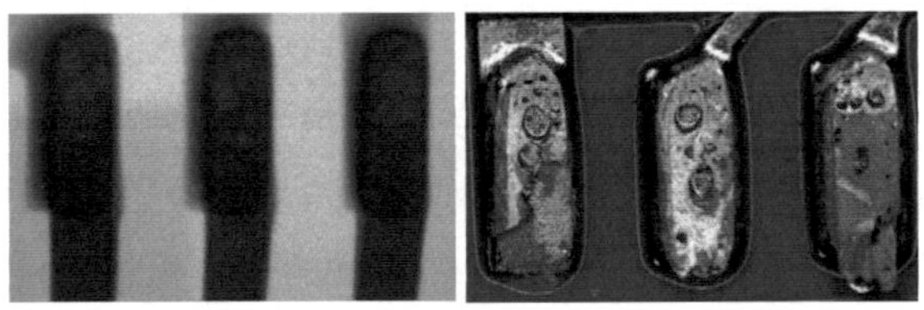

Figure 11. X-ray and fracture photographs of pin void of sample S4#.

By Comparing the pin void ratio of S1# and S2# samples, it can be seen that the void section is mostly circular or elliptical without pin pressure bar, and irregular with pin pressure bar. Both S3# and S4# pin void ratio morphologies are circular or elliptical, which may be that the pin pressure bar fails to fully press the pins. By comparing the pin void ratios of S2# and S3# samples, there is no significant difference between the etched pin and the stamped plane pin.

4 Conclusions

This paper presented an experimental method of analyzing the pin void ratio for high voltage hybrid integrated intelligent power module. The combination of the X-ray, fracture and slice analysis method is utilized in order to effectively address pin void unascertainable problem. The analysis results show that based on the existing stamping plane pins, adjusting the reflow curve has no obvious improvement on the pin void ratio. When the pin pressure bar is not used, the cross section of the pin void is mostly circular or elliptical, and the fracture shows the longitudinal depth of the void. However, the pin void is irregular and the fracture shows that the void is shallow in the longitudinal direction. Additionally, the cross-sectional area of the pin void is larger with the use of the pressure bar than without it, but the cavity volume will be smaller.

References

[1] Tsukizawa, M., Sasaki, H., Lee, J.H., Jeong, J.Y., Bae, H.S. (2018) A new single in-line package of intelligent power modules for low power motor drive applications. In: International Conference on Electric Power and Energy Conversion Systems. Kitakyushu. pp: 1–4.

[2] Li, Y.R.Y., Fu, G.C., Wan B., Yan, X.J., Zhang, W.F., Li, W. (2022) Phase□field modelling of lead□free solder joint void growth under thermal□electrical coupled stress. Journal of Electronic Materials, 51: 259–272.

[3] D.C. Katsis, D.C., van Wyk, J.D. (2003) Void-induced thermal impedance in power semiconductor modules: some transient temperature effects. IEEE Transactions on Industry Applications, 39: 1239–1246.

[4] Fu, S.C., Mei, Y.H., Li X., Lu G. Q. (2017) A high power density multichip phase-leg IGBT module with void-free die attachment using nanosilver paste. In: IEEE Energy Conversion Congress and Exposition. Cincinnati. pp. 3659–3662.
[5] Zhang, J., Zhang, B.B., Liu, Y.Y., Fei, J.M., Wan, C.A. (2014) The improvement of soldering process of new power module packaging material with large soldering area. In: International Conference on Electronic Packaging Technology. Chengdu. pp. 231–234.
[6] Lin, K. L., Yao, S. (2000) Solder thickness variation with respect to soldering parameters components and packaging technologies. IEEE Transactions on Components, Packaging and Manufacturing Technology, Part A: Packaging Technologies, 23: 661–664.
[7] Shishido, N., Hayama, Y., Morooka, W., Hagihara, S., Miyazaki, N. (2019) Application of nonlinear fracture mechanics parameter to predicting wire-liftoff lifetime of power module at elevated temperatures. IEEE Journal of Emerging and Selected Topics in Power Electronics, 7: 1604–1614.
[8] Ito, M., Kato J., Nonaka T. (2015) 3D analysis of flip chip interconnection using FIB-SEM slice and view. In: International Conference on Electronics Packaging and iMAPS All Asia Conference. Kyoto. pp: 102–105.

Text data mining of power based on natural language processing technology

Xin Wang[1,a], Jiming Tian[1,b*], Fei Li[2,c]

[1]Zhejiang Huayun Information Technology Co.,Ltd, Hangzhou, Zhejiang Province,310007,China

[2]Founder International Co.,Ltd,Beijing, 100086,China

[a]1035155968@qq.com,[b*]jimingtian@163.com, [c]lifeia7@163.com

Abstract. This paper first introduces the development of text mining technology, points out the difficulties of text mining technology, and then designs the power text mining based on natural language processing technology Then it introduces the typical application of text data mining technology in 95598 customer service complaints. Finally, it puts forward the application effect and challenge of power text mining.

1. Introduction

Power big data has become a hot research object at present, but only a small amount of data is mined and utilized every year[1]. Generally speaking, power data has the characteristics of large volume, many types, low value density and fast change, and it's harder to use. The low value density of the data means that most of the data are normal data of the power grid, with only a small amount of abnormal data[2].

At present, the rapid development of natural language processing and artificial intelligence[3-5] provides a good technical basis for power text mining, which has its own conditions and broad prospects. However, text mining technology in the domestic power industry is still a new frontier field, and most of the research is still in the exploratory experimental stage, the application benefits have not yet appeared. Compared with the achievements of internet and medicine, the research on text mining in power industry needs to be strengthened.

The existing text mining is mainly for English words, which only needs to be separated by spaces, so the text mining based on English does not need to consider text preprocessing. However, only sentences and paragraphs can be distinguished in Chinese text, and it is relatively difficult to segment Chinese words, which has seriously hindered the development of Chinese power data text mining. In China, the power data is mainly presented in Chinese text, so this paper studies and analyzes the Chinese text mining and analysis of power data using natural language processing technology.

This paper first introduces the development of text mining technology, points out the difficulties of text mining technology, then designs the power text mining based on natural language processing technology, then it introduces the typical application of text data mining technology in 95598 customer service. Finally, it puts forward the application effect and challenge of power text mining.

Content from this work may be used under the terms of the Creative Commons Attribution 3.0 licence. Any further distribution of this work must maintain attribution to the author(s) and the title of the work, journal citation and DOI.
Published under licence by IOP Publishing Ltd

2. Text mining technology overview

2.1. Text mining technology development process
The concept of text mining first appeared in the mid-1980s, it integrated natural language processing and data mining part of the technology and concepts[6, 7]. In the early days, text mining was once questioned and criticized for its scientific nature, but with the development of text mining technology, its application has gradually been recognized and become a research hotspot. In a narrow sense, text mining refers to the process of extracting previously unknown, understandable, and ultimately usable knowledge from large amounts of text data, and using this knowledge to better organize information and solve problems. Text mining includes both text knowledge extraction and text data processing for various applications. At present, the main research directions of text mining are text structure analysis, text semantic analysis, text summarization, text relevance analysis, etc. In the field of social science intelligence, text mining is used to monitor hot topics on Weibo[8], user comment semantic analysis[9], spam classification[10], etc.

2.2. Text mining technology difficult points
Text data belongs to natural language, and its mining involves linguistics, mathematics, computer science, information science, psychological science and cognitive science etc. From the technical point of view, text mining is difficult in both data and method. On the data side, text is unstructured data, has no deterministic form and lacks machine readable semantics, and needs to be transformed into structured data to be further mined. In addition, the text adopts different language, it is difficult to transplant the mining technology simply. On the methods side, although the existing text mining methods have been able to solve some problems, but there are still poor results, such as search engines can not return accurate answers based on the input text. There are still many text mining problems that have not yet been effectively addressed. From the point of view of application, there are many kinds of problems when universal text mining method is used in professional domain, so the adaptability of universal text mining method should be considered.

3. Design of text data mining in power market based on natural language processing technology
The process of power text mining based on NLP technology includes[6, 7, 11]: Data pre-processing, text clustering, text classification.

3.1. Data pre-processing
As far as power text mining is concerned, it deals with a mass of text whose content is composed of human natural language and can not be read and understood directly by computer. In particular, the structure of natural language is too complicated, including grammar, semantics, pragmatics, etc. It is easy to produce ambiguity and needs to be based on contextual information to understand language connotation, direct analysis can easily misguide the end result. Therefore, the key problem of text mining is to solve the problem of structured text representation, which can be read and understood by computer, and reflect the real features of text. Data pre-processing is intended to transform semi structured or unstructured text into structured data text for computer systems to read and understand, thus preparing for data mining. Generally speaking, Chinese text preprocessing includes three parts: text segmentation, feature representation and feature extraction[12].

3.1.1. Text segmentation
A sentence or paragraph is divided into words according to established rules. In English document processing, words are separated by spaces, without the need to design algorithms for text word segmentation. In the text, words can not be directly distinguished from words, although there are obvious segmentation characters in single words, whole sentences and paragraphs. However, Chinese text segmentation is relatively complex, and as an independent research direction, there are many important applications, such as information retrieval, chinese-foreign translation, automatic

summarization, automatic classification and so on. Especially with the rapid growth of network data and the expansion of Chinese information, the importance of Chinese text segmentation has become more and more prominent. At present, there are two Chinese word segmentation methods, dictionary-based word segmentation and annotation-based word segmentation. Dictionary-based segmentation requires the construction of a machine dictionary with sufficient memory, which matches the string to be analyzed with the entry in the dictionary according to a certain strategy, in this paper, the process of word segmentation is regarded as the problem of word tagging in a sentence, and Chinese characters are annotated to form words, that is, word formation.

3.1.2. Feature representation
As a first step in data pre-processing, word segmentation is also the initial stage of data structuring. The unstructured data is an important vehicle for the transformation of data into structured data. Feature representation models include Bool logic model, Vector space model and probability model.

3.1.3. Feature extraction
After feature representation, feature selection, i. e. feature extraction, is needed. The evaluation value of the feature words is calculated based on the specific evaluation function, and the feature words are sorted according to the size to select the desired feature words, also called the evaluation function method. So the key of feature extraction is the design of evaluation function. The most widely used evaluation function includes mutual information, information gain, statistical method and so on.

3.2. Text clustering
After Data pre-processing comes a key part of text mining, which is called text clustering. Power text clustering is an unsupervised classification, which does not have label information and automatically classifies text. After feature extraction and vectorization, the clustering process of text data is the same as that of machine learning[13]. The commonly used clustering algorithms include partition, hierarchy, graph theory and so on. The corresponding representative algorithms are K-means algorithm and DBSCAN algorithm.

3.3. Text classification
The purpose of power text classification is more explicit, which is to supervise the learning process. Text classification is an important part of the task of massive data management, which is to provide the pre-defined category label for the given text and classify the text accurately and efficiently. After Data pre-processing, the text classification methods are also various, the most widely used are Bayes algorithm, K-means neighbor algorithm, decision tree algorithm, neural network and support vector machine algorithm. Deep learning is also suitable for text classification, and the classification effect is good. Power text classification flow[14] as shown in Figure 1.

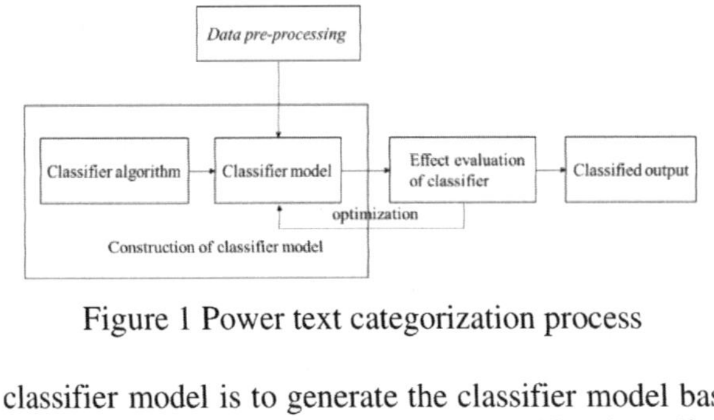

Figure 1 Power text categorization process

The construction of classifier model is to generate the classifier model based on the pre-processed data, and then classify the new text data automatically. The classification effect evaluation of classifier is the key link of power text automatic classification, and the validity of evaluation index is closely related to the reliability of test results. The commonly used evaluation indexes of power text classification include recall rate (W) , precision rate (V) , accuracy rate (W') and mission rate (G) to evaluate the effect of text classification. Is represented by a contingency table, as shown in Table 1.

Table 1 contingency table

	Machine evaluation 'yes'	Machine evaluation 'no'
Manual evaluation 'yes'	A	C
Manual evaluation 'no'	B	D

The relationship is expressed as follows:

$$W = \frac{A}{A+C}, V = \frac{A}{A+B}, G = 1 - W = \frac{C}{A+C}, W' = \frac{A+D}{A+B+C+D} \quad (1)$$

4. Case analysis

Starting with the natural language processing technology, the text of the power complaint form is deeply mined, the text content is analyzed by the word segmentation technology, and the word segmentation result is used as an auxiliary feature to extract, so as to control the main problems of power input, take a targeted and personalized service strategy to improve Customer satisfaction.

4.1. *Dmplementation of text segmentation*

Power text segmentation is the automatic segmentation of words through the computer, the use of big data software Python and Hidden Markov model, the content of power complaints processing segmentation. The results of the participle are shown in table 2.

Table 2 results of power text segmentation

Sentences	text segmentation
Customer call form	Customer, call, morning, a female, employee of the power company, home, inform, mother, electricity, ladder, electricity price,
Customers have been coming in	Customer, last year, voltage, long time: low impact, resident production, solution, the power company, verification processing, power outage
Account number 410724	Account number: 410724, customer company, male, bad attitude, staff, situation, work, call
Customer calls to complaints	Customer, complaint, call, 95598, warranty, repair, phone: inform: Question: Reply: 13838389438
Customer calls to feedback	Customer, call, location, plan, power outage, announcement, time, power outage information, SMS notification, call, phone, site
Customer calls source	Customer, Zhejiang Province, Hangzhou, Xixing Street, Jianghui Road, early stage, line, transformation, construction

4.2. *Feature extraction implementation*

For power text data word segmentation processing, each word can be used as a text identification feature, all features in the power text set frequency[15], as shown in table 3.

Table 3 frequency of occurrence of different characteristics

Customers	Power cut	Evaluation	Wrong connection	Malfunctions	Using electricity
1232	1085	1044	760	96	521
Rush repair	Business Hall	Ammeter	In the same time	Electricity poles	Meter reading
81	691	45	37	31	33
Lead	Today	Because	Around	Recover	Power supply institute
12	8	9	8	72	115

4.3. Implementation of classifier model

Based on the big data algorithm, a classifier model is built for the test data text, by which the data text of the power complaint work order can be automatically classified, and the specific distribution state of hot events in different business can be fully understood, through the analysis of the results, an effective response strategy is designed to achieve power proactive service, improve Customer satisfaction and loyalty, reduce the number of complaints and increase customer stickiness.

4.3.1. Explicit model

Taking power failure as an example, a suspected power failure list and a given specific weight non-power failure list are randomly selected as training sets in the complaint list. The text classification models of decision tree c 5 tree, SVM algorithm, C&R Tree algorithm are established. Recall rate, precision rate, accuracy rate and omission rate are selected as evaluation indexes, after comparing and analyzing the whole effect of the classifier model, the decision tree C5 Tree model is selected as the classifier model of power cut.

4.3.2. Effectt evaluation

Input the power complaint list to be classified and classify it by C5 tree model of decision tree, and evaluate the effect of classification. The result is as shown in table 4.

Table 4 results of evaluation of classification effectiveness

Evaluation index	W	V	W′	G
Evaluation result	93.3%	35.3%	77.8%	11%

As can be seen from table 4, the recall rate of decision tree C5 tree model is about 93.3% , and the omission rate is about 11% , the precision rate is about 35.3% and the accuracy rate is about 77.8% , which means that although all suspected power outage orders can be roughly identified, other hot spot orders are identified as power outage orders, so it is necessary to further optimize and improve the classifier model. In order to improve the accuracy of the segmentation results, the model is optimized based on three levels: enriching and optimizing thesaurus and thesaurus; C5 tree algorithm for tuning parameter decision tree, that is, optimizing the depth of the improved model and the leaf nodes. After adjusting the classifier model, the training set worksheets are input and classified in the model. The results of the classification effectiveness evaluation after model tuning are shown in table 5. From Table 5, it can be seen that the precision and accuracy of the classifier model have been improved obviously, and the mission rate has been increased, but it is still within the acceptable level. Based on the data validation of the training set, the optimized classifier model can more accurately identify the hot event complaint worksheet[16].

Table 5 evaluation results of classification effectiveness after model tuning

Evaluation index	W	V	W′	G
Evaluation result	92.4%	96.2%	93.3%	10.1%

Based on the classification of hot issues complaint worksheets and marketing department business related to each type of business for the impact of Customer satisfaction and degree of complaint, in order to enhance customer satisfaction and reduce customer complaints, and strengthen customer loyalty, so as to promote the long-term and stable development of power supply enterprises.

5. Clonclusion

In summary, this paper studies power text mining and classification based on natural language processing technology, and analyzes the actual effect of power text mining and classifier model with complaint worksheet as an example, using big data analysis technology, segmentation analysis method and so on to mine the text of complaint worksheet, which breaks through the existing restriction of vague customer complaint demand, so as to master the problem of power customer complaint, in order to improve the Customer satisfaction and loyalty, the paper puts forward the strategy of pertinence and Individuation Service, which has the advantages of high accuracy, wide application range and flexible methods The precision and accuracy of the classifier model are very high, and the optimized classifier model can identify the hot event complaint sheet more accurately.

As power industry is highly specialized and requires high accuracy and reliability of text mining models, which requires researchers to have both power domain knowledge and text mining experience, in order to establish a text mining technology system that meets the characteristics and application requirements of power text, we should combine adaptive improvement with independent innovation. At present, the research of power text data mining has attracted much attention, but it is still faced with the puzzle that the demand of mining is not clear and the mining effect is difficult to achieve the expectation. So it is urgent to have the application results of text mining to demonstrate and lead, and inspire more demand of power text mining, attract more people to participate in the research of power text mining technology and application, promote the theory and engineering application of power text mining flourish development.

References

[1] Xue Yusheng, Lai yening, integration of big energy thinking and big data thinking (I) big data and big power data, Automation of Electric Power Systems, 40(2016)1-8.
[2] Zhang Pei, Wu Xiaoyu, he jing-han, big data technology in active distribution network, Electric Power Construction, DOI (2015)52-59.
[3] P.M. Nadkarni, L. Ohno-Machado, W.W. Chapman, Natural language processing: an introduction, Journal of the American Medical Informatics Association, 18 (2011) 544-551.
[4] J. Hirschberg, C.D. Manning, Advances in natural language processing, Science, 349 (2015) 261-266.
[5] X. Qiu, T. Sun, Y. Xu, Y. Shao, N. Dai, X. Huang, Pre-trained models for natural language processing: A survey, Science China Technological Sciences, DOI (2020) 1-26.
[6] C. Dreisbach, T.A. Koleck, P.E. Bourne, S. Bakken, A systematic review of natural language processing and text mining of symptoms from electronic patient-authored text data, International journal of medical informatics, 125 (2019) 37-46.
[7] T.A. Koleck, C. Dreisbach, P.E. Bourne, S. Bakken, Natural language processing of symptoms documented in free-text narratives of electronic health records: a systematic review, Journal of the American Medical Informatics Association, 26 (2019) 364-379.
[8] Ding Zhaoyun, Jia Yan, Zhou Bin, summary of data mining research, Journal of Computer Research and Development, 51(2014)691.
[9] Peng Lan, Zhou Qihai, Qiu Jiangtao, Factor Model Study on the usefulness of consumer online reviews, Library and Information Service, 2011.
[10] Xiaohua Zhang, research on spam filtering based on text categorization, Anhui University, 2017.
[11] F. Zhang, H. Fleyeh, X. Wang, M. Lu, Construction site accident analysis using text mining and natural language processing techniques, Automation in Construction, 99 (2019) 238-248.
[12] Li Huaqiang, Liu haoxue, Han Yanhu, partner selection of Multimodal Transport Alliance, Journal of Highway and Transportation Research and Development, 35(2018)152-158.
[13] Lanlina, Shi Ruisheng, Gou xuerong, the research of learning partner recommendation method based on learning behavior model, Modern Educational Technology, 28(2018)67-73.
[14] Zhao Jinhui, Wang Xuehui, Zhou Yu, partner selection demand-driven Virtual Enterprises, Control and Decision 34(2019)2627-2634.
[15] Jian Jie, Wang Mi Lin, an shi quan, Su Jia Fu, partner selection for Collaborative Product Innovation Alliance, Computer Integrated Manufacturing Systems, Doi (2019)752-763.
[16] Wu Feifei, Li Ruiyu, Huang Lucheng, identification and selection of cross-industry R & D Partners, Journal of Intelligence, 9(2019) .

A Fuzzy Query Method Oriented to Knowledge Graph

Yunxing Liu[1,a*]

[1]College of Computer Science and Technology, Nanjing University of Aeronautics and Astronautics, Nanjing, Jiangsu, China

[*]email: liuyunxing@nuaa.edu.cn

Abstract. With the widespread application of Knowledge Graphs, Knowledge Graph query technology has received widespread attention. The daily use of natural language usually contains fuzzy terms. The use of fuzzy terms in queries is more in line with natural language habits. Therefore, fuzzy questions have been widely discussed and applied in relational data models, and their use in NoSQL data models has also begun to be affected in recent years. RDF (Resource Description Framework) is a representative model of the knowledge graph, and SPARQL is the standard query language of RDF. To realize the fuzzy query of the knowledge graph, this paper introduces the unclear expansion syntax of aggregate function based on summarizing and analyzing the existing SPARQL fuzzy expansion. Supplements the qualitative preference analysis method then proposes and SPARQL1. 1 Corresponding fuzzy SPARQL1.1 (f-SPARQL1.1) simultaneously provides strategies and techniques for rewriting ambiguous queries into classic SPARQL queries. In addition, this article also provides a visual interface for defining the membership functions of fuzzy terms and provides three interactive methods for setting fuzzy membership functions. Compared with the previous fuzzy query methods for RDF models, the method proposed in this paper has more vital fuzzy query ability and more flexible universality. The experimental results show the usability of the method proposed in this paper.

1. Introduction

Google first proposed the Knowledge Graph(KG) in 2012. It usually consists of entities and relationships. Well-known large-scale knowledge graphs such as WikiData, DBpedia, FreeBase. KG has been widely used in various downstream tasks, such as intelligent question answering systems and recommendation systems. In most intelligent studies, knowledge graphs act as knowledge bases and are the brain and core of functions. Therefore, knowledge graph query and exploration tasks as essential means of acquiring knowledge have received extensive attention[1]. The daily use of natural language usually contains ambiguous terms, and it is more common to include vague words in query sentences. For example, when asking about temperature, we typically use "Is it **cold** outside now?" instead of "Is the temperature outside now **lower than 10 degrees**?". The use of vague terms is a unique way of expression for humans, and the use of fuzzy terms in queries is more in line with natural language habits. Therefore, fuzzy queries have been widely discussed in relational database models. With the rise of the Semantic Web, data models such as XML (Extensible Markup Language) and RDF (Resource Describe Framework) have emerged one after another. In recent years, the application of fuzzy queries on NoSQL databases has also received more and more attention, such as [2, 3]. Fuzzy query tasks are usually carried out from two aspects. One is the fuzzification of the data model itself, such as work [4, 5]. The other is to expand the data query language by fuzzy grammar and transform it into a fuzzy set theory. Precise query, such as work[6-8]. RDF is a leading data

Content from this work may be used under the terms of the Creative Commons Attribution 3.0 licence. Any further distribution of this work must maintain attribution to the author(s) and the title of the work, journal citation and DOI.
Published under licence by IOP Publishing Ltd

representation model of the Knowledge Graph. SPARQL is the standard query language of RDF. SPARQL1.1 is an upgraded version after 2013.

This article proposes fuzzy SPARQL1.1(f-SPARQL1 .1) Grammar, used for the fuzzy query of the knowledge graph. This article inherits from the previous work f-SPARQL[6], fp-SPARQL[7], and SPARQLf-p[8] to realize the fuzzy query. The form of FILTER (?X θ FT)[WITH α] was first expressed by f-SPARQL. Both fp-SPARQL and SPARQLf-p inherit the state to implement ambiguous queries. But unlike f-SPARQL, fp-SPARQL defines 13 language intervals and executes preference queries. SPARQLf-p adds support for fuzzy relations while executing preference queries. SPARQLf-p executes preference query and fuzzy relation query. SPARQL1.1 newly introduced the function of value aggregation; through the COUNT, SUM, MIN, MAX and AVG operators, you can count and summarize database instances. In relational databases' past fuzzy query work, fuzzy aggregation operations have also been discussed and researched, such as [9, 10]. The fuzzy query work represented by f-SPARQL does not consider the fuzzy query in the case of aggregation, so this article supplements the fuzzy expansion of the aggregate query, and realizes the use of fuzzy terms to group the data and perform the aggregation operation, and do it on the fuzzy set The ability to aggregate functions.

The fuzzy set theory[11] proposed by Zadeh is the primary theoretical support in the field of fuzzy query, and the unclear membership function is a means to describe fuzzy term sets. Therefore, the unclear membership function plays a vital role in the fuzzy query process. Still, the recent SPARQL fuzzy query expansion method does not pay much attention to establishing the fuzzy membership function. This brings difficulties to the actual establishment and flexible migration of fuzzy queries. Therefore, this article designed a visual interface that defines the membership function required for fuzzy terms to solve this problem. It provides users with automatic, semi-automatic, and custom-setting membership function modes. In this way, the user's operation difficulty is reducmethodand customizing the membership function can set a membership function that is more in line with the user's personality.

The main contributions of this paper are as follows (1) Summarize and inherit the fuzzy extended grammar of the f-SPARQL series, supplement the fuzzy aggregate query grammar, and propose the fuzzy extended grammar f-SPARQL1.1 of SPARQL1.1. At the same time, we provide a method to convert f-SPARQL1.1 into executable SPARQL grammar. (2) we propose three modes for establishing fuzzy membership functions, and realize a visual query interface, providing a more flexible way of establishing membership functions.

2. Related Work

Implementation of the fuzzy query by fuzzy extension of SPARQL language includes [8, 12, 13], some of which are based on fuzzy RDF model, and some are based on accurate RDF model. Compared with the exact RDF model, fuzzy RDF attaches fuzzy membership to RDF triples to express fuzzy semantics. Although accurate RDF models cannot describe fuzzy semantics, most open knowledge graphs are still represented by correct RDF data, and most downstream tasks are still based on accurate RDF models. We explore the fuzzy query method for the accurate RDF data model, but the SPARQL extension work on the fuzzy RDF model still has reference significance.

Aggregate functions have been studied in relational database models. However, for the fuzzification of SPARQL aggregate functions, the FSA-SPARQL[13] team recently focused on this research, but this research is aimed at fuzzy RDF data models. In an inaccurate RDF data model, all of its aggregation functions take into account the fuzzy attributes of triples, while the fuzzy aggregation operation that should be a counting operation becomes the fuzzy base of the fuzzy set, SUM, AVG, MIN, AVG has become a version weighted by the membership. The aggregate function syntax is usually followed by Group by. When discussing aggregate functions, it is often accompanied by a discussion of data grouping. FSA-SPARQL considers grouping data based on fuzzy sets and then compiles them into accurate SPARQL data before execution. Although FSA-SPARQL implements the fuzzy expansion of aggregation and functions, it has a strong dependence on fuzzy RDF models and cannot be directly migrated to accurate RDF data models. Therefore, this article proposes suitable for accurate RDF models. An extended version of the aggregate function of the model. The fuzzy aggregation function proposed by FSA-SPARQL takes membership as a weight to participate in the calculation of the aggregation function. It can combine the characteristics of fuzzy membership to reflect the data, but it loses the original semantics of SUM, AVG, and MIN.

For the SPARQL query work of fuzzy RDF data models, because RDF itself has fuzzy attributes and membership degrees, it is usually not too much to explain the process of establishing fuzzy membership functions. For accurate RDF data models, the establishment of membership functions is relatively important. In the fuzzy expansion of the f-SPARQL series for precise RDF, most of the default fuzzy membership functions are used, but the default membership function is very rigid and does not necessarily meet the individual needs of users. In the past fuzzy query work on precise relational data models, SQLf[10] directly defined fuzzy membership functions in the query language but only supported trapezoidal fuzzy membership functions. Work [14] provides three fuzzy membership function setting methods but only supports the triangular membership function setting, and the user can only change the range and number of intervals of the membership function. Work [10] provides a visual interface that can use commands to set fuzzy membership functions for the fuzzy query of the Neo4j database. However, it is necessary to understand the membership function setting commands. The supported fuzzy membership functions are limited in shape, and the interactive experience is not ideal. Work [15] provides an iPad-based software that can easily define linguistics by hand-drawing on the iPad, and is very flexible. Inspired by the work [14], this paper provides three forms of setting fuzzy membership functions. The difference is that this work can not only set the fuzzy range and the number of domain intervals but also apply different shapes of fuzzy membership functions, not limited to triangles. Preliminaries

2.1. *Fuzzy Set Theory*

Fuzzy set theory[11] was proposed by Zadeh. In classical set theory, an element belongs to a set or does not belong to a set. The fuzzy set theory provides the ability to describe the degree to which an element belongs to a certain set in classical set theory. The degree to which element x belongs to set A is 0 or 1. However, it is impossible to accurately determine whether the element x belongs to set A. In fuzzy set theory, the membership degree of an element is usually calculated by the membership function. The fuzzy set theory proposed by Zadeh uses the value between [0,1] to indicate the degree to which an element belongs to the set instead of using absolute 0 and 1. Define A as the fuzzy set F of U in the universe U, and define the membership function as shown in formula (1):

$$\mu_A(u): U \rightarrow [0,1] \qquad (1)$$

There are some more important concepts in fuzzy set theory, $Supp(A) = \{x \mid x \in U \wedge \mu_A(x) > 0\}$ represents the support of fuzzy set A, which is a set of elements with non-zero membership degree in the fuzzy set, $Core(A) = \{x \mid x \in U \wedge \mu_A(x) = 1\}$ is a group of elements of fuzzy set A that completely belong to fuzzy set A, A_α represents the cut-off set of fuzzy set A, and the membership degrees of elements in A are all greater than α or greater than or equal to α.

2.2. *Fuzzy Terms*

Fuzzy terms are often used in daily life. Fuzzy terms mainly include fuzzy words, modifiers, and fuzzy operators. Fuzzy words refer to words that already contain vague meanings, such as “cold”, “hot”, “expensive” and “cheap”. Most of these are adjectives. Modifiers modify the degree of ambiguity, such as “very”, “special”, “slightly”. The fuzzy word modified by the modifier can be regarded as a compound fuzzy term item, and the fuzzy membership function of this fuzzy item is usually transformed from the membership function of the modified fuzzy word. Fuzzy operators, such as “close to”, “at least”, “not less than”, and “approximately” are usually placed before the precise quantifier, which represents the fuzzification of the precise quantifier.

3. F-sparql1.1

f-SPARQL1.1 inherits the work of the f-SPARQL series. These works are mainly completed in the form of fuzzy expansion of the FILTER constraint. The grammatical details can be understood from the literature [16-18]. Next, we will introduce the work of this article: the related grammar of fuzzy aggregate function and the setting method of fuzzy membership function.

3.1. Syntax Extension

3.1.1. Aggregation for Fuzzy Sets

This article provides fuzzy syntax extension support for the COUNT, SUM, AVG, MIN, MAX aggregate functions, and its extended form is [*AGG_FUNCTION*](?x θ FT) [with α]. Where [*AGG_FUNCTION*](?x θ *FT*) [with α] represents a function of *fCOUNT*, *fSUM*, *fAVG*, *fMIN*, *fMAX*, and *FT* represents a fuzzy term, α is the minimum degree of membership that satisfies the condition. It transforms fuzzy query sentences using fuzzy cut set theory and represents an operator. Table1 explains their meanings respectively. Ordinary aggregation functions can only perform aggregation operations on precise sets. In contrast, the aggregation functions in Table1 can perform aggregation operations on fuzzy sets, calculating the number of entries, average values, maximum and minimum values.

Table 1 Fuzzy extension of SPARQL aggregate function

Syntax	Explanation
fCOUNT(?*x* θ FT) [WITH α]	Returns the number of entries satisfying ?*x* θ FT.
fSUM(?*x* θ FT) [WITH α]	Returns the sum of values satisfying ?x θ FT.
fAVG(?*x* θ FT) [WITH α]	Returns the average of values satisfying ?x θ FT.
fMIN(?*x* θ FT) [WITH α]	Returns the min of the value satisfying ?x θ FT.
fMAX(?*x* θ FT) [WITH α]	Returns the max of the value satisfying ?x θ FT.

3.1.2. Fuzzy Grouping

The aggregate function is built on the basis of GROUP BY. When the query statement does not contain GROUP BY, it actually treats all data as the same group. In the past, GROUP BY was usually followed by the precise attributes of the data. The GROUB BY discussed in this article is followed by fuzzy attributes and classified according to fuzzy terms. For example, "the number of people in various age stages in the data set" should be grouped according to the fuzzy term "young people", "middle-aged people", and "old people". However, the accurate data does not contain the fuzzy terms corresponding to this attribute, so the data needs to be converted into fuzzy terms before classification. For the classified set, fuzzy aggregation operation and ordinary aggregation operation can be performed again. This article defines the grammar shown in Table 2 for fuzzy grouping.

Table 2 Fuzzy extension related to fuzzy grouping

No.	Syntax	No.	Syntax
1	*lingTrans([s, p], value,* α *)*	4	*lingAbstract(s,* p_1,p_2,..., α *)*
2	*lingFew(s,* p_1,p_2,..., α *)*	5	*fgroup by label([s, p], value,* α *)* ...
3	*lingMost(s,* p_1,p_2,..., α *)*		

The function of *lingTrans* in Table 2 is to convert precise values into fuzzy terms. The parameters *s* and *p* are the subject and predicate in the triple *(S, P, O)*. *value* is the data that needs to be converted into fuzzy terms, and a is the threshold. This function uses *s*, *p* to locate the fuzzy membership function group that describes the attributes determined by the subject and predicate, and uses the fuzzy membership function to return all fuzzy terms greater than the threshold. *lingFew* can be similar to *lingMost*. After grouping the fuzzy term attributes determined by (s, p_1, p_2,..., α) in the data set, respectively, it returns the attributes with the largest number and the smallest number in the group, *lingAbstract* is a combination of *lingMost* and *lingFew*, which can return the attributes with the most and least proportions. The function *fgroup by label* can group the column of value according to the fuzzy term corresponding to (*[s, p], value,* α), there can be multiple groups of label functions, which are similar to the ordinary GROUP BY. Using fgroup by can solve the problem "query the average age of each age in the data set."

3.2. Fuzzy Membership Function Setting Method

Fuzzy membership function has a certain degree of objectivity, because fuzzy exists because there is some kind of excess that is not black or white between objective facts, and this kind of excessive restriction causes the fuzzy membership function to conform to certain objective laws. Different people's understanding of the same vague term is subjective. Therefore, the formulation of the membership function should allow subjective adjustments in accordance with certain objective facts. The fuzzy query expansion method of the knowledge graph in this article includes the setting method of fuzzy membership function, and provides three optional setting methods of fuzzy membership function, automatic, semi-automatic, and custom mode.

3.2.1. Auto Mode

The user provides the subject and predicate of the membership function, and the system divides it into three fuzzy sets as "low", "medium" and "high" according to the distribution of the subject and predicate query data in the knowledge graph, and then uses the triangular membership function establishment method to establish Fuzzy membership function. According to the subject and predicate, the maximum and minimum values obtained are *maxV* and *minV* respectively. To establish a fuzzy membership function of "low", "medium" and "high", [*maxV*, *minV*] needs to be divided into six parts. Length *x=(maxValue-minValue)/6*, the range of fuzzy term "low" is [*minValue, minValue+2x*), [*minValue+x, minValue+5x*) is the range of fuzzy term "medium". [*minValue+4x, maxValue)* is the range of the fuzzy term "high".

3.2.2. Semi-automatic Mode

Under the mode of semi-automatic setting of membership functions, users are allowed to specify the range of fuzzy domains, the number of fuzzy term domains, the corresponding fuzzy terms, and the shape of the fuzzy membership function. The system provides membership functions such as triangles and trapezoids. Provide a visual interface. The visual interface is shown in Figure 1.

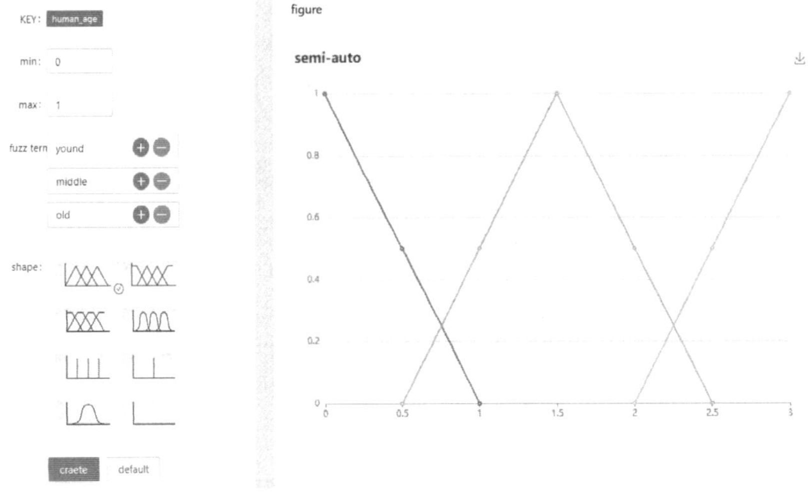

Figure 1 the semi-automatic establishment of fuzzy membership function.

3.2.3. Custom Mode

The use of semi-automatic creation of membership functions can give users a certain degree of freedom, but users need to have a certain understanding of fuzzy set theory, although the form of semi-automatic creation of membership functions provides users with settings for the number of fuzzy intervals, terms and the shape of the membership function. Choice, but there are still some limitations, so this article provides users with a completely manual membership function setting mode. In this mode, the user needs to set the maximum value, minimum value of the membership function and the described fuzzy term, and set the membership function for this fuzzy term by hand. Using this method gives users absolute freedom. Users can fully control the shape of the fuzzy membership function they want. In addition, users can also set the fuzzy membership function by customizing the function expression. Figure 2 shows the interface of the custom fuzzy membership function.

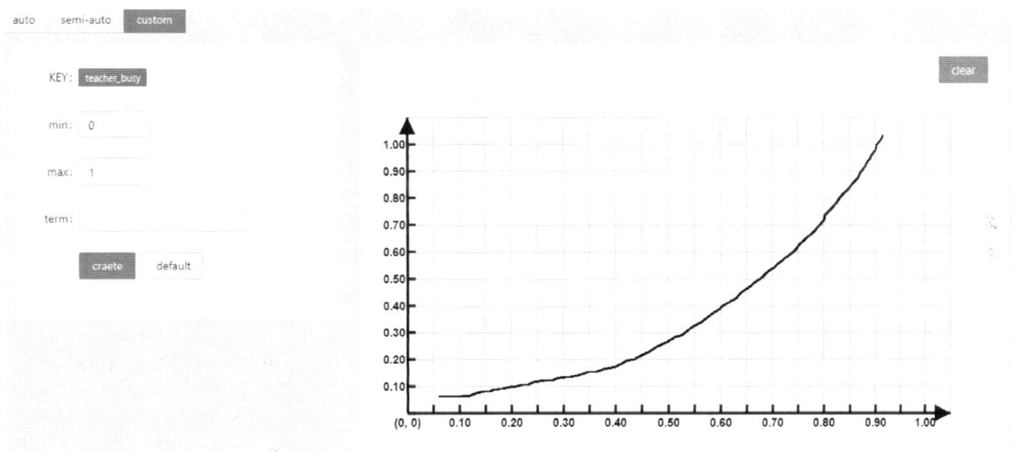

Figure 2 Automatically set fuzzy membership function

4. F-Sparql1.1 Execution Method

This work analyzes f-SPARQL1.1 to realize the fuzzy query process of the knowledge graph. The membership function can be persisted in the database. Part of the f-SPARQL1.1 work is replaced by SPARQL, and the other part is completed after the result of querying the knowledge graph. The next part of this chapter explains the method of converting the syntax extension defined in Chapter 4 into SPARQL1.1.

4.1. Explain The Aggregation Syntax of Fuzzy Sets

Converting fuzzy aggregate functions to SPARQL requires the help of f-SPARQL. This work first parses f-SPARQL1.1 query statements containing fuzzy aggregate functions into f-SPARQL, and then uses f-SPARQL grammar rules to convert fuzzy aggregate functions into The SPARQL statement executed. The grammar of f-SAPRQL conversion to SPARQL can be referred to [6]. Specifically, the fuzzy aggregate function is converted into a general aggregate function, and then the main content of the aggregate function is converted, and the form of FILTER ($?X\ \theta$ FT)[WITH α] is added to the matching main body of f-SPARQL. Algorithm 1 converts the fuzzy aggregation function of f-SPARQL1.1 into details of f-SPARQL.

Algorithm 1 Convert the fuzzy aggregate function

1. fspWithAgg←Input
2. fAgg←{fCOUNT,fMAX,fMIN,fAVG,fSUM}
3. selectPart, wherePart, otherPart←splitFsp(fspWithAgg)
4. items←splitSelect(selectPart)

10. noFuItem←noFuItem.replace('f',")
11. selectPart←selectPart.replace(items[k], noFuItem)
12. wherePart←wherePart + 'FILTER (' +

```
5. for k←1 to len(items) do:
6. if (items[k].contains(fAgg) && lower(items[k])) do:
7. fuzzyTerm←Regex.match(items[k], "\\((.*?)\\)")
8.    threshold←Regex.match(items[k], "with (\d+(\.\d+)?)")
9.    noFuItem←items[k].replace(fuzzyTerm,'*')
      fuzzyTerm + ') WITH ' + threshold
13. end
14. end
15. fsparql←compose(selectPart,
      wherePart, otherPart)
16. return fsparql
```

4.2. *Explain Fuzzy Grouping Grammar*

Functions such as *lingTrans* pave the way for the realization of *Group By*. The lingTrans function corresponds to a set of fuzzy membership functions $\{\mu_A, \mu_B \mu_C, ...\}$ through input parameters [*s, p*]. The *lingTrans* and *lingFew* functions, *lingMost* and *lingAbstract* are all downstream extensions to SPARQ1.1. The algorithm process is to first obtain the parameters [*s,p*] of *lingTrans* expression in the f-SPARQL1.1, Then get a series of fuzzy membership function groups, and calculate the fuzzy membership degree, then filter out all fuzzy terms and membership degrees whose membership degree is greater than α .Get the result $\{(A,d_1),(B,d_2),...\}$ after the above steps. Among them, *A* and *B* represent fuzzy terms, and *d* represents degree of membership. The function *lingAbstract* generates simple abstracts based on fuzzy grouping, grouping by fuzzy terms of the provided characteristics, and analyzing the number of each group. The algorithm flow of *lingAbstract* and *lingMost* and *lingFew* are similar. This chapter will explain the lingMost algorithm.

The input of lingMost is ($s,p_1,p_2,...,$), take the subject in the input parameters of lingMost and other predicates in the parameters to form a tuple of parameters (s, p_1), (s, p_2),... and the threshold as the *lingTrans* function Input, get the return value $R = \{(A,d_1),(B,d_2),...\}$. It should be noted that lingTrans may return multiple fuzzy terms with the same degree of membership. When the lingMost input parameter contains only one predicate pi, the result set after executing the data set is Set(R). Analyze the number of fuzzy terms in each group in the set, and take the largest number of fuzzy terms as the result. When the number of multiple fuzzy terms is the same, the distance of the data belonging to this fuzzy term relative to the fuzzy set Core is calculated respectively, and the data with the shortest distance is selected. If it is still the same, all return. The calculation formula (2) of the distance to the Core is the Euclidean distance between the point with the degree of membership as the coordinate and the point with the value of each dimension being 1.

$$\text{distance} = \sqrt{\sum_{i=1}^{n}(1-d_i)^2} \tag{2}$$

When the *lingMost* input parameter contains multiple predicates, the processing method is different from the above process. When a piece of data e contains multiple predicates $p_1,p_2,...p_n$ multiple sets of fuzzy term results , $R_1,R_2,...R_n$ are obtained. For data *e*, a combination of these terms is needed to express the characteristics of data *e*. That is, the Cartesian product operation is performed on the results of multiple sets of fuzzy terms. $R = R_1 \times R_2 \times ...R_n$. When the result is expressed in fuzzy terms and fuzzy membership as $R_1=\{(term_{11},d_1),(term_{12},d_1)\}, R_2=\{(term_{21},d_2),(term_{22},d_2)\}, R_3=\{(term_{31},d_3)\}$, The result after the Cartesian product operation can be expressed as the result of formula (3). The subsequent processing is similar to the processing with only one predicate, searching for the features that contain the most features of fuzzy terms, and calculating the average distance if the number is the same.

$$\begin{gathered}\{((term_{11},d_1),(term_{21},d_2),(term_{31},d_3)),((term_{12},d_1),(term_{21},d_2),(term_{31},d_3)),\\((term_{11},d_1),(term_{22},d_2),(term_{31},d_3)),((term_{12},d_1),(term_{22},d_2),(term_{31},d_3))\}\end{gathered} \tag{3}$$

The process of *f-group by* is actually similar to the above process, and it is a practical example of grouping in the process of *lingMost*. *f-group* by followed by *label([s, p], value)*, grouped according to fuzzy terms corresponding to [*s, p*]. Fuzzy grouping is different from traditional grouping. In traditional grouping, there is no intersection between groups, which is absolute grouping. However, according to fuzzy term grouping, there may be intersections between groups, because according to fuzzy sets, an element may belong to fuzzy group *A*, or it may belong to the fuzzy term *B*. To achieve fuzzy grouping, first apply the *lingTrans* function to convert the attributes of precise data items into fuzzy terms, and input the parameters in the label into *lingTrans*. The process is consistent with the processing process of the *lingMost* function mentioned above. In order to better demonstrate this process, the data used is shown in Table 3. Assuming that the degree of membership has been calculated according to the membership function. After using the above process to execute the query"Query the average height of each age, the degree of membership is 0.5", the result is {177, old}, {176, middle}, {172, young}.

Table 3 Sample data sheet

Name	Height	D(old)	D(middle)	D(young)
N1	150	0	0	1
N2	175	0	1	0
N3	178	1	0	0
N4	189	0	0	1
N5	176	0.5	0.5	0
N6	177	0	0.5	0.5

5. Experiments

5.1. Experimental Environment and Data Set

In order to verify the effectiveness of the experiment, we will analyze the experiment from the result and efficiency of the fuzzy query of the knowledge graph, and compare and analyze it with f-SPARQL, fp-SPARQL, and SPARQLf-p. This article uses JAVA to convert the f-SPARQL1.1 extension into the downstream processing and upstream extension processing of the SPARQL language, using the RDF data set on Univ-Bench University ontology generated by the LUBM-based data generator uba1.7, The data contains data from 50 universities, including the number of triples 680K, the number of subjects 110K, the number of relations 18, and the number of objects 18K. The operating system used in the experiment is Win-10 64-bit, the processor is Intel(R) Core(TM) i7-8550U CPU @ 1.80GHz, 1.99 GHz, RAM is 16GB.

5.2. Time-consuming analysis of fuzzy query

In this work, there are several functions used for fuzzy grouping to convert the data in the knowledge graph into fuzzy terms. There are two main factors affecting the conversion efficiency. One is the amount of data, and the other is when the function *lingFew* is used. *lingMost*, the number of parameters specified in *lingAbstract*. First consider the execution time of the *lingTrans* function when the amount of returned data increases. The time of this function is related to the fuzzy classification operation. Then consider the execution of the *lingFew, lingMost,* and *lingAbstract* functions with different attributes and different amounts of data. Figure 3 shows the execution time of the *lingTrans* function under different data volumes, and the result is the average result after 100 executions. It can be observed from the icon that the results can still be displayed within 1 second under a million data volume, and the tens of millions of data volume can be processed in about 2 seconds.

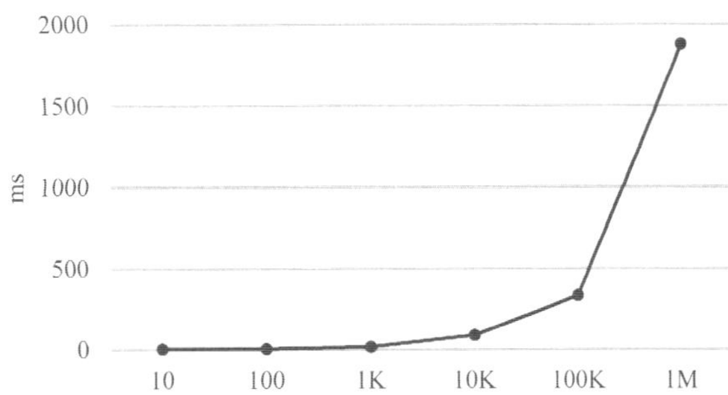

Figure 3 The execution time of the lingTrans function

Then analyze the time consumption of *lingFew*, *lingMost*, and *lingAbstract* with different attribute numbers and data volumes. Figure 4 shows the comparison of the *lingAbstract* function with different data volumes and setting different attribute numbers. In order to avoid the amount of experimental data due to the degree of membership The filter is filtered out and the amount of data changes, and the threshold value in the experiment is set to 0. It can be seen from the figure that as the amount of data and attributes increase, the execution time increases. In the case of 100,000 data, the execution can still be completed within one second, and the execution can still be completed within 10s under the millions of data volume, which reflects the effectiveness of our method. The functions implemented by *lingAbstract* are similar to those implemented by *lingFew* and *lingMost*, so its execution efficiency is almost the same.

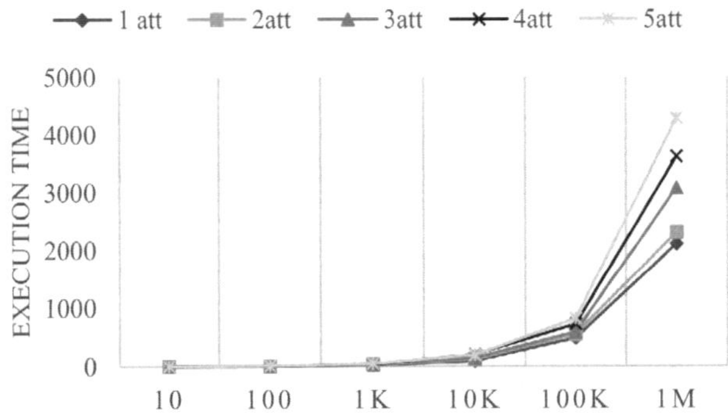

Figure 4 The execution time of the lingAbstract

5.3. Comparison of Fuzzy Membership Function Setting Methods

This chapter uses the same experimental metrics as SPARQLf-p to facilitate comparison, and uses Recall and Precision to analyze the query results. The calculation is shown in formula (4) and formula (5). The query sentence used is the same as the working SPARQLf-p, and the fuzzy membership function defined by it is also used.

$$Recall=\frac{\#relevant\ tuples\ of\ queryresults}{\#relevant\ tuples\ in\ dataset} \tag{4}$$

$$Precision=\frac{\#top\text{-}k\ tuples\ \cap\ relevant\ tuples}{k} \tag{5}$$

The method adopted by SPARQLf-p is to set the fuzzy membership function built-in in the system. When the data changes, it is likely that it no longer meets the user's expectations, and there is a big gap with the form of user-defined membership function. This paper selects 10 users to conduct experiments on fuzzy membership function settings, and compares them with the work of SPARQLf-p using automated and custom methods. After performing the above steps, the average value data of recall and precision of 10 users are obtained as shown in Table 4, where R represents recall and P represents precision.

Table 4 Comparison of recall and precision of different membership function setting methods

	SPARQL*f-p*		**auto**		**custom**	
No.	R	P	R	P	R	P
Q1	0.40	0.40	0.72	0.72	0.95	0.95
Q2	0.60	0.62	0.78	0.78	0.95	0.95
Q3	0.73	0.93	0.76	0.76	0.96	0.96
Q4	0.34	0.34	0.74	0.74	0.98	0.98
Average	0.52	0.57	0.75	0.75	0.96	0.96

From the results in Table 4, it can be analyzed that the membership function set by SPARQLf-p is not applicable to this data, which leads to the problem of not setting preferences Q1, Q2 can not get better results, but Q3 and Q4 due to preferences The existence of the setting can rank the better results higher, because the error of the fuzzy membership function of the working SPARQf-p is mainly to formulate a wider range, so when only the first 50 data are compared, the recall rate is still Can maintain a relatively good level. The automatically set membership function is very close to the user-defined membership function range, so good results have been achieved. Since the user-defined fuzzy membership function is most in line with the user's own thoughts, it is obvious that the user-defined membership function is used. The result of the situation is closest to the user's intention.

6. Conclusion

This paper proposes a fuzzy query method based on the RDF data model of the knowledge graph, which can express query conditions using fuzzy terms in natural language, and can also translate query results into fuzzy terms. This fuzzy query method supplements the fuzzy query expansion for aggregate functions through summary and previous work, and realizes the aggregation operation of fuzzy sets. The expansion is still realized by converting into executable SPARQL language instead of developing new RDF queries. language. We call the above improvement and expansion work f-SPARQL1.1. This article focuses on the method of setting the fuzzy membership function, providing a visual interface and three methods of setting the membership function. Using a custom membership function setting method, the recall rate and precision rate obtained on the LUMB data set have been improved.

References

[1] Lissandrini, M., Pedersen, T. B., Hose, K. and Mottin, D.(2020).Knowledge graph exploration: where are we and where are we going?ACM SIGWEB Newsletter.1-8
[2] Ma, Z., Bai, L. and Yan, L.(2020).Modeling Fuzzy Spatiotemporal Data with XML.Springer Nature,
[3] Castelltort, A. and Martin, T.(2018).Handling scalable approximate queries over NoSQL graph databases: Cypherf and the Fuzzy4S framework.Fuzzy Sets and Systems.348:21-49
[4] Ma, Z., Li, G. and Yan, L.(2018).Fuzzy data modeling and algebraic operations in RDF.Fuzzy Sets and Systems.351:41-63
[5] Mama, R. and Machkour, M.(2021).Fuzzy querying with SQL: Fuzzy view-based approach.Journal of Intelligent & Fuzzy Systems.40:9937-9948
[6] Cheng, J., Ma, Z. M. and Yan, L.(2010).f-SPARQL: A Flexible Extension of SPARQL. In: Database and Expert Systems Applications(DEXA).487–494
[7] Wang, H., Ma, Z. M. and Cheng, J.(2012).fp-Sparql: An RDF fuzzy retrieval mechanism supporting user preference. In: Conference on Fuzzy Systems and Knowledge Discovery(FSKD).443-447
[8] Ma, R., Jia, X., Cheng, J. and Angryk, R. A.(2016).SPARQL queries on RDF with fuzzy constraints and preferences.Journal of Intelligent & Fuzzy Systems.30:183–195
[9] Zadrożny, S. and Kacprzyk, J.(2020).Fuzzy Analytical Queries: A New Approach to Flexible Fuzzy Queries. In: 2020 IEEE International Conference on Fuzzy Systems (FUZZ-IEEE).1–8
[10] Bosc, P., Pivert, O. and Smits, G.(2010).On a Fuzzy Group-By and Its Use for Fuzzy Association Rule Mining. In: Advances in Databases and Information Systems.Berlin, Heidelberg.pp.88--102
[11] Zadeh, L. A.(1996).Fuzzy sets.World Scientific,
[12] Almendros-Jiménez, J. M., Becerra-Terón, A., Moreno, G. and Riaza, J. A.(2021).Flexible Aggregation in FSA-SPARQL. In: International Conference on Fuzzy Systems (FUZZ-IEEE).1-7

[13] Li, G., Li, W. and Wang, H.(2020).Querying Fuzzy RDF Knowledge Graphs Data. In: 2020 IEEE International Conference on Fuzzy Systems (FUZZ-IEEE).1-8
[14] Sarhan, A., Rokne, J. and Alhajj, R.(2018).Integrating flexibility and fuzziness into a question driven query model.Information Sciences.430-431:349-377
[15] Yager, R. R., Reformat, M. Z. and To, N. D.(2019).Drawing on the iPad to input fuzzy sets with an application to linguistic data science.Information Sciences.479:277-291
[16] Yager, R. R., Reformat, M. Z. and To, N. D.(2019).Drawing on the iPad to input fuzzy sets with an application to linguistic data science.Information Sciences.479:277-291

Scatter correction using FBP with data rearrangement for a single full scan CBCT

Guanyu Kang[1,2,a], Yu Han[2,b], Xiaoqi Xi[2,c], Lei Li[2,d], Siyu Tan[2,e], Hongfeng Zhao[3,f] and Bin Yan[2,g*]

[1] Cyberspace Security Institute of Zhengzhou University, Zheng zhou, He nan, China

[2] School of Information Systems Engineering of the University of Information Engineering, Zheng zhou, He nan, China

[3] Dongguan Xinda Optoelectronics Technology Co, Dong guan, Guang dong, China

[a]email: 2271406781@qq.com, [b]email: hy007hy007@126.com, [c]email: 254096709@qq.com, [d]email: leehotline@163.com, [e]email: tan.siyu@hotmail.com, [f]email: ZHF15639140797@126.com

* Corresponding author: [g]email:ybspace@hotmail.com

Abstract. A scattering plate consisting of multiple lead strips is an effective tool for scatter correction and dose reduction in cone-beam CT(CBCT). Existing scattering plate correction methods take advantage of data redundancy by partially blocking cone beam data from a single scan to make up and perform scattering correction. However, for geometric objects with large cone angles during CBCT scans, existing backprojection-filteration(BPF) algorithms cannot accurately reconstruct the object due to data inconsistency and truncation, and may produce additional image artifacts. In order to improve the image quality at large cone angles, this paper proposes a hybrid scatter correction method for filteration-backprojection (FBP) image reconstruction from partially blocked cone beam data using data rearrangement. The proposed method uses a blocker consisting of multiple lead strips that are placed between the light source and the object to provide scatter correction and image reconstruction with beam-blocked data. In combination with a data rearrangement algorithm, the cone-beam projection data is rearranged into a tilted parallel projection data, followed by weighted filtering and inverse projection reconstruction. The experimental results of CBCT show that the proposed method effectively corrects the scattering in each projection and reconstructs the scatter-corrected image from a single scan.

1. Introduction

Scattering artifacts are the main cause of image quality degradation in cone-beam CT(CBCT) imaging, and in general, scattering leads to cupping artifacts and dark streak artifacts that reduce image contrast and resolution [1]. In the past decades, many scatter correction methods have been proposed and developed by researchers. The correction methods are mainly classified into: Monte Carlo (MC) simulation, analytical modeling, and scattering plate correction methods. MC methods estimate the scattering distribution by calculating the photon transport process. MC simulation provides relatively accurate scattering estimates, but the accuracy is usually a trade-off with the amount of computation. Therefore, MC simulations are usually combined with other methods to improve the computational speed. Analytical modeling methods calculate scattering by performing a convolution operation on the

Content from this work may be used under the terms of the Creative Commons Attribution 3.0 licence. Any further distribution of this work must maintain attribution to the author(s) and the title of the work, journal citation and DOI.
Published under licence by IOP Publishing Ltd

scattering kernel of the measured data. Convolution-based methods are computationally efficient, but kernels must be predefined to be used and kernel parameters are sensitive to influence the results. Scattering plate-based methods typically use x-ray beam blockers or scattering plates and can achieve accurate scattering estimates without considering complex physical processes. In 2004, Ning et al [2]. proposed the scattering correction plate method, which takes advantage of this physical property of the strong attenuation of X-rays by lead, and the scattering distribution produced by the object is measured directly by scanning and using interpolation, but a second scan is required, increasing the scanning dose and cost. In 2011, Niu et al [3]. proposed an improved scattering correction plate to take advantage of the redundancy of data during the full scan, constructing a set of alternating left and right lead strips, which can also be acquired by rotating to the left when the right object is obscured, and by reconstructing the left and right sides separately, the whole correction process can be completed with only one scan. In 2018, Liu et al [4]. proposed a scattering correction plate method based on the local filter operator, for the problem of low scattering estimation accuracy of the traditional method, combined with the local filter operator, the reconstruction algorithm is optimized that suppresses the effect of truncation artifacts and improves the scattering correction effect. However, they usually encounter a problem that for scanned objects with large cone angles, it is not possible to obtain enough original signals from the blocked cone beam data for accurate image reconstruction by exploiting redundancy. Due to the discontinuity of the projection data, the conventional FBP algorithm cannot be directly applied to image reconstruction. The existing BPF algorithm used to solve the image reconstruction problem performs the filtering process in the last step of the image domain, and although the reconstruction is more robust to discontinuous data than the FBP algorithm, the reconstruction accuracy is lower compared to the FBP algorithm.

Therefore, this paper uses data rearrangement to propose a hybrid scattering correction method for filtered inverse projection reconstruction coming from partially blocked cone beam data. In this paper, a scattering plate consisting of lead strips arranged at equal intervals is designed and placed between the light source and the measured object, and the lead strips are parallel to the rotation axis direction, and the projection data obtained by scanning will produce discontinuities along the sector direction. By designing the width between the lead strips of the scattering plate, the data redundancy in the middle plane of the cone beam data and the quasi-redundancy in the non-middle plane can be used, and the quasi-redundant data in the non-middle plane can be data rearranged so that sufficient image reconstruction data can be obtained, after which the reconstructed image of the object to be measured is finally reconstructed by weighted filtering and inverse projection reconstruction.

In this paper, we first analyze the reordering algorithm of the cone beam data, then introduce the composition of the proposed scattering correction plate and the process of scattering correction if data redundancy is utilized, and finally demonstrate the feasibility of the proposed method through actual CBCT experiments.

2. Materials and Methods

2.1. Data rearrangement based FBP reconstruction algorithm

The filtered inverse projection reconstruction algorithm based on data rearrangement consists of three main parts: data rearrangement, filtering and inverse projection reconstruction [5].

2.1.1. Data rearrangement.

The geometry of the CBCT system scan is shown in Figure 1. The cone-beam X-ray source emits a beam from S, passes through the object under test, reaches the flat detector and is received, and the trajectory of the X-ray source is a circle of radius R. In order to describe the whole projection process, a virtual detector is introduced parallel to the real detector and located at the central circle point. A three-dimensional coordinate system is established with the center of rotation as the origin, and the set transformation from the real detector to the virtual detector has a certain proportional transformation relationship, i.e. the projection value of any X-ray can be determined on the virtual detector by $p(\beta,m,n)$, where β is the projection angle of the line from the X-ray to the origin to the x-axis, and m and n are its horizontal and vertical coordinates in the coordinate system, respectively.

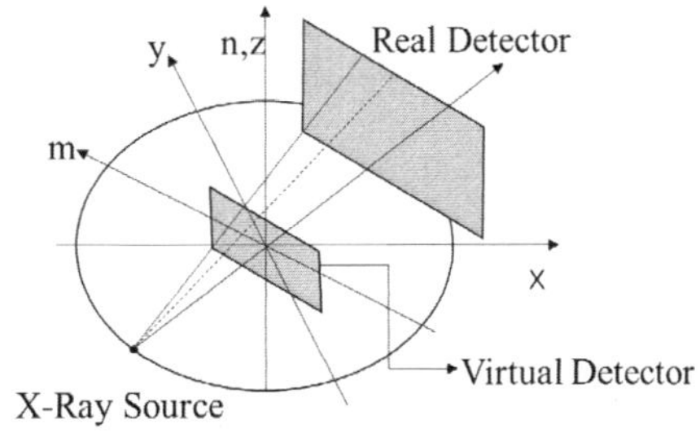

Fig. 1. Cone beam CT source trajectory scan geometry

The CBCT data are first rearranged into tilted ray beams parallel to each other, as shown in Figure 2. Any of the rearranged ray projection values can be represented by $p_v(\gamma,t,s)$, where γ denotes the angle between the parallel beam and the X-axis, and t and s denote the horizontal and vertical coordinates, respectively.

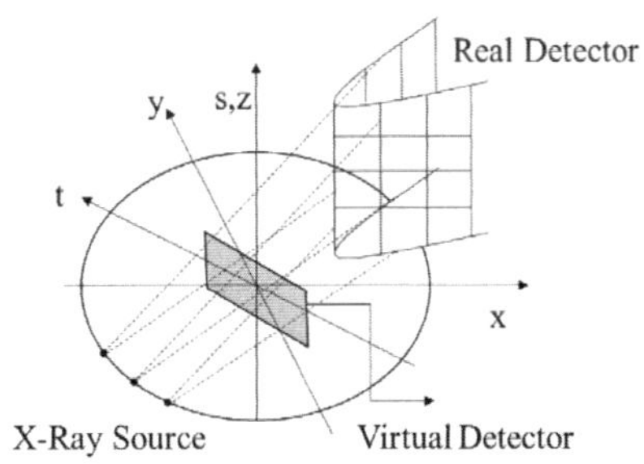

Fig. 2. Rearranged circular trajectory scan geometry

The geometric coordinate transformation relations before and after rearrangement can be shown by Fig. 3, and the following transformation relations exist:

$$\beta = \gamma - \arcsin\frac{t}{R} \tag{1}$$

$$m = \frac{t \bullet r}{\sqrt{R^2 - t^2}} \tag{2}$$

$$n = \frac{s \bullet R^2}{R^2 - t^2} \tag{3}$$

Combining Eqs. (1)-(3), the process of data rearrangement can be expressed as:

$$P_v(\gamma,t,s) = p(\gamma - \arcsin\frac{t}{R}, \frac{t \bullet r}{\sqrt{R^2 - t^2}}, \frac{s \bullet R^2}{R^2 - t^2}) \tag{4}$$

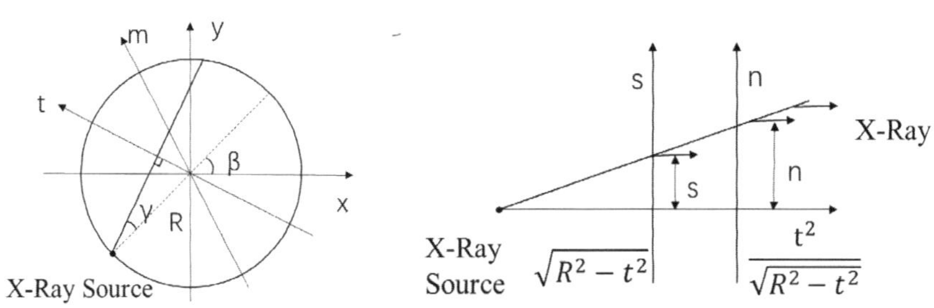

Fig. 3. Geometric diagram of the rearrangement process

2.1.2. Weighted filtering.
FBP algorithm has a simple form with a common tilt filter to filter the rearranged data row by row and do a cone angle weighting. The filtering process can be expressed as follows:

$$P_f(\gamma,t,s)=\{P_v(\gamma,t,s)\times\cos[k(t,s)]\}*g(t) \quad (5)$$

2.1.3. Back projection.
The reconstructed result is obtained by weighting and filtering the projection data after data rearrangement and then back-projecting it. The process of assigning each pixel value on the detector to the beam across all voxels is called back-projection. The inverse projection equation can be expressed as follows:

$$f(x,y,z)=\frac{1}{2}\int_0^{2\pi}P_f(\gamma,t(x,y,\gamma),s(x,y,z,\gamma))d\gamma \quad (6)$$

2.2. Setting of scattering plate
The design and geometric setup of the scattering plate is shown in the left panel of Fig. 4. The lead strips are placed at equal intervals on a thin acrylic plate. The thickness of the strips is about 2 mm, and the width and period of the lead strip strips are 8 mm and 20 mm, respectively, with the length varying according to the actual object size. The lead strips are placed in such a way that they can completely attenuate the X-rays and ensure that the shadow data blocked by the lead strips are scattered values. Considering the penumbra effect and the setup error, we used a strip width less than half the period size of 10.0 mm to ensure the integrity of the object's projected data at complementary angles.

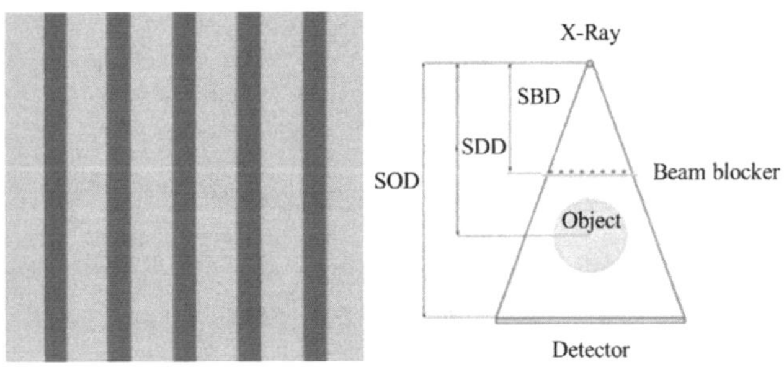

Fig. 4. Geometry of the scattering plate

The scattering plate is placed between the light source and the object and positioned relatively close to the object under test, as shown in the right panel of Figure 4, to ensure a suitable magnification ratio and to keep the red line of the scattering plate and the rotation axis of the light source and the object under test in the same plane to minimize secondary scattering and penumbra effects at the edge of the scattering plate [6].

2.3. Scattering calibration process
Scattering carries mainly low frequency components and is insensitive to small deformations of the internal structure [7]. Since the X-ray photons are almost completely blocked by the scattering plate lead strip, the measured data in the blocked shadow region can be considered as scattering signal. Therefore, the scattering distribution can be estimated from the partially blocked projection data using interpolation methods. x-rays are the same for the same position of the same object at the same initial energy, and the projection data of the object at complementary angles during the circumferential scan are the same. There is redundancy in the scan data during the scanning process [8]. Therefore, the redundancy of the data existing in the scanning process can be exploited to obtain sufficient projection data with only one scan while ensuring that the shielded area can be measured with a sufficiently accurate scattering distribution. Figure 5 shows the scattering correction process.

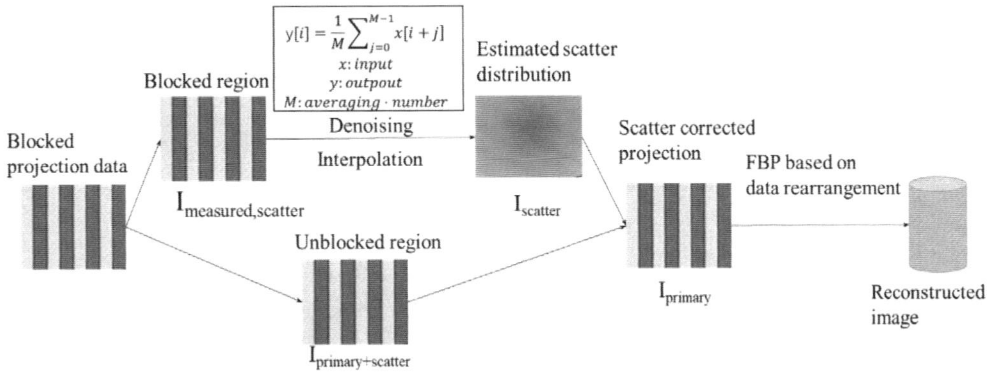

Fig. 5. Scattering calibration process diagram

Scattering estimation is first performed using the central part of the shadows. To avoid the penumbra effect at the edges of the strip shadows, only the data within the central one-third region of each strip shadow is used for estimation [9], and the scattering values to the scattering distribution of the data under the lead strip shadow region are estimated using cubic spline interpolation and subtracted from the original projection data. Then based on the FBP algorithm of data rearrangement, the projection data are rearranged into tilted parallel beam projection data [10], and the projection data are complemented by using the redundancy of the data, and finally the reconstructed results can be obtained by back-projection after weighted filtering of the complemented projection data.

3. Experiments and Results

Fig. 6. Floor 714 of Catphan 700 and projection

The real data in the experiments were obtained by scanning using an industrial CT system in our laboratory. The industrial CT system in our laboratory is a CT system for circular trajectory scanning, which includes a detector, a ray source, a stage, a mechanical system, and computer acquisition and projection software. The source is a 150 kV MICROFOCUS X-ray source L12161-07 and the detector is a THALES PIXIUM RF4343. The Catphan700 body model and the scanned projection images are shown in Figure 6. The sampling interval of the projection is 360° and the scanning interval is 1°. The distance from the light source to the object under test and the distance from the light source to the detector are 750 mm and 969.7 mm, respectively.

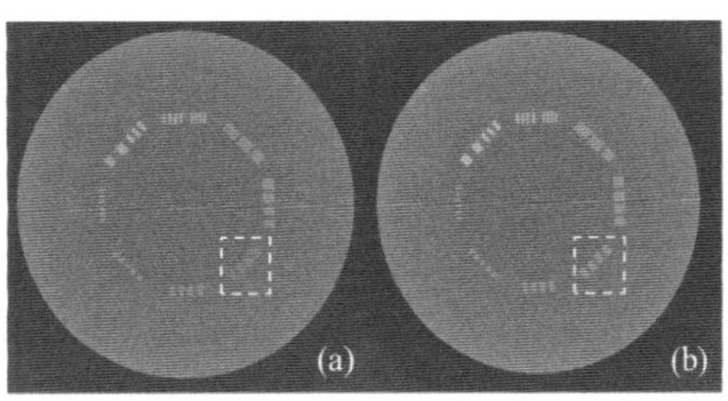

Fig.7. Reconstruction results of aircraft industrial devices (a) Without scattering correction; (b) Correction results of the method in this paper

The experimental results are shown in Figure 7, which is the reconstructed image of the CTP714 layer in the Catphan 700 body model. Figure 7 (a) is the uncorrected reconstructed image, and (b) is the reconstructed image after scatter correction by the method in this paper. Before correction, due to the effect of scattering, the image has low contrast in the yellow marked area, poor image uniformity, and obvious cup artifacts. After correction by the method in this paper, the image is more uniform, the cupping artifacts are reduced, and the edge structure of the yellow marker region of the image is recovered.

The reconstructed images were quantitatively analyzed, and the region of interest (ROI) used for the calculation and the center and edge regions for calculating the object T_{cup} were selected from within the yellow marked area, and the results are shown in Table 6. the scatter correction resulted in an overall improvement in image contrast and CNR, and the decrease in T_{cup} indicated an improvement in cupping artifacts.

Table 1 Comparison table of measurement parameters before and after correction

Numerical Comparison	Evaluation Indicators			
	Contrast	CNR	$T_{cup\%}$	RMSE
Without correction result	7.45	4.56	4.16%	36.54
Correction results of the method in this paper	14.67	13.15	2.89%	20.46

4. Conclusions

In this paper, a new scattering correction method based on scattering plates is proposed, which takes advantage of redundancy to partially block the projection data and deducts the scattering, and then combines the data rearrangement to reconstruct the image from the scattering corrected projection data by the FBP algorithm. Both practical experiments and numerical measurement studies validate and evaluate the proposed method, and successfully demonstrate that the proposed method can estimate the scattering accurately. The proposed method can provide fast and effective CBCT scattering correction for practical applications.

Acknowledgments

National Key Research and Development Program (No.2020YFC1522002)

References

[1] Turan E. Fiber grating spectra[J]. Journal of Lightwave Technology, 1997, 15(8): 1468-1488.
[2] Ning R, Tang X, Conover D, et al. X-ray scatter correction algorithm for cone beam CT imaging[J]. Medical Physics, 2004, 31(5): 67–75.
[3] Niu T, Zhu L, et al. Overview of X-ray Scatter in Cone-beam Computed Tomography and Its Correction Methods[J]. Current Medical Imaging Reviews, 2011, 38(9): 5186-5199.
[4] Liu J. Scatter Correction Method for X-Ray CT: Theory and Preliminary Results[J]. IEEE Transactions on Medical Imaging, 2018, 25(12): 1573-1587.
[5] Zhou K, Tian J, Meng X, et al. A Novel Scatter Correction Method for Cone-Beam Computed Tomography [J]. Journal of Computer Assisted Tomography, 2016, 40(4): 632-640.
[6] Siewerdsen J H, Moseley D J, Bakhtiar B, et al. The influence of antiscatter gridson soft-tissue detectability in cone-beam computed tomography with flat-panel detectors[J]. Medical Physics, 2004, 31(12): 3506-3520.
[7] Xu S, Prinsen P, Wiegert J, et al. Deep residual learning in CT physics: scatter correction for spectral CT[J]. IEEE Access, 2017, 5(99): 1050-1061.
[8] Sisniega A, Zbijewski W, Xu J, et al. High-fidelity artifact correction for cone-beam CT imaging of the brain[J]. Physics in Medicine & Biology, 2015, 60(4): 1415-1439.

[9] Kida S, Nakamoto T, Nakano M, et al. Cone Beam Computed Tomography Image Quality Improvement Using a Deep Convolutional Neural Network[J]. other, 2018, 25(7): 817-827.
[10] Zhang Y, Chen Y, Zhong A, et al. Scatter correction based on adaptive photon path-based Monte Carlo simulation method in Multi-GPU platform[J]. Computer Methods and Programs in Biomedicine, 2020, 53(14): 3841-3861.

Few Shot Object Detection for Tangka Seats Based on Deformable Convolution

Panpan Xue[1,2*], Wenjin Hu[1,2], Guoyuan He[1,3]

[1]Key Laboratory of China's Ethnic Languages and Information Technology of Ministry of Education, Northwest Minzu University, Lanzhou, Gansu 730030, China

[2]School of Mathematics and Computer Science, Northwest Minzu Univsersity, Lanzhou, Gansu 730000, China

[3]National Languages Information Technology, Boulder, Northwest Minzu University, Lanzhou, Gansu, 730000, China

[*]Corresponding author's e-mail: y191730464@stu.xbmu.edu.cn

Abstract. According to deformable target sizes in the detection for Thangka Seats, an optimized few shot Thangka detection model is proposed. The deep residual network based on the deformable convolution is designed to extract the Thangka image features to decrease the missing detections rate. Experimental results show that the proposed method has better performance than the FSOD model on the Thangka dataset. The AP of 2-way 1-shot is 28.5 %, and the AP50 reaches 67.6 %, which increases by 4.7 % and 4.7 %, respectively.

1. Introduction

Tangka, a scroll painting, is a unique artistic expression of Tibetan Buddhism. Due to the rich historical background and national characteristics, it was listed in the name of intangible cultural heritage by the United Nations in 2006[1]. The detection of Tangka Seats can deepen the understanding of the semantic common sense behind Tangka, and bring convenience to the Tangka cultural relics enthusiasts and researchers. Therefore, the detection of Tangka Seats based on machine learning has important academic significance.

Deep convolutional neural networks (CNNs) have achieved success in target detection[2-6], which largely depends on a large number of training data with accurate label annotation. In recent years, few shot object detection has been developed rapidly. Given some base classes with sufficient labels and some new classes with only a small number of samples, the task of few shot object detection is to obtain a model that can detect both base objects and new objects in the test, which will bring convenience to the headwear detection of Thangka. LSTD[7] proposed knowledge transfer regularization and background suppression regularization to promote knowledge transfer from source domain to target domain. Karlinsky et al.[8] converts the Region of Interest (RoI) into the feature vector as the input of distance metric learning sub-network. Singh et al.[9] adopt the fine-tuning method to obtain the detection results by fine-tuning the pre-training model in the target domain. However, all these methods learn category-specific embedding and require fine-tuning for new categories. FSAM[10] proposes a few shot object detection network to detect unseen category objects with only a few training instances. The Attention-RPN and multi-relation detector make full use of the

Content from this work may be used under the terms of the Creative Commons Attribution 3.0 licence. Any further distribution of this work must maintain attribution to the author(s) and the title of the work, journal citation and DOI.
Published under licence by IOP Publishing Ltd

similarity between a small number of training samples and test sets to detect new objects, and suppress the error detection in the background.

In order to decrease the rate of miss detection and false detection and improve the detection accuracy for small-sample Thangka dataset, we propose a deformable few shot object detection method(dc-fsod) to adapt to the detection of small sample Thangka dataset, which introduce deformable convolution(DC)[11] to extract feature in ResNet. This method has better realized the research purpose of real-time object detection in small sample Tangka dataset, improve the accuracy of object detection, decrease miss detection rate and false detection rate, and verify the effectiveness of this method in small sample Tangka image detection.

2. Proposed Method

The dc-fsod model is a small sample object detection network with a weight-shared framework, shown in Figure 1. The model first extracts features of support images and query images through ResNet with DC, and then performs the bounding box regression and localization in the query branch, which is the Faster R-CNN network containing RPN and detector. This framework is utilized to train the matching relationship between support and query features, so that the network can learn common features among same categories. Since RPN only classifies anchor boxes between foreground and background, all foreground anchor boxes are given high confidence, and the ROI sent to the next stage contains many categories unrelated to the support image, which affects the later training of the network. When the support set information is introduced into RPN through attention mechanism, i.e., Attention-RPN, the above problem can be well addressed. The Attention-RPN produce relevant proposals while suppressing proposals in other categories. Finally, the multi-relation detector is adopted to produce an accurate parsing between support and potential boxes in the query.

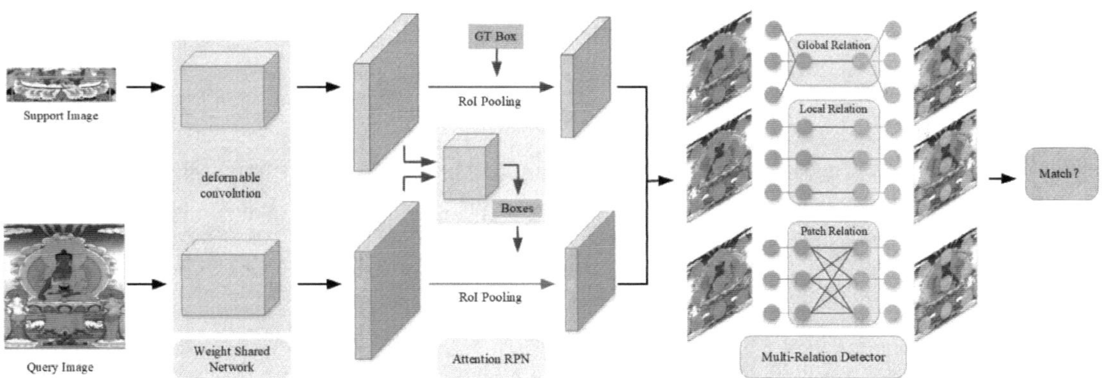

Figure 1 dc-fsod model architecture

2.1. Introduction of DC

In the 2D convolution, for each location p_0 on the output feature map, the eigenvalue calculation formula is

$$G(p_0) = \sum_{p_i \in K} \omega(p_i) f(p_0 + p_i)$$

$$\mathcal{F} = \{(x,y)|(-1,-1),(-1,0),\cdots,(1,1)\} \quad (1)$$

Among them, $i = 1,...,N$, $N=|F|$, ω represents the weight of convolution kernel, f is the input feature map, F represents the output feature map, which defines a 3×3 kernel with dilation 1, p_0 represents central location and p_i enumerates the location in F.

Figure 2 shows the image data of Seats in Thangka Main Portraits. It can be seen that the appearance of Seats are variable and the size is different. Therefore, it is necessary to address the problem of the geometric deformation of the target in the process of object detection in Thangka image. However, the traditional convolution operation does not have scale invariance and rotation invariance, which induce that the fixed geometric shapes are only dealt with, and the geometric transformation of the target can not be adapted.

Figure 2 The Image Data of Seat in Thangka Seats

For this reason, DAI et al. proposed a deformable convolution network, which replaced the standard convolution in the neural network with DC. Numerous studies[12-13] have shown that this method has excellent performance in the field of target detection. In this paper, deformable convolution is introduced into the backbone network ResNet41.

In DC, the regular grid F is augmented with offsets {Δp_i|i=1,...,N}, where N=|F|. Equ.(1) becomes

$$G(p_0) = \sum_{p_i \in k} \varpi(p_i) f(p_0 + p_i + \Delta p_i) \tag{2}$$

where $f(p_0 + p_i + \Delta p_i)$ represents the input eigenvalue of the sampling offset location. As illustrated in Figure 3, DC adopts a parallel convolution layer to learn offset so that the sampling point of the convolution kernel on the input feature map are offset. That is, a direction vector is added to the location of each sampling point, which makes the convolution kernel become arbitrary shape and realizes random sampling near the current location. When the sampling is concentrated in the ROI or the target, the modeling ability for the deformed target is further improved.

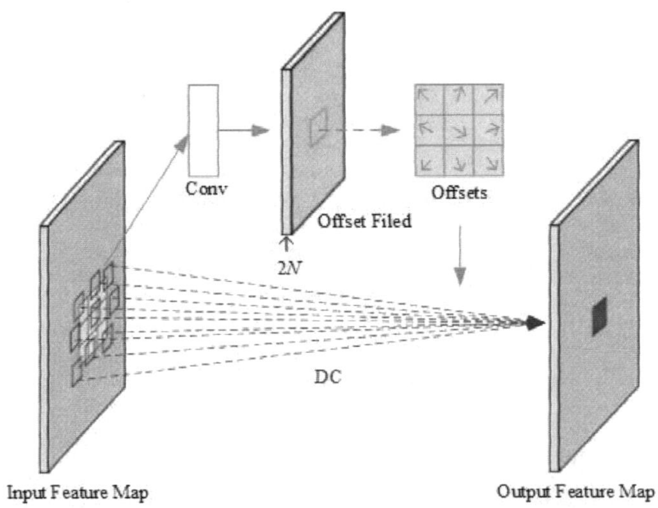

Figure 3 Illustration of 3×3 deformable convolution

3. Experiments

3.1. Experiments Sets

The experiment employs pytorch framework. and is carried out in Ubuntu 16.04 system. All experimental platforms are: Intel (R) Core (TM) 5-1035G1, the main frequency is 1.19GHz, and the memory is 16GB. GPU is used to accelerate the calculation.

3.2. Tangka Dataset

Our research object is Tangka Seats, which is from the Institute of Thangka, School of Art, Northwest Minzu University. There are three categories: padmasana, cushion, sumeruseat, as shown in Figure 4.

Figure 4 Tangka image detection categories

We balance the dataset by means of data enhancement to further improve the detection accuracy. Table 1 shows the number of categories before and after the Thangka image equalization.

Table 1. The number distribution of categories before and after Tangka images equalization.

augmentation	padmasana	sumeruseat	cushion
before	659	89	96
after	747	598	629

3.3. Result Analysis

The method proposed in this paper is to solve the detection problem of Seats in Tangka images. In order to verify the effectiveness of our method on the Tangka dataset, the experiments are carried out in three types of samples.

We train respectively dataset on the FSOD model and our model. In the experiment, we set the learning rate as 10-4, the number of iterations as 70000, and the batch size as 4, and then evaluate the performance of the model on the new testing set.

We construct the few shot learning problem of 2-way 1-shot for the training of small sample data, and 2*1 samples which are input as support set for training the target detection task.Fig.9 shows the test process of 2-way 1-shot and the performance results of the experimental test set evaluation model are shown in Table 2.

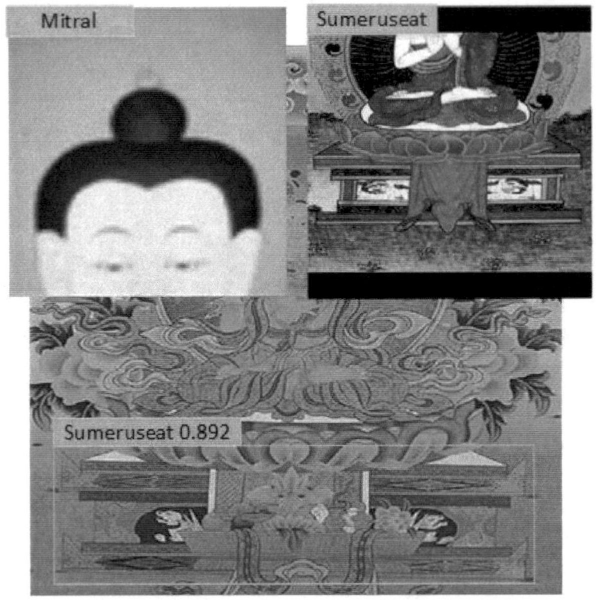

Figure 5 Test results of 2-way 1-shot

Table 2 Evaluation results of Tangka image detection in the 2-way 1-shot.

Method	AP(%)	AP50(%)
FSOD	23.8	63.9
ours	**28.5**	**67.6**

As shown in Table.3, in 1-shot task, our method achieves better results than FSOD model.

4. Conclusion

In order to address the problems of different sizes in the Tangka image detection, we improves the dc-fsod model as follows: DC is introduced to extract the feature of the Tangka image to adapt to the problem of deformable sizes of the Tangka image. The experimental results show that the proposed method has lower miss detection and higher detection accuracy than the FSOD model on the Thangka dataset.

Acknowledgments

This work was supported by the National Natural Science Foundation of China under Grant No. 62061042, 61862057, the Fundamental Research Funds for the Central Universities under Grant [Yxm2021069], [31920200066],[Yxm2021070] the program for innovative research team of SEAC[2018[98]], the Northwest Minzu University Graduate Research Innovation Project [Yxm2021069] and the special fund for talent introduction of Northwest Minzu university.

References

[1] Ge J F. From Temple to Exhibition Hall: Reconstruction of Tangka Viewing Mechanism under the Transformation of Space[J]. Journal Of Lanzhou University(Social Sciences), 2018, 46(2):74-80.

[2] Redmon J, Farhadi A. Yolov3: An incremental improvement[J]. arXiv preprint arXiv:1804.02767, 2018.

[3] Bochkovskiy A, Wang C Y, Liao H. YOLOv4: Optimal Speed and Accuracy of Object Detection[J]. 2020.
[4] Liu W, Anguelov D, Erhan D, et al. Ssd: Single shot multibox detector[C]//European conference on computer vision. Springer, Cham, 2016: 21-37.
[5] Fu C Y, Liu W, Ranga A, et al. Dssd: Deconvolutional single shot detector[J]. arXiv preprint arXiv:1701.06659, 2017.
[6] Ren S, He K, Girshick R, et al. Faster R-CNN: Towards Real-Time Object Detection with Region Proposal Networks[C]. Advances in Neural Information on Processing Systems, 2015.
[7] Chen H，Wang Y L，Wang G Y，et al. LSTD:A Low-shot Transfer Detector for Object Detection［J］. arXiv Preprint，2018，arXiv: 1803．01529.
[8] Karlinsky L， Shtok J， Harary S， et al. RepMet: Representative-Based Metric Learning for Classification and Few-Shot Object Detection[C]// 2019 IEEE/CVF Conference on Computer Vision and Pattern Recognition (CVPR). IEEE, 2019.
[9] Singh P, Varadarajan S, Singh A N, et al. Multi-domain Document Layout Understanding Using Few-Shot Object Detection[M]. 2020.
[10] Fan Q, Zhuo W, Tang C K, et al. Few-Shot Object Detection With Attention-RPN and Multi-Relation Detector[C]// 2020 IEEE/CVF Conference on Computer Vision and Pattern Recognition (CVPR). IEEE, 2020.
[11] Dai J, Qi H, Xiong Y, et al. Deformable Convolutional Networks[C]// The IEEE Conference on Computer Vision and Pattern Recognition(CVPR), 2017:764-773.
[12] Wei R, Song Y, Zhang Y. Enhanced Faster Region Convolutional Neural Networks for Steel Surface Defect Detection[J]. ISIJ international, 2020, 60(3):539-545.
[13] Xu X, Yu T, Hu X, et al. SALMNet: A Structure-Aware Lane Marking Detection Network[J]. IEEE Transactions on Intelligent Transportation Systems, 2020, PP(99):1-12.

Research on intrusion detection method based on feature selection and integrated learning

Qun Liu[1,a], Zhiyong Tong[2,b], Shuiqing Wang[1,c], Ziheng Yang[1,d*]

[1]Electronic Engineering College, Heilongjiang University, Haerbin, Heilongjiang, China

[2]Heilongjiang Provincial Military Command, Haerbin, Heilonhjiang, China

[a]email: lqun9601@163.com, [b]email: tzy1997@163.com,

[c]email: 17863934828@163.com,

*Correspondence: [d]email: yzh@hlju.edu.cn

Abstract. With the introduction of computer technology, network attacks have become more frequent. Some illegal elements may intrude into computers through network attacks to tamper with messages, spread viruses and other destructive behaviors, causing great damage to personal sensitive information, industrial control networks, transaction systems, etc. . For this, this design proposes an improved intrusion detection method based on feature selection and integrated model. The NSL-KDD training data set is used to evaluate the proposed model. First, balance the data categories through the SMOTE-ENN method, and then use feature selection technology and PCA feature extraction technology to reduce the number of irrelevant features and improve the classification accuracy. Finally, using CART as the base classifier, Bagging technology is used to establish an integrated model and an intrusion detection system. Experimental results show that the CART-based Bagging method provides better accuracy, lower false alarm rate and faster model training speed, and the system can detect intrusion attacks with similar attributes and has a certain degree of adaptability.

1. Introduction

With the rapid development of information and network technology, industrial equipment has formed a network, and the high value or absolute number of various types of transactions have caused industry and commerce and their customers to face serious risks. These risks are exacerbated by the increase in the number and maturity of interconnected networks and malicious opponents. Ensuring information security is an important task that must be completed in today's rapid development of the Internet of Things.

Intrusion Detection System (IDS) is an important tool for computer network protection. The potential of intrusion detection technology has a fundamental impact on the performance of intrusion detection systems. IDS prevents possible intrusions by detecting network traffic to ensure the security and integrity of the network information[1]. Recently, IDS systems based on machine learning (ML) have been widely used in network intrusion detection tasks[2]. This article turns to machine learning algorithms for intrusion detection. Machine learning can allow the system to automatically improve performance as experience increases. The decision tree classification algorithm uses the probability of various events to construct a complete decision tree, which is used to assess and predict event risks. The decision tree classification algorithm satisfies the needs of intrusion detection systems very well.

Content from this work may be used under the terms of the Creative Commons Attribution 3.0 licence. Any further distribution of this work must maintain attribution to the author(s) and the title of the work, journal citation and DOI.
Published under licence by IOP Publishing Ltd

It can not only classify attacks and abnormal behaviors, but also actively detect and prevent risks. For the task of intrusion detection, this paper proposes an improved intrusion detection method based on feature selection and integrated model. This method can improve the recognition rate of network attacks and reduce the false alarm rate to a certain extent.

2. Intrusion detection system design

This paper is oriented to the task of intrusion detection. On the NSL-KDD data set, the CART decision tree is used as the base classifier to construct an intrusion detection mechanism through the integrated learning method in machine learning.

2.1. Decision tree

Decision tree is a classifier. The main advantage of decision tree over many other classification techniques is that they generate a set of simple and efficient rules that can be combined with other network protection methods. The core key point affecting the learning of decision tree is how to find the optimal division attribute, we hope that as the splitting process progresses, the data contained in the nodes in the decision tree belongs to the same category to the greatest extent, and then introduces the CART decision tree[3] classification algorithm used in this article.

CART algorithm: CART algorithm is also a decision tree algorithm, and the division of its nodes is based on the Gini index formula (1)

$$Gini(D)=\sum_{k=1}^{|y|}\sum_{k'\neq k} p_k p_{k'} = 1-\sum_{k=1}^{|y|} {p_k}^2 \quad (1)$$

where *Gini(D)* means extracting two samples from data set *D*. The Gini exponent for attribute a can be defined as the formula (2)

$$Gini\_index(D,a)=\sum_{v=1}^{V}\frac{|D^v|}{|D|}Gini(D^v) \quad (2)$$

Finally, in the attribute set A, the attribute with the smallest Gini index after the division is used as the optimal division basis as shown in formula (3)

$$a_* = \arg\min Gini\_index(D,a) \quad (3)$$

where $a \in A$.

2.2. Intrusion detection based on integrated learning

2.2.1. Data balance processing.

There have been many researches on intrusion detection based on machine learning methods, and the most common data used among them is the NSL-KDD dataset[4]. This article uses KDDTrain+ and KDDTest+. The data distribution is shown in Table 1.

It can be konwn from Table 1 that although the NSL-KDD data set has been processed, its category distribution is obviously unbalanced, of which Normal category data accounts for 53.46%, while U2R only has 0.04%. In a real network environment, the normal category of network traffic is also much higher than the attack category. In this case, IDS will focus on the monitoring of normal data flow, thereby improving the accuracy of the overall data flow recognition. Therefore, some attack data traffic may be ignored, which reduces the performance of the protection system. At present, sampling methods such as over-sampling and under-sampling are commonly used to deal with the problem of unbalanced data. Over-sampling methods repeat or synthesize new samples for insufficient classes, while under-sampling will reduce samples of larger classes. In this article, we first use the over-sampling method SMOTE to amplify samples with fewer sample categories[5], and finally use the under-sampling method ENN to remove noise samples, thus forming the SMTE-ENN method.

Table 1 Distribution of cyber attacks

Data Set	Number Statistics					
	Total	Normal	Dos	Probe	U2R	R2L
Training Set	125973	67343	45927	11656	52	995
		53.46%	36.46%	9.25%	0.04%	0.79%
Test Set	22544	9711	7458	2421	200	2654
		43.08%	33.08%	10.74%	0.89%	12.22%

As shown in Figure 1, The sample sizes of Normal, Dos, Probe, R2L, and U2R categories in the original data are 67343, 45927, 11656, 995, and 52, respectively. After the SMOTE-ENN method, the sample sizes are 66877, 67285, 67269, 67277 and 67333, we will conduct research on intrusion detection methods based on decision tree on the basis of this sample.

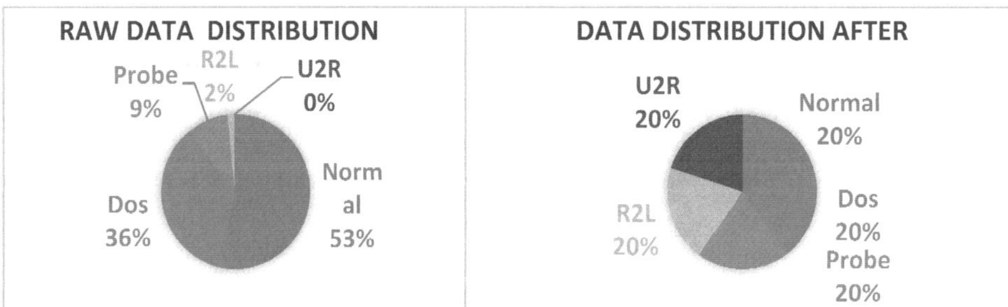

Fig 1 The distribution of training data

2.2.2. PCA feature extraction.

Feature extraction has a wide range of applications in data mining tasks. Generally, before this step is applied to training the model, in order to make the model easier to train, we will adjust the dimensionality of the feature space, generally using dimensionality reduction methods. Feature extraction can not only reduce the dimension of the feature space and the complexity of the task, but also reduce the redundancy of data information, thereby improving the performance of the model. In the face of intrusion detection tasks, this paper introduces the principal component analysis (PCA) algorithm[6] to extract data features.

The PCA algorithm implementation process is as follows

(1) Let input data $X=\{x^1,x^2,....,x^n\}$,where X is n p-dimensional column vectors.

(2) For data centralization, such as formula (4) and formula (5)

$$\mu=\frac{1}{N}\sum_{t=1}^{N}x^t \quad (4)$$

$$x^i=x^i-\mu \quad (5)$$

where $i\in\{1,2,...,n\}$.

(3) Calculate the covariance matrix formula (6)

$$C=\frac{1}{N}XX^T \quad (6)$$

(4) Decompose the eigenvalue formula of covariance matrix C (7)

$$C_{v_i}=\lambda_i v_i \quad (7)$$

where $\lambda_i(i=1,2,...,p)$ is the eigenvalue, and its corresponding eigenvector is $v_i(i=1,2,...,p)$. The eigenvectors corresponding to the first k eigenvalues are used to form the projection matrix $A=\{v_1,v_2,...,v_k\}$, where v_i is the eigenvector corresponding to λ_i.

(5) Finally, project the original sample to the new feature space to obtain the new dimensionality reduction sample formula (8)

$$X' = W^T X \tag{8}$$

Where W^T is a trainable matrix, and X' is n p-dimensional column vectors.

2.2.3. Data balance processing.
At present, the integrated learning method is often used in data min-ing tasks. Ensemble learning is a method that uses multiple classifiers. Its purpose is to improve the robustness of the model and learn how to improve the classification performance from each subclass-ifier. This method subdivides a big problem, and then adopts targeted solutions for each part. The suc-cess of the ensemble method depends on the diversity of the misclassified instances of each subclass-ifier[7]. This paper is oriented to intrusion detection tasks and uses CART as the classifier to constr-uct a Bagging ensemble learning model.

First introduce the Bagging method of building an integrated CART decision tree. Under this method, each decision tree is independently trained through the bootstrap method, and then they will be aggregated through an appropriate method. Usually, we divide the training set into T subsets, and then use each subset to build a decision tree, and finally these independent decision tree are built into an integrated decision tree. From the statistical facts, we need to use diversified training samples to obtain better integrated learning results and improve model performance. For this reason, we often use Bootstrap technology, which constructs T subsets through repeated sampling with replacement. The sample x in each subset may appear repeatedly or not. Finally, each subset will be used to train a specific decision tree.

After the training is over, we use the voting mechanism to combine the predicted output, and the process is shown in Algorithm 1.

Algorithm1 Bagging Algorithm Flow

Input: Training set $D = \{(x_1, y_1), (x_2, y_2), \ldots, (x_m, y_m)\}$;

Base learner $\mathcal{L}$;

Number of training rounds T.

Output: $H(x) = \arg\max \sum_{t-1}^{T} \Pi(h_t(x) = y)$

Begin:

1. **for** t=1,2,…,T **do**
2. $h_t = \mathcal{L}(D, D_{bs})$
3. **end for**

3. Results

This article divides the data set into two types of normal and abnormal data according to sample labels. The following first introduces the classification indicators used: false negative (FN), false positive (FP), true positive (TP) and true negative (TN).

This paper adopts the *accuracy* rate as the formula (9) and the false alarm rate *FAR* as the formula (10) as the evaluation indicators.

$$accuracy = \frac{TP + TN}{TP + FP + TN + FN} \tag{9}$$

$$FAR = \frac{FP}{FP + TN} \tag{10}$$

This paper sorts the importance of features by Gini coefficients, as shown in Figure 2, and the first 35 features are selected to be studied separately, aiming to find a subset that can work with a classifier to produce higher classification accuracy. We use the CART decision tree as the base classifier and use the BAGG integration technology to build the integration model. Using a subset of the NSL-KDD dataset, the 35 feature dataset trained and tested these models. The training and testing phases are performed on two separate datasets (KDDTrain+ and KDDTest+). This paper uses Scikit-learn as the basic tool for algorithm implementation.

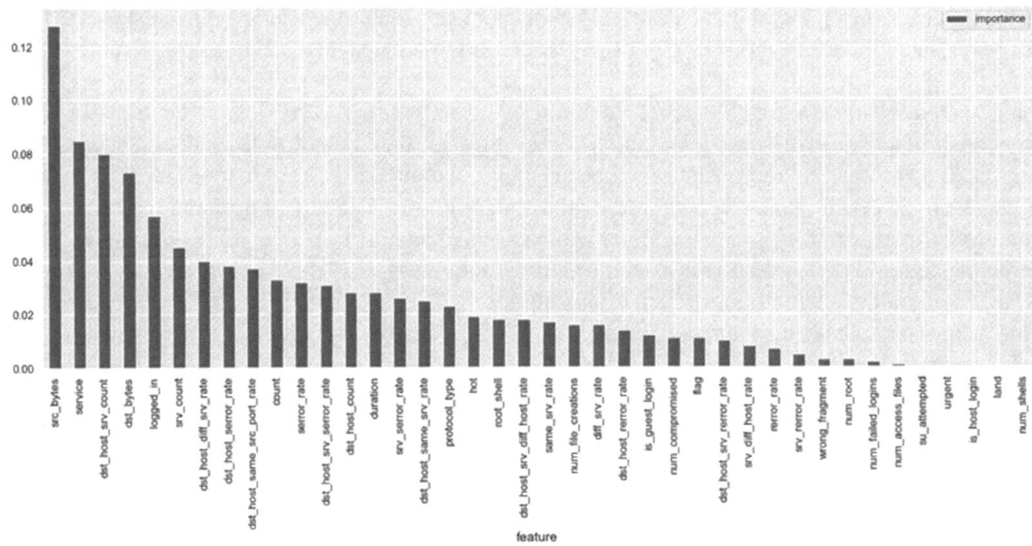

Fig 2 Sort feature importance by Gini

Table 2 shows the experimental results of the ensemble model and other methods proposed in this paper using 35 feature subsets to train and test.

Table 2 experimental results with 35 features

Model	Accuracy(%)	FAR(%)
SVM	80.03	3.88
KNN	78.94	3.01
CART	80.23	3.19
Naïve Bayes	80.87	8.12
Bagging(CART DT)	84.68	1.81

In total, by selecting 35 feature subsets, the classification performance of the Bagging (CART DT) model was greatly improved, and its accuracy increased from 80.51% to 84.68%, and its false positive rate decreased from 2.09% to 1.81%. This is because the Gini coefficients used in the selection feature can also be used as a partition attribute in the decision tree. Therefore, using Gini coefficients for feature selection can help improve the performance of the Bagging (CART DT) model. This makes the Bagging (CART DT) model perform better than other methods on a subset of 35 features. It is worth noting that although some methods currently have a higher accuracy rate than the Bedding (CART DT) model we designed, their false positive rate is also higher than the model we designed. Therefore, the intrusion detection model in this article should not only improve the recognition rate of network attacks, but also reduce the false alarm rate as much as possible to improve the overall performance of IDS.

4. Conclusion

In general, this paper proposes an improved intrusion detection method based on feature selection and integrated model. The NSL-KDD training data set is used to evaluate the proposed model. First, balance the data categories through the SMOTE-ENN method, and then use feature selection technology and PCA feature extraction technology to reduce the number of irrelevant features and improve the classification accuracy. Using CART as the base classifier, Bagging technology is used to build an ensemble model. The experimental results show that the method we designed has a higher recognition rate and a lower false positive rate for network attack data.

References

[1] Hoque M S, Mukit M, Bikas M, et al. An implementation of intrusion detection system using genetic algorithm[J]. arXiv preprint arXiv:1204.1336, 2012.
[2] Biswas S K. Intrusion detection using machine learning: A comparison study[J]. International Journal of pure and applied mathematics, 2018, 118(19): 101-114.
[3] Safavian S R, Landgrebe D. A survey of decision tree classifier methodology[J]. IEEE transactions on systems, man, and cybernetics, 1991, 21(3): 660-674.
[4] Tavallaee M, Bagheri E, Lu W, et al. A detailed analysis of the KDD CUP 99 data set[C]. 2009 IEEE symposium on computational intelligence for security and defense applications. IEEE, 2009: 1-6.
[5] Chawla N V, Bowyer K W, Hall L O, et al. SMOTE: synthetic minority over-sampling technique[J]. Journal of artificial intelligence research, 2002, 16: 321-357.
[6] Hadri A , Chougdali K , Touahni R . Intrusion detection system using PCA and Fuzzy PCA techniques[C]. 2016 International Conference on Advanced Communication Systems and Information Security (ACOSIS). IEEE, 2016.
[7] Aravind M , Kalaiselvi V . Design of an intrusion detection system based on distance feature using ensemble classifier[C]. International Conference on Signal Processing. 2017:1-6.

Topology Modeling for Reconfigurable Scan Networks in IEEE 1687 Networks

Yizhen Xu[1,a], Shouhong Chen[1,2,b], Jun Ma[1,2,c*], Cuifeng Xu[1,2,d], Ke Zeng[1,e]

[1]School of Electronic Engineering and Automation, Guilin University of Electronic Technology, Guilin, Guangxi, China

[2]Guangxi Key Laboratory of Automatic Detecting Technology and Instruments, Guilin University of Electronic Technology, Guilin, Guangxi, China

[a]e-mail: nuonuointernet@163.com, [b]e-mail: cshgl@guet.edu.cn

[c]*e-mail: majun@guet.edu.cn

[d]e-mail: xcf4100@guet.edu.cn, [e]e-mail: zengke_1997@163.com

Abstract—This paper designs an automatic generation method of Reconfigurable Scan Networks (RSNs) directed topology model combined with graph theory. By studying IEEE 1687 chains information, RSNs directed topology model is generated automatically based on the interconnections between on-chip instruments and the corresponding internal register length which are determined on standard Instrument Connectivity Language (ICL) script. The test and result analysis demonstrate the feasibility and effectiveness of the technology, which solves the inefficiency and high error rate of RSNs directed topology model built by artificial analyzing ICL script.

1. Introduction

With the rapid development of integrated circuits (ICs) industry, a large number of embedded instruments have been equipped for testing, configuration, and monitoring [1]. Embedded instruments lead to complex System-on-Chips (SoCs) designed, which often integrates cores of many different functions in one SoCs, and embedded instruments that may incorporate multiple functions in each core [2]. IEEE Std 1687 (iJTAG) adopts a reconfigurable scan chain structure to standardize access to instruments by specifying architecture and description language from test access ports. It introduces the Instrument Connectivity Language (ICL) to describe the network structure, including the connectivity between various components of the network-on-chip architecture and instruments on the scan paths that exist in the network, so that more instruments can be connected to the iJTAG instrument network systematically.

The analysis and operation of embedded instruments on SoCs are more concerned with minimizing the path cost to achieve the best access to them currently [3]. Topological model is an entity description of the topological relationship between the instruments on the SoCs [4], which is the basis for logical analysis that determines the analysis of the best access path of the instrument directly. The establishment of a topology model has a great influence on the efficiency and accuracy of reconfigurable scan chain analysis. According to the description of the ICL script, the scan chain of connected instruments on-chip can be analyzed. However, artificial analysis of reconfigurable scan chains from ICL script description takes a major of time to establish a chain topology model because of the high error probability and

Content from this work may be used under the terms of the Creative Commons Attribution 3.0 licence. Any further distribution of this work must maintain attribution to the author(s) and the title of the work, journal citation and DOI.
Published under licence by IOP Publishing Ltd

inefficiency in manual data processing. Moreover, with the continuous expansion of technology, there are more and more embedded instruments have been integrated into modern SoCs, which causes severe tests on the model built of reconfigurable scan chain topology. As a result, the way of artificial formation for scan chain topology has been unable to meet the requirements for modern testing [5]. Therefore, this paper proposes a novel method of establishing the topology model of RSNs combining with the method of graph theory. By parsing the ICL script through lexical grammar, the connection relationship between the network-on-chip components and corresponding internal register length are obtained to establish the node topology information, which can traverse it to generate a scan network topology model quickly and effectively.

2. Topological Model

SoCs is a kind of micro-component network, which is an abstraction of communication between components [6]. Topology is the basis of SoCs visualization when SoCs have been designed. It not only defines the layout and connection mode of nodes and lines in the network that determines the access path to the instrument but also shows the connection topology of the network, which becomes the basis of logical operation and analysis for any device.

2.1. Reconfigurable scan networks

With the help of Reconfigurable Scan Networks (RSNs), IEEE Std 1687 focuses on the access and interconnection to on-chip instrumentation on SoCs [7]. In RSNs, the scan chain is not fixed and immutable but can be reconfigured to vary to allow the online operation of different scan chains. Fast access to on-chip devices can be achieved again using only those instruments currently needed for access in the path.

To realize the scan chain structure of variable-length, the Segment Insertion Bit (SIB) component is introduced by RSNs to dynamically reconfigure the scan chain, which can be used to access the on-chip instruments. For achieving access to the target instrument, reconfigured scan networks can be done in many different ways, according to the target requirements [8]. Figure 1 shows the general structure of RSNs consisting of SIB and Test Data Register (TDR), on which corresponding instruments for various functions can be mounted on the TDR. In addition, the RSNs also has capabilities that include the IEEE1687 client interface, where connections between the devices are connected from their ports through wires.

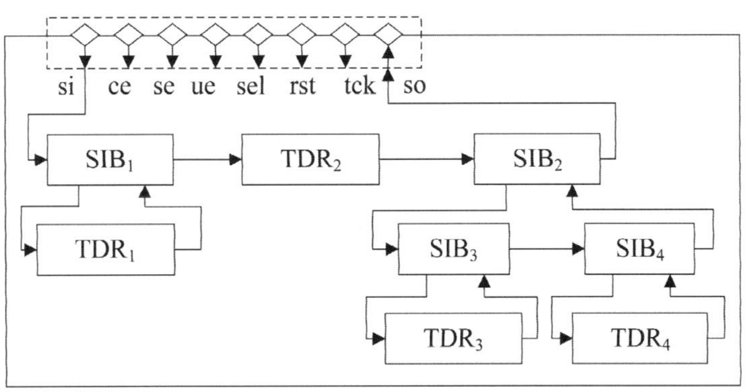

Figure 1. Example of a reconfigurable scan networks

2.2. Digraph model of RSNs

The topology diagram is a simplified representation of the RSNs topology view that reflects the structural relationship between entities. The digraph is an effective tool to describe a project or system, which is widely used to solve practical problems [9]. Since on-chip instruments are accessed in the scan chain network based on the data flow model, hence, the RSNs topology is consistent with the digraph in terms of the abstract topology construction. In the RSNs topology, the nodes of the digraph are used to represent the instruments in the network because each element is associated with a vertex. Meanwhile, the edges of the digraph are used to represent the connection between the instruments in the network, which reflects the flow direction of the test data stream. A digraph that describes the access process can be expressed as:

$$G = (V(G), E(G)) \tag{1}$$

Where:

$$V(G) = \{v_1, v_2, ..., v_n\} \tag{2}$$

$$E(G) = \{e_1, e_2, ..., e_n\} \tag{3}$$

$V(G)$ represents the set of vertexes v_i in the digraph G, which indicates each module in RSNs. $E(G)$ represents the set of edges e_i in the digraph G, which indicates the transmission direction of operation module signal in RSNs, that is, the access direction of on-chip instruments. To express RSNs more significantly, the digraph is weighted [10]. For directed edges, the weight corresponds to the SR length within the module inner the RSNs. The set of weights can be expressed as:

$$A = \{SR_1, SR_2, ..., SR_n\} \tag{4}$$

The RSNs digraph model is built as shown in Figure 2.

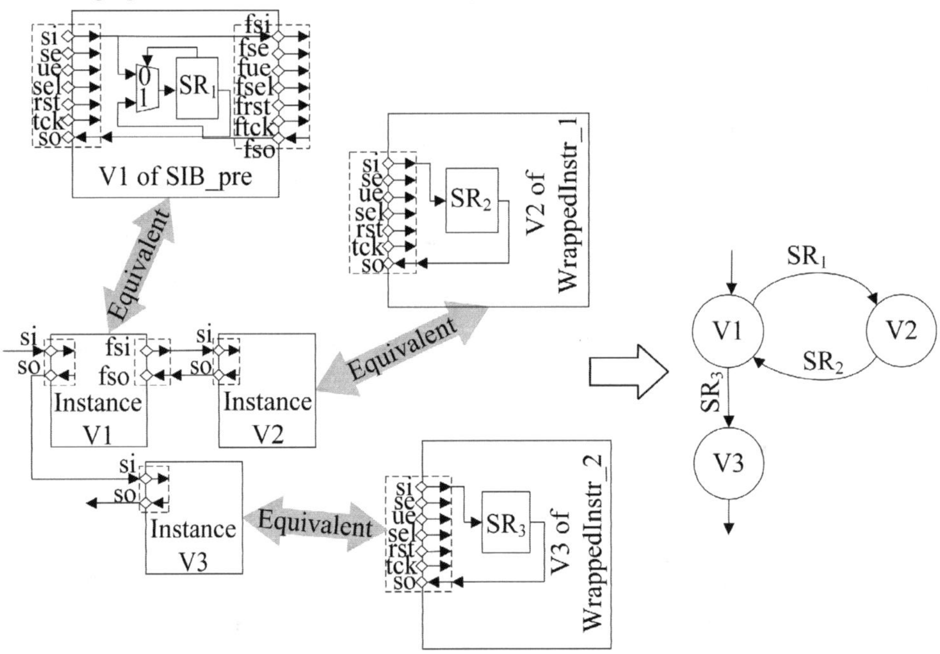

Figure 2. Digraph model of RSNs

3. Automatic Generation Implementation Of Rsns Topology

The topology graph provides a view of the interconnection between network modules. According to the construction rules of RSNs digraph model, the premise of constructing a RSNs topology structure is to get the connection information of nodes. The automatic generation of RSNs topology is based on certain known conditions, using clear methods to enable the program to automatically generate node topology information. Therefore, parsing the ICL script is the key to RSNs topology generation.

3.1. ICL script parsing

The fundamental entity in ICL is called a module. A module in a network is defined by its ports and internal contents which may be any combination of registers, combinational logic, or other child modules. As a result, logical hierarchies may also appear in the networks. As a descriptive language, ICL follows uniform format rules. Generally, the description of an ICL script contains the information about each hierarchy in the logical hierarchy, which includes the top-level module structure and other hierarchical modules. Each module contains the port information of the current module, while other hierarchical modules can be the information of a device simply, including device ports and internal instruments. Other hierarchical modules are instantiated in the top-level module. The ICL script information is decomposed as shown in Figure 3.

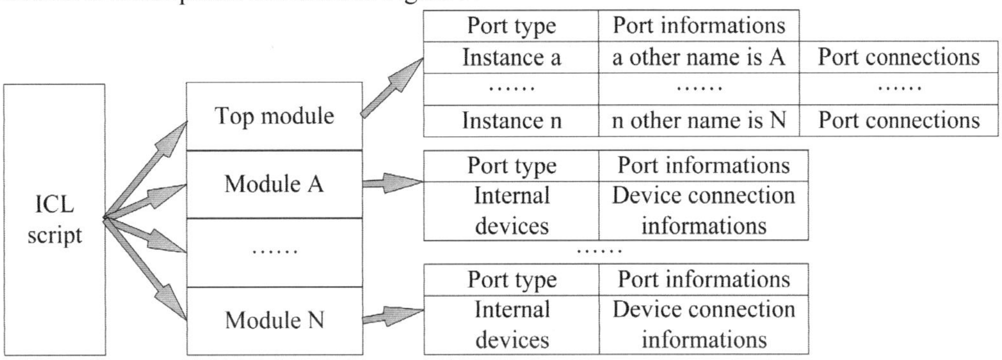

Figure 3. ICL script information decomposition

According to the instrument elements and logical connection on the network described by the ICL script, the script parsing module automatically parses the ICL script by the preset rules. The parsing of node information is completed by lexical analysis, and the acceptable set of lexical units is described by the regular expression [11]. The lexical analyzer gets ICL script characters and checks them, then outputs lexical units that match regular expressions until all characters of the ICL script have been read. The ICL script parsing process is shown in Figure 4.

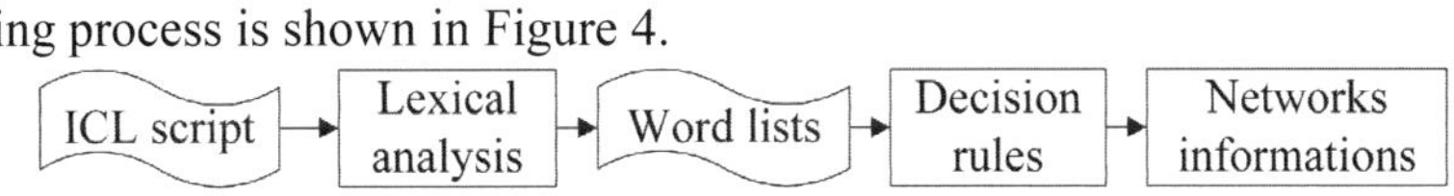

Figure 4. ICL script parsing process

3.2. *RSNs directed topology generation*

The flow chart of automatic generation of RSNs directed topology is shown in Figure 5.

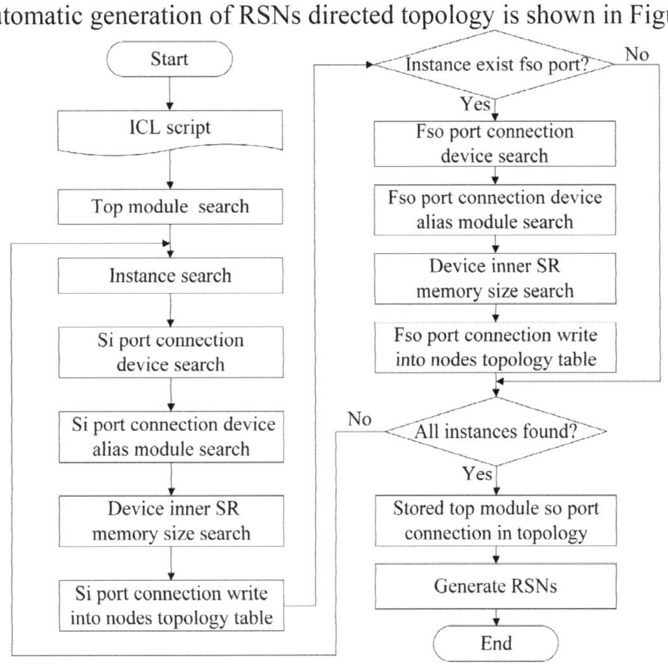

Figure 5. RSNs directed topology generation process

ICL represents the behavior of the networks. Identifying the node information is the key to generating the directed topology of the scan networks. The related information such as node instruments and attributes can be analyzed by parsing the ICL script, also the scan chain is related to Instance. Therefore, the generation of RSNs revolves around the interconnection between Instances. From one node in the topology, that is, Instance, which is the instantiation of another module in the current module, starting to search for nodes connected to its ports to form corresponding directed edges. Moreover, each Instance is an independent module, where its internal SR memory length is assigned to the corresponding edge weight. When all the Instances have been searched, the port connection information of the top-level module is stored in the topology table in which to generate the RSNs directed topology by traversing all the node information.

3.3. Result verification

This paper verifies the aforementioned RSNs topology model and its algorithm with the ICL script given in Appendix E of IEEE Std 1687 in the process of RSNs topology verification. The correlations between the instruments obtained from the analysis of the example ICL script are shown in Table 1.

Table 1. Correlations between instruments in example ICL script

Instance	Instance alias	Instrument connected to si port	Instrument connected to fso port	SR memory size in alias module	Instrument connected to top module ports
SIB1	SIB_pre1	×	WI1	1	si
SIB2	SIB_pre2	SIB1	WI2	1	×
SIB3	SIB_pre3	SIB2	WI3	1	×
SIB4	SIB_pre4	SIB3	SIB5	1	so
SIB5	SIB_pre5	WI4	SIB6	1	×
SIB6	SIB_pre6	WI5	WI6	1	×
WI1	WrappedInstr1	SIB1	×	10	×
WI2	WrappedInstr2	SIB2	×	8	×
WI3	WrappedInstr3	SIB3	×	6	×
WI4	WrappedInstr4	SIB4	×	12	×
WI5	WrappedInstr5	SIB5	×	7	×
WI6	WrappedInstr6	SIB6	×	3	×
Note: "×" indicates that the corresponding connect device does not exist.					

As shown in Table 1, the chains in the hierarchical structure consist of 12 testability on-chip instruments, including SIB and TDR. For ease of analysis, the input port and output port of the top-level module will be displayed as separate nodes. The RSNs topology generated by the directed topology generation algorithm is shown in Figure 6.

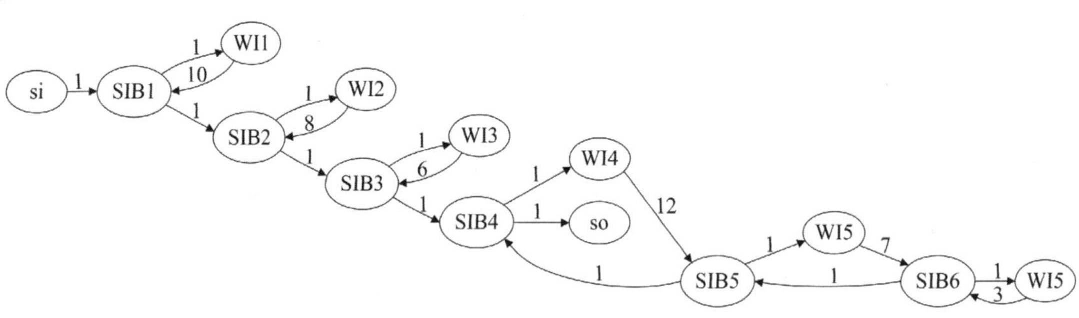

Figure 6. RSNs topology generated by example ICL script

The results shown in Figure 6 not only reflect the chain topology information in RSNs clearly, but also express the internal register length of each structure. RSNs topology allows fast analysis of the SoCs chain and greatly reduces the access time of on-chip devices.

4. Conclusion

This paper proposes a technical scheme to study and design a directed topology model of RSNs. By studying IEEE 1687 chains information and combining it with the digraph model, RSNs directed topology model is generated automatically based on the interconnections between on-chip instruments the corresponding internal register length which are determined on standard ICL script. With universality, this technology shows the connections between on-chip instruments as a whole and simplifies the preparation for boundary-scan tests, as well as saves time and reduces labor cost for the best path selection of on-chip instrument access.

Acknowledgments

This work was supported in part by the Natural Science Foundation of Guangxi Province under Grant 2015GXNSFDA139003, in part by the Guangxi Key Laboratory of Automatic Detecting Technology and Instruments under Grant YQ14115 and YQ15101.

References

[1] Zadegan,F.G., Larsson,E., Jutman,E. (2014) Design, Verification and Application of IEEE 1687[C]. In: 2014 IEEE 23rd Asian Test Symposium (ATS), Hangzhou, pp, 93-100.
[2] Liu,Y., Yan,X.L. (2012) Design of Flexible Scan-chain Based on IEEE P1687[J]. Computer Measurement and Control, 20(9):2357-2359.
[3] Ibrahim,A., Kerkhoff,H.G. (2016) Analysis and design of an on-chip retargeting engine for IEEE 1687 networks[C]. In: 2016 IEEE European Test Symposium (ETS), Amsterdam, pp, 1-6.
[4] Wang,M., Shi,M.C. (2017) Study on Topology Model of Urban Drainage GIS System[J]. Bulletin of Surveying and Mapping, 2017(8):129-134.
[5] Hou,X.N.,Chen,S.H., Yan,X.L. (2018) Research and implementation of SQLite based automatic generation for boundary scan testing chain[J]. Modern Electronics Technique, 41(8):64-67,71.
[6] Benini,L., Bertozzi,D. (2005) Network-on-chip architectures and design methods[J]. IEEE Proceedings-Computers and Digital Techniques. 152(2):261-272.
[7] Thiemann,B., Feiten,L., Raiola,P., et al. (2019) On Integrating Lightweight Encryption in Reconfigurable Scan Networks[C]. In:2019 24th IEEE European Test Symposium (ETS), Baden-Baden, pp, 1-6.
[8] Tšertov,A., Jutman,A., et al. (2016) A suite of IEEE 1687 benchmark networks[C]. In:2016 IEEE International Test Conference (ITC), Fort Worth, pp, 1-10.
[9] Yao,G.H., Zhu,D.M., Ma,S.H. (2009) Approximating the Directed Minimum Degree Spanning Tree of Directed Acyclic Graph[J]. Journal of Computer Research and Development, 46(6):1052-1057.
[10] Fan,G.J., Hou,Y.E., Zhang,Q. (2014) Construction Method of E-R Model of Data Flow Diagram Based on Directed Graph[J]. Computer and Modernization, 2014(6):12-16.
[11] Shen,Y.Q., Yan,G., Tang,J.T., Luo,Y.S. (2017) Test Case Automatically Parsing Based on Lexical and Grammar Analysis[J]. Electric Drive for Locomotives, 2017(1):74-76.

SAR image simulation of sea surface targets

Xianping Cheng[1,a], **Weijie Xia**[1,b*]

[1]College of Electronic Information Engineering, Nanjing University of Aeronautics and Astronautics Nan Jing, Jiang Su, China

[a]email：1044532864@qq.com

[*b]email：nuaaxwj@nuaa.edu.cn

Abstract: SAR(Synthetic Aperture Radar) images of sea targets are of great significance to the military. In view of the high cost of obtaining the measured data of sea targets SAR images, this paper implements a sea target SAR image simulation method based on electromagnetic scattering calculations. This method uses the Phillips spectrum to generate the sea surface geometric model, and combines the target model and the sea surface model; then uses the SBR(Shooting and Bouncing Rays) to calculate the electromagnetic scattering of the sea surface target; finally, the SAR image of the sea surface target is simulated by the RCS(Radar Cross Section) obtained by the calculation. Through the analysis of the results of different sea states and different incident angles, it is shown that this method can complete the SAR image simulation of sea surface targets at different sea states and different angles, and the simulation effect is relatively realistic.

1. Introduction

With more than 70% of the earth's surface area occupied by the ocean, the ocean is of great significance to the military of a country. With the frequent occurrence of maritime disputes around the world in recent years, sea surface SAR image simulation has therefore attracted wide attention from scholars in various countries. The ocean is a complex natural environment, and the motion of the sea surface is also usually difficult to predict due to its irregularity.

The current SAR image simulation researches can be divided into two categories. The first one base on the principle of radar imaging and simulates the process of target signal echo formation, and the focus of this type of method is to simulate the echo signal and use the simulated echo data for SAR image simulation. The second one is based on the scattering characteristics of the target reflected by the SAR image, and the simulated image is obtained by merging the features of the radar image with the scattering coefficient map of the radar image[2][3].

Since the 1950s, many scholars have conducted researches on sea surface SAR imaging and proposed different sea surface SAR image simulation methods. Yu et al. conducted SAR simulation research on shallow sea terrain[4]. Zhang et al. used the scattering model to model the echo and imaged the dynamic sea surface[5]. Zhu et al. established a suitable and feasible ocean surface scattering model based on the two-scale model[6]. Guo et al. studied the impact of sea echo on ships, and simulated the ships with SAR images[7].

The above methods are all aimed at the SAR image simulation of the sea surface or the target. This paper combines the geometric model of the sea surface and the target, and calculates the electromagnetic scattering of the model to realize a SAR image simulation method for the sea surface target.

Content from this work may be used under the terms of the Creative Commons Attribution 3.0 licence. Any further distribution of this work must maintain attribution to the author(s) and the title of the work, journal citation and DOI.
Published under licence by IOP Publishing Ltd

2. Sea surface model simulation

The variability of the ocean makes wave fluctuations difficult to predict. The magnitude of the wind affects the fluctuation of the waves, and as the wind speed increases, the level of the waves also increases. Internationally, the Douglas scale is usually used to describe the level of the waves. Table 1 below is the sea level classification method.

Table 1 Sea State Correspondence Table

Sea state	Wind rating	Wind speed(m/s)	height of the wave(m)
0	0	0~0.3	0
1	1	0.3~1.6	0~0.1
2	2	1.6~3.4	0.1~0.5
3	3~4	3.4~8.0	0.5~1.25
4	5	8.0~10.8	1.25~2.5
5	6	10.8~13.9	2.5~4.0
6	7	13.9~17.2	4.0~6.0

When performing sea surface simulation, the sea surface is usually viewed as consisting of points on its surface. The height of the waves on the sea surface is defined as $h(x,y)$, and assume that the spectrum signal of the known sea surface is $\tilde{h}(\omega_x,\omega_y)$. After the inverse Fourier transform of the spectrum signal, the height information of the sea surface can be known. Since the height of the sea surface changes with time, the height information of the sea surface at time t is:

$$h(\vec{X},t)=\sum \tilde{h}(\vec{k},t)e^{i\vec{k}\cdot\vec{X}} \tag{1}$$

where $\vec{X}=(x,y)$ represents the coordinates of any point on the sea surface, $\vec{k}=(\omega_x,\omega_y)$ represents the frequency domain coordinates, and $\tilde{h}(\vec{k},t)$ represents the spectrum signal, which is calculated as follows:

$$\tilde{h}(\vec{k},t)=\tilde{h}_0(\vec{k})e^{j\omega(k)t}+\tilde{h}_0^*(-\vec{k})e^{-j\omega(k)t} \tag{2}$$

where $\tilde{h}_0(\vec{k})$ and $\tilde{h}_0^*(-\vec{k})$ are conjugate each other, $\tilde{h}_0(\vec{k})$ is calculated as follows:

$$\tilde{h}_0(\vec{k})=\frac{1}{\sqrt{2}}(\xi_r+i\xi_i)\sqrt{P_h(\vec{k})} \tag{3}$$

where ξ_r and ξ are mutually independent random numbers and obey the normal distribution with zero-mean and unit variance. $P_h(\vec{k})$ is the Phillips sea spectrum, which takes the following form[8]:

$$P_h(\vec{k})=A\frac{e^{-1/(kL)^2}}{k^4}\left|\vec{k}\,\square\,\vec{\omega}\right|^2 \tag{4}$$

In the formula, $L=V^2/g$, V is the wind speed, g is the acceleration of gravity, k is the modulus of the vector $\vec{k}$, and A is the Phillips constant.

According to the above method, the sea surface of different sea states is simulated, and the simulation results are shown in the following Fig. 1:

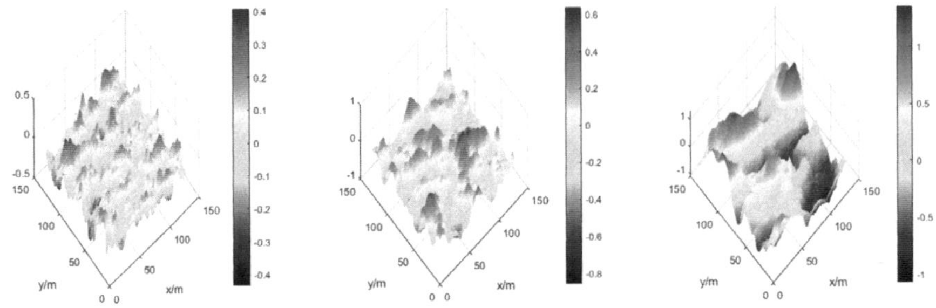

(a) (b) (c)

Fig. 1Three-dimensional schematic diagram of the sea surface.
(a) sea state two;(b) sea state three;(c) sea state four.

Comparing with the above figure, we can find that as the sea state increases, the height of the wave and length of the wave will also increase, and the height of the wave is consistent with the above-mentioned Douglas scale.

3. RCS calculation

In this paper, SBR is used to simulate the RCS of the sea surface target and use the simulation results to simulate the SAR image. The sea surface target is segmented according to the distance direction and azimuth direction, and each resolution unit is treated as a scattering point, and the RCS value of each scattering point can be calculated to simulate the SAR image. The RCS of each scattering point is calculated as follows:

$$\sigma = 4\pi R^2 \frac{|E_s|^2}{|E_i|^2} \tag{5}$$

where E_i represents the intensity of the incident field, E_s represents the intensity of the scattered field, and R represents the distance from the scattering point.

The SBR uses a combination of geometric optics and physical optics to solve the scattering field of the target[9][10]. This method can calculate multiple scattering between complex targets, and its calculation results can meet the actual engineering needs.

The geometric optics method is a high-frequency approximation that represents the energy propagation characteristics of electromagnetic waves, using which the field strength of electromagnetic waves can be calculated. The electric field strength at the observation end can be expressed as the electric field strength at the reference end as:

$$E(s) = E(0)Ae^{-jks} \tag{6}$$

In the formula, $k = 2\pi\lambda$ is the free space wave number, s represents the distance between the observation end and the reference end along the ray, and A represents the diffusion factor, which is 1 in the far-field condition.

The physical optics method uses the Stratton-Chu equation to solve the approximate RCS under high frequency conditions, and its expression is as follows:

$$E_s = \int\left[-j\omega\mu\left(\hat{i}\times H\right)\psi_0 + \left(\hat{i}\times E\right)\times\nabla\psi_0 + \left(\hat{i}\cdot E\right)\nabla\psi_0\right]ds \tag{7}$$

In the formula, ds is the differential of the outer surface of the scattering target, $\hat{i}$ is the normal loss of the scattering surface, E and H respectively represent the total power plant and total magnetic field on the target surface, and $\psi_0 = e^{jk_0 r}/4\pi r$ represents the Green's function of free space. Under far-field conditions, $\nabla\psi_0 \approx jk_0\hat{s}\psi_0$. Assuming the target is an ideal conductor:

$$\hat{i}\times E = 0, \hat{i}\cdot H = 0, \hat{i}\times H = 2\hat{i}\times H^i \tag{8}$$

Under the above conditions, the available scattering field calculation formula is as follows:

$$E_s = jk_0\psi_0 \int \left[\hat{s} \times (\hat{s} \times (\hat{n} \times H)) \exp\left(jk_0 r^{`} (\hat{n} - \hat{s}) \right) \right] ds \quad (9)$$

$\hat{n}$ and $\hat{s}$ indicate the unit vector of the incident direction and the observation direction.

The geometric model generated in this paper is composed of many face elements, and each face element can be regarded as a flat plate. From the above formula, the calculation formula for the RCS of the flat plate is as follows:

$$\sigma = \frac{64\pi}{\lambda^2} a^2 b^2 \cos^2 \phi [\frac{\sin(2k\sin(\phi))}{2k\sin(\phi)}]^2 \quad (10)$$

Among them, a and b represent half the length and width of the plate, and ϕ is the angle between the incident electromagnetic wave and the plane.

4. Experiments

In this paper, the SAR images are simulated for different sea states and different incidence angle information. First, use the Phillips sea spectrum to model the sea surface to generate sea surface models of different sea states, and then the sea surface model and the target model are combined as shown in Fig. 2 below; then the ray-tracing method is used for electromagnetic scattering calculation of the geometric model; finally, the results of electromagnetic scattering calculation are used to simulate the SAR images. The following Table 2 shows the radar operating parameters and the information related to the geometry model of sea surface target.

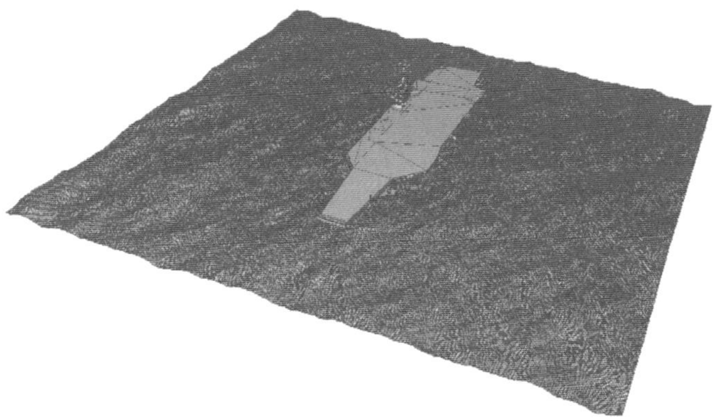

Fig. 2 Geometry model of sea surface target

Table 2 Parameter information table

Parameter name	Parameter size	Parameter name	Parameter size
Pitch angle	30°	Azimuth angle	0~90°
Center frequency	10GHz	Bandwidth	50MHz
Distance sampling point	512	Azimuth sampling point	512
Sea state	0~6	Sea size	512m*512m

4.1. SAR image simulation of different sea states

The following Fig. 3 shows the scatter map of the sea surface target and the SAR image simulation diagram under different sea states, where Fig. 3 (a) indicates the scatter map of the sea surface target, and Fig. 3 (b) indicates the simulated SAR image of the sea surface target. The brighter the figure indicates the greater the scattering intensity. Comparing the simulated results, it can be seen that the scattering intensity of the target is much greater than that of the sea surface, and the surface background features are more obvious as the sea state increases, and the aircraft carrier almost does not change. The outline of the target can be seen from the simulated SAR images, and the places where the target scatters more strongly are mainly distributed on the shell and buildings of the target.

(a)

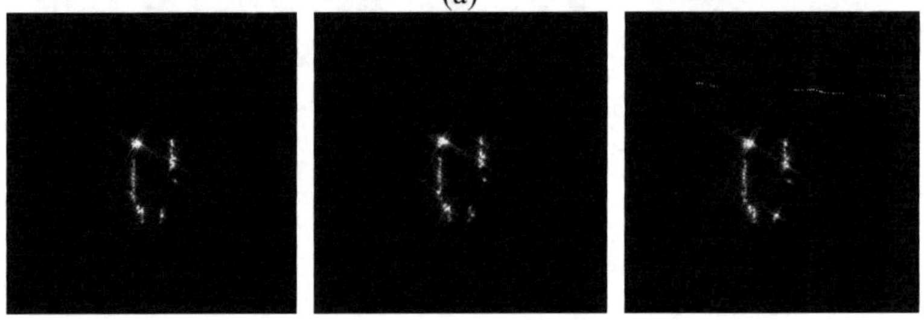

(b)

Fig. 3 Image simulation of sea surface targets in different sea states.
(a) Scattering map of different sea states; (b) SAR images of different sea state.s

4.2. SAR image simulation of different incident angles

The following Fig. 4 shows the simulation of SAR images of sea targets at different incident angles when the sea state is 3. It can be seen from the Fig. 4 below that with the change of the incident angle, the shadow of the ship in the scattering image has also changed, which is consistent with the actual situation, and the scattering intensity of the sea surface has hardly changed. From the simulated SAR image, it can be seen that when the incident angle changes, the scattering intensity of different positions of the ship changes greatly, and the scattering points facing the incident direction will be more scattered.

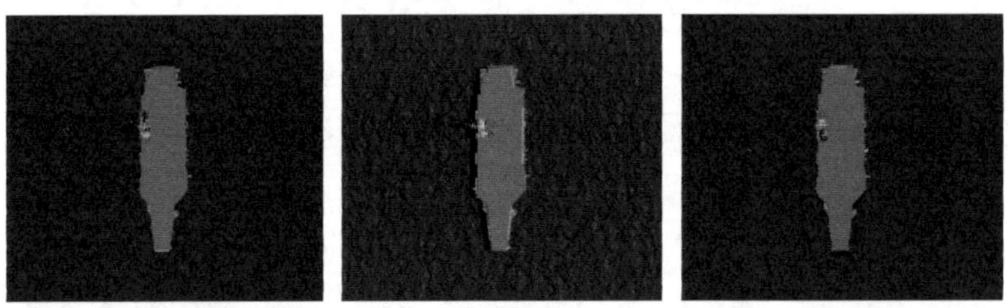

(a)

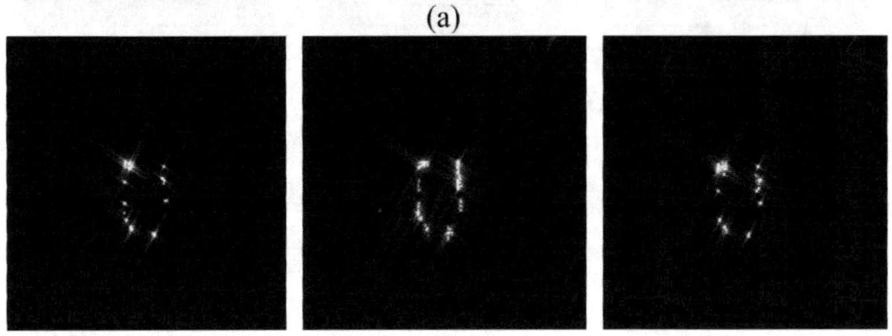

(b)

Fig. 4 Image simulation of sea surface targets with different incident angles.
(a) Scattering diagrams at different incident angles; (b) SAR images with different incident angles.

From the above analysis, it can be seen that the scattering intensity of the sea surface target is greater than the sea surface scattering intensity, and the scattering of the sea surface target is less affected by the sea state as the incident direction of electromagnetic waves changes, the scattering of the sea surface target changes greatly, and the incident angle The change has a greater impact on SAR simulation.

5. Conclusion

In this paper, the SBR is used to calculate the electromagnetic scattering of the geometric model of the sea surface target, and the calculation result is used to simulate the SAR image. In the process of SAR image simulation for different sea states and different incident angles, the SAR image simulation of sea targets is carried out by changing the information of sea state and incident angle. Through analysis and comparison, it can be found that different sea states have little effect on the SAR imaging of sea targets; the difference in incident angle has a greater influence on the imaging of sea targets. The SAR image simulation of sea targets by the method in this paper can simulate the SAR images of sea targets under different sea states and different incident angles.

However, the target is stationary in this paper, and its scattering characteristics will inevitably change with the movement of the sea surface and the target. Therefore, the next step should consider the SAR image simulation of dynamic sea targets.

References

[1] Zhu, M. , Zou, J. , Fei, X. , & Wei, D. . (2012). Imaging simulation of sea surface with sar. Springer Berlin Heidelberg.
[2] Zhang, Q. J. , Wang, X. B. , Liu, W. , Zhu, Y. , & Ding, Z. G. . (2016). Research on echo simulation of geosynchronous SAR. International Radar Conference. IET.
[3] Yoshida, T. , & Rheem, C. K. . (2013). Sar image simulation in the time domain for moving ocean surfaces. Sensors, 13(4), 4450-4467.
[4] Yu, X. , Chong, J. , Hong, W. , & Zhu, M. . (2011). Simulation study on shallow sea topography imaging by along-track interferometric sar. DBLP.
[5] Zhang, M. , Chen, H. , & Yin, H. C. . (2011). Facet-based investigation on em scattering from electrically large sea surface with two-scale profiles: theoretical model. IEEE Transactions on Geoscience & Remote Sensing, 49(6), 1967-1975.
[6] Zhu, M. , Zou, J. , & Hou, J. . (2015). Imaging simulation of sea surface with full polarization SAR. 2015 IEEE 5th Asia-Pacific Conference on Synthetic Aperture Radar (APSAR). IEEE.
[7] Guo, Y. , H Wang, H Ma, Ke, L. , & Xia, Z. . Sar image simulation of ship targets based on multi-path scattering.
[8] Newell, A. C. , & Zakharov, V. E. . (2008). The role of the generalized phillips' spectrum in wave turbulence. Physics Letters A, 372(23), 4230-4233.
[9] Hamel, P. , Adam, J. P. , Kubicke, G. , & Pouliguen, P. . (2012). An improved hybridization technique of geometrical optics physical optics. Antenna Technology and Applied Electromagnetics (ANTEM), 2012 15th International Symposium on. IEEE.
[10] Weinmann, & Frank. (2006). Ray tracing with po/ptd for rcs modeling of large complex objects. IEEE Transactions on Antennas & Propagation.

Research on Mine OFDM Intelligent Receiver Based on Deep Learning

Xuhong Li[1,a], Yong Pan [2,b*], Anyi Wang [3,c], Zhiyuan Feng[4,d]

[1]Xi'an University of science and technology, School of communication and information engineering, Xi'an, Shaanxi, China

[2]Xi'an University of science and technology, School of communication and information engineering, Xi'an, Shaanxi, China

[3]Xi'an University of science and technology, School of communication and information engineering, Xi'an, Shaanxi, China

[4]Xi'an University of science and technology, School of communication and information engineering, Xi'an, Shaanxi, China

[a]email: lixhong105@xust.edu.cn

[b*]email: 1443211944@qq.com

Abstract: coal is an important energy. Coal mining needs a series of equipment, among which communication equipment can ensure the safety of underground workers in the mining process and improve production efficiency. Aiming at the problems of multipath fading and anti-interference of signal transmission in coal mine, researchers propose to optimize the quality of signal transmission in coal mine with OFDM technology. Compared with traditional OFDM technology, this paper proposes a method of mine OFDM intelligent receiver based on deep learning, which aims to use deep learning technology to replace the signal recovery process of the receiver of traditional OFDM technology. The simulation results show that in terms of bit error rate, the signal recovery of the receiver based on deep learning technology is obviously better than that of the traditional OFDM receiver. This will improve the economic benefits of coal mining in the future.

1. Introduction

In the process of coal mining, countless mining accidents have occurred all over the world, resulting in heavy casualties and property losses [1]. Therefore, it is particularly necessary to have reliable communication equipment to ensure personnel safety in production and prevent major mine accidents. With the continuous development of deep learning technology, people begin to explore and study it and apply it to the field of wireless communication, hoping to achieve unexpected results. In the field of wireless communication, the traditional wireless communication system is composed of transmitting part and receiving part.

For OFDM transmission technology, the general principle is to divide the total channel bandwidth into several orthogonal subchannels, and convert the serial high-speed code stream in the channel into parallel low-speed code stream. Then, these low-speed code streams are used to modulate the subcarriers of several mutually orthogonal subchannels to realize the synchronous transmission of signals in these subchannels. At the same time, OFDM can effectively combat the problem of multipath effect. Because

Content from this work may be used under the terms of the Creative Commons Attribution 3.0 licence. Any further distribution of this work must maintain attribution to the author(s) and the title of the work, journal citation and DOI.
Published under licence by IOP Publishing Ltd

of its orthogonality, the receiver can receive the signal separately, so that the interference between subchannels is greatly reduced. Moreover, the sub channels comply with flat fading, so that the inter symbol interference (ISI) no longer exists. Therefore, OFDM technology is suitable for the communication environment such as mine channel [2].

In the traditional wireless communication system, the signal is output at the transmitter. Through the wireless channel, the autocorrelation of the training sequence is used for frame synchronization at the signal receiving end to find the initial position of the signal, and then the cyclic prefix is used to calculate the frequency offset and compensate the frequency offset. Then, the cyclic prefix is removed, the orthogonal carrier demodulation is carried out, and the time domain signal is converted into frequency domain signal by FFT. The pilot is used for channel estimation and channel equalization. Then the signal is demodulated and decoded, and finally the output binary signal is obtained. It is not difficult to find that this series of processes is particularly complex, and the local optimal performance of each module in the receiving part cannot be guaranteed to be the global optimal performance. The error of the preprocessing module may affect the optimization of the subsequent processing module, lead to the accumulation of errors, and affect the performance of the receiver. When an algorithm is designed to simulate this process, it is usually assumed that a module is in an ideal state. For example, the channel fading is specified as Rayleigh model and the noise is additive Gaussian white noise. Therefore, the traditional receiver optimization is the optimal performance under ideal conditions rather than the optimal performance in actual scenarios [3].

This paper proposes a method based on deep learning technology to replace the traditional OFDM technology receiver signal recovery. Firstly, this process is simplified, and there will be no such concern, Because when the convolutional neural network (CNN) in deep learning is used to realize this process, compared with the tradition, it is globally optimized. When the forward network works, it obtains the loss function and back propagation, continuously optimizes the network parameters, and finally obtains the optimal result [4] . At the same time, he does not rely on theoretical modeling and has unity. So as to realize the intelligent recovery and reception of signals.

2. Materials and Methods

2.1 Analysis of Underground Wireless Channel in Coal Mine

Coal mine underground roadway is a limited, narrow and non free space. Due to many underground mechanical equipment, irregular coal mine roadway structure, roadway wall rock parameters and roughness and other factors, radio waves occur reflection, diffraction and scattering when propagating in underground roadway. It makes the downhole channel produce very serious multipath fading effect. This kind of signal fading caused by multipath propagation is called small-scale fading. This paper adopts mine small-scale Nakagami-m fading channel [5]. The probability density is as follows:

$$P(r)=\frac{2}{\Gamma(\mathrm{m})}\left(\frac{m^m}{\Omega^m}\right)r^{2m-1}\exp\left(-\frac{mr^2}{\Omega}\right),r\geq 0,m\geq\frac{1}{2} \tag{1}$$

Where, $\Gamma(\mathrm{m})$ is the gamma function and $\Omega=E(r^2)$ is the signal power; m is called the shape factor or Nakagami constant, which can reflect the severity of multipath signal fading. The expression is:

$$m=\frac{\Omega^2}{E\left[(r^2-\Omega)^2\right]},m\geq\frac{1}{2} \tag{2}$$

2.1.1 Rayleigh Distribution

When the signal propagates in the non visual range in the mine wireless channel, that is, when the m value is 1, the Nakagami-m distribution is transformed into Rayleigh distribution. Rayleigh fading distribution describes the distribution type of flat falling signal of underground wireless channel, and can also describe the time-varying characteristics of multipath component of received signal at the receiver[6]. When two noise signals are orthogonal, the synthetic signal obeys Rayleigh distribution. The probability density function is:

$$P(\mathrm{r}) = \begin{cases} \frac{r}{\delta^2} exp(-\frac{r^2}{2\delta^2}), r \geq 0 \\ 0, r < 0 \end{cases} \tag{3}$$

Where, δ represents the delay spread value of the received signal, and δ^2 represents the time average power of the received signal.

2.1.2 Ricean Distribution

When the signal propagates in the visual range in the mine wireless channel, that is, 2 < m < 4, the Nakagami-m distribution is transformed into Ricean distribution. This distribution can describe the signal with line of sight propagation in the ideal underground channel, and synthesize the multipath components propagating from different paths to the receiver into the static non fading signal. At this time, the envelope distribution of other weak multipath signals attached to the main signal follows the Rician distribution[7]. With the weakening of the main component, the mixed signal changes from Rician distribution to Rayleigh distribution. The probability density function is:

$$P(\mathrm{r}) = \begin{cases} \frac{r}{\delta^2} exp(-\frac{r^2+A^2}{2\delta^2}) I_0 \ (\frac{Ar}{\delta^2}) \ , 0 \leq r \leq \infty, \ 0 \leq \mathrm{A} \\ 0, r < 0 \end{cases} \tag{4}$$

Where, parameter a represents the peak value of the main signal amplitude, parameter k is the rice factor, and the rice distribution is determined by it. When $A \rightarrow> 0, K \rightarrow -\infty dB$, and the amplitude of the main signal decreases, the Rician distribution changes to the Rayleigh distribution. To sum up, the Nakagami-m fading model simulated in MATLAB 2019b software is as follows:

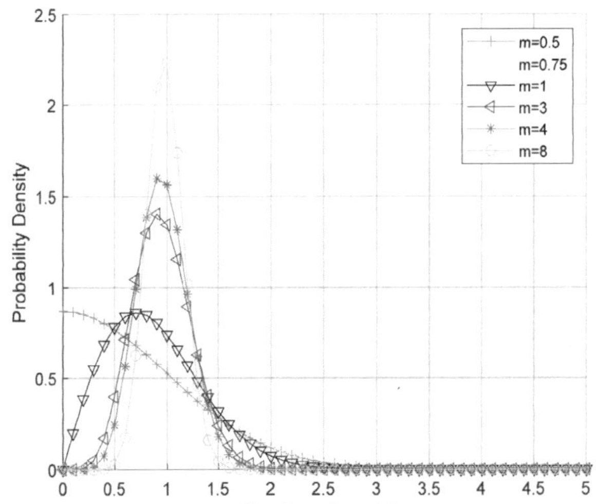

Fig. 1 Nakagami-m function curve corresponding to different m values References

The six curves in the figure correspond to the probability density curves of Nakagami Fading when m = 0.5, 0.75, 1, 3, 4 and 8 respectively. When 0.5 ≤ m < 1, the degree of multipath fading is more serious than Rayleigh fading. When the m value is larger, the probability density curve is closer to the pulse shape, indicating that the fading degree is smaller. in the mine roadway, the nakagami constant m is about 0.75.

2.2 OFDM intelligent receiver model

OFDM technology is mainly multi carrier technology, which can disassemble the channel into multiple flat fading channels to reduce the negative impact of multipath fading on the transmission signal. On this basis, OFDM technology and deep learning technology are organically combined to establish an OFDM wireless communication system based on deep learning, which corresponds to the complex communication environment under the mine. Then, the convolutional neural network (CNN) in deep learning is further introduced to replace the receiving part of the traditional OFDM wireless communication system with convolutional neural network, so as to build an OFDM intelligent receiver. The system model diagram is roughly as follows:

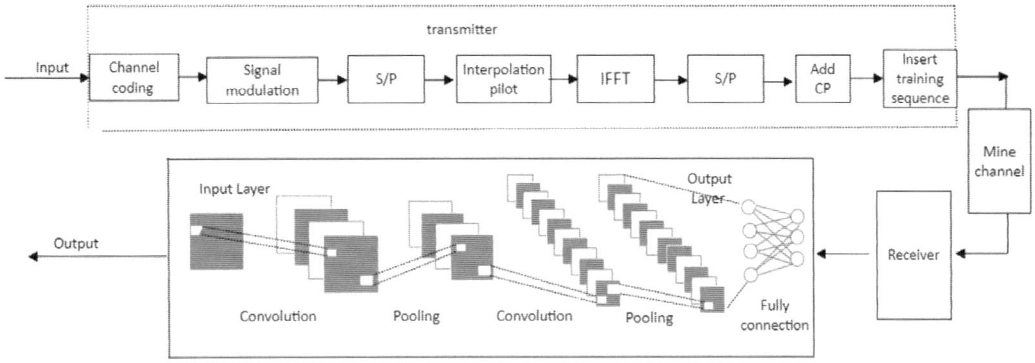

Fig. 2 model diagram of OFDM intelligent receiver

As shown in Figure 2, the transmitter is consistent with the traditional OFDM communication system, that is, the source signal is a series of binary data, which is randomly generated by computer simulation. Firstly, the signal is channel coded to enhance the anti-interference performance of the system, reduce the bit error rate and improve the reliability of signal transmission. Secondly, the signal is modulated, the binary signal is converted into a complex signal, and a certain regular pseudo-random sequence is inserted into the signal as the pilot signal before orthogonal carrier modulation. After orthogonal carrier modulation, a cyclic prefix is inserted in front of each symbol as a guard interval, and finally a training sequence is inserted in front of the frame to achieve frame synchronization at the receiving end. The traditional receiving process has been mentioned in the preface, which will not be repeated here. The OFDM intelligent receiver uses a convolutional neural network to replace the whole signal receiving and processing process, sends the received signal channel into the convolutional neural network, completes the signal receiving process and restores the original information bit stream.

The structure diagram of the convolutional neural network used in this paper is shown in Figure 3. The convolutional neural network mainly processes the input data through convolution, batch normalization, activation function, pooling and other operations. Convolution layer is used to extract data features, batch normalization can speed up the training speed of the model, activation function is used to increase the nonlinearity of the model and pooling layer to reduce the dimension of the data features extracted from convolution layer. Finally, through the data features obtained from the full connection layer, the classification output is realized by using the classification activation function. The depth neural network used in this paper is relatively simple, but it has achieved good signal reception and recovery effect.

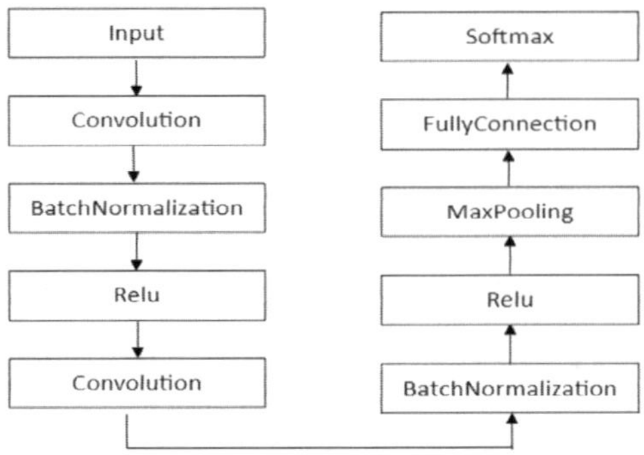

Fig. 3 structure of convolutional neural network

3. Results & Discussion

In this section, an experimental simulation of the proposed OFDM smart receiver is carried out. The simulation is based on the MATLAB 2019b software platform. Firstly, a data set is generated, which including 20000 training data samples and 20000 test data samples, and then the data is send to the built convolutional neural network for training. Finally, the performance of OFDM intelligent receiver is analyzed by bit error rate curve. The BER performance of traditional receiver and intelligent receiver in mine channel with Nakagami constant m = 0.75 and rice fading channel with Nakagami constant m = 3 under BPSK modulation mode is compared and analyzed.

The parameter settings of OFDM wireless communication system transmitter are shown in the table:

Tab. 1 OFDM transmitter parameters

Parameter	value
FFT points	64
Number of subcarriers per symbol	46
Number of OFDM symbols	8
Pilots	8
Cyclic prefix	16
modulation mode	BPSk
Nakagami-m	m=0.75,m=3
SNR	0:8

The training parameters of convolutional neural network are set as shown in the table:

Tab. 2 network training parameters

Parameter	value
Initial Learning rate	0.1
Optimizer	Adam
Batch size	256
epochs	8
Training data	20000
Test data	20000

The experimental platform: Matlab2019b.

Specific experimental steps:

1. Randomly generated binary information bitstream.
2. At the OFDM transmitter, BPSK coding, modulation, pilot insertion, module, IFFT, CP addition and other operations are performed.

3.Through the mine channel with Nakagami constant m = 0.75 and rice fading channel with Nakagami constant m = 3.

4. The traditional OFDM receiver and convolutional neural network are used to replace the OFDM receiver for data reception and recovery.

5. Compared with the original information bit stream, the bit error rate curve is obtained.

6. The bit error rate is analyzed and summarized.

Fig. 4 shows the BER performance curve of traditional OFDM receiver and OFDM receiver based on deep learning in mine channel with Nakagami constant m = 0.75. From the curve, it can be seen that the performance of OFDM receiver based on deep learning is better than that of traditional OFDM receiver. This shows that the use of deep learning technology instead of traditional methods has made remarkable progress.

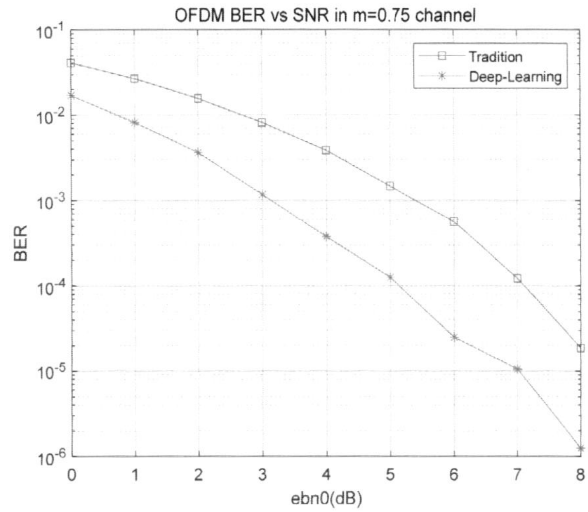

Fig. 4 m=0.75,Tradition vs Deep-Learning

Fig. 5 bit error rate performance curve of traditional OFDM receiver and OFDM receiver based on deep learning in rice channel with Nakagami constant m = 3. From the curve, it can be seen that the performance of OFDM receiver based on deep learning is not as improved as that of traditional receiver in the case of m = 0.75, but it is better than traditional receiver in general, On the other hand, it shows that the deep learning intelligent receiver proposed in this paper is very suitable for wireless communication in coal mine.

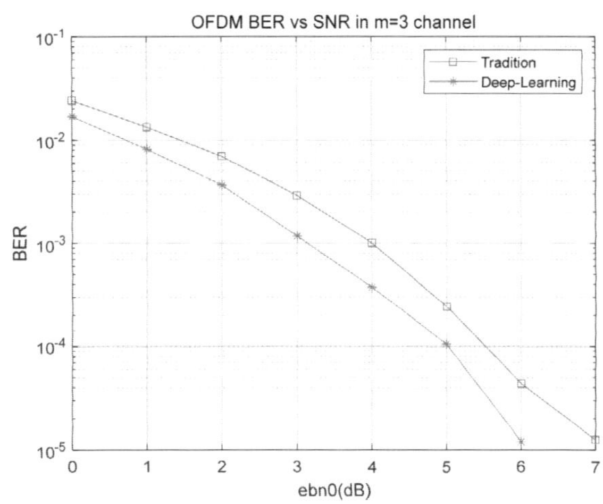

Fig. 5 m=3,Tradition vs Deep-Learning

4. Conclusions

In view of the particularity and complexity of signal propagation environment in coal mine, we analyze the characteristics of wireless channel in coal mine, and propose an OFDM intelligent receiver model based on deep learning. Through the simulation experiment of the intelligent receiver. Compared with the traditional OFDM receiver, the receiver based on CNN has obvious advantages. The OFDM intelligent receiver not only improves the successful reception rate of the signal, optimizes the bit error performance and avoids the accumulation of bit errors. Moreover, it improves the complex problem of traditional step-by-step and module receiving, ensures the communication quality of underground coal mine, which has certain application prospects and good economic benefits. What's more, it is suitable for mine wireless communication.

For the OFDM intelligent receiver model based on deep learning proposed in this paper, the next work can be focused on improving and optimizing the network. It can also be combined with other systems, such as MIMO. In the future, the whole transceiver can be optimized by combining the key technologies in 5G and 6G. It is expected to make greater and greater contributions to coal mining in the future.

References

[1] Zhang Y. Current Situation and Development Trend of Multi System Fusion Data Analysis for Coal Mine Safety Monitoring[J]. International Journal of Computational and Engineering,2020,5(3):

[2] I Kumar,R Mishra. An Efficient ICI Mitigation Technique for MIMO-OFDM System in Time-Varying Channels[J]. Mathematical Modelling and Engineering Problems,2020,7(1):

[3] Zheng S , Chen S , Yang X . DeepReceiver: A Deep Learning-Based Intelligent Receiver for Wireless Communications in the Physical Layer[J]. IEEE Transactions on Cognitive Communications and Networking,2020,PP(99):1-1.

[4] Hao Y,Geoffrey Ye L & Biing-Hwang Juang.(2018).Power of Deep Learning for Channel Estimation and Signal Detection in OFDM Systems.. IEEE Wireless Commun. Letters(1), doi: 10.1109/LWC.2017.2757490

[5] Zhang X Y,Sun Q N & Gao Y.(2020).Spectral efficiency of adaptive OFDM systems over high mobility Nakagami-m fading channels. Systems Science & Control Engineering(1), doi:10.1080/21642583.2020.1854132.

[6] Zhou H,Wang R,Zhao H R & An Z Z.(2021).Application of m-ldpc codes in multipath Rayleigh fading channels. Journal of Physics: Conference Series(1), doi:10.1088/1742-6596/1738/1/012037.

[7] Yuanxue, Xin, Pengfei, Shi, Wenrui, & Tang, et al. (2019). Joint processing of pilot and data for massive mimo systems in ricean fading channels. IEEE Access, 7, 83615-83627.

Harmonic detection technology based on ant colony optimization BP neural network

Jianyu Wu[1,a], Hairong Zou[1,b*]

[1]School of Electrical Engineering,Shanghai Dianji University, Shanghai, 201306, China

[a]email: 1643337538@qq.com

*Hairong Zou: [b]email: zouhr@sdju.edu.cn

Abstrat-At present, ip-iq algorithm based on instantaneous reactive power theory is the most widely used in the field of harmonic detection.However, the ip-iq algorithm based on instantaneous reactive power has coordinate transformation and inverse transformation, which leads to cumbersome calculation.the existence of low-pass filter leads to delay and poor dynamic performance of the algorithm.The harmonic detection based on BP neural network has the advantages of simple calculation, fast detection speed, strong adaptive ability and good dynamic performance.However, the initial weights and thresholds are random, so it is easy for the network to fall into local optimal solution during training, resulting in low detection accuracy. This paper presents a harmonic detection method based on ant colony algorithm optimized BP neural network. ACO-BP, BP and ip-iq algorithms are compared by Simulink simulation. The results show that ACO-BP harmonic detection algorithm is feasible and superior

1. Introduction

With the development of power electronic technology, a large number of power electronic devices are connected to the power grid, which greatly improves the utilization rate of electric energy. However, the access of a large number of power electronic devices also injects a large number of harmonics into the power grid, which seriously affects the power quality of the power grid[1]. As an important part of harmonic control, harmonic detection technology has high research value. In this paper, shunt active power filter is used as a harmonic control device[2].

According to the detection principle, the commonly used harmonic detection methods can be summarized as follows: (1) harmonic detection method based on Fourier transform theory; (2) Harmonic detection method based on instantaneous reactive power theory; (3) Harmonic detection method based on neural network; (4) Harmonic detection method based on wavelet transform theory. Among them, artificial neural network has been widely used because of its advantages in small amount of calculation, accuracy, reliability and adaptive ability. In document [3], hopfield neural network is used to predict the content of each harmonic, which has high prediction accuracy for the whole harmonic, but it is difficult to detect the interharmonic. In document [4], fuzzy neural network is applied to harmonic detection with high accuracy. In document [5], BP neural network is applied to harmonic detection., because the gradient descent method is used as the learning method, the network is easy to fall into the local optimal solution, so the detection result is not very ideal. In this paper, ACO-BP model is used to detect harmonics in power grid, and the feasibility and superiority of this harmonic detection method are verified by building a model in Simulink.

Content from this work may be used under the terms of the Creative Commons Attribution 3.0 licence. Any further distribution of this work must maintain attribution to the author(s) and the title of the work, journal citation and DOI.
Published under licence by IOP Publishing Ltd

2. Harmonic detection methods

2.1 Ip-iq method based on instantaneous reactive power

In 1983, Japanese scholar h. Akagi and others proposed the instantaneous reactive power theory based on time domain and non sinusoidal conditions and the concepts of instantaneous real power P and instantaneous virtual power Q, and then put forward the concepts of ip and iq on this basis[6].

Ip-iq harmonic detection algorithm is widely used because of its high accuracy and real-time. However, the existence of coordinate transformation and inverse transformation leads to cumbersome calculation, and the existence of low-pass filter leads to delay and poor dynamic performance. The principle of ip-iq method is shown in Fig.1.

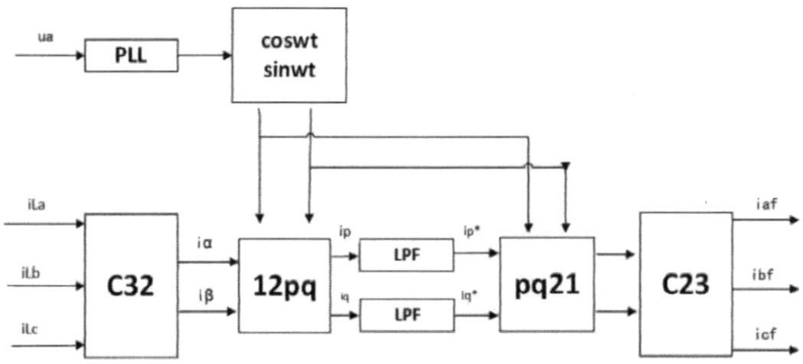

Fig.1 Ip-iq harmonic detection method based on instantaneous reactive power theory

2.2 BP neural network

In recent years, BP neural network has also been developed in the field of harmonic detection because of its strong adaptive ability and good dynamic performance[7-8]. BP neural network can be divided into three layers, namely input layer. The signal is calculated layer by layer from the input layer through the hidden layer.Finally, it is output by the output layer to calculate the error between the output result of the network and the expected output result. If it meets the accuracy requirements set by the network, the result is output. If it does not meet the requirements, the error is back propagated, the error is transmitted to the input layer through the hidden layer, and the weight threshold of each layer of the neural network is adjusted by the gradient descent method. After several cycles, the trained network can meet the preset accuracy requirements or reach the preset maximum training times. Because the gradient descent method is easy to make the BP neural network fall into the local optimal solution and affect the harmonic detection accuracy, ACO-BP harmonic detection method is proposed in this paper.

2.3 ACO-BP

Ant colony algorithm (ACO) is a simulated optimization algorithm to simulate ant foraging behavior. It was first proposed by Italian scholar Dorigo m et al. In 1991 and first used to solve TSP (traveling salesman problem)[9-10]. Then, the basic principle and mathematical model of ant colony algorithm are systematically studied. The principle of ant colony algorithm is: when ants are looking for food, they will release pheromones on their walking path. It can also sense pheromones released by other ants. The shorter the distance, the higher the pheromone content will be. The higher the pheromone content will attract more I ants to walk. Finally, the shortest path will be obtained. When it is applied to the problem of optimal solution, the shortest path corresponds to the optimal solution. The process of Optimizing BP neural network by ant colony algorithm is shown in Fig.2.

Ant colony optimization BP neural network process:

1) Import data and normalize the data.

2) Configure BP neural network parameters to determine the initial weight and threshold of BP neural network.The parameters of ant colony algorithm are initialized, the elements of weight matrix and threshold vector constitute the path coordinates of ant population, and the mean square error is used as the adaptability. The ant coordinates and the highest pheromone are initialized, and the ant colony

algorithm begins to iterat.

3) The pheromone content was calculated according to the ant position.

4) Calculate the highest pheromone and update the position of the optimal ant individual.

5) Update ant location.

6) Perform 3-5 cycle until the requirements are met.

7) Take out the optimal ant position coordinates, get the optimal initial weight and threshold, bring them into BP neural network for training, and further modify the weight and threshold to meet the requirements.

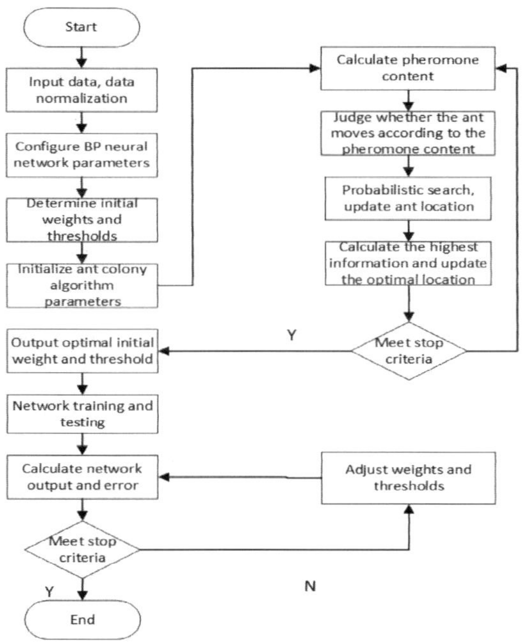

Fig.2 Ant colony optimization BP neural network flow chart

3. Simulation comparison

The load side load currents ia, ib and ic detect the fundamental currents iaf, ibf and icf of the load current through the ip-iq algorithm. The load current ia, ib, ic, fundamental current iaf, ibf, icf and A-phase voltage ua are sampled for 5 cycles, and the first unstable cycle is deleted. The remaining 4 cycles ia, ib, ic are used as the input of ACO-BP model, and iaf, ibf and icf are used as the expected output to construct a 4-9-3 neural network. After the network structure is determined, the transfer functions of the hidden layer and the output layer are selected as the default logsig and purelin functions. The training function is selected as the default trainlm. BP neural network is trained with the same parameters.Using gensim function, the above ACO-BP model and BP model in MATLAB are generated into the corresponding Simulink simulation module. For comparison, put ip-iq, BP and ACO-BP models into the same simulation model, as shown in the Fig.3.

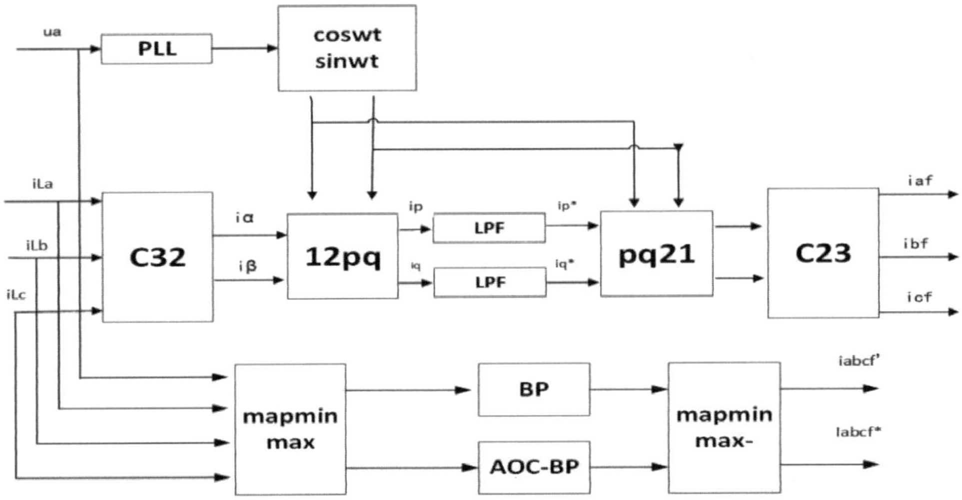

Fig.3 Ip-iq, BP, ACO-BP harmonic detection model

A nonlinear load simulation circuit as shown in Figure 4 is built in Simulink. The power frequency is set to 50 Hz, the operation time of each cycle is 0.02 s, and the system operation time is set to 0.1 s.

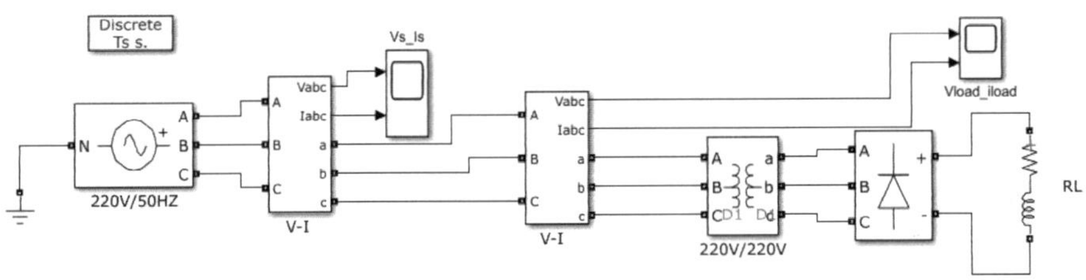

Fig.4 Nonlinear load main circuit

The model in Figure 3 is combined with the model in Figure 4 through the active power filter, and the phase a current at the power side under different states is measured, as shown in Figure 5. In Fig. 5, (a) is the a-phase current waveform of the power supply side without the compensation device, and (b), (c) and (d) are the a-phase current waveform of the power supply side after the ip-iq, BP and ACO-BP models are connected to the compensation of the active power filter. Table 1 shows the current distortion rate of phase a current at the power supply side under the four models.

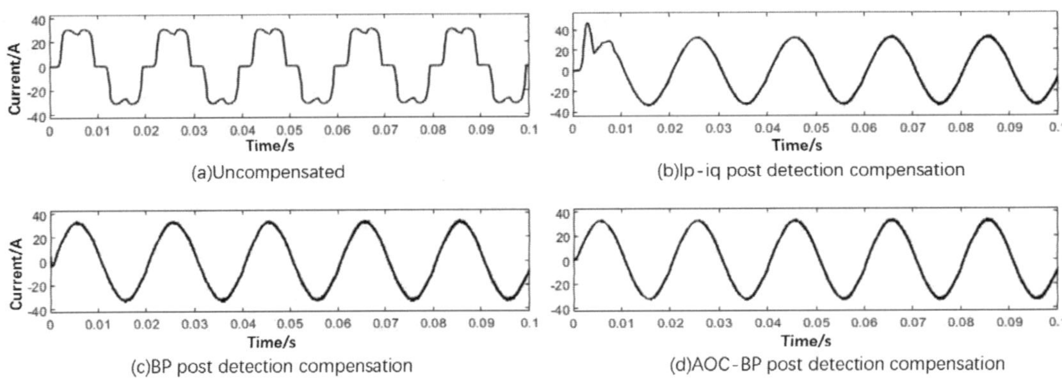

Fig.5 Phase a current at power side

The current waveform in Fig. 5 is analyzed by FFT to obtain the current distortion rate of each model in different time periods. The specific data are shown in Table 1.

Table 1 THD of four models

Category	0-0.02s	0.02-0.04Ss	0.04-0.06s	0.06-0.08s	0.08-0.1s
Uncompwnsated	24.32%	24.35%	24.42%	24.25%	25.12%
ip-iq	18.22%	2.50%	2.84%	3.19%	3.45%
BP	5.23%	3.33%	3.53%	3.67%	3.78%
ACO-BP	4.44%	2.78%	2.97%	3.24%	3.45%

As shown in Figure 5, the whole operation cycle is analyzed. At the stage of 0 ~ 0.02 s, the THD measured by ACO-BP model is the smallest of the four models, and is significantly lower than that of ip-iq model,indicating that the detection speed advantage of ACO-BP model is obvious compared with ip-iq model.In the rest of the time, the current distortion rates of ACO-BP and ip-iq models are almost equal. In the whole running time period, the THD of ACO-BP model is less than that of BP model. To sum up, ACO-BP model has high accuracy in both dynamic and steady state.

4. Conclusions

In this paper, ant colony algorithm is used to optimize the initial weight and threshold of BP neural network, and the trained ACO-BP model is applied to power grid harmonic detection. The simulation results show that the harmonic detection technology based on ACO-bp model has high accuracy and detection speed in both steady-state circuit and dynamic circuit.

References

[1] Michalec, L., Jasiński, M., Sikorski, T., Leonowicz, Z.J., Suresh, V. (2021) Impact of Harmonic Currents of Nonlinear Loads on Power Quality of a Low Voltage Network–Review and Case Study. Energies., 14: 3665-3665.

[2] Gong, J., Li, D.Y., Wang, T.K., Pan, W.H., Ding, X.Z. (2021) A comprehensive review of improving power quality using active power filters. Electric Power Systems Research., 199:

[3] Prakash, C., Alka, S., Madhusudan,S. (2018) Application of Hopfield Neural Network for Harmonic Current Estimation and Shunt Compensation. Electric Power Components and Systems., 46: 290-301.

[4] Lei, W.J., Wang, Y.X., Wang, L,. Cao, H. (2015) A Fundamental Wave Amplitude Prediction Algorithm Based on Fuzzy Neural Network for Harmonic Elimination of Electric Arc Furnace Current. Mathematical Problems in Engineering., 18:

[5] Liu, Y.Y., Zeng, J., Wang, K.L. (2013) A Harmonic Detecting Scheme Based on BP Neural Network. Applied Mechanics and Materials.,2645: 392-392.

[6] Van, B.C., Cong, P.V., Minh ,T.C. (2015) Comparison on Two Harmonic Current Determinate Methods of p-q and ip-iq. International Journal of Scientific Engineering and Technology., 4: 302-305.

[7] Xie, Y.X., Yan, Y.J., Li, X., Ding, T.S., Ma, C. (2021) Fault diagnosis method for scintillation detector based on BP neural network. Journal of Instrumentation., 16:

[8] Zhang, B.D., Jing, Z.P. (2012) An Identification Method of Load Harmonic Current Based on BP Neural Network. In: 2012 IEEE International Conference on Computer Science and Automation Engineering（CSAE 2012）. Beijing. pp. 39-43.

[9] Shi, L.B., Hao, J., Zhou, J.Q., Xu, J.Y. (2003) Ant colony optimization algorithm with random perturbation behavior to the problem of optimal unit commitment with probabilistic spinning reserve determination. Electric Power Systems Research., 69: 295-303.

[10] Zhao, X.P., Yue, Q.F., Pei, J.B., Pu, J.W., Huang, P., Wang, Q. (2021) Ecological Security Pattern Construction in Karst Area Based on Ant Algorithm..International journal of environmental research and public health., 18: 6863-6863.

AUTHOR INDEX

Institute of Physics
Dirac House, Temple Back
Bristol BS1 6BE UK

ISSN: 1742-6588
ISBN 978-1-7138-5483-8

CURRAN ASSOCIATES INC.
proceedings
.com